S. H. IRVINE

Matrix Algebra for Social Scientists

Matrix Algebra
for Social Scientists

PAUL HORST
University of Washington

HOLT, RINEHART AND WINSTON, INC.

New York   Chicago   San Francisco   Toronto   London

To Muriel

Preface

The purpose of this book is, essentially, to introduce students, teachers, and research workers without mathematical training to a branch of mathematics which is becoming increasingly valuable for the analysis of scientific data. It was written primarily for behavioral scientists with no more than a reasonably good working knowledge of high school freshman algebra who have passed an introductory course in statistics. The overriding concern in the selection, organization, and treatment of topics throughout the book is with the collection and processing of experimental data for the purpose of answering questions crucial to the prediction and control of human behavior.

The book is an outgrowth of two different kinds of experience. First, for ten years during and prior to World War II, while engaged in psychological and personnel research work, I developed and adapted the more elementary concepts of matrix algebra for use by computing staffs consisting in large part of persons with no more than a high school education. Second, for seventeen years during and since the war I have taught the principles of simple matrix algebra, together with its applications to the analysis of social science data, to students with only high school mathematics. Much of my treatment is not orthodox from the mathematician's point of view. In general, I have been willing to sacrifice mathematical elegance and rigor for the sake of simplicity and ease of comprehension, although I have not intentionally sacrificed mathematical accuracy. No attempt is made to cover the more advanced concepts of matrix theory. The following are what I regard as among the unique characteristics of the book:

1. It is specifically oriented toward the three major types of analyses used in the social sciences: (1) multiple regression analysis, (2) factor analysis, and (3) analysis of variance.

2. It emphasizes the distinction, so important in the social sciences, between data matrices and derived matrices.

3. It emphasizes the distinction between entities and attributes in the matrix presentation of data. This is frequently a source of confusion to both students and mature scholars in the social sciences.

4. In addition to the conventional notion of matrix multiplication, it develops the concept of matrix multiplication as the sum of major products of vectors. This concept is of great practical value in the analysis of social science data.

5. The concept of rank from a computational point of view and entirely without the aid of determinants is developed. In fact, those elements of matrix algebra essential for the social sciences are presented entirely without the aid of determinants.

6. The concept of nonsingular matrices is minimized, and, instead, primary emphasis is given to the general class of rectangular matrices (basic matrices) whose rank is equal to the smaller dimension of the matrix.

7. The concept of "basic structure" of any real vertical matrix is developed as the triple product from left to right, respectively, of (1) a nonhorizontal matrix orthonormal by columns, (2) a diagonal matrix, and (3) a nonvertical matrix orthonormal by rows. The traditional and narrower related concept of latent roots and vectors of a square matrix is subordinated.

8. The concept of the general inverse of *any* real matrix is defined as that matrix which satisfies the following conditions: if any real matrix is multiplied by its general inverse on either side and the product subtracted from the identity matrix, then the trace of the square of the difference matrix is a minimum. The conventional inverse of a nonsingular matrix is considered as a special case of the general inverse of a nonbasic rectangular matrix, and the importance of this concept for the analysis of data in the social sciences is emphasized.

Chapter 1 develops the point of view that all scientific investigations must be based on matrices of observed data.

Chapter 2 shows how we indicate in symbolic form that tables of observed data are to be considered as matrices to which the powerful methods of matrix algebra may be applied.

Chapter 3 describes the different kinds of matrices that are of special interest and value in the analysis of data.

Chapter 4 through 11 develop, in easy steps, the rules of matrix operations basic to the analysis of large bodies of scientific data.

Chapters 12 and 13 show how the notions of elementary statistics may be generalized and simplified by means of matrix notation to handle the highly complex phenomena of the social sciences.

Chapter 14 introduces a special class of matrices (orthogonal) essential for simplifying the analysis of some of the more complex scientific problems.

Chapters 15 and 16 are concerned with the concept of rank, which is used in science to formalize the "law of parsimony."

Chapters 17 and 18 develop the concept of basic structure, which, because of its simplicity and greater usefulness in the analysis of social science data, is substituted for the more conventional mathematical concepts of latent roots and vectors.

Chapters 19, 20, and 21 develop, in easy stages and for various special cases, the notion of matrix inversion, which is the basis for solving for the unknowns in scientific studies.

Chapter 22 develops and presents applications of the general rank reduction theorem, which is of great importance in the social sciences but which has received little emphasis in conventional texts on matrix algebra.

Chapter 23 expands the conventional mathematical approach to the solution of linear equations to fit the real life problems which scientists are called upon to solve.

The professional mathematician may become impatient with the leisurely pace of Part I and most of Part II. Also, he may experience irritation with the unconventional format and the unorthodox development for many of the proofs. If so, I beg his indulgence, for I can assure him that the traditional approach of the mathematician has negatively conditioned many students to the study of mathematics. It is this large group that I have sought to reassure and to persuade that mathematics is not necessarily frustrating, but can actually be interesting, entertaining, and above all, useful. Reactions to earlier dittoed and mimeographed versions of the text seem to indicate that I have been at least partially successful.

In particular, some mathematicians may become restive with the extended treatments involving matrix transposition and supermatrices. If so, I invite them to become involved for a time in projects in which these concepts are crucial for efficient data analysis operations. It is here that we see how concepts, simple in the abstract, can become surprisingly confusing in application.

I have deliberately avoided the conventional practice of using boldface characters to distinguish matrices from scalar quantities. This reflects my position that the matrix should be the primary operating unit rather than the scalar or vector, which are merely special cases of matrices. I prefer to define scalar quantities verbally whenever any ambiguity arises. In general, I have been much more concerned than the professional mathematician with the development of a functional, as distinguished from a descriptive, symbolic system.

I owe a particular debt of gratitude to Professor Robert M. Thrall, who patiently read the manuscript and made many valuable suggestions. Most of these I was able to utilize to good advantage. There are, however, several points of disagreement between us. A brief discussion of these may throw further light on the essential differences between my approach and that of the professional mathematicians.

The use of subscript notation to indicate the dimensions of a matrix is objectionable to Professor Thrall, whereas I have found it extremely useful in the manipulation of matrix expressions.   It is true that ambiguities can arise, but I find it much better to clarify these for particular situations than to complicate the notation with additional symbols which, among other things, may cause confusion with computer programming notation. Also, in the case of the supermatrix notation in Chapter 11, Professor Thrall points out certain limitations.   I confess that I am not completely satisfied with the notation as I have developed it thus far.   I am thoroughly in agreement, however, with Bodewig that a major objective should be to "make the notation do the work."   If the notation for the multiplication of supermatrices seems complex, it is because data analysis involving supermatrices may be complex and confusing, even though the concepts themselves are simple.

I am particularly indebted to Professor Thrall for calling my attention to the work of E. H. Moore and his students on the "general inverse." This concept arises naturally from the definition of the basic structure of a matrix.   Since my own independent invention of the concept in 1941 I have found it very useful in the analysis of behavioral science data and I have been repeatedly surprised that mathematicians have not generally been familiar with it.

Another significant comment by Professor Thrall suggests unnecessary proliferation resulting from the distinction between horizontal and vertical matrices.   This comment is crucial because it focuses the spotlight on the great gulf between pure mathematicians and the data processors.   In the early work on matrix theory, the interest of mathematicians centered primarily on square matrices.   It is only more recently that general interest has spread to rectangular matrices.   The truth is, of course, that data matrices are rarely square and that problems involving the fundamental concept of rank are inextricably interwoven with considerations of major and minor products of matrices, whether they are data matrices or derived matrices.   In the analysis of data matrices, one cannot help but be impressed with the fundamental roles played by rank and dimensionality.

Finally, a most persuasive suggestion from Professor Thrall was that the Gram-Schmidt orthogonalization process might have been considered in Chapter 14 on orthogonal matrices.   As traditionally presented, this process is developed in scalar notation.   In terms of the matrix terminology and notation progressively developed in previous chapters of this book, it seems to come more naturally in section 10 of Chapter 19 on the inverse of a matrix.   It is true, however, that the professional mathematician who is accustomed to the conventional Gram-Schmidt presentation might be apt to overlook it in this setting.

I also acknowledge the assistance of Helen Ranck, Barbara Jaeger, and Judy Goodstein, who typed the manuscript, and of George Burket, who

checked the exercises and answers.   I, however, assume responsibility for
such errors as may remain.

Finally, I wish to express my deep appreciation for the financial support
provided by the Office of Naval Research Contract Nonr-477(08) and
Public Health Research Grant M-743(C6).

P.H.

Seattle, Washington
1962

Contents

PART III. The Structure of a Matrix

Part I

Simple Matrix Concepts

Chapter 1

Introduction

1.1 What a Matrix Is

It is often true that simple dictionary definitions don't help you much to understand the meaning of words. This is certainly the case for the word "matrix." An abridged Webster's dictionary defines matrix as "a cavity in which anything is formed such as a mold or die." This definition has little to do with the way we shall use the word, even though it is adequate for one of the most common uses. The mathematicians have used *matrix* to mean something quite different. However, since we are going to discuss pretty much the same things as the mathematicians do when they speak of a "matrix," we shall use the same word.

The word *matrix* comes from the Latin, like a good many technical or scholarly words. The plural is therefore the Latin plural, *matrices*. Thus a quite simple concept is called by a rather technical name.

Basically, a matrix, as we shall use the word, is a table of numbers with so many rows and so many columns. Figure 1.1.1 is an example of a matrix.

$$
\begin{array}{ccc}
4 & 3 & 1 \\
6 & 2 & 5 \\
1 & 3 & 4 \\
7 & 9 & 5
\end{array}
$$

FIG. 1.1.1

You will notice that Fig. 1.1.1 has four rows and three columns.

Another example of a matrix is Fig. 1.1.2.

$$
\begin{array}{cccc}
5 & 9 & 12 & 14 \\
6 & 8 & 13 & 15 \\
14 & 3 & 13 & 10 \\
15 & 16 & 8 & 13 \\
14 & 9 & 3 & 11
\end{array}
$$

FIG. 1.1.2

3

You see that Fig. 1.1.2 has five rows and four columns.

You can have as many or as few rows and columns in a matrix as you wish, although usually there are special reasons for deciding how many rows and how many columns you will have. You could, for example, have a matrix with only one row and one column, in which case you would have, of course, only a single number.

The numbers used in a matrix are real numbers, numbers in the ordinary sense. Some of you doubtless know about imaginary and complex numbers. If so, don't expect to find numbers of this kind in the matrices we talk about. If you haven't heard about such numbers, just forget we mentioned them. We shall have very little use for them in this book.

1.2   Where Matrices Come From

The matrices used in the social sciences are of two general kinds. Some matrices come directly from our observations of things and people. These we call data matrices. Other matrices come from data matrices when we apply certain rules to them. These we call derived matrices. When we apply appropriate rules to some data matrices, we may get extremely interesting and useful results.

There is a big difference between our interest in matrices and that of the mathematician. We are very much concerned with the sources of data matrices, and with the practical applications of the results after certain rules have been applied. The mathematician, on the other hand, is usually not very much interested in how the matrix came to be or what is done with the results after the rules are applied. He is chiefly interested in mathematical properties of various types of matrices. Because of this difference of interest, we shall have little or no use for some of the mathematician's rules. On the other hand, we shall find it convenient to emphasize some rules the mathematician uses very little, if at all.

Where do the numbers in our data matrices come from? As you will see, they come from the world in which you live, and from the people who live in it with you. The numbers may have a great deal to do with the hopes and ambitions, the frustrations and successes, of millions of people. For example, in Fig. 1.1.1 each row may represent a person who has applied for a job, and each column may represent his grade in a given school subject, so that the grade of the second person in the third subject would be 5. Or again, in Fig. 1.1.2 the rows might represent common stocks listed on the New York stock exchange, and the four columns might be the highest price of the stock for each of four successive months. For example, the highest price of stock 2 during the third month was 13.

A good example of matrices that are of great interest to hundreds of thousands of people throughout the country may be found on the sports pages of any daily newspaper. Look at

```
 2    1   2
 3    7   2
 0    1   4
11   16   2
 0    0   2
```

FIG. 1.2.1

The five rows in Fig. 1.2.1 represent five different baseball teams and the three columns are respectively runs, hits, and errors.

Another example comes from the weather report in a typical daily newspaper.

```
94   63   .00
96   73   .24
64   49   .12
81   55   .00
72   58   .56
75   60   .00
```

FIG. 1.2.2

In Fig. 1.2.2 each row represents a city in the United States, and the columns represent, in order, the highest temperature, the lowest temperature, and the rainfall in inches for the day. You see, then, that almost any place you look you can find a data matrix, for a data matrix is basically a very simple way of organizing the numbers used to describe the various kinds of information that come to our attention every day.

1.3 What the Rows and Columns Mean

Rows and columns play an important role in both data and derived matrices. You should therefore get to understand at once the simple meanings of rows and columns in data matrices. In the first example in Fig. 1.1.1 the rows represent four different people, and the columns are grades of these persons in three different subjects. In Fig. 1.1.2 the rows represent five different common stocks, and the columns are the highest prices reached in each of four successive months. In Fig. 1.2.1 the rows are five different ball teams and the columns are records for each of three different things that ball teams do. In Fig. 1.2.2 the rows are six different cities, and the columns are four different things about the weather in those cities on a given day. Ordinarily then, you could say that the rows in a data matrix represent things which exist in and of themselves. These we call *entities*.

Now in all four cases the columns represent things that do not exist in and of themselves. We can't have a grade in a school subject unless somebody makes the grade. Neither can we have the highest price in a given month unless it is the highest price of some particular thing like a common stock. In the same way, we can't have a hit or a run unless someone makes that hit or run. Nor can we have a temperature reading or rainfall except at some particular place. We may say then that usually the columns of a primary matrix represent *attributes* of the entities in the rows. These attributes are sometimes called *variables*.

In general then, we may say that in a data matrix the rows represent entities and the columns represent attributes. We could of course let rows be attributes and columns be entities. The plan we have adopted is fairly common practice, however, because usually there are many more entities than attributes in a matrix. The weather report is an example. Usually the number of cities on which weather is reported is much greater than the number of different things reported about the weather. Similarly, the daily stock market report lists hundreds of stocks, but ordinarily only a small number of things are reported about each stock—such as the high, the low, the close, and the number of shares traded. So also a matrix of school grades would usually not include more than six to eight subjects, but it might include dozens or even hundreds of students.

For convenience we have come to construct data matrices with rows as entities and columns as attributes. In the width of an ordinary newspaper column most of the important attributes about a particular subject can be given, but this space would not hold more than a few of the many desired entities. Furthermore, we are used to reading horizontally across the page, and our interest tends to focus on a single entity rather than a single attribute of a lot of entities. We can, therefore, by following our natural reading habits, find all that is recorded about a single entity in one row.

To summarize what we have been saying, a data matrix ordinarily includes more entities than attributes; this is one reason why we choose to let the rows represent entities. Furthermore, to be useful for scientific purposes, a data matrix must in general have more entities than attributes.

You should know, however, that the method we use here is by no means standard practice among scientists. Some writers prefer to let entities be columns and attributes be rows. Others are not consistent; they use both styles of matrix construction. In this book, however, unless we give warning to the contrary, we shall stick to the rule of using rows for entities and columns for attributes in a data matrix.

1.4 Changing the Order of Rows or Columns

One important feature of both entities and attributes is that the order in which they are recorded is usually unimportant with respect to the rules or operations to be applied to them. It does not matter, for example, in Fig.

1.1.1, which row comes first, second, and so on, nor in what order the columns are arranged. For practical purposes it may be convenient to have persons or cities or other entities listed in alphabetical order, but this is not necessary as long as it is clear which row is which person or city. In the same way, in Fig. 1.1.2 it might be convenient to have the attributes, the school subjects, in alphabetical order as Algebra, English, History, and so on, but this is not necessary as long as it is clear what each column is. You can state this principle by saying that rows or columns of a data matrix usually do not represent a scale or *continuum* (plural *continua*).

An example of a matrix which is not a data matrix and in which neither the rows nor the columns are interchangeable is Fig. 1.4.1.

Weight in pounds

		90	91	92	
Height in inches	63	0	1	2	3
	62	0	3	1	4
	61	2	0	1	3
					10

Fig. 1.4.1

This table represents a group of ten boys who range in height from 61 through 63 inches and in weight from 90 through 92 pounds. Each number in the table is the number of boys in the group having a specified weight and height. For example, the number of boys who are 62 inches tall and weigh 91 pounds is three. The number of boys who are 61 inches tall and weigh 92 pounds is only one. But Fig. 1.4.1 is not a data matrix. The rows represent a scale or continuum of height and the columns represent a scale or continuum of pounds. You can easily see, however, what kind of a data matrix Fig. 1.4.1 would come from. Since all the numbers in the table add up to 10, there were ten boys altogether. Therefore, the entities in the data matrix were boys, and it had ten rows. The matrix had two columns or attributes, one of which was height and the other weight.

1.5 Some Exceptions to Interchangeability of Rows or Columns

There can be exceptions to the rule that the order of entities or attributes of data matrices can be changed about. The most important is a matrix in which either the entity or the attribute is some sort of time sequence, such as days or months. In Fig. 1.1.2, for example, the entities are common stocks and the attributes are successive months. While it would be quite possible to put July before June or May before April, it would be hard to think of a good reason for doing so. Again you might have a matrix as in Fig. 1.5.2,

$$
\begin{array}{ccc}
20 & 30 & 33 \\
10 & 15 & 21 \\
19 & 19 & 22 \\
20 & 21 & 22
\end{array}
$$

Fig. 1.5.2

where the rows are students in a typing class and the columns are the number of words per minute in each of three successive weeks. You would not want to put the first week last or the second week first.

Similarly, the entities might be successive time intervals. This could happen if all the numbers for the matrix were obtained from a single person. Suppose, for example, this person were ill in a hospital and each day for two weeks a record was made on each of three attributes—his temperature, blood pressure, and pulse rate. Then you would have a matrix of 14 rows and 3 columns. Each entity, now, would be the individual on each of the 14 days. Here again you would not ordinarily switch the days out of their natural sequences.

A really thorough treatment of the cases in which time sequences can be either entities or attributes would lead us into problems that need not concern us in this book. Therefore, you may simply remember that for the most common data matrices the order of entities can be changed about in any way you please and so also can the order of the attributes.

1.6 Reciprocal Nature of Entities and Attributes

By now it should be clear that merely by looking at a matrix it is impossible to tell whether the rows are the attributes and the columns the entities or the other way around. There is nothing in the matrix form, in itself, to tell you which are entities and which are attributes. In constructing the matrix, you determine this for yourself. It makes no difference, however, in your use of the matrix or the rules you will learn to apply to it, whether rows or columns are entities or attributes. It is well to remember this, so that when you come to some of the applications of matrix algebra to the behavioral sciences you will not be confused. For example, one of these applications has to do with analyzing human personality. A form of matrix algebra known as factor analysis is used for solving such problems. Such work utilizes the concepts discussed in Chapters 17 and 18; you can decide arbitrarily after the analysis which you want to designate as attributes, and which as entities. Unless this is clear, a great deal of time can be wasted in arguing about what rules shall be applied to the data matrix in order to come out with appropriate answers.

1.7 Scientific Predictions and Data Matrices

But what is a matrix used for? In the social sciences, it is a basis for prediction. For a particular group of entities, some attributes may be predicted from others. All scientific predictions are based on the numbers in data matrices and on certain operations or rules applied to them. This is true whether you are making scientific predictions about the weather, what grades you will make if you major in English, how successful you will be as a doctor, how happy your marriage will be, how much oil, if any, can be found on Uncle Ben's old wheat farm, or the price of General Motors common stock by next Christmas. If the predictions of these or any other events you can think of are not based directly or indirectly on data matrices and operations upon them, then you can be sure that the predictions are not scientific.

Ordinarily, we wish to predict certain attributes for specific entities. If your future is in question you are the entity, and the attributes are your English grades, success as a doctor, or marital happiness. Similarly, in our other examples, Uncle Ben's wheat farm is the entity, and the attribute is the amount of oil it can yield; and the entity is General Motors common stock, and the attribute is its price next Christmas.

1.8 The Criterion Attributes

Usually the attributes you want to predict are the kinds of things that mean a great deal to you or someone else. Or, they may be things which over a period of time would turn out to be of some importance to a great many people. It is important to you to know whether you will be a successful doctor or whether your marriage will be a happy one.

Another characteristic of the kinds of attributes we wish to predict is that finding out about them directly could be very costly. Assuming you had the money and could get through medical school, you could try practicing medicine and see how you made out. But if you failed, think of all the money, time, and effort wasted in preparing for a medical career.

The third and most important characteristic of the kind of attributes we wish to predict is the very fact that the measure of that attribute for a particular entity is not available at present. A prediction necessarily deals with the future.

We call the attribute we wish to predict the *criterion attribute* or simply the *criterion* (plural *criteria*).

Now in any scientific prediction study, you may well have two or more criteria. You may want to predict your grades not only in English, but in mathematics, history, and many other subjects, or you may want to predict not only how successful you would be as a doctor, but also as an engineer, an architect, a lawyer, and so on.

1.9   The Predictor Attributes

In a general way you can say, then, that for scientific prediction the columns or attributes in a data matrix are of two kinds—those we want to predict, or the criteria, and those we use to make our predictions. These latter we call *predictor* attributes or variables. In general we are able to find numbers or measures of predictor attributes for each of a group of entities, such as people, sooner and more economically than we can get the numbers for the criterion attributes or columns. See Fig. 1.9.1, for example.

$$
\begin{array}{cc}
21 & 63 \\
22 & 66 \\
23 & 69 \\
24 & 72 \\
25 & 75 \\
26 & 78 \\
\end{array}
$$

Fig. 1.9.1

Here we have a matrix of six rows and two columns. Suppose that each row represents a bookkeeper in a big department store. Suppose also that the first column represents the scores that the six bookkeepers made on an arithmetic test before they were hired for the job, and that the second column represents their weekly salary at the end of the first year of employment. Here the first column in the matrix was available a year before the second column. Suppose you are willing to take as a measure of success on the job the salary a person is making at the end of the first year. Then if you are willing to grant that success on a job is important, and that it is also important to know in advance how successful you will be, you will see that the second column in Fig. 1.9.1 is what we agreed to call a criterion attribute. So also, if you are willing to agree that knowledge of arithmetic is related to competence in a bookkeeper, you see that the first column might be a predictor attribute.

1.10   The Prediction Formula

Now take a good look at the numbers in both columns of Fig. 1.9.1. The second column is three times the first. By examining the table you can draw the conclusion that if you know a person's score on the arithmetic test before employment, you can tell how much he will be earning as a bookkeeper in the department store at the end of the year. Actually, of course, you could never tell exactly how successful a person will be in a job by knowing his score on a test before employment, but many matrices of the sort given in Fig. 1.9.1 have shown that for some jobs a test is a reasonably good indication of later success.

Now look at Fig. 1.10.1.

40	26	184
28	21	140
60	23	212
43	22	174
21	25	142
30	28	172

Fig. 1.10.1

Suppose that the rows represent secretaries in a large business office, and that the last column represents their monthly salary at the end of the first year of employment. This column will then be the criterion attribute. Suppose the first column gives the scores on an arithmetic test taken before the girls were employed, and the second their scores on a reading test. Now if you were good at spotting number relationships, a little study of the figure would reveal to you a rule that relates the first two columns to the third. You could also find this relationship, however, by using some simple rules of matrix algebra. If you applied these rules you would find that if you took two times the arithmetic score of a secretary plus four times her reading score, you would get her monthly salary after one year of employment. Here again, we have greatly oversimplified the matter. It is true, however, that in many cases we can get a better estimate of how successful a person will be on a job if we have several test scores.

You see now that the essential characteristic of the predictor attributes is that they enable you to reconstruct the criterion column. By using the simple rules of matrix algebra, we find how accurately it is possible to reconstruct the criterion column from the predictor columns, and the rules show us exactly how to do it. For example, in Fig. 1.10.1 the application of the approximate rules would show that two times the number in the first column plus four times the number in the second column would give the corresponding number in the third or criterion column.

Thus matrix algebra is helpful when you deal with a matrix of entities and attributes in which one column is a criterion attribute. It helps you to find the numbers to use in multiplying the numbers in each of the other columns, so that when you add the products together for a given entity you get as close to the corresponding number in the criterion column as possible. The numbers you multiply by are called *weights*. For any one attribute, you will always use the same weight for all the entities. For example, in Fig. 1.10.1 if you multiply the arithmetic score by two and the reading score by four for one person, then you must also use two for the arithmetic score of every other person and four for his reading score.

Sometimes matrix algebra will help you discover that no matter what weights you use, it is not possible to reconstruct the numbers for the criterion attribute at all closely. In any case, however, you can find out *how* closely it is possible to reconstruct them—exactly, not at all, or somewhere in between.

1.11   The Bases of Scientific Prediction

But what is the good of reconstructing measures of criterion attributes for different persons or entities when you already know them? The truth is that, for these *particular* persons or entities, reconstructing the criterion attributes does no good at all. Here we must pause to state a principle that is basic, not only to all science, but to all knowledge. The only way to judge what will happen in the future is by what has happened in the past. Other things being equal, the more frequently things have happened in the past, the more sure you can be they will happen in the future. That is, the criterion attributes are a tool of the empirical approach to the search for truth.

Your only purpose in examining a matrix of persons or entities whose criterion attributes you already know is to get weights to use on the predictor attributes of those entities whose criterion attributes you don't know, in order to predict them. This is one of the reasons why a data matrix generally has many more entities than attributes. The entities represent experience, and the more entities there are in a data matrix, the more experience is represented in it. A more technical reason has to do with the statistician's "degrees of freedom." This subject is treated in detail in statistics texts.

1.12   Two Kinds of Entities

We may then call a data matrix with both predictor and criterion attributes or variables a *prediction data matrix*. But by now you have noticed that we also have two kinds of entities. These are, first, the entities for which we have both the predictor attributes and one or more criterion attributes; and second, those entities for which we have only the predictor attributes. Figure 1.12.1 is an example of a prediction data matrix that includes both kinds of attributes and both kinds of entities. It is simply Fig. 1.10.1 to which have been added entries for some girls whose arithmetic and reading scores are known, while their salaries at the end of the first year in the office are not known, since the girls have not yet been employed.

	40	26	184	
	28	21	140	
I	60	23	212	II
	43	22	174	
	21	25	142	
	30	28	172	
	32	22	. . .	
III	37	29	. . .	IV
	56	20	. . .	
	. . .	. . .	. . .	

Fig. 1.12.1

You may very appropriately call Fig. 1.12.1 a complete prediction data matrix. Figure 1.10.1 you may call an experimental prediction data matrix, because it includes only the entities used to determine experimentally the weights to be applied to the predictor attributes of the entities whose criterion attributes are unknown.

Therefore, we call the entities whose criterion attributes are known *experimental entities*, and those whose criterion attributes are to be determined *administrative entities*.

Returning to Fig. 1.12.1, then, you see that a complete prediction data matrix consists of four sections or quadrants. Quadrant I, the upper left, consists of the predictor attributes for the experimental entities. Quadrant II, the upper right, consists of the criterion attribute for the experimental entities. Quadrant III is the predictor attributes for the administrative entities. Quadrant IV must be left blank until you have applied to the numbers in the other three quadrants the appropriate rules of matrix algebra that will enable you to get a prediction of the numbers in this quadrant.

You should also notice in passing that in Fig. 1.12.1 there are blanks at the bottom of Quadrants III and IV. These blanks indicate that usually in a complete prediction matrix the number of administrative entities is not definitely known. Once you have used the experimental part of the matrix to determine the appropriate prediction weights, you may go on applying these weights more or less indefinitely to the predictor attributes of one administrative entity after another in order to predict their criterion score or attribute. Thus, once you found the weights to be applied to arithmetic and reading scores for predicting success as a secretary, you might go on year after year using these weights to predict the success of each applicant for the job.

On the other hand, the number of entities in the experimental part of the matrix must be fixed for any particular research or study. Once you start operating on it with the rules of matrix algebra, you can't very conveniently add or take out rows. You should be sure that you have in it the number and kind of entities that are needed to give results you can have confidence in. The same is true with the number of attributes. Whether or not you have the right number and kind of each depends on a great many things. We can give very little attention to these in this book, but you should know several of them.

The human sciences, such as economics, anthropology, psychology, education, and political science, are of course very complex. To predict things about people, things that are important to them, requires that you take into consideration a great many attributes. Therefore, other things being equal, the more predictor attributes you have in an experimental prediction matrix, the more accurately you can expect to predict the criterion for the applied entities.

For best results in prediction, however, the number of entities in the experimental matrix should be much greater than the number of attributes. The more predictor variables you have, the more entities you will need. You can learn how to tell whether you have enough entities and attributes in an experimental matrix by reading technical books in statistics. What you should know now is that in the human sciences, the experimental prediction matrix should be large if the results are to be useful for prediction.

SUMMARY

1. What a matrix is
   a. *Definition.* A matrix is a table of numbers or symbols that stand for numbers, having rows and columns.
   b. *Example.*

$$\begin{matrix} 1 & 2 & 3 \\ 3 & 4 & 5 \\ 6 & 7 & 2 \end{matrix}$$

2. Source of matrices:
   a. Data matrices come from experimental observations.
   b. Derived matrices come from operations on data matrices.

3. Meaning of rows and columns of data matrices:
   a. Rows usually consist of entities, representing people or things.
   b. Columns usually consist of characteristics or attributes of the entities.

4. Interchangeability of rows and columns:
   a. The serial order of the entities and of the attributes is usually irrelevant.

b. Entities and attributes do not usually represent continua.

c. Dimensions of derived matrices may be continua.

5. Exceptions to interchangeability of rows and columns:
   a. Time intervals for entities are not interchangeable.
   b. Time intervals for attributes are not interchangeable.

6. Reciprocal nature of entities and attributes:
   a. The experimenter makes the distinction between entities and attributes.
   b. The rules of matrix algebra do not depend on a distinction between entities and attributes.

7. Scientific prediction and data matrices:
   a. All scientific predictions must be based on data matrices.
   b. Criterion and predictor attributes are required for scientific prediction.

8. Criterion attributes:
   a. *Definition.* Criterion attributes are those we wish to predict.
   b. Criterion attributes are important as such, are costly to assess, and are not currently available.
   c. A data matrix may have more than one criterion attribute.

9. Predictor attributes:
   a. *Definition.* Predictor attributes are those we predict from.
   b. Predictor attributes are not necessarily important in themselves, are relatively easy to assess, and are currently available.

10. The prediction formulas:
    a. If predictor attributes are to be useful they must have numerical relationships with the criterion attributes or variables.
    b. Matrix algebra enables us to find these relationships.

11. Kinds of entities:
    a. *Definition.* Experimental entities are those from which the prediction formulas are deduced.
    b. *Definition.* Administrative entities are those to which the formulas are applied.

EXERCISES

1. Make up a data matrix from which Fig. 1.4.1 could be derived.

2. Given the data matrix

$$
\begin{array}{cc}
X & Y \\
1 & 1 \\
3 & 5 \\
2 & 3 \\
4 & 7 \\
8 & 15 \\
6 & ? \\
7 & ?
\end{array}
$$

(a) What should the last two numbers in the Y column be?

(b) Write an equation for the relationship of the Y to the X column.

3. Given the matrix

$$
\begin{array}{ccc}
1 & 2 & 3 \\
2 & 5 & 6 \\
3 & 6 & 9
\end{array}
$$

How can this be rewritten by changing only one number so that the rows and columns are proportional?

ANSWERS

1.

Height	Weight
90	61
90	61
91	62
91	62
91	62
91	63
92	61
92	62
92	63
92	63

2.

(a) 11, 13

(b) $Y = 2X - 1$

3.

$$
\begin{array}{ccc}
1 & 2 & 3 \\
2 & 4 & 6 \\
3 & 6 & 9
\end{array}
$$

Chapter 2

The Language of Matrices

It is quite possible to use the kind of algebra you learned as a high school freshman on the data matrices from the behavioral sciences, but since these matrices are usually large, elementary algebra is very tedious and confusing to use. For the analysis of large tables of numbers, matrix algebra is ideally suited. We shall therefore in this book learn how a whole table of numbers may be handled with matrix algebra much as you handled single numbers with elementary algebra.

2.1 Showing that a Table of Numbers Is a Matrix

Methods of designation

We have said that basically a matrix is a table of numbers; now we need to elaborate on this definition. When we call a table of numbers a matrix, we imply that it is subject to a set of rules indicating what can be done with these numbers, and also what cannot be done.

There are several ways in which a table of numbers can be designated a matrix. One way is to put a double ruling on either side of the table. See Fig. 2.1.1.

$$\left\|\begin{array}{ccc} 4 & 3 & 1 \\ 6 & 2 & 5 \\ 1 & 3 & 4 \\ 7 & 9 & 5 \end{array}\right\|$$

Fig. 2.1.1

A second method commonly used to show that a table of numbers is a matrix is to enclose it in parentheses as in Fig. 2.1.2.

$$\begin{pmatrix} 5 & 9 & 12 & 14 \\ 6 & 8 & 13 & 5 \\ 14 & 3 & 2 & 10 \\ 15 & 16 & 8 & 13 \\ 14 & 9 & 3 & 11 \end{pmatrix}$$

Fig. 2.1.2

17

A third method is to enclose the matrix in brackets, as in Fig. 2.1.3.

$$\begin{bmatrix} 4 & 3 & 1 \\ 6 & 2 & 5 \\ 1 & 3 & 4 \\ 7 & 9 & 5 \end{bmatrix}$$

FIG. 2.1.3

All three of these methods—the double ruling, the parentheses, and the brackets—are equally good. In textbooks on matrix algebra, and in published articles which use matrix algebra, you will find that some writers use one method of designation and some use another. In fact, some writers may use more than one of the methods.

As you will see later, a single column or a single row of numbers may be regarded as a special case of a matrix. Once we start applying the rules of matrix algebra, it is important to know when a matrix of this kind should be considered as a row and when it should be considered as a column. In order to save space in printing, it is sometimes customary to write all single column matrices as rows. Then in order to show that a row matrix as printed should really be a column matrix, the row is enclosed in braces as shown in Fig. 2.1.4.

$$\{4 \quad 3 \quad 5 \quad 7\}$$

FIG. 2.1.4

Here the braces mean that the row should really have been written as a column, as in Fig. 2.1.5, except that there wasn't space to write it as a column.

$$\begin{Vmatrix} 4 \\ 3 \\ 5 \\ 7 \end{Vmatrix} \quad \text{or} \quad \begin{bmatrix} 4 \\ 3 \\ 5 \\ 7 \end{bmatrix} \quad \text{or} \quad \begin{pmatrix} 4 \\ 3 \\ 5 \\ 7 \end{pmatrix}$$

FIG. 2.1.5

Now if the one-line matrix in Fig. 2.1.4 should actually have been written as a row, rather than as a column, it would have been written in one of the usual three forms shown in Fig. 2.1.6.

$$\begin{Vmatrix} 4 & 3 & 5 & 7 \end{Vmatrix}$$
$$(4 \quad 3 \quad 5 \quad 7)$$
$$[4 \quad 3 \quad 5 \quad 7]$$

FIG. 2.1.6

One thing you must carefully avoid in designating a table of numbers as a matrix: never use a single line on each side of the table of numbers, as in Fig. 2.1.7.

$$\begin{vmatrix} 4 & 3 & 1 \\ 6 & 2 & 5 \\ 1 & 3 & 4 \end{vmatrix}$$

FIG. 2.1.7

If a matrix has the same number of rows as it has columns, as Fig. 2.1.7 does, the single line on each side of the table of numbers has a very definite meaning to mathematicians. Figure 2.1.7 is not a matrix but a *determinant*. In mathematics, a determinant is a single number and the line on each side of the table means that you multiply various combinations of the numbers together and then add all of these products together to get that single number. In this course we shall not be concerned with how to get a determinant or single number from a square table of numbers. Although most mathematicians and people who use and write about matrix algebra also use determinants in matrix algebra, we shall have no use for determinants in this book. The use of determinants needlessly complicates many of the rules and procedures of matrix algebra. Just remember never to use a single ruling on each side of a table of numbers to show that it is a matrix.

Preferred designation

In this book we shall use brackets around the numbers to show that the tables are matrices. These are more convenient than the double ruling on both sides simply because it is difficult for some people to draw two parallel vertical lines quickly. For matrices that consist of a single row we shall also ordinarily use brackets, as in Fig. 2.1.8.

$$[3 \quad 7 \quad 9]$$

FIG. 2.1.8

In general, it is more practical to use brackets for a matrix that has a large number of rows. Actually it doesn't matter much what method you use as long as you are consistent. In no case, however, shall we use braces to show that numbers that have been written in a single row should really have been written in a column. If the matrix is a row, we shall write it as such and if it is a column, we shall write it as a column.

2.2 Specifying the Size of a Matrix

As you might suppose, the size of a matrix is determined by the number of rows and the number of columns it contains. The number of rows and columns is called the *order* of the matrix. In giving the order of a matrix, it is customary to mention the number of rows first and the number of columns second. For example, Fig. 2.1.1 is a four-by-three matrix because it has four rows and three columns. Figure 2.1.2 is a five-by-four matrix because it has five rows and four columns. In giving the order of a matrix it is customary to write "four by three" as *4 by 3*. A five-by-four matrix would be written 5 by 4. For example, if we had a matrix of scores on 10 different tests for 20 different persons, this would be a 20 by 10 matrix.

The term *dimensions* of a matrix is used by social scientists to mean exactly the same thing as order. The word in this use is not common among mathematicians, however. A matrix, as we use the term, has two dimensions; one of *height* and one of *width*. As you would guess, the height of the matrix is simply the number of rows it has. The width is the number of columns. We mention height first and width second.

It is natural, then, that if the height of a matrix is greater than its width, we call it a *vertical* matrix. A vertical matrix has more rows than columns. Figs. 2.1.1, 2.1.2, 2.1.3, 2.1.5, and 2.1.9 are all vertical matrices. Similarly, if the width of a matrix is greater than its height, we call it a *horizontal* matrix. A horizontal matrix has more columns than rows. Figure 2.1.8 is an example of a horizontal matrix. Another example is Fig. 2.2.1.

$$\begin{bmatrix} 6 & 4 & 3 \\ 7 & 2 & 9 \end{bmatrix}$$

Fig. 2.2.1

Remember, however, that Fig. 2.1.4 is not a horizontal matrix because the braces show that it should really have been written in column form and that it is therefore a vertical matrix. Since most data matrices have more rows than columns, they are typically vertical rather than horizontal. The matrix of 10 test scores for 20 persons would of course be a vertical matrix.

It would be just a coincidence if a data matrix had exactly the same number of rows and columns; and as noted in Chapter 1, such a data matrix would not have much scientific value for prediction purposes because it would have the same number of entities as attributes, whereas it should have many more entities than attributes. However, in operating upon data matrices for the purpose of drawing scientific conclusions from them, we often get derived matrices that have the same number of rows as columns, that is, *square* matrices. In a square matrix, of course, the height is equal to the width. An example is Fig. 2.2.2.

$$\begin{bmatrix} 3 & 7 & 9 \\ 4 & 2 & 6 \\ 8 & 1 & 3 \end{bmatrix}$$

Fig. 2.2.2

Here we have a matrix of three rows and three columns, or a 3 by 3 matrix. You can indicate in several ways that a matrix is square. In giving the order you can simply give it as you would for any other matrix. Thus you would say a 3 by 3 matrix, a 4 by 4 matrix, and so on. Perhaps the simplest and best way to indicate that a matrix is square, if you want to give its size at the same time, is simply to say a third-order matrix, a fourth-order matrix, and so on. The fact that you give only one number in specifying the order means that both dimensions of the matrix are equal to this number.

2.3   The Numbers in a Matrix

Methods of designation

Several different names can be used in referring to any single number within a matrix. The number is sometimes called a *coordinate* of the matrix. A less technical way of referring to these numbers is simply to call them entries. Any particular number in the matrix is called an *entry* in the matrix. A third way of referring to the numbers is to call them *elements*. We shall somewhat arbitrarily choose the word *element* to refer to a single number within the matrix. An element in the matrix of test scores is simply the score of a given person on a particular test.

A specific element of the matrix is designated by referring to the row and column in which it is located. As in specifying order, we mention first the row and then the column. For example, the element in the third row, second column, of Fig. 2.1.2 is 3. The element in the third row, third column, of Fig. 2.2.2 is 3.

Total number of elements

The total number of elements in a matrix is obviously its height times its width, or the product of its two dimensions. The total number of elements in Fig. 2.1.3, for example, is 4×3, or 12. The total number of elements in the matrix of test scores is 200.

The principal diagonal

In speaking of a square matrix, such as Fig. 2.2.2, we give a special name to all elements whose row and column number are the same. There would, of course, be as many of these elements as the order of the matrix. In Fig. 2.2.2 there are three such elements. The element in the first row, first column, is 3, in the second row, second column, it is 2, and in the third row, third column, it is 3. These elements form what we call the *principal diagonal* of the matrix. The principal diagonal of a square matrix consists of the elements that have the same number for both the row and column.

2.4   Using a Letter to Stand for a Number in a Matrix

Numerical subscripts of letter elements

You will recall that in elementary algebra you used letters such as a, b, c, x, y, z, and so on to stand for numbers. We use the same practice in matrix algebra. In developing the rules for matrix operations, it is convenient to let letters stand for the elements of the matrix. The letters may do this in various ways. One method is shown in Fig. 2.4.1.

$$\begin{bmatrix} a_1 & a_2 & a_3 \\ b_1 & b_2 & b_3 \\ c_1 & c_2 & c_3 \end{bmatrix}$$

FIG. 2.4.1

Here we use different letters to represent the different rows of the matrix. The letter a is used to represent numbers in the first row, b, numbers in the second, and c, numbers in the third. Here we have a third-order square matrix. To designate the columns, we use a subscript on each letter. (A subscript is a small letter or number written a little below and to the right of another number or letter.) The subscripts in Fig. 2.4.1 designate the columns in which the elements are located. Thus, if Fig. 2.4.1 represented the numbers in Fig. 2.2.2, b_3 would represent 6 because the element in the second row, third column, of Fig. 2.2.2 is 6, and the element in the second row, third column, of Fig. 2.4.1 is b_3.

Another plan you could use to let letters stand for the elements of a matrix is similar to that in Fig. 2.4.1, but the subscript of the letter indicates the row and the letter itself indicates the column, as in Fig. 2.4.2.

$$\begin{bmatrix} a_1 & b_1 & c_1 \\ a_2 & b_2 & c_2 \\ a_3 & b_3 & c_3 \\ a_4 & b_4 & c_4 \end{bmatrix}$$

FIG. 2.4.2

Here you have a 4 by 3 matrix where all the a's are in the first column, all the b's in the second, and all the c's in the third. Every letter with a subscript of 1 is in the first row, every letter with a subscript of 2 is in the second row, and so on. Figure 2.4.2 might be used to represent the matrix in Fig. 2.1.3. Then a_3 would be the element in the third row, first column, which is the number 1.

The third way to indicate that a letter stands for a number in a matrix is to use a single letter with two subscripts. This principle is illustrated in Fig. 2.4.3.

$$\begin{bmatrix} a_{11} & a_{12} & a_{13} \\ a_{21} & a_{22} & a_{23} \\ a_{31} & a_{32} & a_{33} \\ a_{41} & a_{42} & a_{43} \end{bmatrix}$$

FIG. 2.4.3

Here we have used the letter a to stand for each element in the matrix, but a different pair of subscripts gives the location of each element. The first number of the subscript is the number of the row in which the element is found. The second is the number of the column in which the element is found. Thus, here a_{23} is the element in the second row, third column. You will see that the first number in each element is the same in any given row. The second number of the subscript is the same for every column. Without looking at Fig. 2.4.3, then, you should be able to tell what a_{23} is in Fig. 2.1.3. It is the element in the second row, third column, or 5.

In this book, we shall use this third method to let a letter stand for an element of the matrix. It is more convenient than the methods shown in Figs. 2.4.1 and 2.4.2, because as soon as the subscripts of an element are given, you know at once in which row and which column to find it. On the other hand, if you use the methods of either Fig. 2.4.1 or Fig. 2.4.2, you first have to remember which of the two you decided to use, so as to know whether the letters are rows and the subscripts columns, or the other way around. Even assuming you can remember this, if you use more than three or four letters, you might have trouble deciding promptly whether f, say, is the fifth, sixth, or seventh letter in the alphabet.

You must note, however, that although in general we shall use the double subscript system, where the first subscript means row and the second means column, there will be some exceptions when the order of the subscripts will be reversed. We shall call your attention to these exceptions when the occasions arise.

Letter subscripts of letter elements

You have seen how it is convenient to let letters stand for the elements of a matrix and to use subscript numbers to indicate in which row and column each element is located. Sometimes it is also convenient to use letters for the subscripts, as in Fig. 2.4.4.

$$\begin{bmatrix} a_{11} & a_{12} & \cdots & a_{1m} \\ a_{21} & a_{22} & \cdots & a_{2m} \\ \cdots & \cdots & \cdots & \cdots \\ a_{n1} & a_{n2} & \cdots & a_{nm} \end{bmatrix}$$

FIG. 2.4.4

If the matrix has a large number of rows and columns, as many data matrices do, it might be far too much trouble to write out all of the letter elements with their appropriate number subscripts. Therefore we write the first two elements in the first row and then use a series of three dots (ellipsis) to indicate that all of the other elements in the first row except the last are taken for granted. We write the last element as a_{1m}, where m stands for the number of columns. Here we do not commit ourselves as to what m is. We can let it be anything we please to suit any particular matrix. In the same way, the second row contains the first two elements, then three dots, and finally, the elements a_{2m} to indicate that this is the last of the m elements in the row. Then we have a row of dots the full width of the matrix to show that there are a number of other rows that we simply don't take the time to write in.

Finally, we write in the first two elements of the last row, followed by three dots, and the last element. You will notice that each of these elements in the last row has n as a first subscript. This means that we do not commit ourselves specifically as to the number of rows in the matrix. We let n stand for the number of rows, which can be anything we please. You see then that the last element in the last row has letters for both of its subscripts. The first one indicates the number of rows in the matrix and the last one the number of columns. In general, then, the subscripts of the lower right element of the matrix give you the order or dimensions of the matrix, whether in terms of actual numbers or in terms of letters that stand for numbers.

A system closely resembling that in Fig. 2.4.4 is given in Fig. 2.4.5.

$$\begin{bmatrix} a_{11} & \cdots & a_{1m} \\ \cdots & \cdots & \cdots \\ a_{n1} & \cdots & a_{nm} \end{bmatrix}$$

Fig. 2.4.5

Here we give only four elements in the matrix: the first element in the first row, the last element in the first row, the first element in the last row, and the last element in the last row. Figure 2.4.5 is a little more compact than Figure 2.4.4, but some people find it easier to remember just what the dots in a row mean if the second element in each row, and the second row, are given.

A third variation of the use of letter subscripts is shown in Fig. 2.4.6.

$$\begin{bmatrix} a_{11} & \cdots & a_{ij} & \cdots & a_{lm} \\ \cdots & \cdots & \cdots & \cdots & \cdots \\ a_{i1} & \cdots & a_{ij} & \cdots & a_{im} \\ \cdots & \cdots & \cdots & \cdots & \cdots \\ a_{n1} & \cdots & a_{nj} & \cdots & a_{nm} \end{bmatrix}$$

Fig. 2.4.6

Here we give three rows of the matrix and three elements in each row. The method is similar to Fig. 2.4.5 except that in row 1 we put an element between the first and the last one, with the second subscript given as j. We also put a row between the first and last row, with i as the first subscript. We mean that this is the ith row, or any unspecified row between the first and last one. In the same way, the column between the first and last one means that this is the jth column, or any unspecified column between the first and last one. The intermediate element in the ith row, then, is a_{ij}. We can then refer to the ijth element of the matrix, without specifying which particular row or column the element is in, so as to make general statements about the matrix elements.

In all three examples given in Figs. 2.4.4, 2.4.5, and 2.4.6, you will see that whenever a letter is used for a subscript it has the same meaning as a number. When the letter is in the first position, it refers to the row in which the element is found; in the second position, it refers to the column. In all three methods, the subscripts of the lower right element indicate the dimensions of the matrix. Thus, in each of the three examples, the element a_{nm} means that we have an $n \times m$ matrix, or a matrix with n rows and m columns.

2.5 Using a Single Letter to Stand for an Entire Matrix

What makes matrix algebra so useful in working with large tables of numbers is that we can let a single letter stand for an entire matrix. Figure 2.5.1 is an example.

$$a = \begin{bmatrix} a_{11} & a_{12} & \cdots & a_{1m} \\ a_{21} & a_{22} & \cdots & a_{2m} \\ \cdots & \cdots & \cdots & \cdots \\ a_{n1} & a_{n2} & \cdots & a_{nm} \end{bmatrix}$$

FIG. 2.5.1

Here we let the single letter a without any subscripts represent an entire matrix whose elements are represented by the letter a with subscripts. Another example is Fig. 2.5.2.

$$b = \begin{bmatrix} b_{11} & \cdots & b_{1m} \\ \cdots & \cdots & \cdots \\ b_{n1} & \cdots & b_{nm} \end{bmatrix}$$

FIG. 2.5.2

Here the letter b stands for a matrix whose elements are represented by the letter b with subscripts. You can choose any letter you wish to represent a matrix. In the two examples, we use the same letter to stand for the matrix and for the individual elements in the matrix. This is not necessary, however, for we can use a single letter to represent a matrix whose elements are numbers. We can use the letter c, for example, to stand for the matrix given in Fig. 2.2.2. This we would write as in

$$c = \begin{bmatrix} 3 & 7 & 9 \\ 4 & 2 & 6 \\ 8 & 1 & 3 \end{bmatrix}$$

FIG. 2.5.3

Neither is it necessary that we use the same letter to stand for the matrix even if the elements are letters. We could, for example, write a matrix as in Fig. 2.5.4.

$$F = \begin{bmatrix} c_{11} & c_{12} & c_{13} \\ c_{21} & c_{22} & c_{23} \\ c_{31} & c_{32} & c_{33} \\ c_{41} & c_{42} & c_{43} \end{bmatrix}$$

FIG. 2.5.4

Here we let F stand for a matrix whose elements are represented by the letter c. There is nothing in the rules of matrix algebra to tell us how to select a single letter to stand for an entire matrix. Such decisions must be made on the basis of simplicity or convenience. Other things being equal, you would doubtless prefer to use the plan of Figs. 2.5.1 or 2.5.2, where the letter that stands for the entire matrix is the same as the letter that stands for the elements in the matrix.

Using a single letter to stand for a number greatly simplifies algebraic manipulations and also the description of the computation worksheets required for the analysis of a particular set of data.

Specification of order

If you use a single letter to stand for an entire matrix, then you must be sure to specify the order of the matrix. This can be done in several different ways. You can, for example, simply write down the equation as given in Figs. 2.5.1, 2.5.2, 2.5.3, or 2.5.4; then anyone can tell by looking at the right side of the equation what the order of the matrix is. Once you have done this, you can go ahead using the letter symbol in your operations without bothering to say anything further about its order. Actually, however, you do not even need to write down the equations as given in these examples. You could simply say, for example, let a be an n by m matrix, with n and m standing for any unspecified but different numbers. That is enough to define the order, and then you can proceed to apply the rules of matrix algebra which you will learn on the following pages.

Another method used to indicate the order of a matrix is to use subscripts on the letter which stands for the entire matrix. If you do this, you must state clearly, however, that the subscripts indicate the number of rows and columns in the matrix and that the letter stands for the entire matrix. (Subscripts may be used for various other purposes.) If you adopt this practice, the first subscript will ordinarily represent the number of rows and the second letter the number of columns. For example, you could say, "Let a_{nm} be an n by m matrix." It is perfectly clear then that the symbol a_{nm} refers, not to the element in the nth row and mth column, but to an entire matrix with n rows and m columns. Usually when subscripts are used to indicate the order of a matrix, they will be letters rather than numbers. This, however, is not an absolute rule and we may occasionally make exception to it. The use of letter subscripts to indicate the order of a matrix is of value when we take up supermatrices in Chapter 5, and matrix multiplication in Chapter 9.

SUMMARY

1. Showing that a table of numbers is a matrix:
   a. If a table of numbers is designated as a matrix, it is subject to a well defined set of operations.
   b. A table of numbers may be designated as a matrix by enclosing it in double rulings on each side, parentheses, or brackets. Braces may be used to enclose a row of numbers that is meant to be a column.
   c. The preferred designation in this book is the use of brackets.

2. How to specify the size of a matrix:
   a. The number of rows and columns in a matrix is called its *order*. The number of rows is usually mentioned first.
   b. By dimensions we mean the same thing as order. The *height* is the number of rows and the *width* the number of columns. A *vertical* matrix has more rows than columns. A *horizontal* matrix has more columns than rows.

3. The numbers in a matrix:
   a. A single number in a matrix is called an *element* of the matrix.
   b. It may also be called a *coordinate* or an entry.
   c. When a particular element in the matrix is specified, its row position is usually mentioned first and its column second.
   d. The number of elements in a matrix is the product of its height and width.
   e. The principal diagonal of a square matrix consists of the elements whose row and column positions are the same.

4. Using a letter to stand for a number:
   a. A letter with double numerical subscripts may be used to indicate the

elements in the matrix. The first number in the subscript generally indicates the element's row position and the second its column.

  b. A letter element may have letter subscripts to designate any row and column location.

5. Using a single letter to stand for a matrix:

  a. A single letter used to stand for an entire matrix simplifies algebraic manipulations and the description of computation worksheets.

  b. The order of a matrix represented by a single letter may be indicated verbally, as, for example, n by m, (3 by 4), etc.

  c. A single letter standing for a matrix may carry double subscripts to indicate its order; usually the first subscript is its height and the second its width.

EXERCISES

1. Which of the following is not a matrix?

$$\begin{bmatrix} 2 & 3 & 4 \\ 3 & 5 & 1 \\ 2 & 6 & 7 \end{bmatrix} \quad \begin{pmatrix} 2 & 3 & 4 \\ 3 & 5 & 1 \\ 2 & 6 & 7 \end{pmatrix} \quad \begin{vmatrix} 2 & 3 & 4 \\ 3 & 5 & 1 \\ 2 & 6 & 7 \end{vmatrix} \quad \begin{Vmatrix} 2 & 3 & 4 \\ 3 & 5 & 1 \\ 2 & 6 & 7 \end{Vmatrix}$$
$$\quad A \qquad\qquad B \qquad\qquad C \qquad\qquad D$$

2. What are the heights of matrices whose orders are as follows:

  (a) 4 by 3   (b) 7 by 10   (c) 20 by 4   (d) (n by m)

3. How many elements do each of the matrices in 2 above have?

4. Which of the following matrices have principal diagonals?

$$\begin{bmatrix} 1 & 4 & 3 \\ 2 & 1 & 7 \end{bmatrix} \quad \begin{bmatrix} 2 & 4 \\ 3 & 1 \\ 1 & 7 \end{bmatrix} \quad \begin{bmatrix} 1 & 5 & 7 \\ 2 & 4 & 6 \\ 3 & 2 & 1 \end{bmatrix} \quad \begin{bmatrix} 1 & 7 & 8 & 4 \\ 2 & 5 & 6 & 9 \\ 7 & 4 & 3 & 2 \end{bmatrix} \quad \begin{bmatrix} 2 & 3 \\ 8 & 1 \end{bmatrix}$$
$$\quad A \qquad\qquad B \qquad\qquad C \qquad\qquad\quad D \qquad\qquad\quad E$$

5. Given

$$b = \begin{bmatrix} 2 & 4 & 3 & 5 \\ 7 & 6 & 4 & 2 \\ 8 & 5 & 9 & 4 \\ 3 & 1 & 2 & 7 \\ 6 & 8 & 4 & 10 \end{bmatrix}$$

What are the numerical values of the following elements: b_{13}, b_{24}, b_{53}, b_{42}, b_{31}?

6. Given

$$a = \begin{bmatrix} a_{11} & \cdots & a_{1m} \\ \cdots & \cdots & \cdots \\ a_{n1} & \cdots & a_{nm} \end{bmatrix} \qquad \begin{aligned} m &= 3 \\ n &= 4 \end{aligned}$$

What is the height and width of a?

7. A_{rs} is a matrix. Without further information what would you take to be its height and width?

ANSWERS

1. C

2. (a) 4   (b) 7   (c) 20   (d) n

3. (a) 12   (b) 70   (c) 80   (d) nm

4. C and E

5. 3, 2, 4, 1, 8

6. Height 4, Width 3

7. Height r, Width s

Chapter 3

Kinds of Matrices

There are various kinds of matrices that are particularly important in the behavioral sciences. We shall list them for ready reference.

3.1 The General Rectangular Matrix

You are already familiar with the rectangular matrix, in which the height and the width, or the number of rows and the number of columns, is generally not the same. As you have seen, most of the data we get from life situations are of the kind that can be arranged in the form of a rectangular matrix. Also, most matrices, to be useful for scientific prediction, must have more rows than columns, or more entities than attributes. While most data matrices are rectangular, it does not follow that all rectangular matrices are necessarily data matrices. Many derived matrices, that is, matrices obtained by applying the rules of matrix algebra to data matrices, may also be rectangular.

3.2 The Square Matrix

Usually we think of a rectangle as being larger in one direction than the other. You have seen, however, that we can have a special case of a rectangular matrix in which the height and the width are the same. This is a square matrix, one having the same number of rows as columns. In Fig. 3.2.1 two square matrices are shown; one has two rows and columns and the other four rows and columns.

$$\begin{bmatrix} 2 & 3 \\ 1 & 7 \end{bmatrix} \quad \begin{bmatrix} 2 & 4 & 1 & 7 \\ 3 & 2 & 4 & 1 \\ 6 & 7 & 4 & 2 \\ 4 & 3 & 2 & 8 \end{bmatrix}$$

Fig. 3.2.1

If we use letters to stand for the elements of a square matrix and double number subscripts to indicate the row and column for each element, then the last subscript for all elements in the last column will be the same as the first subscript for all elements in the last row. Figure 3.2.2 is an example of a square matrix of this kind.

$$\begin{bmatrix} a_{11} & a_{12} & a_{13} \\ a_{21} & a_{22} & a_{23} \\ a_{31} & a_{32} & a_{33} \end{bmatrix}$$

FIG. 3.2.2

You see that, in this third-order square matrix, 3 is the last subscript for each element in the last column, and 3 is also the first subscript for each element in the last row. Notice that in a square matrix both subscripts of the lower right-hand element are equal. In this case the subscripts are both 3. Another example of a square matrix with letter elements is given in Fig. 3.2.3.

$$\begin{bmatrix} a_{11} & a_{12} & \cdots & a_{1n} \\ a_{21} & a_{22} & \cdots & a_{2n} \\ \cdots & \cdots & \cdots & \cdots \\ a_{n1} & a_{n2} & \cdots & a_{nn} \end{bmatrix}$$

FIG. 3.2.3

Here the last subscript for each element in the last column is n, and the first subscript of each element in the last row is also n. For the last, or lower right-hand element, both subscripts are n.

As you saw in Chapter 2, Section 2, if we let a single letter such as a represent the entire square matrix, we can indicate that the matrix is square by saying that a is an nth-order square matrix, or simply, a is an nth-order matrix. If we decide to let subscripts indicate the order of the matrix, we would let a_{nn} be a square matrix of order n. We must be sure to state this definition or the symbol might be mistaken for the last element in the last row of a square matrix.

3.3 Symmetric Square Matrices

A special kind of square matrix that is useful in the analysis of data in the behavioral sciences is the *symmetric matrix*. This is usually a derived rather than a data matrix. A symmetric matrix may be defined as one in which the corresponding elements of corresponding rows and columns are equal. It is shown in Fig. 3.3.1.

$$\begin{bmatrix} 11 & 4 & 3 \\ 4 & 15 & 10 \\ 3 & 10 & 17 \end{bmatrix}$$

FIG. 3.3.1

Look at row 1 and the corresponding column, which is column 1. The second element in row 1 is 4 and the corresponding, or second, element in column 1 is also 4. In column 2, you see that the elements are, in order, 4, 15, and 10. In the corresponding row, that is, the second row, you see that the elements are also 4, 15, and 10.

Subscript designation

If you choose to indicate a symmetric matrix by using a letter for the element, and double subscripts to give its position, you may do so in several ways. Fig. 3.3.2 shows one way.

$$\begin{bmatrix} a_{11} & a_{12} & a_{13} \\ a_{12} & a_{22} & a_{23} \\ a_{13} & a_{23} & a_{33} \end{bmatrix}$$

FIG. 3.3.2

Notice here that the subscripts for the element in the second row, first column, are in the same order as the subscripts for the element in the first row, second column. In both cases, the subscripts are, in order, 1, 2: The rule for writing these subscripts is simple. First write the elements with their subscripts in the principal diagonal. Then write the elements with their subscripts as you would normally, above and to the right of the principal diagonal. Finally write in the subscripts for the elements below the principal diagonal in just the reverse order from which you would normally write them.

This method of writing subscripts is one of the exceptions to the rule given in Section 2.4, that the first subscript indicates the row and the second subscript indicates the column.

It is common practice to use the system of subscripts shown in Fig. 3.3.2 to indicate that a matrix is symmetric. We may, however, stick to the original rule, as in Fig. 3.2.3. Then, to show that the matrix is symmetric, we say that $a_{ij} = a_{ji}$. This means that the element in the ith row and the jth column is the same as the element in the jth row and the ith column. For example, in Fig. 3.3.2 the element in the first row, third column, is the same as the element in the third row, first column.

A very satisfactory method of letting a single letter stand for a symmetric matrix is simply to say, for example, let x be a symmetric matrix. Since a symmetric matrix is clearly defined, there would be no question as to what is meant by this statement.

Writing the symmetric matrix

Very frequently, you will find that when a symmetric matrix of numbers is written down, the numbers above the principal diagonal are omitted, and it is assumed that you know how to fill them in. Or, just the reverse practice may be followed; the numbers below the principal diagonal may be omitted. Even though this practice is frequent, it is usually not recommended. First, it is much more convenient in practical applications to have all of the numbers in the table written in. Second, if the numbers are not written in, they may reasonably be assumed to be 0. As we shall see, there is a special kind of matrix in which the numbers, either above or below the principal diagonal, *are* 0. It would be a serious mistake to take a symmetric matrix for a matrix of this type.

A convenient way to write down a symmetric matrix of numbers is first to write the numbers in the principal diagonal and those above and to the right. (These numbers are often called the *supradiagonal* elements.) You can then fill in the matrix below the principal diagonal by starting with the first column and copying, in order down the column, each number in the first row. Then you go to the second column, and beginning immediately below the second diagonal, you copy in order each of the numbers in the row to the right of the second diagonal. (Numbers below the diagonal are termed *infradiagonal* elements.) In this way you complete each of the unfinished columns in the matrix so that it will, in fact, be a symmetric matrix, as you can well verify.

The most common examples of symmetric matrices used in the anaylsis of data from the behavioral sciences are the statistician's *correlation* and *covariance* matrices. The nondiagonal elements in the symmetric matrix are correlation coefficients or covariances. If you have had a course in statistics, you are already familiar with these terms. You will learn more about them in Chapters 12 and 13.

3.4   The Diagonal Symmetric Matrix

You have seen that the square matrix is a special case of the rectangular matrix, and the symmetric matrix is a special case of the square matrix. There are also special cases of the symmetric matrix. The most general of these is the *diagonal matrix*. In a diagonal matrix, all elements except those in the principal diagonal are 0 (Fig. 3.4.1).

$$\begin{bmatrix} 15 & 0 & 0 \\ 0 & 12 & 0 \\ 0 & 0 & 9 \end{bmatrix}$$

Fig. 3.4.1

Another way to define a diagonal matrix is to say that every element is 0 except those whose row and column numbers are equal. In Fig. 3.4.1 only the elements in the first row, first column; in the second row, second column; and in the third row, third column, are different from 0.

Methods of writing the diagonal matrix

You may quite properly write a diagonal matrix with no entries whatever in the nondiagonal positions, as in Fig. 3.4.2.

$$\begin{bmatrix} 5 & & \\ & 12 & \\ & & 9 \end{bmatrix}$$

Fig. 3.4.2

If a matrix is written in the form given in Fig. 3.4.2, all elements in the nondiagonal positions are understood to be zero.

We may write a diagonal matrix with letter elements in the same way, as in Fig. 3.4.3.

$$\begin{bmatrix} a_{11} & & & \\ & a_{22} & & \\ & & \cdots & \\ & & & a_{nn} \end{bmatrix}$$

Fig. 3.4.3

This is a diagonal matrix of order n. You see that for each element the first and second subscripts are the same.

It is not necessary, however, to use double subscripts when you let letters stand for the elements in a diagonal matrix. You may use single subscripts as in Fig. 3.4.4.

$$\begin{bmatrix} a_1 & & & \\ & a_2 & & \\ & & \cdots & \\ & & & a_n \end{bmatrix}$$

Fig. 3.4.4

It is perfectly clear that Fig. 3.4.4 is a diagonal matrix in which a_1 is the first diagonal element, a_2 the second, and so on.

A common example of a diagonal matrix used frequently in the analysis of data from the human sciences has for its elements the *standard deviations* of the attributes. These are indicated by the small Greek letter sigma, σ. A diagonal matrix made up of standard deviations is given in Fig. 3.4.5.

$$\begin{bmatrix} \sigma_1 & & \\ & \sigma_2 & \\ & & \sigma_3 \end{bmatrix}$$

Fig. 3.4.5

Single letter designation of a diagonal matrix

We may let a single letter stand for a diagonal matrix as well as any other kind of matrix. But, since the diagonal matrix has so many interesting and useful properties, and since it is used so frequently in the analysis of data in the behavioral sciences, a special, easily recognized symbol for it is useful. Therefore, we usually let some form of the letter d represent a diagonal matrix. The following symbols are commonly used: D, d, Δ, δ.

Various subscripts serve to indicate the different diagonal matrices we are working with. In Fig. 3.4.6, 3.4.7, and 3.4.8 subscripts are used on the letter D to indicate various diagonal matrices.

$$D_a = \begin{bmatrix} a_1 & & & \\ & a_2 & & \\ & & \cdots & \\ & & & a_n \end{bmatrix}$$

FIG. 3.4.6

$$D_b = \begin{bmatrix} b_1 & & \\ & b_2 & \\ & & b_3 \end{bmatrix}$$

FIG. 3.4.7

$$D_\sigma = \begin{bmatrix} \sigma_1 & & & \\ & \sigma_2 & & \\ & & \sigma_3 & \\ & & & \sigma_4 \end{bmatrix}$$

FIG. 3.4.8

Here Fig. 3.4.6 is an nth-order diagonal matrix, whose diagonal elements are a_1, a_2, etc. In Fig. 3.4.7 we let D_b equal a diagonal matrix whose diagonal elements are b_1, b_2, and b_3. The subscript σ to D in Fig. 3.4.8 indicates a diagonal matrix each of whose elements is a standard deviation. Such matrices are useful in deriving correlation matrices from covariance matrices.

3.5 Scalar Diagonal Matrices

A special case of a diagonal matrix is the *scalar matrix*, a diagonal matrix all of whose diagonal elements are equal (Fig. 3.5.1).

$$\begin{bmatrix} 3 & 0 & 0 \\ 0 & 3 & 0 \\ 0 & 0 & 3 \end{bmatrix}$$

FIG. 3.5.1

Here every diagonal element is 3. We can also use letter elements to indicate a scalar matrix. Then we omit.the subscripts on the diagonal elements to show that they are all equal. In Fig. 3.5.2 letters are used to designate the elements of a scalar matrix.

$$\begin{bmatrix} a & 0 & 0 \\ 0 & a & 0 \\ 0 & 0 & a \end{bmatrix}$$

Fig. 3.5.2

3.6   The Identity Matrix

A special case of a scalar matrix is the *identity matrix*. The identity matrix is a scalar matrix all of whose diagonal elements are 1. Since it is still a special case of a diagonal matrix, all of the nondiagonal elements are 0. Figure 3.6.1 is an example of a third-order identity matrix.

$$\begin{bmatrix} 1 & 0 & 0 \\ 0 & 1 & 0 \\ 0 & 0 & 1 \end{bmatrix}$$

Fig. 3.6.1

When a single letter is to stand for an identity matrix, we may use the capital letter I. In this book, I always stands for an identity matrix. Not everyone uses the capital letter I for the identity matrix; sometimes the number 1 is used to represent it. The number 1 has a more convenient use, however, for another kind of matrix, which will be considered soon. There-fore we shall not use the number 1 for the identity matrix, even though some writers do.

The identity matrix serves the same purposes in matrix algebra as the number 1 does in ordinary algebra, or what we shall now call *scalar* algebra. Scalar algebra is simply the algebra that deals with single numbers rather than tables of numbers.

You can indicate the order of an identity matrix in several ways. One is simply to state the order by saying, let I be a fourth-order identity matrix. Sometimes, in the equations you will be using, it is easy to tell the order by the position in which the identity matrix appears. In such a case it is not necessary to specify the order. It is also sometimes convenient to specify the order by the use of a subscript. For example, I_3 would be a third-order identity matrix. You could represent Fig. 3.6.1 then by I_3; or, in general, you could let I_n be an nth-order identity matrix.

3.7   The Sign Matrix

The sign matrix is of particular importance in research on personality traits. It makes possible designation of an attribute, such as a personality trait, in terms of its opposite. Thus the plus number may mean introversion and the minus number extroversion.

The sign matrix is similar to the identity matrix, except that some of the 1's in the diagonal may carry minus signs. A sign matrix is shown in Fig. 3.7.1.

$$\begin{bmatrix} -1 & 0 & 0 \\ 0 & 1 & 0 \\ 0 & 0 & -1 \end{bmatrix}$$

Fig. 3.7.1

The sign matrix is seldom treated as such in standard mathematical books and articles on matrices. Nevertheless, because it is useful in analyzing data from the behavioral sciences, we give it this special name, using the small letter i to represent it. A subscript can be used to indicate the order of the sign matrix as well as the identity matrix. Unless otherwise specified, however, we shall state the order in words and reserve subscripts to indicate the particular sign matrix we are talking about. Figure 3.7.2, for example, uses i_a to indicate a second-order sign matrix whose first diagonal element is -1 and whose second is 1.

$$i_a = \begin{bmatrix} -1 & \\ & 1 \end{bmatrix}$$

Fig. 3.7.2

Figure 3.7.3, on the other hand, uses b as a subscript, so that i_b is a matrix whose first diagonal element is 1 and whose next two diagonal elements are -1.

$$i_b = \begin{bmatrix} 1 & & \\ & -1 & \\ & & -1 \end{bmatrix}$$

Fig. 3.7.3

Note that we have not used 0's in the nondiagonal positions in either Fig. 3.7.2 or 3.7.3. It is not necessary to use 0's either for sign matrices or for identity matrices, even though we did use them for clarity in Fig. 3.6.1.

3.8   Skew Symmetric Matrices

Another special kind of square matrix is the *skew symmetric matrix*. In a skew symmetric matrix, the principal diagonal is 0, and corresponding elements above and below it are of equal absolute value but are opposite in sign. Another way of saying this is that a_{ij} is equal to $-a_{ji}$. An example of a skew symmetric matrix is Fig. 3.8.1.

$$\begin{bmatrix} 0 & -3 & 4 \\ 3 & 0 & -6 \\ -4 & 6 & 0 \end{bmatrix}$$

FIG. 3.8.1

Here you see that the element in the first row, second column, is -3, and the element in the second row, first column, is plus 3. The element in the first row, third column, is 4, and the element in the third row, first column, is -4. The skew symmetric matrix plays an important role in certain psychological and sociological scaling techniques.

3.9   Triangular Matrices

Upper triangular matrix

A derived matrix that is useful in the behavioral sciences is a square matrix in which all elements below the principal diagonal are equal to 0. This is an *upper triangular matrix* (Fig. 3.9.1).

$$\begin{bmatrix} 2 & 5 & 7 \\ 0 & 3 & 6 \\ 0 & 0 & 7 \end{bmatrix} \quad \text{or} \quad \begin{bmatrix} 2 & 5 & 7 \\ & 3 & 6 \\ & & 7 \end{bmatrix}$$

FIG. 3.9.1

Note that in the left-hand matrix of Fig. 3.9.1, 0's are written in below the principal diagonal element. In the right-hand matrix the corresponding spaces are left blank. This is simply another way of writing an upper triangular matrix. Where no values are written in below the principal diagonal, we assume that these are 0.

You will recall that we discouraged the common practice of omitting the elements in a symmetric matrix from either above or below the principal diagonal, because the omitted elements might be assumed to be 0's. Therefore if the elements *are* omitted below the principal diagonal, we assume that the matrix is an upper triangular matrix and not a symmetric matrix. Another way of defining an upper triangular matrix is to say that an element is zero if its row number is greater than its column number.

Lower triangular matrix

As you might well guess, we can also have a triangular matrix in which all elements *above* the principal diagonal are zero. This is a *lower triangular matrix*, shown in Fig. 3.9.2.

$$\begin{bmatrix} 6 & 0 & 0 \\ 3 & 4 & 0 \\ 5 & 2 & 6 \end{bmatrix} \quad \text{or} \quad \begin{bmatrix} 6 & & \\ 3 & 4 & \\ 5 & 2 & 6 \end{bmatrix}$$

FIG. 3.9.2

For the lower triangular matrix also we may write in the zeros above the principal diagonal or omit them. In a lower triangular matrix an element is zero if its column number is greater than its row number.

If we let letters stand for the numbers in a triangular matrix, we ordinarily omit the zero entries, as in Fig. 3.9.3.

$$\begin{bmatrix} a_{11} & a_{12} & a_{13} \\ & a_{22} & a_{23} \\ & & a_{33} \end{bmatrix} \quad \begin{bmatrix} a_{11} & & \\ a_{21} & a_{22} & \\ a_{31} & a_{32} & a_{33} \end{bmatrix}$$

FIG. 3.9.3

The left-hand matrix is an upper triangular matrix and the right-hand one is a lower triangular matrix. A single letter may designate either an upper or lower triangular matrix. We may say simply, let a be an upper or lower triangular matrix, as the case may be; or, for an upper triangular matrix, we may say that a is a matrix such that a_{ij} is 0 if i is greater than j. For a lower triangular matrix, we may say that a is a matrix such that a_{ij} is 0 if j is greater than i. In the behavioral sciences, triangular matrices are used in intermediate stages of the solutions of prediction problems.

3.10 Partial Triangular Matrices

Upper partial triangular matrix

Another kind of matrix that is very common and useful in the behavioral sciences is the *partial triangular matrix*. A partial triangular matrix is not a square matrix. As we have two types of triangular matrices, so we have two types of partial triangular matrices, namely the upper partial triangular matrix and the lower partial triangular matrix. An upper partial triangular matrix is shown in Fig. 3.10.1.

$$\begin{bmatrix} 2 & 5 & 7 & 9 & 4 & 8 \\ 0 & 3 & 6 & 5 & 5 & 2 \\ 0 & 0 & 7 & 8 & 4 & 9 \end{bmatrix}$$

FIG. 3.10.1

This is a horizontal matrix in which an element is 0 if its row number is larger than its column number. In an upper partial triangular matrix all elements in the lower left hand corner are 0.

Lower partial triangular matrix

A lower partial triangular matrix is one in which an element is 0 if its column number is greater than its row number. Figure 3.10.2 is a lower partial triangular matrix.

$$\begin{bmatrix} 6 & 0 & 0 \\ 3 & 4 & 0 \\ 5 & 2 & 6 \\ 2 & 8 & 9 \\ 9 & 5 & 8 \end{bmatrix}$$

FIG. 3.10.2

A lower partial triangular matrix is always a vertical matrix. The 0 elements are in the upper right hand corner of the matrix. Like the triangular matrix, the partial triangular matrix typically arises in solutions to prediction or estimation problems that involve a number of predictor and criterion variables.

3.11   The Zero-One Matrix

The general case

One general kind of matrix used extensively in the behavioral sciences is the *zero-one* matrix. Each element in this matrix is either 0 or 1 (Fig. 3.11.1).

$$\begin{bmatrix} 0 & 1 & 1 & 0 \\ 0 & 0 & 1 & 0 \\ 1 & 1 & 1 & 0 \\ 0 & 0 & 1 & 1 \\ 1 & 1 & 0 & 1 \end{bmatrix}$$

FIG. 3.11.1

Many data matrices are zero-one matrices. For a simple example, suppose that the entities or rows in a zero-one matrix are people and the columns or attributes are the separate items in an objective test or examination. If a person answers an item correctly, the element corresponding to that person and that item is 1. If he answers an item incorrectly, the element corresponding to that person and the incorrectly answered item is 0. For another example of a zero-one matrix that you might get from sociological data, suppose the entities are different families. Suppose the attributes or columns represent items indicative of the families' standard of living, such as automobiles, telephones, electric refrigerators, electric dishwashers, television sets, and so on. If a family possessed one of these items, you would put a 1 in the appropriate column for that family. If it did not possess the item you would put a 0. There are many other ways in which zero-one matrices may arise from attributes describing or characterizing large numbers of people or entities. The zero-one type of matrix is extremely useful not only because in it data can be represented simply, but also because operations with it are relatively simple.

The permutation matrix

Several kinds of zero-one matrices are not data matrices but are nevertheless useful and important. One of these is the *permutation* matrix. It serves a most important function in applications of matrix algebra to experimental data. It is a square matrix; each column in the matrix has a single 1 and all the other elements in the column are 0. Similarly one of the elements in a row is 1 and all the rest of the elements in the row are 0. A permutation matrix is shown in Fig. 3.11.2.

$$\begin{bmatrix} 0 & 1 & 0 & 0 \\ 0 & 0 & 1 & 0 \\ 1 & 0 & 0 & 0 \\ 0 & 0 & 0 & 1 \end{bmatrix}$$

FIG. 3.11.2

You see that in this example each row has only one 1 in it, as does each column. Notice that according to the definition, the identity matrix is a special case of a permutation matrix. The identity matrix is one in which all nondiagonal elements are 0 and all diagonal elements are 1. This of course means that each row and each column has a single 1 in it and all of the other elements are 0. Since the permutation matrix is so generally useful in matrix algebra, we use the special symbol, π (Greek letter pi), to represent it.

You recall that the order in which both entities and attributes are arranged may be more or less arbitrary. The permutation matrix may be used to rearrange the order of either or both. In general it is used to rearrange rows or columns of any matrix.

3.12 The e_{ij} Matrix

A special case of a zero-one matrix is one all of whose elements are 0 except one and that is simply the number 1. This rather peculiar matrix has some practical uses in isolating and rearranging specific rows or columns in a matrix. An e_{ij} matrix is given in Fig. 3.12.1.

$$\begin{bmatrix} 0 & 0 & 0 & 0 \\ 0 & 0 & 1 & 0 \\ 0 & 0 & 0 & 0 \\ 0 & 0 & 0 & 0 \\ 0 & 0 & 0 & 0 \end{bmatrix}$$

FIG. 3.12.1

Here we have a 5 by 4 matrix in which all elements are 0 except the element in the second row, third column. This element is 1.

To designate a particular e_{ij} matrix we use numbers for the subscripts i and j to indicate the row and column in which the unit element is found. For example, Fig. 3.12.1 is an e_{23} matrix.

3.13 The Zero (Null) Matrix

Just as in scalar algebra, or for that matter, ordinary arithmetic, zero is a number, so also in matrix algebra there is a zero matrix. As you would guess, a zero matrix is one in which every element is 0, as in Fig. 3.13.1.

$$\begin{bmatrix} 0 & 0 & 0 \\ 0 & 0 & 0 \\ 0 & 0 & 0 \\ 0 & 0 & 0 \end{bmatrix}$$

FIG. 3.13.1

Here we have a 4 by 3 zero matrix. Another name more commonly used by mathematicians for a zero matrix is the *null* matrix. Since this term has come into such general usage, we shall also use it.

From the definition of a null matrix you see that it can be of any order. Ordinarily we indicate a null matrix by a simple 0 and specify its order. For example, we say let 0 be a 3 by 4 or n by m null matrix.

3.14 Vectors

A matrix may consist of only one row or one column. This form, called a *vector*, has many important uses in scientific prediction. The two general varieties of vectors are of course the *column vector*, a matrix of only one column, and the *row vector*, a matrix of only one row. Figure 3.14.1 is a column vector.

$$\begin{bmatrix} 9 \\ 12 \\ 13 \end{bmatrix}$$

FIG. 3.14.1

Figure 3.14.2 is a row vector.

$$[10 \quad 12 \quad 13 \quad 14]$$

FIG. 3.14.2

The order of a vector is specified simply by indicating the number of its elements. For example, Fig. 3.14.1 is a third-order column vector. Fig. 3.14.2 is a fourth-order row vector.

The vector is often the type of matrix in which we are most interested in problems of scientific prediction. For example, in Chapter 1, you recall, we talked about the weights to be applied to the predictor attributes in order to reconstruct or predict the criterion attribute. These weights may be represented as a vector in which the weight for each attribute is an element of the vector.

When letters are to stand for the elements of a vector, we need only one subscript, since there is only one row or column. In Fig. 3.14.3 the elements of a row vector are represented by letters.

$$[a_1 \quad a_2 \quad a_3]$$

FIG. 3.14.3

In Fig. 3.14.4, the elements of a column vector are letters.

$$\begin{bmatrix} b_1 \\ b_2 \\ b_3 \\ b_4 \end{bmatrix}$$

FIG. 3.14.4

This is a fourth-order column vector.

We use the same method to indicate that a vector is of order n that we used for other matrices. On the left in Fig. 3.14.5 is an nth order row vector, and on the right, an nth order column vector.

$$[a_1 \quad a_2 \quad \ldots \quad a_n] \qquad \begin{bmatrix} b_1 \\ b_2 \\ \ldots \\ b_n \end{bmatrix}$$

FIG. 3.14.5

It is not necessary, however, to write in the second element of these vectors. To simplify the notation, we can write these vectors including only the first and last elements, as in Fig. 3.14.6.

$$[a_1 \quad \ldots \quad a_n] \quad \begin{bmatrix} b_1 \\ \ldots \\ b_n \end{bmatrix}$$

FIG. 3.14.6

Since the vector has so many special properties, and since it is convenient to recognize it wherever it is used, we ordinarily use a distinctive symbol for it, namely V or v. To indicate different vectors, we can use different subscripts. For example, in Fig. 3.14.6 we can represent the right-hand vector by V_b.

3.15   The Unit Vector

We may have a vector in which all the elements are 1 or unity. This is called the *unit vector*. The unit vector is extremely useful in the analysis of experimental data to sum rows or columns of a matrix. It may be either a row or column vector. In Fig. 3.15.1, row and column unit vectors are given.

$$[1 \quad 1 \quad 1 \quad 1 \quad 1] \quad \begin{bmatrix} 1 \\ 1 \\ 1 \end{bmatrix}$$

FIG. 3.15.1

The unit column vector is indicated simply by the number 1, as in Fig. 3.15.2.

$$1 = \begin{bmatrix} 1 \\ 1 \\ 1 \end{bmatrix}$$

FIG. 3.15.2

To show a unit vector is a row, we put a prime after the 1 as in Fig. 3.15.3.

$$1' = [1 \quad 1 \quad 1]$$

FIG. 3.15.3

You will see in Chapter 4 why we use the prime for this purpose.

To indicate the order of a unit vector, we may use a subscript. For example, we could use a subscript 3 to indicate a unit vector of order 3 as in

$$1_3 = \begin{bmatrix} 1 \\ 1 \\ 1 \end{bmatrix}$$

FIG. 3.15.4

Ordinarily, however, the way in which the vector is used serves to indicate its order.

3.16   The Sign Vector

The *sign vector* is similar to a unit vector, but one or more of its elements is minus 1, as shown in Fig. 3.16.1.

$$[1 \quad -1 \quad -1 \quad 1]$$

FIG. 3.16.1

Sign vectors may be either row or column vectors. They are useful in factor analysis, and in general, where some rows or columns are to be added and others subtracted.

3.17   The Zero-One Vector

As a special case of a zero-one matrix, we may also have *zero-one vectors*. In a zero-one vector each element is either 0 or 1, as in Fig. 3.17.1.

$$\begin{bmatrix} 0 \\ 1 \\ 0 \\ 1 \\ 1 \end{bmatrix}$$

FIG. 3.17.1

A variety of zero-one vector is a vector all of whose elements are 0 except one, and that is unity. This we call an e_i *vector*, or, sometimes, an *elementary* vector. In Fig. 3.17.2, row and column e_i vectors are given. The e_i vector is used extensively in the analysis of data from the behavioral sciences. It serves to segregate rows, columns, and elements of a matrix. (This is explained in Section 10.17.)

$$[0 \quad 1 \quad 0 \quad 0] \qquad \begin{bmatrix} 0 \\ 0 \\ 1 \end{bmatrix}$$

FIG. 3.17.2

The subscript i in the symbol e_i indicates that the ith element is unity. To specify exactly the location of the unity element in the vector, we use a number for a subscript. Thus, in Fig. 3.17.2, we have on the left a fourth-order e_2 vector and on the right a third-order e_3 column vector.

3.18   The Zero Vector

A zero or null vector is actually a null matrix having only one row or column. Therefore all elements in the null vector are 0. Row and column null vectors are given in Fig. 3.18.1.

$$[0 \quad 0 \quad 0 \quad 0] \qquad \begin{bmatrix} 0 \\ 0 \\ 0 \end{bmatrix}$$

FIG. 3.18.1

We indicate the null vector simply by 0, defining the symbol as a vector to distinguish it from a matrix. The order of the null vector may be specified by a single subscript on the 0 or by a statement of the order. For example, we can say, let 0_3 be a third-order column vector. Ordinarily, however, the way in which the null vector is used will make clear its order. Then the order need not be specified.

3.19   The Scalar Quantity

A matrix with only one row and one column is called a *scalar quantity*, or simply a scalar. So you see the letters you encountered in ordinary algebra are matrices of one row and one column, or scalar quantities. In the same way, the numbers you learned about in arithmetic are matrices of one row and one column, or scalars.

To distinguish between a scalar quantity and a symbol for a general rectangular matrix, some texts print scalars in italics and the matrix symbol in bold face. In the case of hand written notations there seems to be no uniform practice. Ordinarily you can make the difference clear simply by defining certain letters as standing for scalar quantities, rather than matrices. This is the method used in this text.

In matrix algebra in general and in this book, we shall use scalar quantities very freely along with the matrices we employ.

SUMMARY

1. The general rectangular matrix:
    a. The height and width are different.
    b. The data matrix is usually rectangular and typically, though not necessarily, vertical.

2. The square matrix:
    a. The square matrix is rarely a data matrix.
    b. In the behavioral sciences it is usually derived from a data matrix.

3. Symmetric matrices:
   a. A symmetric matrix is a special case of a square matrix, in which corresponding elements of corresponding rows and columns are equal.
   b. In a symmetric matrix with elements a_{ij}, we have $a_{ij} = a_{ji}$.
   c. A symmetric matrix is sometimes written with only the principal diagonal and the elements either above or below it. This is incorrect and can lead to confusion with a triangular matrix.

4. The diagonal matrix:
   a. The diagonal matrix is a special case of a symmetric matrix in which all nondiagonal elements are zero.
   b. It may be written without entries in the off diagonal positions.
   c. The diagonal elements of a diagonal matrix may be written unambiguously with a single subscript to indicate their position.
   d. In this book the following symbols will usually be used to indicate diagonal matrices: D, d, Δ, δ.

5. The scalar matrix:
   a. This is a special case of a diagonal matrix in which all the diagonal elements are equal.
   b. The diagonal elements may all be indicated with the same letter without subscripts.

6. The identity matrix:
   a. This is a special case of the scalar matrix in which the diagonal elements are unity.
   b. In this book the identity matrix will be indicated by I, never by 1 as in some texts.
   c. The identity matrix serves the same function in matrix algebra as 1 in scalar algebra.
   d. The order of a diagonal matrix may be indicated verbally or by a subscript.

7. The sign matrix:
   a. This differs from the identity matrix only in that some of the diagonal elements may be -1.
   b. It is indicated in the book by i.

8. The skew symmetric matrix:
   a. This matrix has zeros in the main diagonal, and corresponding elements above and below the diagonal are equal in absolute magnitude and opposite in sign.
   b. In a skew symmetric with elements a_{ij}, we have $a_{ij} = -a_{ji}$.

9. Triangular matrices:
   a. A square matrix with all 0 elements above the main diagonal is called a lower triangular matrix.

    b. A square matrix with all zeros below the main diagonal is called an upper triangular matrix.

10. Partial triangular matrices:

    a. A vertical matrix with 0 for each element whose column number is greater than its row number is called a lower partial triangular matrix.

    b. A horizontal matrix with 0 for each element whose row number is greater than its column number is called an upper partial triangular matrix.

11. Zero-one matrices:

    a. This is a matrix all of whose elements are either 0 or 1.

    b. A permutation matrix is a special case of a zero-one matrix in which each row and column has a single 1 in it. It is therefore square.

12. The e_{ij} matrix has all 0 elements save one which is 1.

13. The null matrix:

    a. This is a matrix all of whose elements are 0.

    b. It is indicated by 0 and the dimensionality must be specified.

    c. The null matrix serves the same function in matrix algebra as zero does in scalar algebra.

14. Vectors:

    a. This is a special kind of matrix with only one row or column.

    b. The order of a vector is indicated by a single number which is the number of elements in it. However, it is also necessary to specify whether it is a row or column.

    c. A letter designation for a vector need carry only a single subscript to indicate its position in the vector.

15. The unit vector:

    a. This vector has unity for all its elements.

    b. A column unit vector is indicated by 1 and a row by 1'.

    c. The order of a unit vector may be indicated verbally or by means of a subscript, or may be evident from the context in which the vector is used.

16. The sign vector: Each element is either $+1$ or -1.

17. The zero-one vector:

    a. This is a special case of a zero-one matrix having only one row or column.

    b. The e_i vector is a special case of a zero-one vector having only a single 1 and that in the ith position.

18. The null vector:

    a. This is a vector all of whose elements are 0.

    b. It is indicated by 0 together with the specifications of its order, verbal or by subscript.

19. Scalar quantity

    a. This is a matrix with only one row and column—the kind of number familiar to most persons.

    b. It may be distinguished from a symbol for a general rectangular matrix by a particular type style or, for a particular problem, by verbal definition of notation.

EXERCISES

Given the following matrices:

1.
$$\begin{bmatrix} 2 & 3 & 4 & 1 \\ 2 & 1 & 3 & 2 \\ 4 & 2 & 1 & 5 \end{bmatrix}$$

2.
$$\begin{bmatrix} 3 & 4 & 1 \\ 0 & 2 & 5 \\ 0 & 0 & 6 \end{bmatrix}$$

3.
$$\begin{bmatrix} 4 \\ 3 \\ 7 \end{bmatrix}$$

4.
$$\begin{bmatrix} a_{11} & \cdots & a_{1n} \\ a_{n1} & \cdots & a_{1n} \end{bmatrix}$$

5.
$$\begin{bmatrix} 2 & 0 & 0 \\ 4 & 7 & 0 \\ 3 & 1 & 4 \end{bmatrix}$$

6. $[7 \quad 9 \quad 2]$

7.
$$\begin{bmatrix} 4 & 3 & 2 \\ 3 & 5 & 1 \\ 2 & 1 & 3 \end{bmatrix}$$

8.
$$\begin{bmatrix} 1 & 3 & 4 & 7 & 2 \\ 0 & 2 & 4 & 3 & 1 \\ 0 & 0 & 8 & 7 & 6 \end{bmatrix}$$

9.
$$\begin{bmatrix} 1 \\ 1 \\ 1 \end{bmatrix}$$

10.
$$\begin{bmatrix} 15 & 0 & 0 \\ 0 & 3 & 0 \\ 0 & 0 & 4 \end{bmatrix}$$

11.
$$\begin{bmatrix} 1 & 0 & 0 \\ 2 & 4 & 0 \\ 5 & 3 & 7 \\ 7 & 6 & 2 \\ 8 & 5 & 1 \end{bmatrix}$$

12. $[0 \quad 1 \quad 0 \quad 0]$

13.
$$\begin{bmatrix} 3 & 0 & 0 \\ 0 & 3 & 0 \\ 0 & 0 & 3 \end{bmatrix}$$

14.
$$\begin{bmatrix} 0 & 1 & 1 & 0 \\ 1 & 0 & 1 & 0 \\ 1 & 0 & 1 & 1 \\ 0 & 1 & 1 & 1 \\ 0 & 0 & 1 & 0 \end{bmatrix}$$

15. $[1 \quad 1 \quad -1 \quad 1]$

16.
$$\begin{bmatrix} 1 & 0 & 0 \\ 0 & 1 & 0 \\ 0 & 0 & 1 \end{bmatrix}$$

17.
$$\begin{bmatrix} 0 & 1 & 0 \\ 0 & 0 & 1 \\ 1 & 0 & 0 \end{bmatrix}$$

18.
$$\begin{bmatrix} 0 \\ 0 \\ 0 \end{bmatrix}$$

19.
$$\begin{bmatrix} 1 & 0 & 0 \\ 0 & -1 & 0 \\ 0 & 0 & 1 \end{bmatrix}$$

20.
$$\begin{bmatrix} 0 & 1 & 0 \\ 0 & 0 & 0 \\ 0 & 0 & 0 \end{bmatrix}$$

21. 4

22.
$$\begin{bmatrix} 0 & 2 & -4 \\ -2 & 0 & 3 \\ 4 & -3 & 0 \end{bmatrix}$$

23.
$$\begin{bmatrix} 0 & 0 \\ 0 & 0 \\ 0 & 0 \end{bmatrix}$$

For each of the following terms, give the number of the matrix above that corresponds to it. Give one and only one number for each letter.

(a) Rectangular

(b) Lower triangular

(c) Unit vector

(d) Square

(e) Upper partial triangular

(f) e_i vector

(g) Symmetric

(h) Lower partial triangular

(j) Sign vector

(k) D

(l) Zero-one matrix

(m) Null vector

(n) Scalar matrix

(o) π

(p) Scalar

(q) I

(r) e_{ij}

(s) i

(t) Null matrix

(u) Skew symmetric

(v) Upper triangular

(w) Row vector

(x) Column vector

ANSWERS

(a) 1, (b) 5, (c) 9, (d) 4, (e) 8, (f) 12, (g) 7, (h) 11, (j) 15, (k) 10, (l) 14, (m) 18, (n) 13, (o) 17, (p) 21, (q) 16, (r) 20, (s) 19, (t) 23, (u) 22, (v) 2, (w) 6, (x) 3

Chapter 4

The Transpose of a Matrix

The concept of the transpose of a matrix is usually dealt with in one sentence or at most one paragraph in the majority of texts on matrix algebra. As a consequence of this cursory treatment, mathematicians not infrequently become confused when they attempt to use the transpose in the solution of problems. Unless you thoroughly master the concept and its implications for the analysis of experimental data, you may well make errors costing many man-hours or days of computational labor, or many hundreds of dollars in high-speed computor time.

It is advisable first to consider two related concepts: the *equality* of matrices and the *natural order* of a matrix.

4.1 Equality of Matrices

In matrix algebra, as in scalar algebra, the concept of equality is highly important. Very simply, two matrices are equal to each other if, and only if, each element of one is equal to the corresponding element of the other. Study Fig. 4.1.1.

$$\begin{bmatrix} a_{11} & a_{12} \\ a_{21} & a_{22} \\ a_{31} & a_{32} \end{bmatrix} = \begin{bmatrix} b_{11} & b_{12} \\ b_{21} & b_{22} \\ b_{31} & b_{32} \end{bmatrix}$$

$$a \qquad\qquad b$$

FIG. 4.1.1

The matrix on the left we call a, and that on the right we call b. Then to say $a = b$ means that element a_{11} is equal to element b_{11}, a_{12} is equal to b_{12}, a_{21} is equal to b_{21}, and so on.

According to this definition of equal matrices, two matrices cannot be equal to each other unless thay are of the same order. Study Fig. 4.1.2.

$$\begin{bmatrix} 4 & 3 \\ 2 & 1 \\ 3 & 4 \end{bmatrix} \qquad \begin{bmatrix} 4 & 2 & 3 \\ 3 & 1 & 4 \end{bmatrix}$$

FIG. 4.1.2

Here we have two matrices; the first is a 3 by 2 matrix, the second a 2 by 3 matrix. The elements in the first column of the left-hand matrix are, in order, the same as the elements in the first row of the right-hand matrix. Similarly, the elements in the second column of the left-hand matrix are, in order, the same as the elements in the second row of the right-hand matrix. Nevertheless, the two matrices cannot be said to be equal. They are not of the same order, nor is the ijth element of one in general equal to the ijth element of the other.

4.2   The Natural Order of a Matrix

Although mathematicians do not talk about the natural order of a matrix, the idea is very useful in applying matrix algebra to large masses of data in the social sciences and elsewhere.

Natural order of data matrices

For data matrices, we shall say that the elements are in their *natural order* when the entities are rows and the attributes are columns. Usually, therefore, the natural order of a data matrix has more rows than columns; it is usually a vertical matrix.

Natural order of other rectangular matrices

In dealing with a rectangular matrix for which the entities and attributes are not specified, we usually take the natural order to be the vertical form of the matrix. Since a vector is a special case of a rectangular matrix, we can regard the natural order of a vector as the column form. In this book, unless we specifically state otherwise, the column vector is regarded as the natural order of the vector.

The partial triangular matrix, described in Section 3.10 is a rectangular matrix. We may say that the natural order of a partial triangular matrix is the lower partial triangular matrix. You recall that a lower partial triangular matrix is vertical. Thus the natural order of a partial triangular matrix is shown in Fig. 4.2.1.

$$\begin{bmatrix} 2 & 0 & 0 \\ 3 & 4 & 0 \\ 1 & 6 & 2 \\ 3 & 5 & 4 \\ 3 & 7 & 1 \end{bmatrix}$$

FIG. 4.2.1

Natural order of square matrices

There is only one kind of square matrix for which we specify a natural order. This is the square triangular matrix. We call the lower triangular matrix the natural order of the triangular matrix. The reason for this you have probably guessed. It is because the triangular matrix is a special case of the partial triangular matrix. Since the lower partial triangular matrix is regarded as the natural order, we shall, for the sake of uniformity, regard its special case, the lower square triangular matrix, also as the natural order. In Fig. 4.2.2 are shown two examples of the natural order of a triangular matrix.

$$\begin{bmatrix} 1 & 0 & 0 \\ 2 & 3 & 0 \\ 3 & 1 & 1 \end{bmatrix} \quad \begin{bmatrix} a_{11} & & \\ a_{21} & a_{22} & \\ a_{31} & a_{32} & a_{33} \end{bmatrix}$$

FIG. 4.2.2

You see, therefore, that in the natural order of a triangular matrix the first subscript is never smaller than the second.

In speaking of the square matrix we usually have no way of specifying the natural order. Very rarely is a data matrix a square matrix. Furthermore, in operations on data matrices or other derived matrices, we have very little use for square matrices in general. Several special types of square matrices are used; these will be discussed in later chapters. One, however, with which you are already familiar (Section 3.11) is the permutation matrix, a zero-one matrix with a 1 in each row and column, all the remaining elements being 0. We do not define the natural order of a permutation matrix.

Other types of square matrices with which you are already familiar are the general symmetric matrix and various special cases of it—the diagonal matrix, the scalar matrix, the identity matrix, and the sign matrix. We do not define the natural order of any form of symmetric matrix, since writing rows for columns or vice versa would not change the value of a symmetric matrix.

4.3 What the Transpose of a Matrix Is

By this time you may have guessed what we mean by the transpose of a matrix. One matrix is called the transpose of another if the rows of the first are written as the columns of the second. In Fig. 4.1.2, the second or right-hand matrix is the transpose of the first. You will see that the rows of the first form the corresponding columns of the second. Another example of a matrix and its transpose is given in Fig. 4.3.1.

$$\begin{bmatrix} 3 & 2 \\ 2 & 1 \\ 7 & 9 \end{bmatrix} \quad \begin{bmatrix} 3 & 2 & 7 \\ 2 & 1 & 9 \end{bmatrix}$$

FIG. 4.3.1

Here the second matrix is the transpose of the first. Of course, we can also say that the first is the transpose of the second. Notice that if the rows of one matrix are the columns of another, then also the columns of the first matrix must be the rows of the other. These are two different ways of saying the same thing.

Another way of defining the transpose of a matrix is to say that the a_{ij}th element of the matrix is the a_{ji}th element of the transpose. For example, in Fig. 4.3.1, suppose that i is 3 and j is 1. Then, a_{ij} would be a_{31}, which in the left-hand matrix is 7; a_{ji} would be a_{13}, which in the right-hand matrix is also 7.

It should be clear then that the order of the transpose of a matrix is the reverse of the order of the original matrix. If a matrix is of order 3 by 4, then its transpose is of order 4 by 3; or, in general, if the order of a matrix is n by m, then the order of the transpose is m by n.

4.4   The Natural Order and the Transpose

Since the natural order of a matrix is usually vertical, the transpose of its natural order is usually horizontal. Also, since in the natural order of a data matrix the entities are usually rows and the columns attributes, we think of the transpose of a data matrix as ordinarily having rows for attributes and columns for entities. In general, therefore, we think of the transpose of a data matrix as being usually horizontal.

4.5   The Symbol for the Transpose

When a letter stands for a matrix

When we let a single letter stand for a matrix, it is customary to put a prime after it to refer to the transpose of the matrix. Figure 4.5.1 illustrates how we ordinarily let the unprimed letters stand for the original matrix and the primed letters for its transpose.

$$a = \begin{bmatrix} 4 & 3 \\ 2 & 1 \\ 7 & 9 \end{bmatrix} \qquad a' = \begin{bmatrix} 4 & 2 & 7 \\ 3 & 1 & 9 \end{bmatrix}$$

Fig. 4.5.1

You see now why in Chapter 3, Section 15, we let the number 1 stand for a column unit vector and $1'$ stand for the unit row vector.

Letters with double subscripts for elements

If we let letters with double subscripts stand for the elements of the matrix, we sometimes reverse our subscript procedure to indicate the transpose of a matrix. For example, if a is given as on the left of Fig. 4.5.2, then its transpose a' would be given as on the right.

$$a = \begin{bmatrix} a_{11} & a_{12} \\ a_{21} & a_{22} \\ a_{31} & a_{32} \end{bmatrix} \qquad a' = \begin{bmatrix} a_{11} & a_{21} & a_{31} \\ a_{12} & a_{22} & a_{32} \end{bmatrix}$$

Fig. 4.5.2

$$a = \begin{bmatrix} a_{11} & a_{12} & \cdots & a_{1m} \\ a_{21} & a_{22} & \cdots & a_{2m} \\ \cdots & \cdots & \cdots & \cdots \\ a_{n1} & a_{n2} & \cdots & a_{nm} \end{bmatrix} \qquad a' = \begin{bmatrix} a_{11} & a_{21} & \cdots & a_{n1} \\ a_{12} & a_{22} & \cdots & a_{n2} \\ \cdots & \cdots & \cdots & \cdots \\ a_{1m} & a_{2m} & \cdots & a_{nm} \end{bmatrix}$$

Fig. 4.5.3

You see that on the right of Fig. 4.5.2 the first subscript refers to the column in which the element is found and the second refers to the row. This is another of the exceptions to the rule given in Chapter 2, by which the first figure in the subscript represents the row and the second figure represents the column.

In general, if we let a stand for the left-hand matrix in Fig. 4.5.3, then a', its transpose, would stand for the right-hand matrix. Notice particularly in Fig. 4.5.3 that the lower right-hand element in the matrix a is the same as the lower right-hand element in a'. This element in both cases is a_{nm}. However, in the first matrix n refers to the row and m to the column, whereas in the second matrix n refers to the column and m to the row.

When subscripts are used to indicate order

You may recall that in Chapter 2 we said the order of a matrix is sometimes indicated by double subscripts to a letter that stands for the entire matrix. If a matrix has n rows and m columns, we may simply let a_{nm} represent this entire matrix. If we do use this double subscript notation, the transpose of the matrix is one with m rows and n columns. We would indicate this relationship as shown in Fig. 4.5.4.

$$a'_{nm} = a_{mn}$$

Fig. 4.5.4

Here you see that putting a prime after a letter has the effect of interchanging the subscripts of the matrix. We could translate Fig. 4.5.4 into words by saying that the transpose of a matrix with n rows and m columns is a matrix with m rows and n columns.

4.6 The Transpose of a Transpose

It is easy to see that if you find the transpose of a matrix and then you find the transpose of that transpose, you get the original matrix right back. We say then that the transpose of a transpose is the original matrix. (This seems perfectly obvious, but sometimes it is easy to forget when you are working with matrix equations in a mechanical way without thinking about what all the different symbols mean.) We can state this principle in the form of an equation, as in Fig. 4.6.1.

$$[a']' = a$$

Fig. 4.6.1

So that you will get used to using symbols in matrix equations, we shall proceed as follows:

$$a' = b \qquad (4.6.1)$$

$$b' = a \qquad (4.6.2)$$

$$[a']' = a \qquad (4.6.3)$$

We start with Eq. (4.6.1). This says that the transpose of a is equal to b. Equation (4.6.2) says that the transpose of b is equal to a. If we substitute the left-hand side of Eq. (4.6.1) for b into the left-hand side of Eq. (4.6.2), we have Eq. (4.6.3), which is the same as Fig. 4.6.1. A simple way of stating the rule is that two primes cancel each other.

4.7 The Transposes of Common Kinds of Matrices

It will be convenient to discuss together relationships between various types of matrices and their transposes.

Rectangular matrix

As we have already seen, the transpose of a vertical matrix is a horizontal matrix. The transpose of a horizontal matrix is a vertical matrix. Clearly the transpose of a square matrix has to be a square matrix.

Triangular matrix

The rules for partial triangular and triangular matrices are equally obvious. The transpose of an upper triangular matrix is a lower triangular matrix, and the transpose of a lower triangular matrix is an upper triangular. Similarly the transpose of an upper partial triangular matrix is a lower partial triangular matrix, and the transpose of a lower partial triangular matrix is an upper partial triangular matrix.

Symmetric matrix

The transpose of any symmetric matrix is the matrix itself, since the row of a symmetric matrix is the same as the corresponding column. If, therefore, we let S be *any* symmetric matrix, we can express this rule by Eq. (4.7.1). If we let D be *any* diagonal matrix, we can write Eq. (4.7.2), which can be represented in expanded notation by Fig. 4.7.1.

$$S = S' \qquad (4.7.1)$$

$$D = D' \qquad (4.7.2)$$

$$\begin{bmatrix} d_1 & 0 & 0 \\ 0 & d_2 & 0 \\ 0 & 0 & d_3 \end{bmatrix}' = \begin{bmatrix} d_1 & 0 & 0 \\ 0 & d_2 & 0 \\ 0 & 0 & d_3 \end{bmatrix}$$

FIG. 4.7.1

The transpose of a scalar matrix is the same scalar matrix. An illustration is Fig. 4.7.2.

$$\begin{bmatrix} 3 & 0 & 0 \\ 0 & 3 & 0 \\ 0 & 0 & 3 \end{bmatrix}' = \begin{bmatrix} 3 & 0 & 0 \\ 0 & 3 & 0 \\ 0 & 0 & 3 \end{bmatrix}$$

FIG. 4.7.2

The transpose of an identity matrix is the same identity matrix. This can be written as in Eq. (4.7.3) or Fig. 4.7.3.

$$I' = I \qquad (4.7.3)$$

$$\begin{bmatrix} 1 & 0 & 0 \\ 0 & 1 & 0 \\ 0 & 0 & 1 \end{bmatrix}' = \begin{bmatrix} 1 & 0 & 0 \\ 0 & 1 & 0 \\ 0 & 0 & 1 \end{bmatrix}$$

FIG. 4.7.3

The transpose of a sign matrix is the same sign matrix. This can be written as in Eq. (4.7.4) or as in Fig. 4.7.4.

$$i' = i \qquad (4.7.4)$$

$$\begin{bmatrix} 1 & 0 & 0 \\ 0 & -1 & 0 \\ 0 & 0 & -1 \end{bmatrix}' = \begin{bmatrix} 1 & 0 & 0 \\ 0 & -1 & 0 \\ 0 & 0 & -1 \end{bmatrix}$$

FIG. 4.7.4

The transpose of a square null matrix is the same null matrix. This is illustrated by Fig. 4.7.5.

$$\begin{bmatrix} 0 & 0 & 0 \\ 0 & 0 & 0 \\ 0 & 0 & 0 \end{bmatrix}' = \begin{bmatrix} 0 & 0 & 0 \\ 0 & 0 & 0 \\ 0 & 0 & 0 \end{bmatrix}$$

FIG. 4.7.5

The scalar quantity (Section 3.19) is a symmetric matrix with only one row and column—that is, a single number. Of course, if the row were written as a column, or the column as a row, this matrix would not be altered. Therefore, the transpose of a scalar quantity is the same scalar quantity. This is so obvious that you may wonder why it should be stated so carefully. Sometimes, however, this simple and obvious rule is very useful in simplifying complex solutions or proofs. We can express this rule by

$$a'_{ij} = a_{ij} \tag{4.7.5}$$

where a_{ij} simply means the element in the ith row and jth column of a matrix, no matter what the size of the matrix. To use a simple ordinary number instead of the letter, we could write the same rule as

$$3' = 3 \tag{4.7.6}$$

This equation merely says that the transpose of the number 3 is the number 3.

4.8   The Transpose of a Vector

You will not have to learn anything new to understand the transpose of a vector. A vector, as you know, is simply a special case of a matrix. If we say that V is a vector, it will be understood to be a column vector, unless otherwise specified. When V or any other letter is used to represent a vector, a prime used after it usually makes it a row vector. Unless otherwise specified, if you see a prime after the symbol for a vector in this book, you can be sure that it is a row vector. This practice is by no means universal, however, and in other books or articles using matrix algebra you may find that the unprimed symbol means a row and the primed symbol a column vector.

It must also now be clear that the transpose of a row vector is a column vector. Figure 4.8.1 illustrates the use of the unprimed and the primed letter to indicate a column and a row vector respectively.

$$V = \begin{bmatrix} 2 \\ 4 \\ 3 \end{bmatrix} \qquad V' = \begin{bmatrix} 2 & 4 & 3 \end{bmatrix}$$

FIG. 4.8.1

We indicate the transpose of special kinds of vectors in the same way. For example, Fig. 4.8.2 illustrates an e_i vector (Section 2.17) and its transpose.

$$e_i = \begin{bmatrix} 1 \\ 0 \\ 0 \\ 0 \end{bmatrix} \qquad e_i' = \begin{bmatrix} 1 & 0 & 0 & 0 \end{bmatrix}$$

FIG. 4.8.2

Figure 4.8.3 illustrates a sign vector (Section 2.15) and its transpose.

$$l = \begin{bmatrix} 1 \\ -1 \\ -1 \\ 1 \end{bmatrix} \qquad l' = \begin{bmatrix} 1 & -1 & -1 & 1 \end{bmatrix}$$

FIG. 4.8.3

The null vector (Section 2.18) and its transpose are illustrated in Fig. 4.8.4.

$$0 = \begin{bmatrix} 0 \\ 0 \\ 0 \end{bmatrix} \qquad 0' = \begin{bmatrix} 0 & 0 & 0 \end{bmatrix}$$

FIG. 4.8.4

SUMMARY

1. Equality of matrices:
   a. Two matrices are said to be equal if and only if all corresponding elements are equal.
   b. Therefore matrices cannot be equal unless their orders are the same.
2. Natural order of a matrix:
   a. *Data matrices.* In this book the natural order of a data matrix is said to be such that rows are entities and columns attributes.
   b. *Other rectangular matrices.* Matrices that are not data matrices are usually considered in natural order if they are vertical. The natural order of a vector, unless otherwise specified, is the column form.
   c. *Square matrices.* The only square matrix to be assigned a natural order is the triangular matrix. Its natural order is taken as the lower triangular form.
3. The transpose of a matrix:
   a. The transpose of a matrix is one in which the rows are written as columns or the columns as rows: therefore the ijth element of the matrix becomes the jith element of the transpose.

    b. The order of the transpose of a matrix is therefore the reverse of the order of the matrix.

4. The natural order and the transpose.

    a. The transpose of the natural-order data matrix usually has attributes for rows and entities for columns.

    b. The transpose of a natural-order rectangular matrix is usually a horizontal matrix.

5. The symbol for the transpose:

    a. If a letter stands for a matrix it is primed to indicate its transpose: a' is the transpose of a.

    b. If the natural order is defined, the letter representing it is unprimed to indicate it is the natural order.

    c. The double subscript notation in which the first subscript stands for column and the second for row may be reversed to indicate the transpose.

6. The transpose of a transpose:

    a. If the transpose of a matrix is transposed the original matrix is restored.

    b. This is indicated by $[a']' = a$.

7. Transposes of common kinds of matrices:

    a. The transpose of a vertical matrix is horizontal and vice versa.

    b. The transpose of a square matrix is a square matrix.

    c. The transpose of a lower triangular matrix is an upper triangular matrix and vice versa. The same rule holds for partial triangular matrices.

    d. The transposes of all symmetric matrices are the same as the original matrices. This includes: general symmetric, diagonal, scalar, identity, and sign matrices. Obviously it also includes scalar quantities.

8. The transpose of a vector:

    a. The column vector is written without the prime.

    b. Its transpose is a row vector.

    c. The primed symbol for a vector means a row vector unless otherwise specified.

EXERCISES

1. Write the transposes of each of the matrices given in the exercises for Chapter 3.

2. Give the numbers of those whose transposes are vertical.

3. Give the numbers of those whose transposes are different but whose orders are unchanged.

4. Give the numbers of those whose transposes are horizontal.

5. Give the numbers of those whose transposes leave the matrix unchanged.

ANSWERS

2. (1), (6), (8), (12), (15)

3. (2), (4), (5), (17), (20), (22)

4. (3), (9), (11), (14), (18), (23)

5. (7), (10), (13), (16), (19), (21)

Chapter 5

The Supermatrix

We have seen that the operation called transposition, though not needed in scalar algebra, is important in matrix algebra. Another operation used only in matrix algebra gives rise to what are called *supermatrices*. This operation, like transposition, is simple and frequently is given only summary treatment by the mathematicians. It is, however, essential in the construction of experimental designs in the behavioral sciences, and its competent utilization in such research can be very helpful. Therefore we shall develop in detail the concept of the supermatrix.

5.1 Definition of a Supermatrix

The supermatrix and its elements

Before trying to define a supermatrix, let us designate the kind of matrices we have been talking about up to now as *simple matrices*. By a simple matrix we mean a matrix each of whose elements is just an ordinary number or a letter that stands for the number. In other words, the elements of a simple matrix are scalars, or scalar quantities.

A supermatrix, on the other hand, is one whose elements are themselves matrices with elements that can be either scalars or other matrices. In general, in the kind of supermatrices we shall deal with in this book the matric elements have only scalars for their elements. Suppose we define four matrices as in Fig. 5.1.1.

$$a_{11} = \begin{bmatrix} 2 & 4 \\ 3 & 7 \end{bmatrix} \qquad a_{12} = \begin{bmatrix} 5 & 1 \\ 3 & 2 \end{bmatrix}$$

$$a_{21} = \begin{bmatrix} 6 & 1 \\ 7 & 5 \end{bmatrix} \qquad a_{22} = \begin{bmatrix} 4 & 7 \\ 2 & 5 \end{bmatrix}$$

Fig. 5.1.1

You will notice that we are using number subscripts on each of the letters that stand for entire matrices. This is one reason why we discouraged the

62

use of number subscripts to indicate order when a single letter was used to stand for a matrix (Section 2.5). You see now we want to use number subscripts to stand for separate matrices. The reason we have chosen the particular number subscripts for each matrix in Fig. 5.1.1 will become clear if we let a stand for an entire matrix whose elements are themselves these four matrices, as in Fig. 5.1.2.

$$ a = \begin{bmatrix} a_{11} & a_{12} \\ a_{21} & a_{22} \end{bmatrix} $$

FIG. 5.1.2

Note that we are using the same rule of double subscripts that we used for simple matrices. The first subscript indicates the row in which the matric element is found and the second subscript indicates the column.

We can write out the matrix a in Fig. 5.1.2 in terms of the original matrix elements in Fig. 5.1.1. This we do in Fig. 5.1.3.

$$ a = \left[\begin{array}{cc|cc} 2 & 4 & 5 & 1 \\ 3 & 7 & 3 & 2 \\ \hline 6 & 1 & 4 & 7 \\ 7 & 5 & 2 & 5 \end{array} \right] $$

FIG. 5.1.3

Here the elements are divided vertically and horizontally by thin lines. If the lines were not used, the figure would be read as a simple matrix.

Thus far we have referred to the elements in a supermatrix as matric elements. It is perhaps more usual to call the elements in a supermatrix *submatrices*. We speak of the submatrices within a supermatrix.

Order and the supermatrix

The order of a supermatrix is defined in the same way as that of a simple matrix. The height of a supermatrix is the number of rows of submatrices in it. The width of a supermatrix is the number of columns of submatrices in it.

As you may have guessed, all submatrices within a given row must have the same number of rows; likewise all submatrices within a given column must have the same number of columns. A diagrammatic illustration of this rule is Fig. 5.1.4.

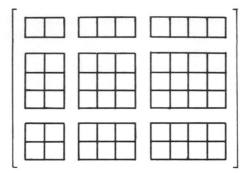

FIG. 5.1.4

In the first row of rectangles we have one row of squares for each rectangle; in the second row of rectangles we have three rows of squares for each rectangle; and in the third row of rectangles we have two rows of squares for each rectangle. Similarly for the first column of rectangles we have two columns of squares for each rectangle. For the second column of rectangles we have three columns of squares for each rectangle, and for the third column of rectangles, we have four columns of squares for each rectangle.

One thing should now be clear from the definition of a supermatrix. The superorder of a supermatrix tells us nothing about the simple order of the matrix from which it was obtained by partitioning. Furthermore, the order of a supermatrix tells us nothing about the orders of the submatrices within that supermatrix.

5.2 Partitioned Matrices

It is always possible to construct a supermatrix from any simple matrix that is not a scalar quantity.

The supermatrix from the simple matrix

The process of constructing a supermatrix from a simple matrix is called *partitioning*. A simple matrix may be partitioned by dividing or separating the matrix between certain specified columns, and then between certain specified rows. Or the procedure may be reversed. The division may be made first between rows and then between columns. To illustrate, suppose we have a simple matrix as in Fig. 5.2.1.

$$a = \begin{bmatrix} 4 & 3 & 2 & 7 \\ 3 & 6 & 1 & 4 \\ 2 & 1 & 5 & 2 \\ 7 & 4 & 2 & 7 \end{bmatrix}$$

FIG. 5.2.1

This is a 4 by 4 matrix with numbers for elements. Now let us draw a thin line between the first and second columns. This gives us the matrix in Fig. 5.2.2.

$$a = \begin{bmatrix} 4 & 3 & 2 & 7 \\ 3 & 6 & 1 & 4 \\ 2 & 3 & 5 & 2 \\ 7 & 4 & 2 & 7 \end{bmatrix}$$

FIG. 5.2.2

Actually now the figure may be regarded as a supermatrix with two matric elements forming one row and two columns. However, we shall next divide the matrix between rows 2 and 3 to give us Fig. 5.2.3.

$$a = \begin{bmatrix} 4 & 3 & 2 & 7 \\ 3 & 6 & 1 & 4 \\ \hline 2 & 1 & 5 & 2 \\ 7 & 4 & 2 & 7 \end{bmatrix}$$

FIG. 5.2.3

It is now a second-order supermatrix, with two rows and two columns of submatrices. If we define the submatrices as in Fig. 5.2.4, we can write them simply as in Fig. 5.2.5.

$$a_{11} = \begin{bmatrix} 4 \\ 3 \end{bmatrix} \qquad a_{12} = \begin{bmatrix} 3 & 2 & 7 \\ 6 & 1 & 4 \end{bmatrix}$$

$$a_{21} = \begin{bmatrix} 2 \\ 7 \end{bmatrix} \qquad a_{22} = \begin{bmatrix} 1 & 5 & 2 \\ 4 & 2 & 7 \end{bmatrix}$$

FIG. 5.2.4

$$a = \begin{bmatrix} a_{11} & a_{12} \\ a_{21} & a_{22} \end{bmatrix}$$

FIG. 5.2.5

The elements now are the submatrices defined in Fig. 5.2.4, and therefore, Fig. 5.2.5 is, in terms of letters, the supermatrix given in Fig. 5.2.3.

According to the methods we have just illustrated, a simple matrix can be partitioned into a supermatrix in any way that happens to suit our purposes.

5.3 The Natural Order and the Supermatrix

You will recall that usually in working with simple matrices we take the natural order to be the vertical form of the matrix. In working with the supermatrix, we have no simple rule for determining its natural order. The natural order of a supermatrix is usually determined by the natural order of the corresponding simple matrix. Furthermore, we are not usually concerned with the natural order of the submatrices within a supermatrix. Here again, if it becomes necessary or convenient to define the natural orders of these submatrices, we take their natural orders to be their orders in the natural order of the simple matrix from which they were constructed by partitioning.

5.4 Symmetric Partitioning

By the symmetric partitioning of a matrix we mean that the rows and columns are partitioned in exactly the same way. If the matrix is partitioned between the first and second columns and between the third and fourth columns, then to be symmetrically partitioned it must also be partitioned between the first and second rows and the third and fourth rows.

According to this rule of symmetric partitioning only square simple matrices can be symmetrically partitioned. An example of a symmetrically partitioned matrix is given in Fig. 5.4.1.

$$
\left[
\begin{array}{c|cc|c|c}
4 & 3 & 2 & 7 & 9 \\
\hline
3 & 6 & 1 & 4 & 8 \\
2 & 1 & 5 & 7 & 7 \\
\hline
7 & 4 & 2 & 7 & 7 \\
\hline
9 & 8 & 7 & 3 & 6
\end{array}
\right]
$$

Fig. 5.4.1

Here you see that the matrix has been partitioned between columns one and two, three and four, and four and five. It has also been partitioned between rows one and two, three and four, and four and five.

One fact you should notice in Fig. 5.4.1 (if you have not already reached this conclusion yourself) is that some of the elements in a supermatrix may be scalar quantities. Here the first element in the first row of submatrices is 4, or a single number. This is also true of the elements in the first row and the third and fourth columns. You can also find other matric elements in Fig. 5.4.1 that are scalars. Notice also that some of the submatrices are row vectors and some are column vectors.

While it is possible to partition symmetrically any square simple matrix, this type of partitioning is usually performed on simple matrices that are themselves symmetric. That is, there is nothing in matrix algebra itself to cause us to prefer the symmetric partitioning of symmetric simple matrices rather than general square matrices, but it usually happens that in the analysis of data, the matrices we can usefully partition symmetrically *are* symmetric simple matrices.

5.5 The Symmetric Partitioning of a Symmetric Simple Matrix

The general case

Let us take a fourth-order symmetric matrix and partition it between the second and third rows and also the second and third columns, as in Fig. 5.5.1.

$$a = \left[\begin{array}{cc|cc} 4 & 3 & 2 & 7 \\ 3 & 6 & 1 & 4 \\ \hline 2 & 1 & 5 & 2 \\ 7 & 4 & 2 & 7 \end{array}\right]$$

FIG. 5.5.1

We can represent this matrix as a supermatrix with letter elements. The submatrices are then as shown in Fig. 5.5.2.

$$a_{11} = \begin{bmatrix} 4 & 3 \\ 3 & 6 \end{bmatrix} \qquad a_{12} = \begin{bmatrix} 2 & 7 \\ 1 & 4 \end{bmatrix}$$

$$a_{21} = \begin{bmatrix} 2 & 1 \\ 7 & 4 \end{bmatrix} \qquad a_{22} = \begin{bmatrix} 5 & 2 \\ 2 & 7 \end{bmatrix}$$

FIG. 5.5.2

We can therefore write the matrix a with letter elements representing the submatrices. This we do in Fig. 5.5.3.

$$a = \begin{bmatrix} a_{11} & a_{12} \\ a_{21} & a_{22} \end{bmatrix}$$

FIG. 5.5.3

Notice that the diagonal elements in Fig. 5.5.3 are a_{11} and a_{22}. If you look at Fig. 5.5.2, you will see that a_{11} is a symmetric matrix and a_{22} is also a symmetric matrix.

Furthermore the nondiagonal elements in Fig. 5.5.3 are the matrices a_{12} and a_{21}. Referring again to Fig. 5.5.2, you see that a_{21} is the transpose of a_{12}.

We therefore have two very simple rules about the matric elements of a symmetrically partitioned symmetric simple matrix. The first of these is that the diagonal submatrices of the supermatrix are all symmetric matrices. The second is that the matric elements below the diagonal are the transposes of the corresponding elements above the diagonal. Figure 5.5.4 is an example of a third-order supermatrix obtained from a symmetric partitioning of a symmetric simple matrix.

$$a = \begin{bmatrix} a_{11} & a_{12} & a_{13} \\ a'_{12} & a_{22} & a_{23} \\ a'_{13} & a'_{23} & a_{33} \end{bmatrix}$$

FIG. 5.5.4

This figure illustrates the second rule. The elements below the diagonal have subscripts exactly the same and in the same order as their corresponding elements above the diagonal. However, to indicate the transposes, the elements below the diagonal have primes after them while those above do not. For example, the element in the first row and second column is a_{12}, whereas the element in the second row and first column is a'_{12}. We can write the general expression for a symmetrically partitioned symmetric matrix as in Fig. 5.5.5.

$$a = \begin{bmatrix} a_{11} & a_{12} & \dots & a_{1n} \\ a'_{12} & a_{22} & \dots & a_{2n} \\ \dots & \dots & \dots & \dots \\ a'_{1n} & a'_{2n} & \dots & a_{nn} \end{bmatrix}$$

FIG. 5.5.5

If we were writing either Fig. 5.5.4 or Fig. 5.5.5, however, the reader could not tell that they are symmetrically partitioned symmetric simple matrices. unless we also specified that a_{11}, a_{22}, and so on, are symmetric submatrices. If in Fig. 5.5.5, for example, we said let a_{ii} equal a'_{ii}, then the reader would know that a symmetric matrix had been symmetrically partitioned.

The diagonal simple matrix

If we want to indicate a symmetrically partitioned simple diagonal matrix, we could write this as in Fig. 5.5.6.

$$D = \begin{bmatrix} D_1 & 0 & \dots & 0 \\ 0' & D_2 & \dots & 0 \\ \dots & \dots & \dots & \dots \\ 0' & 0' & \dots & D_n \end{bmatrix}$$

FIG. 5.5.6

Here we would have to specify the order of the diagonal submatrices D_1, D_2 to D_n. The prime on each 0 below the diagonal means that its order is reversed from the order of the corresponding null matrix above the diagonal. Figure 5.5.6 is a precise way of indicating a symmetrically partitioned diagonal matrix or *superdiagonal matrix*. However, we ordinarily need not go to the trouble of writing in the zeros. There will usually be no misunderstanding if we write the matrix as in Fig. 5.5.7.

$$D = \begin{bmatrix} D_1 & & & \\ & D_2 & & \\ & & \cdots & \\ & & & D_n \end{bmatrix}$$

FIG. 5.5.7

The difference between Figs. 5.5.6 and 5.5.7 is that in the latter we have omitted the zeros or null submatrices.

The identity matrix

The procedure for partitioning an identity matrix is shown in Fig. 5.5.8.

$$I = \begin{bmatrix} I_s & \\ & I_t \end{bmatrix}$$

FIG. 5.5.8

The subscripts s and t indicate the number of rows and columns in the first and second identity matrices respectively. It is often convenient to use these subscripts to indicate the order of the identity submatrices.

Although it is possible to perform nonsymmetric partitioning of symmetric simple matrices, the occasions when this is useful or convenient in the analysis of data are very rare. In general, symmetric simple matrices are subjected only to symmetric partitioning.

5.6 General Diagonal Supermatrices

You have seen in the preceding section how we may partition a simple diagonal matrix to get a superdiagonal matrix. In this case the matric diagonal elements are themselves simple diagonal matrices. In certain experimental designs we may require superdiagonal matrices whose diagonal elements may be any type of simple matrix. They may be square, horizontal, vertical, triangular, or symmetric. They may be row or column vectors of either general or special types, such as unit, e_i, or sign vectors. What the elements are depends on the particular experimental design and the mathematical model chosen for the analysis of the data.*

---

*For applications of a number of different superdiagonal matrices see Horst, Paul, "Generalized Canonical Correlations and Their Applications to Experimental Data," *J. Clin. Psychol.*, Monog. Supp. No. 14 (October 1961).

Figure 5.6.1 is an example of a general superdiagonal matrix whose matric diagonal elements are rectangular matrices.

$$D_m = \left[\begin{array}{cc|ccc} 1 & 3 & 0 & 0 & 0 \\ 2 & 1 & 0 & 0 & 0 \\ 4 & 3 & 0 & 0 & 0 \\ \hline 0 & 0 & 2 & 1 & 5 \\ 0 & 0 & 4 & 3 & 2 \end{array}\right]$$

FIG. 5.6.1

It may be written symbolically as in Fig. 5.6.2.

$$D_m = \begin{bmatrix} m_1 & 0 \\ 0 & m_2 \end{bmatrix}$$

FIG. 5.6.2

Notice that in the partitioning, the simple orders of the matric diagonal elements determine the simple orders of the null matric elements. (The subscripts here do not indicate order.)

An example of a superdiagonal matrix with vector elements, which can be useful in certain experimental designs, is given in Fig. 5.6.3.

$$D_1 = \left[\begin{array}{c|c|c} 1 & 0 & 0 \\ 1 & 0 & 0 \\ \hline 0 & 1 & 0 \\ 0 & 1 & 0 \\ 0 & 1 & 0 \\ \hline 0 & 0 & 1 \\ 0 & 0 & 1 \end{array}\right]$$

FIG. 5.6.3

Here the diagonal elements are simply column unit vectors. Figure 5.6.3 may be written symbolically as in Fig. 5.6.4.

$$D_1 = \begin{bmatrix} 1_1 & 0 & 0 \\ 0 & 1_2 & 0 \\ 0 & 0 & 1_3 \end{bmatrix}$$

FIG. 5.6.4

The subscripts on the diagonal unit vectors in Fig. 5.6.4 indicate their positions in the matrix D_1 and not the simple orders of the vectors.

5.7 The Partial Triangular Matrix as a Supermatrix

You are already familiar with the partial triangular matrix. It is sometimes convenient to regard it as a supermatrix. This we do by partitioning it between the triangular part of the matrix and the rest of the matrix. In this way we have two submatrices, one of which is a square triangular matrix and the other a rectangular matrix. If we partition an upper partial triangular matrix into two submatrices in this manner, we have a 1 by 2 supermatrix. The first element in the row is an upper triangular matrix and the second element is a rectangular matrix. An upper partial triangular matrix partitioned in this fashion is shown in Fig. 5.7.1.

$$\begin{bmatrix} 3 & 4 & 7 & 6 & 8 \\ 0 & 2 & 1 & 2 & 5 \\ 0 & 0 & 5 & 4 & 9 \end{bmatrix}$$

FIG. 5.7.1

If we partition a lower partial triangular matrix in the manner just described, we have a 2 by 1 supermatrix. The first element in this supermatrix is a square lower triangular matrix and the second element is a rectangular matrix. A lower partial triangular matrix partitioned in this way is given in Fig. 5.7.2.

$$\begin{vmatrix} 2 & 0 & 0 \\ 9 & 4 & 0 \\ 8 & 3 & 6 \\ \hline 5 & 2 & 9 \\ 4 & 7 & 3 \end{vmatrix}$$

FIG. 5.7.2

It is convenient at times to use letters to indicate the submatrices of a partial triangular matrix. In Fig. 5.7.3 we represent Fig. 5.7.1 in this way.

$$B' = [T' \mid a]$$

FIG. 5.7.3

Since we have agreed that the natural order shall be that of the lower partial triangular matrix, we use a prime on the B in Fig. 5.7.3 to show that it is the transpose of the lower partial triangular matrix. On the right side of the equation we use T' for the square triangular submatrix. The prime shows that it is an upper triangular matrix. We use an a to indicate the rectangular submatrix. We do not use a prime after the letter a because it just happens that in Fig. 5.7.1 this submatrix has more rows than columns. Even though we have no hard and fast rules for indicating the natural order in a supermatrix, it is often convenient to prime horizontal matric elements, particularly in the case of partitioned partial triangular matrices. The separate submatrices of Fig. 5.7.1 that have been used in the right side of Fig. 5.7.3 are shown in Fig. 5.7.4.

$$T' = \begin{bmatrix} 3 & 4 & 7 \\ 0 & 2 & 1 \\ 0 & 0 & 5 \end{bmatrix} \qquad a = \begin{bmatrix} 6 & 8 \\ 2 & 5 \\ 4 & 9 \end{bmatrix}$$

Fig. 5.7.4

In the same way, we can indicate with letters the submatrices in Fig. 5.7.2 as in Fig. 5.7.5.

$$T = \begin{bmatrix} 2 & 0 & 0 \\ 9 & 4 & 0 \\ 8 & 3 & 6 \end{bmatrix} \qquad a' = \begin{bmatrix} 5 & 2 & 9 \\ 4 & 7 & 3 \end{bmatrix}$$

Fig. 5.7.5

Using the symbols as defined here, we can write the same matrix as in Fig. 5.7.6.

$$C = \begin{bmatrix} T \\ - \\ a' \end{bmatrix}$$

Fig. 5.7.6

Here you see that for the C on the left we do not use a prime because we regard the lower partial triangular matrix as the natural order. On the right side, the first submatrix is a lower triangular matrix which we indicate without a prime. The lower submatrix, a', has a prime because in Fig. 5.7.5 we see that this matrix is horizontal rather than vertical.

It would be quite possible in Fig. 5.7.1 to have, say, 7 columns instead of only 2. In this case it would be perfectly appropriate to use a' instead of a in Fig. 5.7.3 to show that the matrix is horizontal. Similarly in Fig. 5.7.2 it would be quite possible for the matrix below to have more rows than columns, in which case we could use, in Fig. 5.7.6, a for the lower submatrix instead of a', to show that it was vertical.

In most analyses of prediction data matrices, the partial triangular matrix plays an important role. Usually, the order of the triangular submatrix is the same as the number of predictor attributes in the prediction matrix while the number of columns in the rectangular part of the upper partial triangular matrix is the same as the number of criterion attributes. Frequently, this number is only one and in many cases it is less than the number of predictor attributes. For computational convenience, we usually work with the upper partial triangular matrix rather than the lower; and therefore the standard form of the rectangular part is usually as given in Fig. 5.7.1 or Fig. 5.7.3. Therefore the number of rows in the rectangular submatrix is usually equal to the number of predictor attributes while the number of columns is usually equal to the number of criterion attributes.

5.8 Supervectors

Type I column supervectors

Since a vector is a special case of a matrix with only one row or one column, we shall define a simple vector as we did a simple matrix. A *simple vector* is a vector each of whose elements is a scalar. It is useful, however, to define a number of different types of *supervectors*. The first of these types results from the partitioning of a simple vector. This we call a *type I* supervector. The elements of a type I column supervector would themselves be column vectors. For example, Fig. 5.8.1 is a type I column supervector that has two column vector elements, each of which in turn has two scalar elements.

$$V = \begin{bmatrix} 1 \\ 3 \\ \cdots \\ 4 \\ 6 \end{bmatrix}$$

FIG. 5.8.1

The terms we use are analogous to those used in working with matrices; we call the vector elements of a supervector *subvectors*. Thus, in Fig. 5.8.1 the first subvector of the vector V consists of the column elements 1 and 3. The second subvector consists of the column elements 4 and 6. We may also write a column vector in terms of its column subvectors as in

$$V = \begin{bmatrix} V_1 \\ V_2 \\ \cdots \\ V_n \end{bmatrix}$$

FIG. 5.8.2

In this figure V_1, V_2, and V_n are all column subvectors of the column vector V. Each V_i column subvector may have as many elements as we please, depending on the particular problem we are working with. In particular, one or more of these column subvectors may have only one element.

Type I row supervectors

We may also have a type I row supervector, each of whose elements is a row vector, as in Fig. 5.8.3.

$$V' = [1 \quad 3 \mid 4 \quad 6]$$

FIG. 5.8.3

In this figure, the first row subvector has the elements 1 and 3 and the second row subvector has the elements 4 and 6. You will see that on the left V has the prime to show that it is a row vector, since we consider the natural order to be the column vector. We can also use letters to stand for the row subvector elements of a type I row supervector as in Fig. 5.8.4.

$$V' = [V_1' \quad V_2' \quad \ldots \quad V_n']$$

FIG. 5.8.4

Here we have a type I row supervector with n row subvectors. You will notice that we use primes after each of the subvectors to show that they too are row vectors.

5.9   Matrices Expressed as Supervectors

We have discussed the type I supervector, which can be obtained by the partitioning of simple vectors. This partitioning, if applied to column vectors, yields column subvectors; and if applied to row vectors, yields row subvectors. We may also consider column supervectors whose elements are row vectors, and row supervectors whose elements are column vectors. It is easy to see that vectors of this type when expressed in simple form are matrices. Such vectors we shall call *type II* supervectors. Let us see in greater detail how we may express a matrix as a type II supervector. We start with Eq. (5.9.1).

$$a = \begin{bmatrix} a_{11} & a_{12} & \ldots & a_{1m} \\ a_{21} & a_{22} & \ldots & a_{2m} \\ \ldots & \ldots & \ldots & \ldots \\ a_{n1} & a_{n2} & \ldots & a_{nm} \end{bmatrix} \tag{5.9.1}$$

Type II column supervectors

Equation (5.9.1) is the general expression for an n by m matrix in expanded form. Now suppose we consider each row of the matrix in Eq. (5.9.1) as a separate row vector. Suppose we let a different symbol represent each one of these row vectors as shown in

$$
\begin{aligned}
a'_{1.} &= [a_{11} \quad a_{12} \quad \ldots \quad a_{1m}] \\
a'_{2.} &= [a_{21} \quad a_{22} \quad \ldots \quad a_{2m}] \\
\ldots & \\
a'_{n.} &= [a_{n1} \quad a_{n2} \quad \ldots \quad a_{nm}]
\end{aligned} \tag{5.9.2}
$$

We can now substitute the left side of the equations in (5.9.2) into the right side of Eq. (5.9.1) to get

$$
a = \begin{bmatrix} a'_{1.} \\ a'_{2.} \\ \ldots \\ a'_{n.} \end{bmatrix}_m \tag{5.9.3}
$$

You see now that the right side of Eq. (5.9.3) is a column vector each of whose elements is a row vector. Equation (5.9.3) then is one way of writing a matrix as a column supervector whose elements are rows.

Now let us discuss the meaning of the symbols used for the elements in the right side of Eq. (5.9.3). First, each element is primed to show that it is a row vector. Second, the first subscript in each element indicates the row in the original matrix from which the row vector is taken. Third, a dot is used in the second subscript position of each element to show that the element has a number of columns in it. Fourth, the subscript n for the last element means that the matrix has n rows. Fifth, the subscript m outside the brackets, or the *external subscript* as it is called, indicates the number of columns in the matrix, or the number of elements in each of the row vectors within the brackets.

Later on in this book we shall frequently refer to a row of elements in a matrix as a row vector of the matrix. We shall speak of the first, the second, or the ith row vector of a matrix. In the same way we shall also speak of a column of elements in a matrix as a column vector of the matrix.

Type II row supervectors

We shall next see that a matrix may also be expressed as a row supervector with elements that are column vectors. First we go back to Eq. (5.9.1) and consider each of its column vectors. We let each of these column vectors be indicated by the symbols given in

$$
a_{.1} = \begin{bmatrix} a_{11} \\ a_{21} \\ \ldots \\ a_{n1} \end{bmatrix} \qquad a_{.2} = \begin{bmatrix} a_{12} \\ a_{22} \\ \ldots \\ a_{n2} \end{bmatrix} \qquad a_{.m} = \begin{bmatrix} a_{1m} \\ a_{2m} \\ \ldots \\ a_{nm} \end{bmatrix} \tag{5.9.4}
$$

Now if we substitute the symbols on the left side of Eq. (5.9.4) into the matrix on the right side of Eq. (5.9.1), we have

$$
a = [a_{.1} \quad a_{.2} \quad \ldots \quad a_{.m}]_n \tag{5.9.5}
$$

Equation (5.9.5) expresses the matrix a as a row supervector whose subvectors are the column vectors of the original matrix given in Eq. (5.9.1).

Be sure now that you understand clearly the meaning of the symbols on the right side of Eq. (5.9.5). First, each of the elements in the row is left unprimed. The fact that primes are *not* used after these elements shows that they are column vectors. Second, the first subscript instead of the second is a dot. It is used to show that the row elements, rather than the column elements, vary. Third, the second subscript after each element indicates the column of the matrix from which the vector is taken. Fourth, the second subscript of the last element is m and indicates the number of columns in the matrix. The subscript outside the brackets, n, or the external subscript, shows the number of rows in the matrix.

Notice then that we may equally well write a matrix as a column supervector whose elements are rows, as we did in Eq. (5.9.3), or as a row supervector whose elements are columns, as we did in Eq. (5.9.5). Therefore from Eqs. (5.9.3) and (5.9.5), we can write

$$
a = \begin{bmatrix} a'_{1.} \\ a'_{2.} \\ \cdots \\ a'_{n.} \end{bmatrix}_m = [a_{.1} \quad a_{.2} \quad \cdots \quad a_{.m}]_n \tag{5.9.6}
$$

Equation (5.9.6), then, illustrates two very useful ways in which a matrix can be expressed as a type II supervector.

5.10   The Equality of Supermatrices

If you look back at Eq. (5.9.6) in Section 5.9, you see that it violates the rule for equality which we gave for simple matrices. There we said that two matrices are equal if and only if each element of one was equal to the corresponding element of the other. This rule implied, of course, that the two matrices were of equal order, because otherwise, for each element in one, there could not be a corresponding element in the other. But you will see that in Eq. (5.9.6) the two matrices on the right which are said to be equal to each other do not even have the same order. The first is a column supervector while the second is a row supervector. Since supervectors are special cases of supermatrices, you can see that two supermatrices may be equal even though they do not have the same super order. In any case, however, two supermatrices are equal if they have the same order and if each submatrix in one is equal to the corresponding submatrix in the other. For submatrices to be equal, they must of course be of the same order and their corresponding elements must be equal.

In general, two supermatrices may be equal if they are of the same order even though their corresponding matric elements are not equal. This may not be obvious at first, but you will readily perceive it if you look at Figs. 5.10.1 and 5.10.2.

$$a_{11} = 3 \qquad a_{12} = [4 \quad 2 \quad 1]$$

$$a_{21} = \begin{bmatrix} 4 \\ 7 \\ 8 \end{bmatrix} \qquad a_{22} = \begin{bmatrix} 6 & 1 & 3 \\ 2 & 1 & 5 \\ 9 & 4 & 6 \end{bmatrix}$$

FIG. 5.10.1

$$b_{11} = \begin{bmatrix} 3 & 4 \\ 4 & 6 \end{bmatrix} \qquad b_{12} = \begin{bmatrix} 2 & 1 \\ 1 & 3 \end{bmatrix}$$

$$b_{21} = \begin{bmatrix} 7 & 2 \\ 8 & 9 \end{bmatrix} \qquad b_{22} = \begin{bmatrix} 1 & 5 \\ 4 & 6 \end{bmatrix}$$

FIG. 5.10.2

Suppose now in Fig. 5.10.3 we let a be a supermatrix made up of the submatrices in Fig. 5.10.1 and b a supermatrix made up of the submatrices in Fig. 5.10.2.

$$a = \begin{bmatrix} a_{11} & a_{12} \\ a_{21} & a_{22} \end{bmatrix} \qquad b = \begin{bmatrix} b_{11} & b_{12} \\ b_{21} & b_{22} \end{bmatrix}$$

FIG. 5.10.3

Now let us substitute the right-hand sides of the equations in Fig. 5.10.1 into the first matrix of Fig. 5.10.3. We also substitute the right-hand sides of the equations in Fig. 5.10.2 into the second, or b matrices of Fig. 5.10.3. The result is

$$a = \left[\begin{array}{c|ccc} 3 & 4 & 2 & 1 \\ \hline 4 & 6 & 1 & 3 \\ 7 & 2 & 1 & 5 \\ 8 & 9 & 4 & 6 \end{array} \right] \qquad b = \left[\begin{array}{cc|cc} 3 & 4 & 2 & 1 \\ 4 & 6 & 1 & 3 \\ \hline 7 & 2 & 1 & 5 \\ 8 & 9 & 4 & 6 \end{array} \right]$$

FIG. 5.10.4

You see that the corresponding scalar elements for matrix a and matrix b are identical.

We can now state the necessary and sufficient condition for two supermatrices to be equal. Two supermatrices are equal if and only if their corresponding simple forms are equal. As you see in Fig. 5.10.4, if the partitioning lines were removed from the matrices, the resulting simple matrices would be identical.

5.11   Vectors Whose Elements Are Matrices

We have considered vectors whose elements are also vectors, and we have seen that there are two general types. We may also have supervectors whose elements are matrices. These we call *type III* supervectors. The partial triangular matrices written in supermatrix form in Figs. 5.7.3 and 5.7.6 are examples of supervectors whose elements are themselves matrices. You may wonder whether, if we have a vector whose elements are matrices, we would use the primed symbol when the submatrices are in a row and the unprimed if the submatrices are in a column. To be consistent with the practice used in the case of simple vectors, we would have to follow this rule. However, it is usually more convenient to let the simple order of a matrix determine whether to use the prime rather than the order of the supervector that represents the matrix.

One interesting and common example of a type III supervector is a prediction data matrix having both predictor and criterion attributes. If such a matrix is partitioned between these two sets of attributes it becomes a second-order type III supervector. For purposes of computation and analysis it is often useful to regard it as such.

SUMMARY

1. Definition of a supermatrix:
    a. Definition of a simple matrix: a matrix whose elements are scalar quantities is called a simple matrix to distinguish it from a supermatrix.
    b. A supermatrix is one in which one or more of its elements are themselves simple matrices. All matric elements in a given row must have the same number of simple rows and all matric elements in a given column must have the same number of simple columns.
    c. The order of the supermatrix is the number of matric rows and columns.

2. Partitioned matrices: Supermatrices may be constructed from simple matrices by partitioning between certain rows and columns.

3. The natural order of a supermatrix: This must be defined, if at all, in terms of the natural order of its simple form.

4. Symmetric partitioning:
    a. This is accomplished by partitioning between the corresponding rows and columns.
    b. Only square matrices can therefore be partitioned symmetrically.

5. Symmetrically partitioned symmetric simple matrices:
    a. All diagonal matric elements are symmetric simple matrices.
    b. Corresponding off-diagonal matric elements are mutual transposes.

6. The general diagonal supermatrix:

    a. This may have, for matric diagonal elements, any kind of simple matrices, including vectors and scalar quantities.

    b. All off-diagonal matric elements will be null simple matrices.

7. The partial triangular matrix as a supermatrix: This matrix may be partitioned to yield a triangular matrix and a rectangular matrix.

8. Type I supervectors:

    a. A partitioned simple vector will be called a type I supervector.

    b. A column type I supervector has simple column vectors for elements.

    c. A row type I supervector has simple row vectors for elements.

9. Matrices expressed as supervectors with vector element:

    a. Such vectors will be called type II supervectors.

    b. A matrix may be expressed as a column supervector whose elements are row vectors.

    c. A matrix may be expressed as a row supervector whose elements are column vectors.

10. Equality of supermatrices:

    a. Two supermatrices are equal if and only if their simple forms are equal.

11. Type III supervectors:

    a. Vectors one or more of whose elements are matrices will be called type III vectors.

    b. A type III supervector may be constructed by partitioning a simple matrix either between rows or between columns but not both.

EXERCISES

1. Given the supermatrix with matric elements

$$a = \begin{bmatrix} a_{11} & a_{12} & a_{13} \\ a_{21} & a_{22} & a_{23} \\ a_{31} & a_{32} & a_{33} \\ a_{41} & a_{42} & a_{43} \end{bmatrix}$$

Assume: a_{11} is [1 by 2], a_{12} and a_{13} are of widths 3 and 4 respectively; a_{21}, a_{31}, and a_{41} are of heights 3, 2, and 1 respectively. What are the orders of a_{22}, a_{33}, a_{32}, and a_{43} respectively?

2. What does the order of a supermatrix tell us about the order of its corresponding simple matrix?

3. Suppose we have an [8 by 6] simple matrix. How may we partition it to have a [2 by 3] supermatrix all of whose matric elements have the same simple order?

4. If a matrix has been symmetrically partitioned, what do we know about its simple order?

5. Given the symmetrically partitioned symmetric simple matrix

$$a = \begin{bmatrix} a_{11} & a_{12} & a_{13} \\ a_{21} & a_{22} & a_{23} \\ a_{31} & a_{32} & a_{33} \end{bmatrix}$$

  (a) What kind of matrices are a_{11}, a_{22}, a_{32}?
  (b) What are the relationships between a_{12} and a_{21}, a_{13} and a_{31}, a_{23} and a_{32}?

6. What is the meaning of the primes below the diagonal in the diagonal super-matrix?

$$D = \begin{bmatrix} D_1 & 0 & 0 \\ 0' & D_2 & 0 \\ 0' & 0' & D_3 \end{bmatrix}$$

7. If the subscripts in the super identity matrix below indicate order, what is the simple order of the matrix?

$$I = \begin{bmatrix} I_2 & 0 & 0 \\ 0' & I_3 & 0 \\ 0' & 0' & I_1 \end{bmatrix}$$

8. Given the matrix

$$T' = \begin{bmatrix} 3 & 4 & 7 & 6 & 8 \\ 0 & 2 & 1 & 2 & 5 \\ 0 & 0 & 5 & 4 & 9 \end{bmatrix}$$

  (a) How would you partition it so that one of the submatrices is a triangular matrix?
  (b) Is it in the natural or transposed order?

9. If V and V' are type I supervectors, what are (a) the elements of V, and (b) the elements of V'?

10. If (b) is a type II row supervector, what are its elements?

11. If (b) is a type II column supervector, what are its elements?

ANSWERS

1. [3 by 3], [2 by 4], [2 by 3], [1 by 4]

2. Nothing

3. Partition between the 2nd and 3rd and the 4th and 5th columns, and between the 4th and 5th rows.

4. It is square.

5. (a) They are all symmetric.
   (b) One is the transpose of the other.

6. Corresponding null submatrices above and below the diagonal have reversed orders.

7. (b)

8. (a) Between 3rd and 4th columns
   (b) Transposed order

9. (a) Column vectors
   (b) Row vectors

10. Column vectors

11. Row vectors

Chapter 6

The Transpose
of a Supermatrix

6.1 General Matrix with Matrix Elements

You can readily deduce the rules for finding the transpose of any super-matrix simply by observing what happens to the submatrices of a simple matrix when it is partitioned. Look at Fig. 6.1.1.

$$A = \left[\begin{array}{cc|cc} 1 & 6 & 1 & 8 \\ 2 & 3 & 3 & 7 \\ 4 & 5 & 4 & 6 \\ \hline 7 & 2 & 9 & 3 \\ 3 & 4 & 7 & 5 \\ \hline 3 & 1 & 2 & 8 \end{array}\right]$$

FIG. 6.1.1

Here we start with a 6 by 4 simple matrix and partition it into a 3 by 2 supermatrix. Then we take the transpose of the matrix A in Fig. 6.1.1 and get a matrix as in Fig. 6.1.2.

$$A' = \left[\begin{array}{ccc|cc|c} 1 & 2 & 4 & 7 & 3 & 3 \\ 6 & 3 & 5 & 2 & 4 & 1 \\ \hline 1 & 3 & 4 & 9 & 7 & 2 \\ 8 & 7 & 6 & 3 & 5 & 8 \end{array}\right]$$

FIG. 6.1.2

Suppose now we represent each of the submatrices in Fig. 6.1.1 by letter symbols with double number subscripts as in Fig. 6.1.3.

$$a_{11} = \begin{bmatrix} 1 & 6 \\ 2 & 3 \\ 4 & 5 \end{bmatrix} \qquad a_{12} = \begin{bmatrix} 1 & 8 \\ 3 & 7 \\ 4 & 6 \end{bmatrix}$$

$$a_{21} = \begin{bmatrix} 7 & 2 \\ 3 & 4 \end{bmatrix} \qquad a_{22} = \begin{bmatrix} 9 & 3 \\ 7 & 5 \end{bmatrix}$$

$$a_{31} = \begin{bmatrix} 3 & 1 \end{bmatrix} \qquad a_{32} = \begin{bmatrix} 2 & 8 \end{bmatrix}$$

FIG. 6.1.3

Thus we can write the matrix A in terms of the letter elements that stand for the matrices in Fig. 6.1.3. This gives us Fig. 6.1.4.

$$A = \begin{bmatrix} a_{11} & a_{12} \\ a_{21} & a_{22} \\ a_{31} & a_{32} \end{bmatrix}$$

FIG. 6.1.4

Next let us take the transposes of the matrices in Fig. 6.1.3. This gives us the matrices in Fig. 6.1.5. Now compare the submatrices in Fig. 6.1.5 with those in Fig. 6.1.2.

$$a_{11}' = \begin{bmatrix} 1 & 2 & 4 \\ 6 & 3 & 5 \end{bmatrix} \qquad a_{12}' = \begin{bmatrix} 1 & 3 & 4 \\ 8 & 7 & 6 \end{bmatrix}$$

$$a_{21}' = \begin{bmatrix} 7 & 3 \\ 2 & 4 \end{bmatrix} \qquad a_{22}' = \begin{bmatrix} 9 & 7 \\ 3 & 5 \end{bmatrix}$$

$$a_{31}' = \begin{bmatrix} 3 \\ 1 \end{bmatrix} \qquad a_{32}' = \begin{bmatrix} 2 \\ 8 \end{bmatrix}$$

FIG. 6.1.5

You will see that the three matrices on the left in Fig. 6.1.5 are respectively the three submatrices in the first matric row of the matrix in Fig. 6.1.2. Similarly, you will see that the three matrices on the right side of Fig. 6.1.5 are respectively the three submatrices in the second matric row of the matrix in Fig. 6.1.2. We can therefore write the matrix in Fig. 6.1.2 in terms of letter elements with double subscripts as indicated in Fig. 6.1.6.

$$A' = \begin{bmatrix} a_{11}' & a_{21}' & a_{31}' \\ a_{12}' & a_{22}' & a_{32}' \end{bmatrix}$$

FIG. 6.1.6

Rules for the transpose of a supermatrix

If you now compare Figs. 6.1.4 and 6.1.6, you can easily discover the rules for writing the transpose of a supermatrix whose elements are letters with double number subscripts. First, the order of the supermatrix is reversed. For example, Fig. 6.1.4 is a 3 by 2 supermatrix, whereas Fig. 6.1.6 is a 2 by 3 supermatrix. Second, a row of the original matrix becomes a column of the transposed matrix, or, what amounts to the same thing, a column of the original matrix becomes a row of the transposed matrix. Third, as a consequence of writing rows for columns and vice versa, the order of the subscripts is changed, so that in the transpose the first subscript indicates the column in which the matric element is found, and the second element indicates the row. These three rules for writing the transpose of a supermatrix whose elements are letters with double subscripts are the same as those for the transpose of a simple matrix. A fourth rule, not the same, is that each element in the transpose is written with a prime, as in Fig. 6.1.6.

We can incorporate these general principles for any order of supermatrix in Figs. 6.1.7 and 6.1.8.

$$A = \begin{bmatrix} a_{11} & a_{12} & \ldots & a_{1m} \\ a_{21} & a_{22} & \ldots & a_{2m} \\ \ldots & \ldots & \ldots & \ldots \\ a_{n1} & a_{n2} & \ldots & a_{nm} \end{bmatrix}$$

Fig. 6.1.7

$$A' = \begin{bmatrix} a'_{11} & a'_{21} & \ldots & a'_{n1} \\ a'_{12} & a'_{22} & \ldots & a'_{n2} \\ \ldots & \ldots & \ldots & \ldots \\ a'_{1m} & a'_{2m} & \ldots & a'_{nm} \end{bmatrix}$$

Fig. 6.1.8

In Fig. 6.1.7 we have an n by m supermatrix. In Fig. 6.1.8 we have the transpose of this supermatrix, an m by n supermatrix obtained from Fig. 6.1.7 by writing rows for columns and taking the transpose of each of the elements. To summarize, then, the transpose of a supermatrix is obtained by writing rows for columns and taking the transpose of each element.

You will see, of course, that Figs. 6.1.7 and 6.1.8 could also illustrate the transpose of a simple matrix, because a simple matrix is actually a special case of a supermatrix, in which the elements are scalars rather than matrices. As you recall, the transpose of a scalar is the scalar itself, so that Fig. 6.1.8 could represent a simple matrix. In that case, there would of course be no point in adding the primes to the elements, since they would be equal to the original scalar elements.

6.2   The Transpose of a Symmetrically Partitioned Symmetric Simple Matrix

Now let us see how the rules for getting the transpose of a supermatrix work out for a supermatrix that has been obtained from a symmetrically partitioned symmetric simple matrix. As you recall from Section 5.5, if we write the submatrices of such a matrix as letter elements with number subscripts, we have Eq. (6.2.1).

$$a = \begin{bmatrix} a_{11} & a_{12} & \ldots & a_{1m} \\ a'_{12} & a_{22} & \ldots & a_{2m} \\ \ldots & \ldots & \ldots & \ldots \\ a'_{1m} & a'_{2m} & \ldots & a_{mm} \end{bmatrix} \tag{6.2.1}$$

Here we have a supermatrix of order m whose infradiagonal elements are the transposes of the corresponding supradiagonal elements and, as you recall, the diagonal elements are themselves symmetric matrices. Now let us write the transpose of a in Eq. (6.2.1) by writing rows for columns and writing the transpose of each element. This gives us Eq. (6.2.2).

$$a' = \begin{bmatrix} a'_{11} & (a'_{12})' & \ldots & (a'_{1m})' \\ a'_{12} & a'_{22} & \ldots & (a'_{2m})' \\ \ldots & \ldots & \ldots & \ldots \\ a'_{1m} & a'_{2m} & \ldots & a'_{mm} \end{bmatrix} \tag{6.2.2}$$

Notice that the infradiagonal elements of Eq. (6.2.2) are exactly the same as the corresponding elements in Eq. (6.2.1). However, the diagonal elements of Eq. (6.2.2) are the transposes of the diagonal elements in Eq. (6.2.1). Also, the supradiagonal elements in Eq. (6.2.2) have the same subscripts as the corresponding elements in Eq. (6.2.1), but they are modified by the addition of parentheses and primes. Remember, however, that the diagonal matric elements of a symmetric matrix, symmetrically partitioned, are symmetric matrices and thus are not altered by transposition. Therefore we have

$$a'_{11} = a_{11} \qquad a'_{22} = a_{22} \qquad \text{etc.} \tag{6.2.3}$$

Recall also that the transpose of a transpose is the original matrix. Therefore we write

$$(a'_{12})' = a_{12} \qquad (a'_{13})' = a_{13} \qquad (a'_{ij})' = a_{ij} \tag{6.2.4}$$

If now in Eq. (6.2.2) on the right we substitute the right-hand side of Eqs. (6.2.3) and (6.2.4), we get Eq. (6.2.5).

$$a' = \begin{bmatrix} a_{11} & a_{12} & \ldots & a_{1m} \\ a'_{12} & a_{22} & \ldots & a_{2m} \\ \ldots & \ldots & \ldots & \ldots \\ a'_{1m} & a'_{2m} & \ldots & a_{mm} \end{bmatrix} \tag{6.2.5}$$

But you will see that the right-hand side of Eq. (6.2.5), the transpose, is exactly the same as the right-hand side of Eq. (6.2.1), the supermatrix itself. This is, of course, as it should be, since we started with a symmetrically partitioned symmetric simple matrix.

We have then the simple rule that the transpose of a symmetrically partitioned simple symmetric matrix is the matrix itself. You might have seen this at a glance without going through all the steps given in Eqs. (6.2.1) through (6.2.5). However, these steps show how, by means of simple rules and well-defined symbols, we can arrive at certain conclusions. Sometimes the conclusions are very obvious; the use of rules and symbols seems unnecessary. At other times the conclusions are not easily drawn, but they can be arrived at by careful use of the same simple rules and symbols.

6.3   The Transpose of a Diagonal Supermatrix

The simple diagonal matrix

At this point in your study of matrices, it should be very easy for you to see that the transpose of a symmetrically partitioned diagonal matrix is simply the original diagonal supermatrix itself. Nevertheless, let us apply the simple rules we have learned and use letter symbols to prove this fact. We let Eq. (6.3.1) represent the diagonal supermatrix.

$$
D = \begin{bmatrix} d_1 & & & \\ & d_2 & & \\ & & \cdots & \\ & & & d_n \end{bmatrix} \tag{6.3.1}
$$

Each of the elements in the right side of this equation is itself a diagonal matrix. To find the transpose of D here, we simply write rows for columns and take the transpose of each element. This gives Eq. (6.3.2).

$$
D' = \begin{bmatrix} d_1' & & & \\ & d_2' & & \\ & & \cdots & \\ & & & d_n' \end{bmatrix} \tag{6.3.2}
$$

But now we remember that the transpose of a diagonal matrix is the diagonal matrix itself, so that we can write

$$
d_1' = d_1 \qquad d_2' = d_2 \qquad \text{etc.} \tag{6.3.3}
$$

If we substitute the right sides of the equations in Eq. (6.3.3) for the elements in the right side of Eq. (6.3.2), we have Eq. (6.3.4).

$$
D' = \begin{bmatrix} d_1 & & & \\ & d_2 & & \\ & & \cdots & \\ & & & d_n \end{bmatrix} \tag{6.3.4}
$$

The right-hand side of Eq. (6.3.4) is exactly the same as Eq. (6.3.1). Therefore, by the use of the symbols and simple rules you have learned up to this point, you have proved that the transpose of a diagonal supermatrix is the diagonal supermatrix itself. Note, however, that if the diagonal elements in Eq. (6.3.1) were other than diagonal simple matrices, we could not remove the primes on the right of Eq. (6.3.2).

The scalar and the identity matrix

Special cases of the diagonal supermatrix are the scalar supermatrix and the identity supermatrix. Since it is obvious that the transpose of each of these is the matrix itself, we need not go through the proof. We show that the transpose of a super identity matrix is equal to the matrix itself by Eq. (6.3.5).

$$\begin{bmatrix} I_1 & & & \\ & I_2 & & \\ & & \cdots & \\ & & & I_n \end{bmatrix}' = \begin{bmatrix} I_1 & & & \\ & I_2 & & \\ & & \cdots & \\ & & & I_n \end{bmatrix} \tag{6.3.5}$$

6.4 The Transpose of a Type I Supervector

You recall that in partitioning a simple column vector we get a column vector whose elements are column vectors, and in partitioning a simple row vector we get a row vector whose elements are row vectors. Suppose V is a column supervector obtained from partitioning a simple column vector as in Fig. 6.4.1.

$$V = \begin{bmatrix} 3 \\ 2 \\ - \\ 1 \\ 4 \\ - \\ 6 \\ 5 \end{bmatrix}$$

FIG. 6.4.1

Then its transpose would be given as in Fig. 6.4.2.

$$V' = [3 \quad 2 \mid 1 \quad 4 \mid 6 \quad 5]$$

FIG. 6.4.2

If we let the subvectors be represented as in Fig. 6.4.3, then Figs. 6.4.1 and 6.4.2 may be written as letters as in the left and right equations of Fig. 6.4.4, respectively.

$$V_1 = \begin{bmatrix} 3 \\ 2 \end{bmatrix} \qquad V_2 = \begin{bmatrix} 1 \\ 4 \end{bmatrix} \qquad V_3 = \begin{bmatrix} 6 \\ 5 \end{bmatrix}$$

FIG. 6.4.3

$$V = \begin{bmatrix} V_1 \\ V_2 \\ V_3 \end{bmatrix} \qquad V' = [V_1' \quad V_2' \quad V_3']$$

Fig. 6.4.4

Or, if there are n subvectors in V, we may indicate the supervector and its transpose by Fig. 6.4.5.

$$V = \begin{bmatrix} V_1 \\ V_2 \\ \ldots \\ V_n \end{bmatrix} \qquad V' = [V_1' \quad V_2' \quad \ldots \quad V_n']$$

Fig. 6.4.5

Thus we use the same rules to get the transpose of a type I supervector as we do to get the transpose of a general supermatrix. We write row for column and prime the elements. We cannot reverse the subscripts, of course, since each subvector has only one subscript.

6.5   The Transpose of a Simple Matrix Written as a Type II Supervector

The column type II supervector

You recall that we can write a matrix as a column supervector whose elements are row vectors, or as a row vector whose elements are column vectors. Let us discuss first the column supervector whose elements are row vectors. Using the method given in Section 5.9, we write this as in Eq. (6.5.1).

$$a = \begin{bmatrix} a_{1.}' \\ a_{2.}' \\ \ldots \\ a_{n.}' \end{bmatrix}_m \qquad (6.5.1)$$

This is the way we indicate a matrix of order n by m if we want to write it as a column supervector with the rows of the matrix as the row subvectors. Now to indicate the transpose of a, we simply write row for column and take the transpose of the elements, as in

$$a' = [(a_{1.}')' \quad (a_{2.}')' \quad \ldots \quad (a_{n.}')']_m \qquad (6.5.2)$$

But remember that two primes cancel each other. Therefore, in terms of the notation used for the row vectors of a matrix, we would have

$$(a_{1.}')' = a_{1.} \qquad (a_{2.}')' = a_{2.} \qquad (a_{n.}')' = a_{n.} \qquad (6.5.3)$$

We could write a single equation for Eq. (6.5.3), using the first subscript as i to show that it means any element of the supervector, as in

$$(a'_{i.})' = a_{i.}. \qquad (6.5.4)$$

If then we substitute the right side of Eq. (6.5.3) or Eq. (6.5.4) for the elements in the right side of Eq. (6.5.2), we have

$$a' = [a_{1.} \quad a_{2.} \quad \ldots \quad a_{n.}]_m \qquad (6.5.5)$$

You see now that in Eq. (6.5.5) any element $a_{i.}$ is a row of the original matrix, written in column form in the transpose.

The row type II supervector

Now consider a matrix written as a supervector whose elements are column vectors. This, as shown in Section 5.9, can be written as

$$a = [a_{.1} \quad a_{.2} \quad \ldots \quad a_{.m}]_n \qquad (6.5.6)$$

Therefore to write the transpose of a given by Eq. (6.5.6), we write the row of elements on the right as a column and prime each of these elements, as in Eq. (6.5.7).

$$a' = \begin{bmatrix} a'_{.1} \\ a'_{.2} \\ \ldots \\ a'_{.m} \end{bmatrix}_n \qquad (6.5.7)$$

On the right side of the equation, we now have the column vectors of the matrix a written in row form. We can now say that $a'_{.i}$ is the ith column (that is, any unspecified column) of a written in row form in a'.

Thus Eqs. (6.5.5) and (6.5.7) show that there are two different ways to write the transpose of a matrix as a supervector. Therefore we can write Eq. (6.5.8).

$$a' = [a_{1.} \quad a_{2.} \quad \ldots \quad a_{n.}]_m = \begin{bmatrix} a'_{.1} \\ a'_{.2} \\ \ldots \\ a'_{.m} \end{bmatrix}_n \qquad (6.5.8)$$

The meaning of the notation

The meaning of all the symbols used in Eqs. (6.5.1), (6.5.5), (6.5.6), and (6.5.7) must be perfectly clear to you if you are to understand the rules we develop in the following chapters. Let us enumerate the specific things you should remember about the notation used in these equations. These are as follows:

(1) a is usually, but not necessarily, a vertical matrix.
(2) a' is usually, but not necessarily, a horizontal matrix.

(3) a' is always the transpose of a.

(4) a is always the transpose of a'.

(5) a can be expressed as either a column supervector with row elements or a row supervector with column elements.

(6) a' can be expressed as either a column supervector with row elements or a row supervector with column elements.

(7) $a'_{i.}$ is always a row vector, each element of which is the corresponding element of the ith row of a.

(8) $a_{i.}$ is always a *column* vector, each element of which is the corresponding element of the ith *row* of a.

(9) $a_{.i}$ is always a *column* vector, each element of which is the corresponding element of the ith *column* of a.

(10) $a'_{.i}$ is always a *row* vector, each element of which is the corresponding element of the ith *column* of a.

(11) $a'_{i.}$ and $a'_{.i}$ are always row vectors even though a' may be a vertical matrix.

(12) $a_{i.}$ and $a_{.i}$ are always column vectors even though a may be a horizontal matrix.

(13) The letter m or n means the number of rows in the matrix if it is inside column brackets.

(14) The letter m or n means the number of columns in the matrix if it is inside row brackets.

(15) The letter m or n means the number of columns in the matrix if it is outside the column brackets.

(16) The letter m or n means the number of rows if it is outside the row brackets.

To summarize the supervector notation for a simple matrix and its transpose, we have then the very important equations (6.5.9) and (6.5.10).

$$a = \begin{bmatrix} a'_{1.} \\ a'_{2.} \\ \cdots \\ a'_{n.} \end{bmatrix}_m = [a_{.1} \quad a_{.2} \quad \cdots \quad a_{.m}]_n \qquad (6.5.9)$$

$$a' = [a_{1.} \quad a_{2.} \quad \cdots \quad a_{n.}]_m = \begin{bmatrix} a'_{.1} \\ a'_{.2} \\ \cdots \\ a'_{.m} \end{bmatrix}_n \qquad (6.5.10)$$

Be sure you understand exactly what these equations mean. They are fundamental to most computational procedures involving matrices, whether desk calculators are employed or whether programs are written for high-speed electronic computers.

6.6 Transpose of a Type III Supervector

As you have learned (Section 5.11), a type III supervector is a special case of a supermatrix with only one row or one column of matric elements. Therefore the transpose of a supervector with matric elements introduces nothing new. We simply apply the rules of writing row for column or column for row and priming the elements. (Since for a type III supervector we use the prime in accordance with the simple order of the matrix rather than the order of the supervector, we do not necessarily write the row form with a prime.) If we let a supervector a be represented by Eq. (6.6.1), where a_1 and a_2 are matrices,

$$a = [a_1 \quad a_2] \tag{6.6.1}$$

then the transpose, a', would be given by Eq. (6.6.2).

$$a' = \begin{bmatrix} a_1' \\ a_2' \end{bmatrix} \tag{6.6.2}$$

On the other hand, if a supervector b were in column form with matric elements such as those in Eq. (6.6.3), then the transpose, b', would be given by Eq. (6.6.4).

$$b = \begin{bmatrix} b_1 \\ b_2 \end{bmatrix} \tag{6.6.3}$$

$$b' = [b_1' \quad b_2'] \tag{6.6.4}$$

A special and frequently used example of a supervector with matric elements is a partial triangular matrix. Suppose we have an upper partial triangular matrix as in Fig. 6.6.1.

$$T' = \begin{bmatrix} 3 & 2 & 4 & \vline & 5 & 2 \\ 0 & 6 & 1 & \vline & 3 & 5 \\ 0 & 0 & 2 & \vline & 1 & 6 \end{bmatrix}$$

FIG. 6.6.1

Let us then define submatrices as in Fig. 6.6.2.

$$t' = \begin{bmatrix} 3 & 2 & 4 \\ 0 & 6 & 1 \\ 0 & 0 & 2 \end{bmatrix} \qquad a = \begin{bmatrix} 5 & 2 \\ 3 & 5 \\ 1 & 6 \end{bmatrix}$$

FIG. 6.6.2

Now, because t' here is simply a matrix of the first three columns of Fig. 6.6.1, and a is a matrix of the last two columns of Fig. 6.6.1, we can write

$$T' = [t' \quad | \quad a] \tag{6.6.5}$$

Then the transpose of T' in Eq. (6.6.5) is simply T, or is given by Eq. (6.6.6)

$$T = \begin{bmatrix} t \\ a' \end{bmatrix} \tag{6.6.6}$$

In Eq. (6.6.6), we have simply written column for row from Eq. (6.5.15) and have taken the transposes of the two matric elements. If we take the transposes of the matrices in Fig. 6.6.2, we have Fig. 6.6.3.

$$t = \begin{bmatrix} 3 & 0 & 0 \\ 2 & 6 & 0 \\ 4 & 1 & 2 \end{bmatrix} \qquad a' = \begin{bmatrix} 5 & 3 & 1 \\ 2 & 5 & 6 \end{bmatrix}$$

Fig. 6.6.3

If now we substitute in Eq. (6.6.6) the right-hand side of the equations given in Fig. 6.6.3, we have Fig. 6.6.4. This is obviously the transpose of Fig. 6.6.1.

$$T = \left[\begin{array}{ccc} 3 & 0 & 0 \\ 2 & 6 & 0 \\ 4 & 1 & 2 \\ \hline 5 & 3 & 1 \\ 2 & 5 & 6 \end{array} \right]$$

Fig. 6.6.4

Several important points should be noted in Eqs. (6.6.5) and (6.6.6). We are at liberty to use the prime or not, as we please, with any matric element, or any symbol for a type III supervector. However, once we have decided which symbols to prime and which to leave unprimed, then in the transpose of the supervector, we must remove the primes from the symbols that were primed and prime the symbols that were unprimed.

You will note that these conclusions were anticipated in Section 5.7.

SUMMARY

1. Transpose of the general supermatrix with matric elements:
   a. Rows are written as columns or columns as rows.
   b. Subscripts of matric elements are reversed.
   c. Each matric element is primed.

2. Transpose of symmetrically partitioned symmetric simple matrix: This must yield the same supermatrix when transposed.

3. Transpose of a diagonal supermatrix:
   a. *The general case.* The diagonal matric elements are transposed.
   b. If simple diagonal matrices comprise the diagonal matric elements, the transpose is the same as the original.

4. Transpose of type I supervectors:
   a. The column supervector becomes a row with row vector elements.
   b. The row supervector becomes a column with column vector elements.

5. Transpose of simple matrix expressed as type II supervector:
   a. The column supervector becomes a row whose row vector elements are transposed to column vector elements.
   b. The row supervector becomes a column whose column vector elements are transposed to row vector elements.

6. Transpose of type III supervector:
   a. The column vector becomes a row with matric elements transposed.
   b. The row vector becomes a column with matric elements transposed.

EXERCISES

1. If a_1. has the sequence of elements 4, 6, 7, 5
      a_2. has the sequence of elements 3, 9, 2, 8
      a_3. has the sequence of elements 5, 6, 4, 2

   write a'.

2. If b'_1 has the sequence of elements 7, 3, 6, 9
      b'_2 has the sequence of elements 8, 2, 4, 2
      b'_3 has the sequence of elements 4, 1, 9, 7

   write b.

3. If c'_1. has the sequence of elements 7, 4, 6, 5
      c'_2. has the sequence of elements 9, 3, 2, 1
      c'_3. has the sequence of elements 3, 9, 5, 7

   write c.

4. If $f_{.1}$ has the sequence of elements 6, 4, 3
      $f_{.2}$ has the sequence of elements 5, 2, 1
      $f_{.3}$ has the sequence of elements 9, 3, 8

   write f'.

5. Let A be a zero-one matrix where the entities are five rats, the attributes four, and the elements successes or failures, thus,

$$A' = \begin{bmatrix} 1 & 0 & 0 & 1 & 0 \\ 0 & 1 & 0 & 1 & 0 \\ 1 & 0 & 1 & 0 & 1 \\ 1 & 1 & 1 & 1 & 1 \end{bmatrix}$$

Write out each of the following vectors: (a) A_1.   (b) A'_4 (c) A_2.   (d) A'_5.
(e) A'_3 (f) A_4.

6. Let B be a zero-one matrix where the entities are four people, the attributes five test items, and the elements success or failure on the item, thus

$$B = \begin{bmatrix} 1 & 1 & 1 & 1 & 1 \\ 1 & 1 & 0 & 1 & 0 \\ 1 & 0 & 1 & 0 & 1 \\ 1 & 1 & 1 & 0 & 0 \end{bmatrix}$$

Write out each of the following vectors: (a) $B'_{.5}$ (b) $B_{2.}$ (c) $B_{.4}$ (d) $B'_{3.}$ (e) $B'_{1.}$ (f) $B_{.5}$.

7. Indicate whether each of the following in simple form is a matrix or a vector:

(a) $\begin{bmatrix} x'_{1.} \\ x'_{2.} \\ \cdots \\ x'_{s.} \end{bmatrix}$ (b) $\begin{bmatrix} x_{1.} \\ x_{2.} \\ \cdots \\ x_{s.} \end{bmatrix}$ (c) $[x_{1.} \quad x_{2.} \quad \cdots \quad x_{s.}]$ (d) $\begin{bmatrix} x_{.1} \\ x_{.2} \\ \cdots \\ x_{.r} \end{bmatrix}$

(e) $[x_{.1} \quad x_{.2} \quad \cdots \quad s_{.r}]$ (f) $[x'_{1.} \quad x'_{2.} \quad \cdots \quad x'_{s.}]$

(g) $\begin{bmatrix} x'_{.1} \\ x'_{.2} \\ \cdots \\ x'_{.r} \end{bmatrix}$

(h) $[x'_{.1} \quad x'_{.2} \quad \cdots \quad x'_{.r}]$

ANSWERS

1. $a' = \begin{bmatrix} 4 & 3 & 5 \\ 6 & 9 & 6 \\ 7 & 2 & 4 \\ 5 & 8 & 2 \end{bmatrix}$
2. $b = \begin{bmatrix} 7 & 8 & 4 \\ 3 & 2 & 1 \\ 6 & 4 & 9 \\ 9 & 2 & 7 \end{bmatrix}$
3. $c = \begin{bmatrix} 7 & 4 & 6 & 5 \\ 9 & 3 & 2 & 1 \\ 3 & 9 & 5 & 7 \end{bmatrix}$

4. $f' = \begin{bmatrix} 6 & 4 & 3 \\ 5 & 2 & 1 \\ 9 & 3 & 8 \end{bmatrix}$

5. (a) $A_{1.} = \begin{bmatrix} 1 \\ 0 \\ 1 \\ 1 \end{bmatrix}$

(b) $A'_{.4} = [1 \quad 1 \quad 1 \quad 1 \quad 1]$

(c) $A_{2.} = \begin{bmatrix} 0 \\ 1 \\ 0 \\ 1 \end{bmatrix}$

(d) $A'_{5.} = [0 \quad 0 \quad 1 \quad 1]$ (e) $A'_{.3} = [1 \quad 0 \quad 1 \quad 0 \quad 1]$

(f) $A_{4.} = \begin{bmatrix} 1 \\ 1 \\ 0 \\ 1 \end{bmatrix}$

6. (a) $B'_{.5} = [1 \quad 0 \quad 1 \quad 0]$

(b) $B_{2.} = \begin{bmatrix} 1 \\ 1 \\ 0 \\ 1 \\ 0 \end{bmatrix}$

(c) $B_{.4} = \begin{bmatrix} 1 \\ 1 \\ 0 \\ 0 \end{bmatrix}$

(d) $B'_{3.} = [1 \quad 0 \quad 1 \quad 0 \quad 1]$

(e) $B'_{1.} = [1 \quad 1 \quad 1 \quad 1 \quad 1]$

(f) $B_{.5} = \begin{bmatrix} 1 \\ 0 \\ 1 \\ 0 \end{bmatrix}$

7. (a) matrix (b) vector (c) matrix (d) vector (e) matrix (f) vector (g) matrix (h) vector

Part II

Simple Matrix Computations

Chapter 7

Addition and Subtraction of Matrices

7.1 Addition of Simple Matrices

The sum of two matrices

As in arithmetic and ordinary algebra, so also in matrix algebra we have the operation of addition. In arithmetic any two numbers can be added together whether they are positive or negative. However, two matrices can be added if and only if they have the same order. The reason for this will be clear when we define the sum of two matrices. The sum of two matrices is a matrix, each element of which is the sum of corresponding elements of the two matrices being added together.

The sum of two vertical matrices is shown in Fig. 7.1.1.

$$\begin{bmatrix} 4 & 6 \\ 3 & 7 \\ 2 & 5 \end{bmatrix} + \begin{bmatrix} 3 & 2 \\ 6 & 4 \\ 3 & 6 \end{bmatrix} = \begin{bmatrix} 4+3 & 6+2 \\ 3+6 & 7+4 \\ 2+3 & 5+6 \end{bmatrix} = \begin{bmatrix} 7 & 8 \\ 9 & 11 \\ 5 & 11 \end{bmatrix}$$

Fig. 7.1.1

The sum of horizontal matrices is shown in Fig. 7.1.2.

$$\begin{bmatrix} 6 & 2 & 4 \\ 3 & 9 & 7 \end{bmatrix} + \begin{bmatrix} 2 & 5 & 8 \\ 7 & 4 & 3 \end{bmatrix} = \begin{bmatrix} 6+2 & 2+5 & 4+8 \\ 3+7 & 9+4 & 7+3 \end{bmatrix} = \begin{bmatrix} 8 & 7 & 12 \\ 10 & 13 & 10 \end{bmatrix}$$

Fig. 7.1.2

We can also add a diagonal matrix to a square matrix of the same order, as in Fig. 7.1.3.

$$\begin{bmatrix} 4 & 0 & 0 \\ 0 & 3 & 0 \\ 0 & 0 & 5 \end{bmatrix} + \begin{bmatrix} 6 & 3 & 2 \\ 3 & 5 & 4 \\ 2 & 4 & 7 \end{bmatrix} = \begin{bmatrix} 4+6 & 3 & 2 \\ 3 & 3+5 & 4 \\ 2 & 4 & 5+7 \end{bmatrix} = \begin{bmatrix} 10 & 3 & 2 \\ 3 & 8 & 4 \\ 2 & 4 & 12 \end{bmatrix}$$

Fig. 7.1.3

Since vectors are special cases of matrices, they are added in the same way. Figure 7.1.4 shows the sum of two third-order column vectors.

$$\begin{bmatrix} 4 \\ 6 \\ 2 \end{bmatrix} + \begin{bmatrix} 3 \\ 9 \\ 7 \end{bmatrix} = \begin{bmatrix} 4+3 \\ 6+9 \\ 2+7 \end{bmatrix} = \begin{bmatrix} 7 \\ 15 \\ 9 \end{bmatrix}$$

Fig. 7.1.4

From the definition of the sum of two matrices, it is obvious that the order of a sum of two matrices is the same as the order of the matrices that were added together.

The sum of any number of matrices

As you may have guessed, we can take the sum of any number of matrices. However, all of those to be added must have the same order. Each element in the sum is the sum of the corresponding elements of the individual matrices that are added. Figure 7.1.5 shows the sum of three second-order matrices.

$$\begin{bmatrix} 6 & 5 \\ 4 & 3 \end{bmatrix} + \begin{bmatrix} 2 & 4 \\ 3 & 5 \end{bmatrix} + \begin{bmatrix} 7 & 9 \\ 6 & 4 \end{bmatrix} = \begin{bmatrix} 6+2+7 & 5+4+9 \\ 4+3+6 & 3+5+4 \end{bmatrix} = \begin{bmatrix} 15 & 18 \\ 13 & 12 \end{bmatrix}$$

Fig. 7.1.5

In general, suppose we have matrices $a, b, \ldots z$ of order n by m defined by Eqs. (7.1.1), (7.1.2), and (7.1.3) respectively. The sum of these matrices is then as given in Eq. (7.1.4).

$$a = \begin{bmatrix} a_{11} & a_{12} & \cdots & a_{1m} \\ a_{21} & a_{22} & \cdots & a_{2m} \\ \cdots & \cdots & \cdots & \cdots \\ a_{n1} & a_{n2} & \cdots & a_{nm} \end{bmatrix} \qquad (7.1.1)$$

$$b = \begin{bmatrix} b_{11} & b_{12} & \cdots & b_{1m} \\ b_{21} & b_{22} & \cdots & b_{2m} \\ \cdots & \cdots & \cdots & \cdots \\ b_{n1} & b_{n2} & \cdots & b_{nm} \end{bmatrix} \qquad (7.1.2)$$

$$z = \begin{bmatrix} z_{11} & z_{12} & \cdots & z_{1m} \\ z_{21} & z_{22} & \cdots & z_{2m} \\ \cdots & \cdots & \cdots & \cdots \\ z_{n1} & z_{n2} & \cdots & z_{nm} \end{bmatrix} \qquad (7.1.3)$$

$$a + b + \ldots + z =$$
$$\begin{bmatrix} a_{11} + b_{11} + \ldots + z_{11} & a_{12} + b_{12} + \ldots + z_{12} & a_{1m} + b_{1m} + \ldots + z_{1m} \\ a_{21} + b_{21} + \ldots + z_{21} & a_{22} + b_{22} + \ldots + z_{22} & a_{2m} + b_{2m} + \ldots + z_{2m} \\ \cdots \cdots \cdots & \cdots \cdots \cdots & \cdots \cdots \cdots \\ a_{n1} + b_{n1} + \ldots + z_{n1} & a_{n2} + b_{n2} + \ldots + z_{n2} & a_{nm} + b_{nm} + \ldots + z_{nm} \end{bmatrix}$$

$$(7.1.4)$$

7.2 Some Simple Laws of Addition

We shall now give two simple laws of addition so familiar and obvious that you may wonder why we bother to state them. The reason will become evident later, when you find that one of the laws you have taken for granted in arithmetic and scalar algebra does not hold for matrix algebra.

The associative law

The first law is the *associative law* of addition. An example of the associative law in arithmetic is given by the simple equations in Fig. 7.2.1.

$$2 + 3 + 5 = (2 + 3) + 5 = 5 + 5 = 10$$
$$2 + 3 + 5 = 2 + (3 + 5) = 2 + 8 = 10$$

Fig. 7.2.1

You see that in the first equation in Fig. 7.2.1, we added the first two numbers, $2 + 3$, to get 5, and then added this sum to the last number, 5, to get 10. In the second row, we first added the last two numbers, $3 + 5$, to get 8, and then added the first number, 2, to 8 to get 10. If we take a little more complicated example, where we have four numbers in the sum, there are still more ways in which we can get the sum of the four numbers, as shown in Fig. 7.2.2.

$$(4 + 1) + 2 + 6 = (5 + 2) + 6 = 7 + 6 = 13$$
$$4 + (1 + 2) + 6 = 4 + (3 + 6) = 4 + 9 = 13$$
$$4 + 1 + (2 + 6) = (4 + 1) + 8 = 5 + 8 = 13$$
$$(4 + 1) + 2 + 6 = 5 + (2 + 6) = 5 + 8 = 13$$

FIG. 7.2.2

In general then, if we have three or more elements to be added, we can reduce the total number of elements to be added by one if we first take the sum of any two adjacent numbers. From this new series, we can further reduce the number to be added by one if again we add any two adjacent numbers. We can continue this process until we get the sum of all of the numbers. The associative law says that irrespective of which two adjacent elements we take first, which two second, and so on, we will always get the same total for all the elements.

The associative law of addition for matrices can be illustrated if we define matrices a, b, and c, as in Fig. 7.2.3.

$$a = \begin{bmatrix} 3 & 4 \\ 4 & 6 \end{bmatrix} \qquad b = \begin{bmatrix} 2 & 1 \\ 2 & 5 \end{bmatrix} \qquad c = \begin{bmatrix} 3 & 2 \\ 4 & 1 \end{bmatrix}$$

FIG. 7.2.3

Now Figs. 7.2.4 and 7.2.5 show how the associative law of addition applies for the sum of the matrices a, b, and c in Fig. 7.2.3.

$$a + b + c = \begin{bmatrix} 3+2 & 4+1 \\ 4+2 & 6+5 \end{bmatrix} + \begin{bmatrix} 3 & 2 \\ 4 & 1 \end{bmatrix} = \begin{bmatrix} 5 & 5 \\ 6 & 11 \end{bmatrix} + \begin{bmatrix} 3 & 2 \\ 4 & 1 \end{bmatrix} = \begin{bmatrix} 8 & 7 \\ 10 & 12 \end{bmatrix}$$

Fig. 7.2.4

$$a + b + c = \begin{bmatrix} 3 & 4 \\ 4 & 6 \end{bmatrix} + \begin{bmatrix} 2+3 & 1+2 \\ 2+4 & 5+1 \end{bmatrix} = \begin{bmatrix} 3 & 4 \\ 4 & 6 \end{bmatrix} + \begin{bmatrix} 5 & 3 \\ 6 & 6 \end{bmatrix} = \begin{bmatrix} 8 & 7 \\ 10 & 12 \end{bmatrix}$$

Fig. 7.2.5

In Fig. 7.2.4, we added first a and b, and then to this sum we added c. In Fig. 7.2.5, we added first b and c, and then added this sum to a. The results are of course the same as you would expect them to be in arithmetic. The associative law of addition for three numbers or matrices can be stated in the form of Eq. (7.2.1).

$$(a + b) + c = a + (b + c) \tag{7.2.1}$$

As a child studying arithmetic, you learned to use the associative law of addition to check your sums. For example, if you had a column of numbers as in Fig. 7.2.6, you learned to get the sum as in Fig. 7.2.7.

$$\begin{array}{c} 3 \\ 5 \\ 7 \\ 4 \\ \hline ? \end{array}$$

Fig. 7.2.6

$$(3 + 5) + 7 + 4 = (8 + 7) + 4 = 15 + 4 = 19$$

Fig. 7.2.7

Then you would check this sum as in Fig. 7.2.8.

$$3 + 5 + (7 + 4) = 3 + (5 + 11) = 3 + 16 = 19$$

Fig. 7.2.8

As a third-grader, however, you would have been utterly bewildered if the teacher had told you to check your sum by using the associative law of addition.

The commutative law

A second simple law of addition is the *commutative law*. An example of the commutative law in arithmetic for addition of two numbers is given in Fig. 7.2.9.

$$3 + 4 = 7 \qquad 4 + 3 = 7$$

FIG. 7.2.9

For three numbers, an example of the commutative law of addition is given in Fig. 7.2.10.

$$2 + 5 + 3 = 10 \qquad 2 + 3 + 5 = 10 \qquad 5 + 2 + 3 = 10$$
$$5 + 3 + 2 = 10 \qquad 3 + 5 + 2 = 10 \qquad 3 + 2 + 5 = 10$$

FIG. 7.2.10

Here we have six different ways in which to arrange the order of the numbers 2, 5, and 3. No matter in what order we arrange them, they add up to 10 in every case. Here again, this seems elementary, but when we take up the multiplication of matrices, you will learn that the commutative law does not hold. Figures 7.2.9 and 7.2.10 show how the commutative law of addition holds for scalar quantities. From the rule for the addition of matrices, it should be easy to see that the commutative law for the addition of matrices will hold also. Figures 7.2.11 and 7.2.12 illustrate the commutative law for the addition of matrices.

$$\begin{bmatrix} 3 & 4 \\ 4 & 6 \end{bmatrix} + \begin{bmatrix} 2 & 1 \\ 2 & 5 \end{bmatrix} = \begin{bmatrix} 3+2 & 4+1 \\ 4+2 & 6+5 \end{bmatrix} = \begin{bmatrix} 5 & 5 \\ 6 & 11 \end{bmatrix}$$

FIG. 7.2.11

$$\begin{bmatrix} 2 & 1 \\ 2 & 5 \end{bmatrix} + \begin{bmatrix} 3 & 4 \\ 4 & 6 \end{bmatrix} = \begin{bmatrix} 2+3 & 1+4 \\ 2+4 & 5+6 \end{bmatrix} = \begin{bmatrix} 5 & 5 \\ 6 & 11 \end{bmatrix}$$

FIG. 7.2.12

In Fig. 7.2.12, we have on the left side of the equation the same two matrices we have in Fig. 7.2.11. The only difference is that the order of the two matrices is reversed. Nevertheless, as any child can plainly see, the sum is the same, no matter in which of the two possible orders they are added.

The commutative law of addition for two matrices can be expressed as

$$a + b = b + a \qquad (7.2.2)$$

This equation holds whether we are talking about scalar quantities or matrices. For three scalars or matrices, the commutative law of addition is given by the six sets of equations in Fig. 7.2.13.

$$a + b + c = e \qquad a + c + b = e \qquad b + a + c = e$$
$$b + c + a = e \qquad c + a + b = e \qquad c + b + a = e$$

FIG. 7.2.13

We can then summarize the commutative law of addition by saying that no matter in what order a set of matrices is arranged, their sum will always be the same.

7.3 Subtraction of Simple Matrices

As you may well suppose, the subtraction of one matrix from another follows the same principle as the addition of two matrices. First, a necessary and sufficient condition that one matrix can be subtracted from another is that the two matrices have the same order. Second, each element of the difference matrix is the difference between the corresponding elements of the two matrices. An example of how one 2 by 3 matrix is subtracted from another 2 by 3 matrix is shown in Fig. 7.3.1.

$$\begin{bmatrix} 4 & 6 & 3 \\ 9 & 2 & 11 \end{bmatrix} - \begin{bmatrix} 5 & 9 & 2 \\ 8 & 3 & 5 \end{bmatrix} = \begin{bmatrix} 4-5 & 6-9 & 3-2 \\ 9-8 & 2-3 & 11-5 \end{bmatrix} = \begin{bmatrix} -1 & -3 & 1 \\ 1 & -1 & 6 \end{bmatrix}$$

FIG. 7.3.1

In general then, we can illustrate the difference between two $n \times m$ matrices, using letter elements, by Fig. 7.3.2.

$$\begin{bmatrix} b_{11} & b_{12} & \cdots & b_{1m} \\ b_{21} & b_{22} & \cdots & b_{2m} \\ \cdots & \cdots & \cdots & \cdots \\ b_{n1} & b_{n2} & \cdots & b_{nm} \end{bmatrix} - \begin{bmatrix} c_{11} & c_{12} & \cdots & c_{1m} \\ c_{21} & c_{22} & \cdots & c_{2m} \\ \cdots & \cdots & \cdots & \cdots \\ c_{n1} & c_{n2} & \cdots & c_{nm} \end{bmatrix} =$$

$$\begin{bmatrix} b_{11} - c_{11} & b_{12} - c_{12} & \cdots & b_{1m} - c_{1m} \\ b_{21} - c_{21} & b_{22} - c_{22} & \cdots & b_{2m} - c_{2m} \\ \cdots & \cdots & \cdots & \cdots & \cdots & \cdots \\ b_{n1} - c_{n1} & b_{n2} - c_{n2} & \cdots & b_{nm} - c_{nm} \end{bmatrix}$$

FIG. 7.3.2

7.4 Addition and Subtraction of Simple Matrices

As a scalar quantity may be the result of both addition and subtraction, so a matrix may be made up of the sums and differences of a number of matrices. Each element of such a matrix would of course be made up of the sums and differences of the corresponding elements of the other matrices. An illustration is Fig. 7.4.1.

$$\begin{bmatrix} 2 & 3 \\ 5 & 7 \end{bmatrix} - \begin{bmatrix} 1 & 7 \\ 4 & 3 \end{bmatrix} - \begin{bmatrix} 3 & 6 \\ 7 & 5 \end{bmatrix} + \begin{bmatrix} 4 & 8 \\ 7 & 8 \end{bmatrix} =$$

$$\begin{bmatrix} 2-1-3+4 & 3-7-6+8 \\ 5-4-7+7 & 7-3-5+8 \end{bmatrix} = \begin{bmatrix} 2 & -2 \\ 1 & 7 \end{bmatrix}$$

FIG. 7.4.1

We may appropriately regard subtraction as a special case of addition in which all of the signs of the elements of one of the matrices have been changed and the resulting matrix added to the other. Then it is easy to see that both the associative and the commutative laws of addition will hold for the sums and differences of a number of matrices.

7.5 The Addition of Supermatrices

From what you know about the sums and differences of simple matrices, you could readily make up corresponding rules for supermatrices. Thus the sum of two or more supermatrices is a supermatrix whose matric elements are the sum of the corresponding matrix elements of the individual matrices. The difference between two supermatrices is a supermatrix whose matric elements are the differences of the corresponding matric elements of the two matrices.

So, again, we may have a supermatrix made up of the sums and differences of any number of supermatrices. It is defined in the same way as the corresponding simple matrix, except that the elements are themselves matrices.

Conformability requirements

Supermatrices, like simple matrices, cannot be added or subtracted unless they have the same order. Even though supermatrices have the same order, however, it may still be impossible to add or subtract them. Not only must the order of the supermatrices be the same, but also the order of the corresponding matric elements of the supermatrices must be the same. This principle is illustrated in Figs. 7.5.1, 7.5.2, and 7.5.3.

$$a_{11} = \begin{bmatrix} 3 & 2 \\ 4 & 6 \end{bmatrix} \qquad a_{12} = \begin{bmatrix} 7 \\ 1 \end{bmatrix}$$

$$a_{21} = \begin{bmatrix} 4 & 3 \end{bmatrix} \qquad a_{22} = \begin{bmatrix} 5 \end{bmatrix}$$

FIG. 7.5.1

$$b_{11} = \begin{bmatrix} 7 \end{bmatrix} \qquad b_{12} = \begin{bmatrix} 1 & 3 \end{bmatrix}$$

$$b_{21} = \begin{bmatrix} 7 \\ 6 \end{bmatrix} \qquad b_{22} = \begin{bmatrix} 4 & 1 \\ 2 & 3 \end{bmatrix}$$

FIG. 7.5.2

Suppose we make up supermatrices from these figures as shown in Fig. 7.3.5.

$$a = \begin{bmatrix} a_{11} & a_{12} \\ a_{21} & a_{22} \end{bmatrix} \qquad b = \begin{bmatrix} b_{11} & b_{12} \\ b_{21} & b_{22} \end{bmatrix}$$

Fig. 7.5.3

It is clear that a and b in Fig. 7.5.3 are both second-order square supermatrices, but here we cannot add together the corresponding matrix elements of a and b because they do not have the same order. For example, a_{11} is a 2 by 2 matrix and b_{11} is a 1 by 1 matrix or scalar, and similarly, a_{12} is a 2 by 1 matrix whereas b_{12} is a 1 by 2 matrix. Therefore, it is clear that two or more supermatrices may be added or subtracted if and only if they are of the same order and their corresponding elements are also of the same order. (We sometimes call the order of a supermatrix the *superorder*.)

7.6   The Transposes of Sums and Differences of Matrices

The rules for transposes of sums and differences of matrices are very simple. The transpose of a sum of two matrices is equal to the sum of the transpose of the two matrices. Figure 7.6.1 is an illustration of this principle.

$$\left(\begin{bmatrix} 1 & 2 \\ 3 & 4 \end{bmatrix} + \begin{bmatrix} 2 & 1 \\ 1 & 2 \end{bmatrix} \right)' = \begin{bmatrix} 3 & 3 \\ 4 & 6 \end{bmatrix}' = \begin{bmatrix} 3 & 4 \\ 3 & 6 \end{bmatrix}$$

$$\begin{bmatrix} 1 & 2 \\ 3 & 4 \end{bmatrix}' + \begin{bmatrix} 2 & 1 \\ 1 & 2 \end{bmatrix}' = \begin{bmatrix} 1 & 3 \\ 2 & 4 \end{bmatrix} + \begin{bmatrix} 2 & 1 \\ 1 & 2 \end{bmatrix} = \begin{bmatrix} 3 & 4 \\ 3 & 6 \end{bmatrix}$$

Fig. 7.6.1

In symbolic form, this is

$$[a + b]' = a' + b' \tag{7.6.1}$$

The transpose of the difference between two matrices is the difference between their transposes. An illustration is Fig. 7.6.1.

$$\left(\begin{bmatrix} 1 & 2 \\ 3 & 4 \end{bmatrix} - \begin{bmatrix} 2 & 1 \\ 1 & 2 \end{bmatrix} \right)' = \begin{bmatrix} -1 & 1 \\ 2 & 2 \end{bmatrix}' = \begin{bmatrix} -1 & 2 \\ 1 & 2 \end{bmatrix}$$

$$\begin{bmatrix} 1 & 2 \\ 3 & 4 \end{bmatrix}' - \begin{bmatrix} 2 & 1 \\ 1 & 2 \end{bmatrix}' = \begin{bmatrix} 1 & 3 \\ 2 & 4 \end{bmatrix} - \begin{bmatrix} 2 & 1 \\ 1 & 2 \end{bmatrix} = \begin{bmatrix} -1 & 2 \\ 1 & 2 \end{bmatrix}$$

Fig. 7.6.2

This rule is expressed symbolically as

$$(a - b)' = a' - b' \tag{7.6.2}$$

The transpose of a sum and difference of matrices is equal to the sum and difference of the transposes of the matrices. You can readily construct examples to illustrate this principle. The rule is

$$(a + b - c)' = a' + b' - c' \tag{7.6.3}$$

The same rules that hold for the transposes of the sums and differences of simple matrices also hold for the sums and differences of supermatrices. You must remember, of course, that when you take the transpose of a supermatrix you not only write rows for columns, but you also take the transpose of each matric element of the supermatrix.

7.7 Sums and Differences of Special Kinds of Matrices

The sums and differences of certain kinds of matrices have special properties that must be kept in mind.

Diagonal matrices

You see at once that the sums and differences of diagonal matrices are a diagonal matrix. If we let D with a subscript indicate various diagonal matrices of the same order, this principle can be readily stated.

$$D_a + D_b - D_c - D_e = D_f \tag{7.7.1}$$

This is so obvious that no proof is needed. Since all nondiagonal elements are 0, their sums and differences are 0.

Scalar matrices

The sums and differences of scalar matrices are scalar matrices. Figure 7.7.1 illustrates this principle.

$$\begin{bmatrix} 3 & & \\ & 3 & \\ & & 3 \end{bmatrix} + \begin{bmatrix} 4 & & \\ & 4 & \\ & & 4 \end{bmatrix} - \begin{bmatrix} 2 & & \\ & 2 & \\ & & 2 \end{bmatrix} = \begin{bmatrix} 5 & & \\ & 5 & \\ & & 5 \end{bmatrix}$$

FIG. 7.7.1

Symmetric matrices

The sums and differences of symmetric matrices are symmetric matrices, as in Fig. 7.7.2.

$$\begin{bmatrix} 3 & 2 & 1 \\ 2 & 4 & 2 \\ 1 & 2 & 5 \end{bmatrix} + \begin{bmatrix} 2 & 1 & 1 \\ 1 & 2 & 1 \\ 1 & 1 & 2 \end{bmatrix} - \begin{bmatrix} 3 & 2 & 1 \\ 2 & 2 & 1 \\ 1 & 1 & 1 \end{bmatrix} = \begin{bmatrix} 2 & 1 & 1 \\ 1 & 4 & 2 \\ 1 & 2 & 6 \end{bmatrix}$$

FIG. 7.7.2

If we let the letter S with various subscripts indicate various symmetric matrices of equal order, the rule is

$$S_a = S_b + S_c - S_d = S_e \qquad (7.7.2)$$

The sum of a square matrix and its transpose

Only a square matrix can be added to its transpose. This rule is evident because a necessary and sufficient condition for addition of two simple matrices is that they have the same order. Obviously, unless a matrix is square, its transpose cannot have the same order as the original, since the order of a matrix and its transpose are reversed.

The sum of a square matrix and its transpose is a symmetric matrix. Not only is the sum a symmetric matrix, but the principal diagonal of the sum is twice the principal diagonal of the original matrix. Let us first take a simple example of the sum of a third-order matrix and its transpose as in Fig. 7.7.3.

$$\begin{bmatrix} 3 & 2 & 2 \\ 1 & 4 & 1 \\ 2 & 3 & 5 \end{bmatrix} + \begin{bmatrix} 3 & 1 & 2 \\ 2 & 4 & 3 \\ 2 & 1 & 5 \end{bmatrix} = \begin{bmatrix} 6 & 3 & 4 \\ 3 & 8 & 4 \\ 4 & 4 & 10 \end{bmatrix}$$

Fig. 7.7.3

Here you see that the sum of the matrix and its transpose is actually symmetric, and also that the diagonal elements of the sum are twice the corresponding diagonal elements of the original matrix.

Let us now prove these two rules for the general case of any square matrix. First we write the equation of the sum of a square matrix and its transpose in expanded form as in Eq. (7.7.3).

$$\begin{bmatrix} a_{11} & a_{12} & \cdots & a_{1n} \\ a_{21} & a_{22} & \cdots & a_{2n} \\ \cdots & \cdots & \cdots & \cdots \\ a_{n1} & a_{n2} & \cdots & a_{nn} \end{bmatrix} + \begin{bmatrix} a_{11} & a_{21} & \cdots & a_{n1} \\ a_{12} & a_{22} & \cdots & a_{n2} \\ \cdots & \cdots & \cdots & \cdots \\ a_{1n} & a_{2n} & \cdots & a_{nn} \end{bmatrix} = \begin{bmatrix} c_{11} & c_{12} & \cdots & c_{1n} \\ c_{21} & c_{22} & \cdots & c_{2n} \\ \cdots & \cdots & \cdots & \cdots \\ c_{n1} & c_{n2} & \cdots & c_{nn} \end{bmatrix}$$

$$(7.7.3)$$

If we use single letters to stand for the matrices in this equation, we have

$$a + a' = c \qquad (7.7.4)$$

Next if we add corresponding elements of the two matrices on the left side of Eq. (7.7.3) we get

$$\begin{bmatrix} a_{11} + a_{11} & a_{12} + a_{21} & \cdots & a_{1n} + a_{n1} \\ a_{21} + a_{12} & a_{22} + a_{22} & \cdots & a_{2n} + a_{n2} \\ \cdots & \cdots & \cdots & \cdots \\ a_{n1} + a_{1n} & a_{n2} + a_{2n} & \cdots & a_{nn} + a_{nn} \end{bmatrix} = \begin{bmatrix} c_{11} & c_{12} & \cdots & c_{1n} \\ c_{21} & c_{22} & \cdots & c_{2n} \\ \cdots & \cdots & \cdots & \cdots \\ c_{n1} & c_{n2} & \cdots & c_{nn} \end{bmatrix}$$

$$(7.7.5)$$

If we now consider any nondiagonal element in the ith row and jth column for each side of Eq. (7.7.4), we may write

$$a_{ij} + a_{ji} = c_{ij} \tag{7.7.6}$$

Considering next the element in the jth row and ith column for each side of (7.7.5), we have

$$a_{ji} + a_{ij} = c_{ji} \tag{7.7.7}$$

But remembering that addition is commutative, we can write

$$a_{ji} + a_{ij} = a_{ij} + a_{ji} \tag{7.7.8}$$

Next we substitute the right side of Eq. (7.7.8) for the left side of Eq. (7.7.7) and get

$$a_{ij} + a_{ji} = c_{ji} \tag{7.7.9}$$

You see now that the left sides of Eq. (7.7.6) and (7.7.9) are exactly the same. Therefore the right side of Eq. (7.7.6) and (7.7.9) must also be equal. Hence, we write

$$c_{ij} = c_{ji} \tag{7.7.10}$$

According to Eq. (7.7.10), the ijth element of the matrix c is equal to the jith element. You recall that this is one definition of a symmetric matrix. Therefore Eqs. (7.7.6) to (7.7.10) prove that the sum of a square matrix and its transpose is symmetric.

The proof is even simpler that the diagonal elements of the sum of a square matrix and its transpose are twice the corresponding diagonal elements of the original matrix. Equating any corresponding diagonal elements of left and right sides of Eq. (7.7.5), we have simply

$$a_{ii} + a_{ii} = c_{ii} \tag{7.7.11}$$

Or

$$2a_{ii} = c_{ii} \tag{7.7.12}$$

Equations (7.7.11) and (7.7.12) thus prove that the diagonal of the sum of a square matrix and its transpose is twice the diagonal of the original matrix.

The difference between a matrix and its transpose

The difference between a square matrix and its transpose is a skew symmetric matrix. An illustration is Fig. 7.7.4.

$$\begin{bmatrix} 3 & 4 & 2 \\ 6 & 1 & 0 \\ 1 & 9 & 5 \end{bmatrix} - \begin{bmatrix} 3 & 6 & 1 \\ 4 & 1 & 9 \\ 2 & 0 & 5 \end{bmatrix} = \begin{bmatrix} 0 & -2 & 1 \\ 2 & 0 & -9 \\ -1 & 9 & 0 \end{bmatrix}$$

Fig. 7.7.4

Here you see that the matrix to be subtracted is the transpose of the first. You also see that in the difference matrix all the diagonals are 0 and corresponding nondiagonal terms are opposite in sign.

We can easily generalize these two rules for any square matrix. First we write in expanded form the difference between a square matrix and its transpose as in Eq. (7.7.13).

$$\begin{bmatrix} a_{11} & a_{12} & \dots & a_{1n} \\ a_{21} & a_{22} & \dots & a_{2n} \\ \dots & \dots & \dots & \dots \\ a_{n1} & a_{n2} & \dots & a_{nn} \end{bmatrix} - \begin{bmatrix} a_{11} & a_{21} & \dots & a_{n1} \\ a_{12} & a_{22} & \dots & a_{n2} \\ \dots & \dots & \dots & \dots \\ a_{1n} & a_{2n} & \dots & a_{nn} \end{bmatrix} = \begin{bmatrix} c_{11} & c_{12} & \dots & c_{1n} \\ c_{21} & c_{22} & \dots & c_{2n} \\ \dots & \dots & \dots & \dots \\ c_{n1} & c_{n2} & \dots & c_{nn} \end{bmatrix}$$

$$(7.7.13)$$

In compact or symbolic notation, we can write the equation as

$$a - a' = c \tag{7.7.14}$$

For Eq. (7.7.14), we now want to prove that Eqs. (7.7.15) and (7.7.16), representing the two rules, are true.

$$c_{ii} = 0 \tag{7.7.15}$$

$$c_{ij} = -c_{ji} \tag{7.7.16}$$

First we subtract each element in the second matrix on the left side of Eq. (7.7.13) from the corresponding element in the first matrix on the left of (7.7.13) and get Eq. (7.7.17).

$$\begin{bmatrix} a_{11} - a_{11} & a_{12} - a_{21} & \dots & a_{1n} - a_{n1} \\ a_{21} - a_{12} & a_{22} - a_{22} & \dots & a_{2n} - a_{n2} \\ \dots & \dots & \dots & \dots \\ a_{n1} - a_{1n} & a_{n2} - a_{2n} & \dots & a_{nn} - a_{nn} \end{bmatrix} = \begin{bmatrix} c_{11} & c_{12} & \dots & c_{1n} \\ c_{21} & c_{22} & \dots & c_{2n} \\ \dots & \dots & \dots & \dots \\ c_{n1} & c_{n2} & \dots & c_{nn} \end{bmatrix}$$

$$(7.7.17)$$

If now from Eq. (7.7.17), we equate the element in the ith row and jth column of the left matrix to the element in the ith row and jth column in the right matrix we get

$$a_{ij} - a_{ji} = c_{ij} \tag{7.7.18}$$

If we equate the element in the jth row and ith column of the left matrix to the element in the jth row and ith column of the right matrix, we get

$$a_{ji} - a_{ij} = c_{ji} \tag{7.7.19}$$

If we change the signs of all terms in Eq. (7.7.19) we get

$$-a_{ji} + a_{ij} = -c_{ji} \tag{7.7.20}$$

Remembering that the commutative law holds for both addition and subtraction, we can change the order of the terms on the left side of Eq. (7.7.20) to get

$$a_{ij} - a_{ji} = -c_{ji} \qquad (7.7.21)$$

But notice that the left sides of Eqs. (7.7.18) and (7.7.21) are exactly the same. Therefore we can equate the right side of the two equations to get

$$c_{ij} = -c_{ji} \qquad (7.7.22)$$

Now Eq. (7.7.22) is exactly the same as Eq. (7.7.16), the second rule for a square matrix. Therefore Eqs. (7.7.18) to (7.7.21) prove that if from a square matrix we subtract its transpose, we get a matrix in which the ijth element is the negative of the jith element.

The proof is even simpler that the diagonal elements of the difference between a square matrix and its transpose are 0. If we equate the ith diagonal element in the left matrix of Eq. (7.7.17) to the ith diagonal element in the right side, we have

$$a_{ii} - a_{ii} = c_{ii} \qquad (7.7.23)$$

But any number subtracted from itself is 0; therefore, we have

$$a_{ii} - a_{ii} = 0 \qquad (7.7.24)$$

If now we substitute the right side of Eq. (7.7.24) into the left side of Eq. (7.7.23), we have, after transposing, Eq. (7.7.24), which we wished to prove.

$$c_{ii} = 0 \qquad (7.7.24)$$

Since this is exactly the same as Eq. (7.7.14), Eqs. (7.7.22) through (7.7.24) prove that the principal diagonal of the difference between a square matrix and its transpose is 0.

SUMMARY

1. Addition of simple matrices:
   a. The sum of two matrices is defined as a matrix each element of which is the sum of the corresponding elements of the two. If $A + B = C$, with elements A_{ij}, B_{ij}, and C_{ij}, respectively, then $A_{ij} + B_{ij} = C_{ij}$.
   b. Therefore only matrices of equal order can be added together, and the order of a sum is equal to the order of the matrices.
   c. The sum of any number of matrices is a matrix each element of which is the sum of corresponding elements of all the matrices.

2. Some simple laws of addition:
   a. The associative law: $(a + b) + c = a + (b + c)$.
   b. The commutative law: $a + b = b + a$.

3. Subtraction of simple matrices:
   a. Subtraction may be regarded as a special case of addition where the signs of all elements in the matrix to be subtracted are reversed.
   b. Then both the associative and commutative laws of addition hold.

4. Addition and subtraction:
   a. Any number of matrices may be combined by addition of some and subtraction of others.
   b. The associative and commutative laws hold for addition and subtraction of any number of matrices.

5. Addition and subtraction of supermatrices:
   a. The sum or difference of two supermatrices is a supermatrix each of whose matric elements is the sum or difference of corresponding elements of the two matrices.
   b. Supermatrices may be added or subtracted; therefore, if and only if they have the same superorder and corresponding matric elements are of the same order.

6. The transpose of a sum or difference:
   a. The transpose of a sum or difference of simple matrices is equal to the sum of their transposes, that is, $(a + b)' = a' + b'$.
   b. The transpose of a sum or difference of supermatrices is equal to the sum or difference of their transposes.

7. Sums and differences of special kinds of matrices:
   a. The sum of any number of diagonal matrices is a diagonal matrix.
   b. The sum of any number of scalar matrices is a scalar matrix.
   c. The sum of any number of symmetric matrices is a symmetric matrix.
   d. The sum of a square matrix and its transpose is a symmetric matrix.
   e. The difference between a square matrix and its transpose is a skew symmetric matrix.

EXERCISES

1. If

$$a + b = c$$

   and a is of order 40 by 3,
   (a) What is the order of b?
   (b) What is the order of c?

2. If

$$a + b = f$$

   and

$$b + c = g$$

what law is illustrated by the following equation?

$$a + g = f + c$$

3. Illustrate the commutative law of addition by writing the following equation in as many ways as possible:

$$a + b + c = g$$

4. If A is a supermatrix and

$$A + B = C$$

(a) what do we know about B?   (b) What do we know about C?

5. Write

$$f = (a + b - c)'$$

without the parentheses.

6. If D_a and D_b are diagonal matrices and

$$D_a + D_b = C$$

what kind of matrix is C?

7. If α and β are scalar matrices and

$$\alpha + \beta = \gamma$$

what kind of matrix is γ?

8. If S_1 and S_2 are symmetric matrices and

$$S_1 - S_2 = b$$

what kind of matrix is b?

9. If S is a symmetric matrix and D a diagonal matrix, and

$$D + S = K$$

what kind of matrix is K?

10. If

$$x + x' = y$$

(a) What do you know about x?   (b) What do you know about y?

11. If

$$x - x' = y$$

what do you know about y?

12. If Z is skew symmetric and

$$Z + Z' = W$$

what do you know about W?

13. If u is skew symmetric and

$$u - u' = M$$

what do you know about M?

ANSWERS

1. (a) 40 by 3   (b) 40 by 3

2. The associative law

3. $a + c + b = g$ $b + a + c = g$ $b + c + a = g$ $c + a + b = g$
   $c + b + a = g$

4. (a) The superorder of B is the same as A, and corresponding matric elements of A and B have the same order
   (b) The superorder of C is the same as A, and corresponding matric elements of A and C have the same order

5. $f = a' + b' - c'$

6. A diagonal matrix

7. A scalar matrix

8. A symmetric matrix

9. A symmetric matrix

10. (a) It is square   (b) It is symmetric

11. It is skew symmetric

12. It is a null matrix

13. $M = 2u$

Chapter 8

Vector Multiplication

Like scalar algebra and arithmetic, matrix algebra allows not only the operations of addition and subtraction, but also the operation of multiplication. Multiplication in matrix algebra, however, is not quite as simple as multiplication in scalar algebra. All multiplication in matrix algebra can be described in terms of the multiplication of vectors by one another. Therefore, we shall first consider the multiplication of vectors before we take up the multiplication of matrices in general. Two kinds of vector multiplication will be described, since each is convenient for defining matrix multiplication and for indicating computational procedures with both desk calculators and electronic computers.

You must note, however, that our treatment of vector and matrix multiplication in this and the following chapter differs from the traditional mathematical presentation. Since matrix notation is much more economical for the handling of behavioral science data than scalar notation we shall dispense as rapidly as possible with scalar notation and use it only rarely. The treatment in this and the following chapter is designed to help you take advantage as completely and quickly as possible of the simplicity and power of matrix notation.

8.1 The Minor Product of Two Vectors

The first kind of vector multiplication we shall consider is called the *minor product* of two vectors. We shall also refer to this product as their *scalar product;* the two terms will be used interchangeably. Figure 8.1.1 shows the minor product of two third-order vectors.

$$[3 \quad 4 \quad 7] \begin{bmatrix} 2 \\ 6 \\ 1 \end{bmatrix} = (3 \times 2) + (4 \times 6) + (7 \times 1) = 37$$

Fig. 8.1.1

115

The factors of a minor product

We speak of the factors of a product in vector multiplication just as we speak of the factors in the product of scalar multiplication. Therefore, in Fig. 8.1.1 we have two factors on the left side. The first of these factors is a third-order row vector and is called the *prefactor*. The second is a third-order column vector and is called the *postfactor*. From the right side of the equation in Fig. 8.1.1, you see that we have obtained the minor product of the two vectors by taking the product of corresponding elements in the two factors and adding these products to get a single number. Thus the first element in the first factor is 3 and the first element in the second factor is 2. The second element in the first factor is 4 and the second element in the second factor is 6. The third element in the first factor is 7 and the third element in the second factor is 1. The product of the two vectors is 37.

You should notice then that in multiplication for a minor or scalar product of two vectors, the prefactor must be a row and the postfactor must be a column. Therefore, in speaking generally of the minor or scalar product of two vectors, we say that a row vector is *postmultiplied* by a column vector, or that a column vector is *premultiplied* by a row vector. The rule is that in a scalar product of two vectors, the prefactor must always be the row vector and the postfactor must always be a column vector.

Notation for the minor product

The minor product of two vectors may now be defined as the sum of the products of corresponding elements of the two vectors. In general, if the vectors V_a and V_b are given as in Fig. 8.1.2, their minor or scalar product is defined by Eq. (8.1.1).

$$V_a = \begin{bmatrix} a_1 \\ a_2 \\ \ldots \\ a_n \end{bmatrix} \qquad V_b = \begin{bmatrix} b_1 \\ b_2 \\ \ldots \\ b_n \end{bmatrix}$$

FIG. 8.1.2

$$V_a'V_b = \begin{bmatrix} a_1 & a_2 & \ldots & a_n \end{bmatrix} \begin{bmatrix} b_1 \\ b_2 \\ \ldots \\ b_n \end{bmatrix} = a_1b_1 + a_2b_2 + \ldots + a_nb_n \qquad (8.1.1)$$

From the definition of the minor product of two vectors, it is clear that the minor product is always a scalar quantity. It is always a single number or letter that stands for a single number. Furthermore, from the definition, the vectors must be of the same order. The scalar or minor product can be found for any two vectors of the same order.

8.2 The Transpose of a Minor Product of Two Vectors

Since the minor product of two vectors is a scalar quantity, then as you would suppose, the transpose of the minor product of two vectors is the same scalar quantity. This can be expressed as

$$(V_a'V_b)' = V_a'V_b \qquad (8.2.1)$$

It is also easy to prove that the transpose of the scalar product of two vectors is equal to the product of their transposes in reverse order. This rule can be represented by

$$V_a'V_b = V_b'V_a \qquad (8.2.2)$$

Equation (8.2.2) says that if we postmultiply the row form of a V_a vector by the column form of a V_b vector, we get the same results as if we postmultiply the row form of the V_b vector by the column form of the V_a vector. This is very easy to prove from the rule of scalar multiplication of vectors. First we go back to Fig. 8.1.2 and write

$$V_b'V_a = [b_1 \quad b_2 \quad \ldots \quad b_n] \begin{bmatrix} a_1 \\ a_2 \\ \ldots \\ a_n \end{bmatrix} = b_1a_1 + b_2a_2 + \ldots + b_na_n \qquad (8.2.3)$$

Now let us compare, term for term, the last expression (the one after the last equals sign) in Eq. (8.1.1) with the last expression in Eq. (8.2.3). Here we have a_1b_1 for the first term in Eq. (8.1.1) as against b_1a_1 for the first term in Eq. (8.2.3). The second terms are respectively a_2b_2 and b_2a_2 and so on to the last term. In other words, the factors in the products of corresponding terms are simply reversed. But remember that each product has scalar factors, that is, it has simply the elements of the vectors as factors. As you well know, in scalar algebra, $A \times B = B \times A$, or in arithmetic $3 \times 4 = 4 \times 3$. Therefore, the last expressions in Eq. (8.1.1) and Eq. (8.2.3) must be equal. If the last expressions are equal, the first expressions must also be equal; therefore we can write

$$V_a'V_b = V_b'V_a \qquad (8.2.4)$$

Now you can see that Eq. (8.2.4) is exactly the same as Eq. (8.2.2). We have therefore proved that the transpose of a minor product of two vectors is equal to the product of their transposes in reverse order.

Although for demonstration we have proved this rule formally by means of equations, you can see that it must be true merely by using common sense. Let us look at the product of a row vector by a column vector when the elements are numbers as in Fig. 8.2.1.

$$[3 \quad 2 \quad 5] \begin{bmatrix} 1 \\ 6 \\ 2 \end{bmatrix} = (3 \times 1) + (2 \times 6) + (5 \times 2) = 3 + 12 + 10 = 25$$

Fig. 8.2.1

Now look at the product of the transposes of the two vectors on the left in Fig. 8.2.1 in reverse order, as in Fig. 8.2.2.

$$[1 \quad 6 \quad 2] \begin{bmatrix} 3 \\ 2 \\ 5 \end{bmatrix} = (1 \times 3) + (6 \times 2) + (2 \times 5) = 3 + 12 + 10 = 25$$

Fig. 8.2.2

If you are familiar with correlation coefficients, you know that in order to calculate a correlation coefficient, we must get a sum of products of scalar quantities which is usually written ΣXY. Having learned how to take the scalar product of two vectors you can therefore write a sum of cross products in vector notation. Suppose we have a column of x values and a column of y values for which we want to get the sum of cross products. We represent the columns of x and y values respectively as in Fig. 8.2.3.

$$x = \begin{bmatrix} x_1 \\ x_2 \\ \cdots \\ x_n \end{bmatrix} \qquad y = \begin{bmatrix} y_1 \\ y_2 \\ \cdots \\ y_n \end{bmatrix}$$

Fig. 8.2.3

If you have studied elementary statistics you know that ΣXY can be written as

$$\Sigma XY = X_1 Y_1 + X_2 Y_2 + \ldots + X_n Y_n \tag{8.2.5}$$

But we can also write the scalar product of two vectors as

$$X'Y = X_1 Y_1 + X_2 Y_2 + \ldots + X_n Y_n \tag{8.2.6}$$

You see that the right side of Eq. (8.2.5) and Eq. (8.2.6) are exactly the same. Therefore the left sides must also be exactly the same and we can write

$$\Sigma XY = X'Y \tag{8.2.7}$$

Equation (8.2.7) shows how to write the sum of cross products in terms of the scalar product of two vectors; and knowing that the scalar product of two vectors is equal to the product of their transposes in reverse order, we can also write

$$\Sigma XY = Y'X \tag{8.2.8}$$

8.3 The Major Product of Two Vectors

The second kind of product of two vectors we shall consider is one not commonly discussed in books on matrix algebra, but very useful in the behavioral sciences. We shall call it the *major product* of two vectors. A numerical example is shown in Fig. 8.3.1.

$$\begin{bmatrix} 2 \\ 4 \\ 3 \end{bmatrix} \begin{bmatrix} 1 & 3 & 2 & 4 \end{bmatrix} = \begin{bmatrix} 2 \times 1 & 2 \times 3 & 2 \times 2 & 2 \times 4 \\ 4 \times 1 & 4 \times 3 & 4 \times 2 & 4 \times 4 \\ 3 \times 1 & 3 \times 3 & 3 \times 2 & 3 \times 4 \end{bmatrix} = \begin{bmatrix} 2 & 6 & 4 & 8 \\ 4 & 12 & 8 & 16 \\ 3 & 9 & 6 & 12 \end{bmatrix}$$

FIG. 8.3.1

Now if you study this equation carefully, you will make a number of important observations. First, it is similar to the equation for the minor or scalar product of two vectors in that it has a prefactor and a postfactor. The vector on the left is the prefactor and the vector on the right is the postfactor. In the equation for the minor product of two vectors, however, the prefactor was a row vector and the postfactor a column vector. For the major product of two vectors, we have just the reverse; the prefactor is a column vector and the postfactor is a row vector. Therefore, to get the major product of two vectors we postmultiply a column vector by a row vector or we premultiply the row vector by a column vector.

If you examine the matrices following the first and second equality signs in Fig. 8.3.1, you see that we have formed the major product of the two vectors in a very simple manner. For the first row of the product matrix each element in the postfactor is multiplied by the first element in the prefactor. For the second row each element in the postfactor is multiplied by the second element in the prefactor. For the third row each element in the postfactor is multiplied by the third element in the prefactor.

Notation for major product

If we go back to the vectors V_a and V_b we can illustrate the general rule for getting the major product of two vectors by Eq. 8.3.1.

$$V_a V_b' = \begin{bmatrix} a_1 \\ a_2 \\ \cdots \\ a_n \end{bmatrix} \begin{bmatrix} b_1 & b_2 & \cdots & b_m \end{bmatrix} = \begin{bmatrix} a_1 b_1 & a_1 b_2 & \cdots & a_1 b_m \\ a_2 b_1 & a_2 b_2 & \cdots & a_2 b_m \\ \cdots & \cdots & \cdots & \cdots \\ a_n b_1 & a_n b_2 & \cdots & a_n b_m \end{bmatrix} \tag{8.3.1}$$

You see from Eq. (8.3.1) that the major product of two vectors is a matrix whose number of rows is equal to the order of the prefactor and whose number of columns is equal to the order of the postfactor. Each element of a row vector of the product matrix is obtained by multiplying each element of the postfactor by the element in the prefactor corresponding to that row. Similarly, each column vector of the product matrix is obtained by multiplying each element in the prefactor by the element in the postfactor corresponding to that column.

In general, if we indicate the major product of two vectors by the symbol C, we can write Eq. (8.3.1) symbolically as

$$V_a V_b' = C \tag{8.3.2}$$

The element in the ith row and jth column of C is the product of the ith element of the prefactor and the jth element of the postfactor. Thus

$$C_{ij} = a_i b_j \tag{8.3.3}$$

Order characteristics of the major product

It is clear then from the definition of the major product of two vectors that there are no restrictions on the orders of the two vectors. For the scalar product of two vectors, we know that both factors must be of the same order, so that corresponding elements can be multiplied together. For the major product of two vectors, there is no such necessity. But, while the minor product is always a scalar quantity, the major product of two vectors is never a scalar quantity but rather a matrix. The number of rows in this matrix is determined by the column vector, or prefactor, and the number of columns is determined by the row vector, or postfactor. Thus, in summary, the order of a matrix product of two vectors is given by the order of the factors and in the same sequence. For example, a fourth-order column vector times a third-order row vector gives a 4 by 3 matrix.

8.4   The Transpose of a Major Product of Two Vectors

We shall now prove that the transpose of the major product of two vectors is equal to the product of the transposes of the two vectors in reverse order. We can express this rule by

$$(V_a V_b')' = V_b V_a' \tag{8.4.4}$$

From the rules for forming major product of two vectors, we can write Eq. (8.4.5).

$$V_b V_a' = \begin{bmatrix} b_1 \\ b_2 \\ \cdots \\ b_m \end{bmatrix} \begin{bmatrix} a_1 & a_2 & \cdots & a_n \end{bmatrix} = \begin{bmatrix} b_1 a_1 & b_1 a_2 & \cdots & b_1 a_n \\ b_2 a_1 & b_2 a_2 & \cdots & b_2 a_n \\ \cdots & \cdots & \cdots & \cdots \\ b_m a_1 & b_m a_2 & \cdots & b_m a_n \end{bmatrix} \tag{8.4.5}$$

Now suppose we write Eq. (8.4.5) compactly as

$$V_b V_a' = F \tag{8.4.6}$$

From Eq. (8.4.5) you can see that the element in the jth row and ith column of the product is the jth element from the b vector and the ith element from the a vector. Therefore, we can write

$$F_{ji} = b_j a_i \qquad (8.4.7)$$

But since the factors of a scalar product of two scalar quantities can be reversed, Eq. (8.4.7) can be written as

$$F_{ji} = a_i b_j \qquad (8.4.8)$$

Now notice that the right side of Eq. (8.4.8) is exactly the same as the right side of Eq. (8.3.3). Therefore, the left sides of the two equations must also be equal, and we can write

$$C_{ij} = F_{ji} \qquad (8.4.9)$$

But remember that the element in the ith row and jth column of a matrix is the same as the element in the jth row and ith column of its transpose; therefore, because of Eq. (8.4.9) we can write

$$C' = F \qquad (8.4.10)$$

Now let us go back to Eq. (8.3.2) and take the transpose of both sides of the equation to get

$$(V_a V_b')' = C' \qquad (8.4.11)$$

Next we substitute the left side of Eq. (8.4.11) in the right side of Eq. (8.4.9) and get

$$(V_a V_b')' = F \qquad (8.4.12)$$

Then we substitute the left side of Eq. (8.4.6) in the right side of Eq. (8.4.12) to get

$$(V_a V_b')' = V_b V_a' \qquad (8.4.13)$$

Now Eq. (8.4.13) is exactly the same as Eq. (8.4.4), which is what we set out to prove. Equation (8.4.4) states that the transpose of the major product of two vectors is equal to the product of the transposes of the two vectors in reverse order. This theorem is proved by Eq. (8.4.5) through (8.4.13). Actually, it would not be necessary to go through all of these steps for you to see that the statement is true. You could simply see by inspection that the last matrix in Eq. (8.3.1) is the transpose of the last matrix in Eq. (8.4.5). However, it is well to learn early how, by using simple rules, step by step, we can prove very important and sometimes complicated theorems with a minimum of mental effort and confusion.

8.5 The Product of a Vector and Its Transpose

Of particular interest in the analysis of behavioral science data is the product of a vector and its transpose. There are two such products, a column vector premultiplied by its transpose and a column vector post-multiplied by its transpose. The first we call the *minor product moment* of a vector and the second the *major product moment*.

The minor product moment

The minor product moment of a vector is obviously a scalar quantity since it is a special case of the minor product of vectors. More specifically it is simply the sum of squares of the elements of the vector, as you can see from Fig. 8.5.1.

$$[1 \quad 3 \quad 2] \begin{bmatrix} 1 \\ 3 \\ 2 \end{bmatrix} = 1 + 3^2 + 2^2 = 1 + 9 + 4 = 14$$

Fig. 8.5.1

In general symbolic form, this is Eq. (8.5.1).

$$V_a'V_a = [a_1 \quad a_2 \quad \ldots \quad a_n] \begin{bmatrix} a_1 \\ a_2 \\ \ldots \\ a_n \end{bmatrix} = a_1^2 + a_2^2 + \ldots + a_n^2 \tag{8.5.1}$$

Since the minor product moment of a vector is a scalar quantity it is obviously equal to its transpose.

The major product moment

In getting the major product moment of a vector and its transpose, we postmultiply the column form of the vector by its row form, or conversely we premultiply the row form of the vector by its column form. We shall prove that the major product moment of a vector is a symmetric matrix. First, we write out the major product of the vector and its transpose as in Eq. (8.5.2).

$$V_a V_a' = \begin{bmatrix} a_1 \\ a_2 \\ \ldots \\ a_n \end{bmatrix} [a_1 \quad a_2 \quad \ldots \quad a_n] = \begin{bmatrix} a_1^2 & a_1a_2 & \ldots & a_1a_n \\ a_2a_1 & a_2^2 & \ldots & a_2a_n \\ \ldots & \ldots & \ldots & \ldots \\ a_na_1 & a_na_2 & \ldots & a_n^2 \end{bmatrix} \tag{8.5.2}$$

If we indicate the major product moment of the vector by C, we can write Eq. (8.5.3).

$$C = \begin{bmatrix} a_1^2 & a_1a_2 & \ldots & a_1a_n \\ a_2a_1 & a_2^2 & \ldots & a_2a_n \\ \ldots & \ldots & \ldots & \ldots \\ a_na_1 & a_na_2 & \ldots & a_n^2 \end{bmatrix} \tag{8.5.3}$$

Then from Eqs. (8.5.2) and (8.5.3) we get

$$V_a V_a' = C \tag{8.5.4}$$

Now from Eq. (8.5.2) we see that the ijth element is $a_i a_j$ and the jith element is $a_j a_i$. We have, therefore,

$$C_{ij} = a_i a_j \tag{8.5.5}$$

$$C_{ji} = a_j a_i \tag{8.5.6}$$

But since the factors on the right side of Eq. (8.5.6) are scalars, they can be reversed to give us

$$C_{ji} = a_i a_j \tag{8.5.7}$$

You see now that the right sides of Eqs. (8.5.5) and (8.5.7) are exactly the same; therefore, the left sides must also be the same, and we have

$$C_{ij} = C_{ji} \tag{8.5.8}$$

Now Eq. (8.5.8) says that the element in the ith row and jth column is the same as the element in the jth row and ith column. This, you remember, is the definition of a symmetric matrix. Therefore, Eqs. (8.5.2) through (8.5.8) prove that the major product moment of a vector is a symmetric matrix.

In the analysis of data from the human sciences we shall have many occasions to use the major product moments of vectors. An example is Fig. 8.5.2.

$$\begin{bmatrix} 4 \\ 3 \\ 1 \end{bmatrix} [4 \quad 3 \quad 1] = \begin{bmatrix} 16 & 12 & 4 \\ 12 & 9 & 3 \\ 4 & 3 & 1 \end{bmatrix}$$

FIG. 8.5.2

It is easy to see from a numerical example like this that the major product must be a symmetric matrix; however, it is good practice for you to prove the theorem in simple steps for the general case, regardless of what particular numbers are used as the elements of the vector. The major product moment of a vector is used extensively in calculating covariance matrices from data matrices.

8.6 Undefined Products of Vectors

We have described two different kinds of products of vectors — the scalar, or minor, product and the major product. To get the minor product of two vectors, we postmultiply a row vector by a column vector. To get the major product of two vectors, we postmultiply a column vector by a row vector. To avoid confusion in the use of matrix algebra as a sort of automatic problem-solving machine, it is well to know what possible vector products have no meaning within the scheme of matrix algebra. We shall, therefore, illustrate the two kinds of products which might be found but which are not meaningful for practical use. The first of these is the product of a column vector and a column vector. By neither of the two methods of vector multiplication can a column vector be multiplied by a column vector. To emphasize the impossibility of such a thing, we give

$$\begin{bmatrix} 3 \\ 2 \\ 4 \end{bmatrix} \times \begin{bmatrix} 4 \\ 3 \\ 1 \end{bmatrix} = \text{nonsense} \qquad \begin{bmatrix} a_1 \\ a_2 \\ a_3 \end{bmatrix} \times \begin{bmatrix} b_1 \\ b_2 \\ b_3 \end{bmatrix} = \text{nonsense}$$

Fig. 8.6.1

Figure 8.6.2 shows in compact notation that the product of a column vector by a column vector has no meaning.

$$V_a V_b = \text{nonsense}$$

Fig. 8.6.2

Neither can a row vector be multiplied by a row vector to yield a useful product. The impossibility of such a product is illustrated by

$$[1 \quad 2 \quad 5] \times [2 \quad 3 \quad 7] = \text{nonsense} \qquad [a_1 \quad a_2 \quad a_3] \times [b_1 \quad b_2 \quad b_3] = \text{nonsense}$$

Fig. 8.6.3

The impossibility of the product of a row vector by a row vector is illustrated in compact notation by

$$V_a' V_b' = \text{nonsense}$$

Fig. 8.6.4

You see, therefore, that in the product of two vectors, whether a minor or major product, one of the factors must always be a row and the other always a column. You cannot have two column factors or two row factors. Consequently, unless otherwise specified, one of the factors will always have a prime and the other will always be unprimed. This must be so because, unless otherwise specified, an unprimed vector is always a column vector and a primed vector is always a row vector. It has been mentioned that this practice is not consistently followed by all books or articles on matrix algebra; however, we shall follow the practice very closely.

8.7   Vector Multiplication and the Commutative Law

We have already seen that by the commutative law of addition we mean $a + b = b + a$, or, in numbers, $3 + 4 = 4 + 3$. The commutative law of multiplication in arithmetic or scalar algebra means that $ab = ba$, or similarly, $3 \times 4 = 4 \times 3$. But the commutative law of multiplication does not hold for the product of two vectors. What you have taken for granted about numbers since early childhood is not true about the special kinds of matrices we call vectors. You can see very readily that vector multiplication is not commutative. Suppose we define two different vectors as in Fig. 8.7.1.

$$V_a = \begin{bmatrix} 1 \\ 3 \\ 2 \end{bmatrix} \qquad V_b = \begin{bmatrix} 4 \\ 2 \\ 5 \end{bmatrix}$$

FIG. 8.7.1

Then the minor product of the two vectors is shown in Fig. 8.7.2.

$$V_a' V_b = \begin{bmatrix} 1 & 3 & 2 \end{bmatrix} \begin{bmatrix} 4 \\ 2 \\ 5 \end{bmatrix} = 4 + 6 + 10 = 20$$

FIG. 8.7.2

The minor product of the two vectors is a scalar quantity or 20. But if we reverse the order of the factors, or as we say, if we commute the factors, we get Fig. 8.7.3.

$$V_b V_a' = \begin{bmatrix} 4 \\ 2 \\ 5 \end{bmatrix} \begin{bmatrix} 1 & 3 & 2 \end{bmatrix} = \begin{bmatrix} 4 & 12 & 8 \\ 2 & 6 & 4 \\ 5 & 15 & 10 \end{bmatrix}$$

FIG. 8.7.3

That is, we get a major product. The two products shown in Figs. 8.7.2 and 8.7.3 cannot possibly be equal because the first one is a scalar quantity and the second one is a matrix.

Similarly, if we commute the factors in a major product of two vectors, we get a minor product, or scalar quantity. Furthermore, it is not possible to commute the factors in a major product unless the two vectors are of the same order. Also, since commuting the factors in the scalar or minor product of two vectors gives the major product of two vectors, it follows that in so doing we always get a square matrix. This, of course, is true because in a minor product the two vectors must be of the same order. If the two vectors are of the same order, their major product must have the same number of rows as columns and hence be a square matrix.

We can show in compact notation that vector multiplication is non-commutative in Fig. 8.7.4.

$$V_a V_b' \neq V_b' V_a$$
$$V_b V_a' \neq V_a' V_b$$

FIG. 8.7.4

The equals sign with a line drawn through it means that the expressions on the right and left are not equal. This symbol is sometimes read as "distinct from." Thus the first line of Fig. 8.7.4 would read, "$V_a V_b'$ is distinct from $V_b' V_a$."

8.8   The Minor Product of Type I Supervectors

Characteristics of the minor product

From what we know about partitioned vectors or type I supervectors, it is fairly easy to deduce three simple rules about their minor products. First, in multiplication with type I supervectors, as with simple vectors, both factors must be of the same order to result in a minor product. Second, corresponding subvectors of the supervectors must also have the same order. Third, the minor product of two type I supervectors is a scalar quantity. These rules become obvious if we give the minor product of two type I supervectors in terms of their vector elements. As in multiplication of simple vectors, the prefactor in the minor product is a row vector and the postfactor is a column vector. Two supervectors are given in Fig. 8.8.1.

$$V_a = \begin{bmatrix} V_{a_1} \\ V_{a_2} \\ \dots \\ V_{a_n} \end{bmatrix} \qquad V_b = \begin{bmatrix} V_{b_1} \\ V_{b_2} \\ \dots \\ V_{b_n} \end{bmatrix}$$

FIG. 8.8.1

Then the minor product of these two supervectors can be illustrated by Eq. (8.8.1).

$$V_a' V_b = [V_{a_1}' \quad V_{a_2}' \quad \dots \quad V_{a_n}'] \begin{bmatrix} V_{b_1} \\ V_{b_2} \\ \dots \\ V_{b_n} \end{bmatrix} = V_{a_1}' V_{b_1} + V_{a_2}' V_{b_2} + \dots + V_{a_n}' V_{b_n} \tag{8.8.1}$$

Here we have used the same rules for the minor product as we did in multiplying simple vectors; that is, we took the product of corresponding elements of the two vectors and added them together. You should notice in Eq. (8.8.1) that in the middle expression all of the subvectors of V_a' are on the left of all of the subvectors of V_b. Similarly, in the last expression in Eq. (8.8.1), each V_a' subvector is on the left of each V_b subvector. Since now the elements of the supervectors are themselves vectors, you are not free to interchange the order of the factors in the products of two elements as you were in multiplying simple vectors. This is because, as you have learned, vector multiplication is not commutative.

You will also see from the right side of Eq. (8.8.1) why corresponding vector elements in the two vectors must be of the same order. Each term in the right side of Eq. (8.8.1) is itself a minor product of two vectors. As we know, both factors in a minor product must have the same number of elements. Notice, of course, that there is no restriction on the number of elements in the subvectors within the supervector. That is, V_{a_1} can have any number of elements, V_{a_2} any number of elements, and so on, as long as V_{a_1} has the same number of elements as V_{b_1}, V_{a_2} the same number as V_{b_2}, and so on. To make Eq. (8.8.1) more meaningful in terms of actual numbers, let us take the instance of two second-order type I supervectors whose subvectors are as given in Fig. 8.8.2.

$$V_{a_1} = \begin{bmatrix} 2 \\ 3 \end{bmatrix} \qquad V_{b_1} = \begin{bmatrix} 2 \\ 1 \end{bmatrix}$$

$$V_{a_2} = \begin{bmatrix} 1 \\ 7 \\ 5 \end{bmatrix} \qquad V_{b_2} = \begin{bmatrix} 2 \\ 1 \\ 3 \end{bmatrix}$$

FIG. 8.8.2

We can now write the minor product of the two supervectors as in Fig. 8.8.3.

$$V_a'V_b = [2 \quad 3 \mid 1 \quad 7 \quad 5] \begin{bmatrix} 2 \\ 1 \\ - \\ 2 \\ 1 \\ 3 \end{bmatrix} = [2 \quad 3] \begin{bmatrix} 2 \\ 1 \end{bmatrix} + [1 \quad 7 \quad 5] \begin{bmatrix} 2 \\ 1 \\ 3 \end{bmatrix} =$$

$$[2 \times 2 + 3 \times 1] + [1 \times 2 + 7 \times 1 + 5 \times 3] = 7 + 24 = 31$$

FIG. 8.8.3

The transpose of the minor product

We shall next prove that the transpose of the minor product of two type I supervectors is equal to the product of their transposes in reverse order. What we shall prove is stated compactly as

$$V_a'V_b = V_b'V_a \tag{8.8.2}$$

We note first that the left side of Eq. (8.8.2) is expressed in terms of the vector elements of the two supervectors in Eq. (8.8.1). In the same way, we can express the right side of Eq. (8.8.2) in terms of the vector elements of the two supervectors by

$$V_b'V_a = [V_{b_1}' \quad V_{b_2}' \quad \cdots \quad V_{b_n}'] \begin{bmatrix} V_{a_1} \\ V_{a_2} \\ \cdots \\ V_{a_n} \end{bmatrix} = V_{b_1}'V_{a_1} + V_{b_2}'V_{a_2} + \cdots + V_{b_n}'V_{a_n}$$

$$\tag{8.8.3}$$

Now compare the right side of Eq. (8.8.1) with the right side of Eq. (8.8.3). Notice that each term on the right in Eq. (8.8.3) is a minor product of two vectors. Remember also that a minor product of two vectors is equal to the product of their transposes in reverse order. For the ith term on the right in Eq. (8.8.3), we could therefore write

$$V'_{b_i} V_{a_i} = V'_{a_i} V_{b_i} \qquad (8.8.4)$$

Now because of Eq. (8.8.4), we can write Eq. (8.8.3) in the form of

$$V'_b V_a = V'_{a_1} V_{b_1} + V'_{a_2} V_{b_2} + \ldots + V'_{a_n} V_{b_n} \qquad (8.8.5)$$

Now notice that the right side of Eq. (8.8.5) is exactly the same as the right side of Eq. (8.8.1). Therefore, the left sides of the two equations must be equal to each other, and we can write

$$V'_a V_b = V'_b V_a \qquad (8.8.6)$$

But you see that Eq. (8.8.6) is exactly the same as Eq. (8.8.2), which is what we set about to prove. Again, as in some of our earlier proofs, you could argue from common sense or see intuitively that the minor product of two type I supervectors would have to be the same as the minor product of their transposes in reverse order, without going through the steps from Eq. (8.8.1) through Eq. (8.8.6). It is well, however, to learn to use these simple mechanical routine procedures to prove the obvious so that you will be able to use them to prove things which are not at all obvious.

We said at the beginning of this section that the minor products of two type I supervectors exist if, and only if, they are of the same order and their corresponding vector elements are of the same order. Strictly speaking, this is true; however, you will have guessed already that if in a computation you have two type I supervectors, you can reduce them to simple form to get a minor product or a scalar, even though the supervectors are not of the same order. You will recall that this principle holds in addition of supermatrices in general. We do not violate or contradict the rule, however, by such an exception because when we reduce the supervectors to simple form we no longer have supervectors.

An example of how the scalar product of type I supervectors is used in the analyses of data from the social sciences can be given. Suppose we have given a number of different tests to entering freshmen at a university. We enter these test scores on sheets or rosters. We use the basic principle described in Chapter 1 for a data matrix. Each column on the sheet represents a different test and each row on the sheet represents an entering freshman. But suppose we had many more entering freshmen than we could get on a single sheet, so that to enter all of the scores of all of the freshmen we have to have n sheets. Now, having entered the scores, we want to find the correlation between the two tests in the first two columns of the sheets. This means we must get the sum of cross products for these two scores for all the people on n sheets. Actually, what we would probably do if we had a calculating machine is to get the sum of cross products for the first sheet, then for the second, and so on, until we had a sum of cross products for each of the sheets. We would then add them all together to get the total sum of cross products for all of the entering freshmen on the two tests. To express in terms of matrix algebra what we are doing in this illustration, we let x in Fig. 8.8.4 represent a type I column supervector of the scores on the first test, and y represent a similar supervector of the scores on the second test.

$$x = \begin{bmatrix} x_1 \\ x_2 \\ \cdots \\ x_n \end{bmatrix} \qquad y = \begin{bmatrix} y_1 \\ y_2 \\ \cdots \\ y_n \end{bmatrix}$$

Fig. 8.8.4

Each element in the supervector is a vector of the scores on a single sheet. Then Eq. (8.8.7) represents the sum of cross products of all of the sheets for the two tests.

$$x'y = x_1'y_1 + x_2'y_2 + \ldots + x_n'y_n \tag{8.8.7}$$

Each term on the right side of Eq. (8.8.7) represents the sums of cross products for a single sheet.

8.9 The Major Product of Type I Supervectors

The major product of type I supervectors differs in several important respects from the minor product. Here, as in multiplication for the major product of simple vectors, the prefactor must be a column vector and the postfactor a row vector. Here also the two vectors need not have the same order. Furthermore, there are no restrictions on the orders of the subvectors in either of the factors of a major product of type I supervectors. A fourth characteristic is that the major product is a supermatrix. The rule for forming the submatrices of the major product of two supervectors is the same as that for forming the elements of the major product of simple vectors. The first row of matric elements of the product matrix is formed by multiplying each element of the postfactor by the first element in the prefactor.

Equation (8.9.1) illustrates how we form the major product of type I supervectors.

$$V_a V_b' = \begin{bmatrix} V_{a_1} \\ V_{a_2} \\ \vdots \\ V_{a_n} \end{bmatrix} [V_{b_1}' \quad V_{b_2}' \quad \cdots \quad V_{b_m}'] = \begin{bmatrix} V_{a_1}V_{b_1}' & V_{a_1}V_{b_2}' & \cdots & V_{a_1}V_{b_m}' \\ V_{a_2}V_{b_1}' & V_{a_2}V_{b_2}' & \cdots & V_{a_2}V_{b_m}' \\ \cdots & \cdots & \cdots & \cdots \\ V_{a_n}V_{b_1}' & V_{a_n}V_{b_2}' & \cdots & V_{a_n}V_{b_m}' \end{bmatrix}$$

$$\tag{8.9.1}$$

You see that in the right side of Eq. (8.9.1) each element in the matrix is the major product of subvectors. Therefore, the elements in this major product are themselves matrices.

It should be clear from Eq. (8.9.1) that the supermatrix has the same number of rows as the prefactor and the same number of columns as the postfactor. Furthermore, it should also be clear that the height of a matrix element is determined by the order of the subvector from the prefactor and the width by the order of the subvector from the postfactor. Let us consider a numerical example of the major product of two type I supervectors. We let the subvectors be given as in Fig. 8.9.1.

$$V_{a_1} = \begin{bmatrix} 2 \\ 3 \end{bmatrix} \qquad V_{b_1} = \begin{bmatrix} 3 \\ 1 \\ 2 \end{bmatrix}$$

$$V_{a_2} = \begin{bmatrix} 2 \\ 4 \\ 1 \end{bmatrix} \qquad V_{b_2} = [2]$$

$$V_{b_3} = \begin{bmatrix} 2 \\ 3 \\ 5 \\ 1 \end{bmatrix}$$

Fig. 8.9.1

Then the major product will be as shown in Fig. 8.9.2.

$$V_a V_b' = \begin{bmatrix} 2 \\ 3 \\ \hline 2 \\ 4 \\ 1 \end{bmatrix} [3\ 1\ 2\ |\ 2\ |\ 2\ 3\ 5\ 1] =$$

$$\begin{bmatrix} \begin{bmatrix} 2 \\ 3 \end{bmatrix}[3\ 1\ 2] & \begin{bmatrix} 2 \\ 3 \end{bmatrix}[2] & \begin{bmatrix} 2 \\ 3 \end{bmatrix}[2\ 3\ 5\ 1] \\ \begin{bmatrix} 2 \\ 4 \\ 1 \end{bmatrix}[3\ 1\ 2] & \begin{bmatrix} 2 \\ 4 \\ 1 \end{bmatrix}[2] & \begin{bmatrix} 2 \\ 4 \\ 1 \end{bmatrix}[2\ 3\ 5\ 1] \end{bmatrix} = \left[\begin{array}{ccc|c|cccc} 6 & 2 & 4 & 4 & 4 & 6 & 10 & 2 \\ 9 & 3 & 6 & 6 & 6 & 9 & 15 & 3 \\ \hline 6 & 2 & 4 & 4 & 4 & 6 & 10 & 2 \\ 12 & 4 & 8 & 8 & 8 & 12 & 20 & 4 \\ 3 & 1 & 2 & 2 & 2 & 3 & 5 & 1 \end{array}\right]$$

Fig. 8.9.2

In this figure V_a is a second-order column supervector, and V_b' is a third-order row supervector. The first element in V_a is of order 2 and the second element of order 3; therefore, all submatrices in the first row of the product have two rows and all in the second row have three rows. The orders of the three subvectors in the postfactor are respectively 3, 1, and 4; therefore, the number of columns in the first, second, and third matrix columns of the product is 3, 1, and 4, respectively.

8.10 The Transpose of the Major Product of Two Type I Supervectors

Again we shall prove by means of simple equations what may already be obvious to you. The transpose of the major product of two type I supervectors is equal to the product of their transposes in reverse order. This rule is given as

$$(V_a V_b')' = V_b V_a' \tag{8.10.1}$$

First we will write out the elements of the supermatrix for the right side of Eq. (8.10.1) in Eq. (8.10.2).

$$V_b V_a' = \begin{bmatrix} V_{b_1} V_{a_1}' & V_{b_1} V_{a_2}' & \cdots & V_{b_1} V_{a_n}' \\ V_{b_2} V_{a_1}' & V_{b_2} V_{a_2}' & \cdots & V_{b_2} V_{a_n}' \\ \cdots & \cdots & \cdots & \cdots \\ V_{b_m} V_{a_1}' & V_{b_m} V_{a_2}' & \cdots & V_{b_m} V_{a_n}' \end{bmatrix} \tag{8.10.2}$$

Here we have simply used the general rule for obtaining each element of the product matrix in terms of the elements of the factors. Next, let us go back to Eq. (8.9.1) and take the transpose of both the right and left sides of this equation. Remember that to take the transpose of this supermatrix, we write rows as columns and take the transpose of each element. This we do in Eq. (8.10.3).

$$(V_a V_b')' = \begin{bmatrix} (V_{a_1} V_{b_1}')' & (V_{a_2} V_{b_1}')' & \cdots & (V_{a_n} V_{b_1}')' \\ (V_{a_1} V_{b_2}')' & (V_{a_2} V_{b_2}')' & \cdots & (V_{a_n} V_{b_2}')' \\ \cdots & \cdots & \cdots & \cdots \\ (V_{a_1} V_{b_m}')' & (V_{a_2} V_{b_m}')' & \cdots & (V_{a_n} V_{b_m}')' \end{bmatrix} \tag{8.10.3}$$

Now notice that each element of the supermatrix on the right side of Eq. (8.10.3) is the transpose of the major product of two vectors. Remember also that the transpose of the major product of two simple vectors is the product of the transposes in reverse order. For the ijth matrix element of Eq. (8.10.3), we have therefore

$$(V_a V_b')' = V_b V_a' \tag{8.10.4}$$

Substituting the right side of Eq. (8.10.4) for each of the elements on the right side of Eq. (8.10.3) we get Eq. (8.10.5).

$$(V_a V_b')' = \begin{bmatrix} V_{b_1} V_{a_1}' & V_{b_1} V_{a_2}' & \cdots & V_{b_1} V_{a_n}' \\ V_{b_2} V_{a_1}' & V_{b_2} V_{a_2}' & \cdots & V_{b_2} V_{a_n}' \\ \cdots & \cdots & \cdots & \cdots \\ V_{b_m} V_{a_1}' & V_{b_m} V_{a_2}' & \cdots & V_{b_m} V_{a_n}' \end{bmatrix} \tag{8.10.5}$$

But now notice that the right side of Eq. (8.10.5) is exactly the same as the right side of Eq. (8.10.2). Therefore, the left sides must also be the same and we can write

$$(V_a V_b')' = V_b V_a' \qquad (8.10.6)$$

Now Eq. (8.10.6) is exactly the same as Eq. (8.10.1), which we started out to prove.

The major product of type I supervectors plays an important role in the calculation of covariance matrices where attributes are partitioned into predictor and criterion sets.

8.11 Special Vector Products

Certain vector products, both minor and major, are of special interest because they occur so frequently. Some of the most common of these we shall now consider.

The minor product moment

The minor product of a vector and its transpose is the sum of squares of its elements. This we have already indicated in Section 8.5, but since the rule is so generally used in the analysis of behavioral science data it will be repeated here. Another example is given in Fig. 8.11.1.

$$\begin{bmatrix} 3 & 4 & 7 \end{bmatrix} \begin{bmatrix} 3 \\ 4 \\ 7 \end{bmatrix} = 3^2 + 4^2 + 7^2 = 9 + 16 + 49 = 74$$

Fig. 8.11.1

We repeat another form of the general rule in Eq. (8.11.1).

$$V_X' V_X = \begin{bmatrix} X_1 & X_2 & \ldots & X_n \end{bmatrix} \begin{bmatrix} X_1 \\ X_2 \\ \ldots \\ X_n \end{bmatrix} = X_1^2 + X_2^2 + \ldots + X_n^2$$

In the summation notation commonly used in elementary statistics, Eq. (8.11.1) can be written in the form of

$$V_X' V_X = \Sigma X^2 \qquad (8.11.2)$$

Products involving the unit vector

The minor product of the unit vector and a general vector is a scalar that is the sum of the elements of the general vector. It does not matter whether we use the unit vector as a prefactor or a postfactor. An illustration of this is Fig. 8.11.2.

$$[1 \quad 1 \quad 1]\begin{bmatrix} 3 \\ 4 \\ 7 \end{bmatrix} = [3 \quad 4 \quad 7]\begin{bmatrix} 1 \\ 1 \\ 1 \end{bmatrix} = 3 + 4 + 7 = 14$$

FIG. 8.11.2

Equation (8.11.3) shows this rule in algebraic form.

$$[1 \quad 1 \quad \ldots \quad 1]\begin{bmatrix} X_1 \\ X_2 \\ \ldots \\ X_n \end{bmatrix} = [X_1 \quad X_2 \quad \ldots \quad X_n]\begin{bmatrix} 1 \\ 1 \\ \ldots \\ 1 \end{bmatrix} = X_1 + X_2 + \ldots + X_n \tag{8.11.3}$$

To use the summation notation, we simply write Eq. (8.11.4) to show that the minor product of a vector and the unit vector is the sum of the elements in the vectors.

$$1'V_X = V'_X 1 = \Sigma X \tag{8.11.4}$$

It is clear then that when you take a sum of experimental values you are getting the minor product of a vector of observations and the unit vector.

The minor product moment of a unit vector is a scalar equal to the order of the vector. This is obvious from Fig. 8.11.3.

$$[1 \quad 1 \quad 1]\begin{bmatrix} 1 \\ 1 \\ 1 \end{bmatrix} = [1 + 1 + 1] = 3$$

$$[1 \quad 1 \quad 1 \quad 1]\begin{bmatrix} 1 \\ 1 \\ 1 \\ 1 \end{bmatrix} = [1 + 1 + 1 + 1] = 4$$

FIG. 8.11.3

If, as given in Section 3.15, we indicate the unit vector by the number 1 and if we use a subscript to indicate its order, this principle can be expressed algebraically as

$$1'_n 1_n = n \tag{8.11.5}$$

In Eq. (8.11.5) the subscript n means that the unit vector has n elements, and Eq. (8.11.5) simply says that when the unit row vector is postmultiplied by the unit column vector, we get a scalar quantity n that is equal to the order of the vector.

The minor product moment of a sign vector is also equal to its order. This can be easily verified since the elements of a sign vector are either $+1$ or -1. Their squares are therefore each 1.

If we take the major product of the unit vector by another vector, we get a very simple type of matrix. If a row vector is premultiplied by a column unit vector, the product is a matrix in which each row is equal to the row vector. An example of this is Fig. 8.11.4.

$$\begin{bmatrix} 1 \\ 1 \\ 1 \end{bmatrix} \begin{bmatrix} 3 & 4 & 7 \end{bmatrix} = \begin{bmatrix} 3 & 4 & 7 \\ 3 & 4 & 7 \\ 3 & 4 & 7 \end{bmatrix}$$

FIG. 8.11.4

One use of this product is in data analysis involving means. If from a matrix of measures we subtract from each measure of an attribute its mean for all the entities, we are actually subtracting from the data matrix the major product of a vector of means and a unit vector.

The general rule is expressed in algebraic form by Eq. (8.11.6).

$$1V_a' = \begin{bmatrix} a_1 & a_2 & \ldots & a_n \\ a_1 & a_2 & \ldots & a_n \\ \ldots & \ldots & \ldots & \ldots \\ a_1 & a_2 & \ldots & a_n \end{bmatrix} \tag{8.11.6}$$

The major product of a general column vector postmultiplied by a unit vector obviously is a matrix in which all column vectors are equal to the prefactor. This is illustrated by Fig. 8.11.5.

$$\begin{bmatrix} 4 \\ 3 \\ 5 \end{bmatrix} \begin{bmatrix} 1 & 1 & 1 \end{bmatrix} = \begin{bmatrix} 4 & 4 & 4 \\ 3 & 3 & 3 \\ 5 & 5 & 5 \end{bmatrix}$$

FIG. 8.11.5

In algebraic form this rule is given by Eq. (8.11.7).

$$V_a 1' = \begin{bmatrix} a_1 & a_1 & \ldots & a_1 \\ a_2 & a_2 & \ldots & a_2 \\ \ldots & \ldots & \ldots & \ldots \\ a_n & a_n & \ldots & a_n \end{bmatrix} \tag{8.11.7}$$

The major product of two unit vectors is a matrix each of whose elements is 1, or unity. This is shown in Fig. 8.11.6.

$$\begin{bmatrix} 1 \\ 1 \\ 1 \end{bmatrix} \begin{bmatrix} 1 & 1 & 1 & 1 \end{bmatrix} = \begin{bmatrix} 1 & 1 & 1 & 1 \\ 1 & 1 & 1 & 1 \\ 1 & 1 & 1 & 1 \end{bmatrix}$$

FIG. 8.11.6

As in the general case, the order of the major product is determined by the orders of the two factors, the rows being determined by the prefactor and columns by the postfactor.

Products involving e_i vectors

The minor product moment of an e_i vector is equal to 1. Remember that an e_i vector is a vector all of whose elements are 0 except one and that is unity. You can easily see why this rule holds, irrespective of the order of the vector, from Fig. 8.11.7.

$$[0 \quad 1 \quad 0 \quad 0] \begin{bmatrix} 0 \\ 1 \\ 0 \\ 0 \end{bmatrix} = 0 \times 0 + 1 \times 1 + 0 \times 0 + 0 \times 0 = 1$$

FIG. 8.11.7

The general rule is

$$e_i'e_i = 1 \tag{8.11.8}$$

The minor product of any e_i' vector by some other e_j vector is always 0. This is illustrated by Fig. 8.11.8.

$$[0 \quad 1 \quad 0 \quad 0] \begin{bmatrix} 1 \\ 0 \\ 0 \\ 0 \end{bmatrix} = 0 \times 1 + 1 \times 0 + 0 \times 0 + 0 \times 0 = 0$$

FIG. 8.11.8

In algebraic form we can state the rule as

$$e_i'e_j = 0 \tag{8.11.9}$$

where i is different from j.

The scalar product of an e_i vector and any other vector is a scalar quantity that is the ith element of the vector. Thus

$$e_i'V = V'e_i = V_i \tag{8.11.10}$$

A numerical example is given in Fig. 8.11.9.

$$[0 \quad 1 \quad 0 \quad 0] \begin{bmatrix} 5 \\ 4 \\ 6 \\ 2 \end{bmatrix} = 4$$

FIG. 8.11.9

The minor product of any e_i vector by a unit vector is unity. Thus

$$e_i'1 = 1'e_i = 1 \tag{8.11.11}$$

It is shown in numerical form in Fig. 8.11.10.

$$[0 \quad 1 \quad 0 \quad 0] \begin{bmatrix} 1 \\ 1 \\ 1 \\ 1 \\ 1 \end{bmatrix} = 1$$

FIG. 8.11.10

The major product of an e_i vector and an e'_j vector is an e_{ij} matrix. You will recall that an e_{ij} matrix is one all of whose elements are 0 except one, which is unity. Figure 8.11.11 shows the major product of an e_i vector and an e'_j vector. You will notice in Fig. 8.11.11 that the position of the 1 in the prefactor determines its row in the matrix, whereas the position of the 1 in the postfactor determines the column position in the matrix.

$$\begin{bmatrix} 0 \\ 1 \\ 0 \\ 0 \end{bmatrix} [1 \quad 0 \quad 0] = \begin{bmatrix} 0 & 0 & 0 \\ 1 & 0 & 0 \\ 0 & 0 & 0 \\ 0 & 0 & 0 \end{bmatrix}$$

FIG. 8.11.11

Notice also that, as in the general case of the major product of two vectors, there is no restriction on the orders of the two factors. Equation (8.11.12) is the algebraic form of the rule for the major product of an e_i and an e'_j vector.

$$e_i e'_j = e_{ij} \tag{8.11.12}$$

Here the equation states that if an e_i column vector is postmultiplied by an e_j row vector, we get an e_{ij} matrix whose unit element is in the ith row and jth column. We therefore have the obvious rule that any e_{ij} matrix may be expressed as the major product of an e_i vector by an e'_j vector.

SUMMARY

1. The minor product of two vectors:
   a. The left factor is a row vector and is called the prefactor. The right factor is a column vector and is called a postfactor. The column vector is said to be premultiplied by the row vector, and the row vector is said to be postmultiplied by the column vector.
   b. The minor product is a scalar quantity that is the sum of products of corresponding elements of the two vectors:

$$V'_a V_b = [a_1 \quad a_2 \quad \ldots \quad a_n] \begin{bmatrix} b_1 \\ b_2 \\ \ldots \\ b_n \end{bmatrix} = a_1 b_1 + a_2 b_2 + \ldots + a_n b_n$$

c. The minor product is sometimes called scalar product.

d. The minor product of two vectors is defined only for vectors of equal order.

2. Transpose of minor product:

a. It is equal to the minor product since it is a scalar.

b. It is equal to the product of the transposes in reverse order, that is,

$$V_a' V_b = V_b' V_a$$

3. The major product of two vectors:

a. A column vector is postmultiplied by a row vector, or a row vector is premultiplied by a column vector, that is, $V_a V_b'$ is a major product.

b. In a major product of vectors having no zero elements the product is a matrix in which the ijth element is the product of the ith element of the prefactor and the jth element of the postfactor.

$$V_a V_b' = \begin{bmatrix} a_1 \\ a_2 \\ \cdots \\ a_n \end{bmatrix} [b_1 \quad b_2 \quad \cdots \quad b_m] = \begin{bmatrix} a_1 b_1 & a_1 b_2 & \cdots & a_1 b_m \\ a_2 b_1 & a_2 b_2 & \cdots & a_2 b_m \\ \cdots & \cdots & \cdots & \cdots \\ a_n b_1 & a_n b_2 & \cdots & a_n b_m \end{bmatrix}$$

c. Rows are all proportional to one another, as are all columns.

d. The order of the major product is the height of the prefactor and the width of the postfactor. For a major product of two vectors no restrictions are placed on the order of either vector.

4. Transpose of major product:

a. It is equal to the product of the transposes in reverse order.

b. $(V_a V_b')' = V_b V_a'$

5. Product of a vector by its transpose:

a. The minor product of a vector by its transpose we call the minor product moment of the vector. It is the sum of squares of the elements of the vector.

b. The major product of a vector by its transpose we call the major product moment. It is a symmetric matrix.

6. Undefined products of vectors:

a. A row vector cannot be multiplied by a row vector nor a column vector by a column vector.

b. One of the factors *must* be a row and the other *must* be a column.

7. The commutative law:

a. The factors in a major product of vectors can be commuted only if they are of the same order; the product is then a scalar rather than a matrix

b. The factors of a minor product of vectors can be commuted but the product is then a matrix rather than a scalar.

c. Vector multiplication is noncommutative, that is,

$$V_a'V_b \neq V_bV_a'$$

even if $V_a'V_b$ is defined.

8. Minor products of type I supervectors:
   a. The minor product of two type I supervectors can be expressed if they are of the same super order and their corresponding subvectors are of the same order.
   b. The minor product of two type I supervectors is the sum of the minor products of corresponding subvectors.

9. Major products of type I supervectors:
   a. No restrictions are placed on the superorders of either factor or on the orders of their subvectors.
   b. The ijth matric element of the product is the product of the ith column subvector of the prefactor postmultiplied by the jth row subvector of the postfactor.
   c. The product is a supermatrix in which the number of matric rows is the superorder of the prefactor and the number of matric columns is the superorder of the postfactor.

10. Transpose of the major product of type I supervectors: It is equal to the product of the transposes in reverse order.

11. Special vector products:
   a. *The unit vector.*
      (1) the minor product of a unit vector with a general vector is the sum of the elements in the latter.
      (2) The minor product moment of a unit vector is the order of the vector.
      (3) The major product of a unit vector postmultiplied by a general vector is a matrix all of whose rows are equal.
      (4) The major product of a unit vector premultiplied by a general vector is a matrix all of whose columns are equal.
      (5) The major product of two unit vectors is a matrix all of whose elements are 1.
   b. *e_i vectors.*
      (1) The minor product:
         (a) With a unit vector, it is 1:

$$e_i'1 = 1'e_i = 1$$

         (b) With a general vector, it is the ith element of the vector:

$$e_i'V = V'e_i = V_i$$

         (c) With itself or minor product moment, it is 1:

$$e_i'e_i = 1$$

(2) The major product:

    (a) With general vector, if e_i is the prefactor, it is a matrix all of whose rows are 0 except the ith, which is the vector; if e_i is the postfactor, it is a matrix all of whose columns are 0 except the ith, which is the vector.

    (b) With another e'_i vector, it is a matrix whose elements are all 0 except the ijth, which is 1.

EXERCISES

1. (a) Express as the product moment of a vector

$$4 + 36 + 25 = 65$$

    (b) What kind of product moment is this?

2. Express as the product of a vector and a unit vector in two ways

$$1 + 3 + 4 + 2 = 10$$

3. Express as the product of a vector and unit vector

$$a = \begin{bmatrix} 3 & 2 & 1 & 4 \\ 3 & 2 & 1 & 4 \\ 3 & 2 & 1 & 4 \end{bmatrix}$$

4. Express as the product of a vector and unit vector

$$b = \begin{bmatrix} 4 & 4 & 4 \\ 6 & 6 & 6 \\ 2 & 2 & 2 \\ 3 & 3 & 3 \end{bmatrix}$$

5. Express the number 4 as the product of unit vectors.

6. Express as the product of unit vectors

(a) $x = \begin{bmatrix} 1 & 1 & 1 & 1 & 1 \\ 1 & 1 & 1 & 1 & 1 \\ 1 & 1 & 1 & 1 & 1 \end{bmatrix}$ (b) $y = \begin{bmatrix} 1 & 1 & 1 \\ 1 & 1 & 1 \\ 1 & 1 & 1 \\ 1 & 1 & 1 \end{bmatrix}$

7. Express the number 1 as the product of e_i vectors.

8. Express in two ways the scalar 0 as the product of an e_1 and e_2 vector.

9. Given the matrices

$$w = \begin{bmatrix} 0 & 0 & 0 & 0 \\ 0 & 0 & 1 & 0 \\ 0 & 0 & 0 & 0 \\ 0 & 0 & 0 & 0 \\ 0 & 0 & 0 & 0 \end{bmatrix} \qquad y = \begin{bmatrix} 0 & 0 & 0 & 0 \\ 1 & 0 & 0 & 0 \\ 0 & 0 & 0 & 0 \end{bmatrix}$$

(a) Express x as the product of e_i vectors using numerical subscripts.
(b) Do the same for y.
(c) What are the orders respectively of the pre- and postfactors of x?
(d) Of y?

10. (a) Express in two ways the number 1 as the product of a unit vector and an e_2 vector
(b) What kind of products are these?

11. (a) Express as the product of vectors

$$a = \begin{bmatrix} 0 & 2 & 0 \\ 0 & 3 & 0 \\ 0 & 1 & 0 \\ 0 & 4 & 0 \end{bmatrix}$$

(b) What kind of product is it?

12. (a) Express as the product of vectors

$$b = \begin{bmatrix} 0 & 0 & 0 \\ 0 & 0 & 0 \\ 2 & 3 & 5 \end{bmatrix}$$

(b) What kind of product is it?

13. (a) Express as a product moment of vectors

$$a = \begin{bmatrix} 1 & 2 & 3 \\ 2 & 4 & 6 \\ 3 & 6 & 9 \end{bmatrix}$$

(b) What kind of product moment is it?

14. Given e_i is a third-order vector and

$$e_1'V = 2 \qquad e_2'V = 3 \qquad e_3'V = 1$$

What is V'?

15. e_1, e_2, e_3 and e_4 are all fourth-order vectors
(a) Express the sum of their major product moments by the appropriate symbol with subscript to indicate order
(b) What is the sum of their minor product moments?

ANSWERS

1.
(a) $[2 \quad 6 \quad 5] \begin{bmatrix} 2 \\ 6 \\ 5 \end{bmatrix} = 65$ (b) Minor product moment

2.
$[1 \quad 1 \quad 1 \quad 1] \begin{bmatrix} 1 \\ 3 \\ 2 \\ 4 \end{bmatrix} = 10$ $[1 \quad 3 \quad 2 \quad 4] \begin{bmatrix} 1 \\ 1 \\ 1 \\ 1 \end{bmatrix} = 10$

3.

$$a = \begin{bmatrix} 1 \\ 1 \\ 1 \end{bmatrix} [3 \quad 2 \quad 1 \quad 4]$$

4.

$$b = \begin{bmatrix} 4 \\ 6 \\ 2 \\ 3 \end{bmatrix} [1 \quad 1 \quad 1]$$

5.

$$[1 \quad 1 \quad 1 \quad 1] \begin{bmatrix} 1 \\ 1 \\ 1 \\ 1 \end{bmatrix} = 4$$

6.

(a) $x = \begin{bmatrix} 1 \\ 1 \\ 1 \end{bmatrix} [1 \quad 1 \quad 1 \quad 1 \quad 1]$ (b) $y = \begin{bmatrix} 1 \\ 1 \\ 1 \\ 1 \end{bmatrix} [1 \quad 1 \quad 1]$

or

(a) $x = 1_3 1_5'$ (b) $y = 1_4 1_3'$

7. $1 = e_1' e_1$

8. $e_1' e_2 = 0$ $e_2' e_1 = 0$

9. (a) $x = e_2 e_3'$ (b) $y = e_2 e_1'$ (c) 5, 4   (d) 3, 4

10. (a) $e_2' 1 = 1$ $1' e_2 = 1$ (b) Minor products (scalar products)

11.

(a) $a = \begin{bmatrix} 2 \\ 3 \\ 1 \\ 4 \end{bmatrix} [0 \quad 1 \quad 0]$ (b) major

12.

(a) $b = \begin{bmatrix} 0 \\ 0 \\ 1 \end{bmatrix} [2 \quad 3 \quad 5]$ (b) major

13.

(a) $a = \begin{bmatrix} 1 \\ 2 \\ 3 \end{bmatrix} [1 \quad 2 \quad 3]$ (b) major

14. $V' = [2 \quad 3 \quad 1]$

15. (a) I_4 (b) 4

Chapter 9

Matrix Multiplication

9.1 Matrices Expressed as Type II Supervectors

The *a* matrix as a type II supervector

Before considering the subject of matrix multiplication, we shall review the notation for matrices expressed as type II supervectors (Section 5.9). The rules for vector multiplication given in Chapter 8, can be applied to matrices in general if the matrices are regarded as supervectors. First, let us review how the matrix a may be expressed as a type II supervector in row form.

$$a = [a_{.1} \quad a_{.2} \quad \ldots \quad a_{.m}]_n \tag{9.1.1}$$

You will recall that the $a_{.i}$ elements on the right are the column vectors of the matrix a, that the internal subscript m means that there are m columns, and the external subscript n means there are n rows. You also recall that we can write a matrix as a type II column supervector as in Eq. (9.1.2).

$$a = \begin{bmatrix} a'_{1.} \\ a'_{2.} \\ \ldots \\ a'_{n.} \end{bmatrix}_m \tag{9.1.2}$$

Remember that the elements on the right of Eq. (9.1.2) are the row vectors of the matrix a, that the internal subscript n means that there are n rows and the external subscript m means there are m columns.

The transpose of Eq. (9.1.1) is then Eq. (9.1.3).

$$a' = \begin{bmatrix} a'_{.1} \\ a'_{.2} \\ \ldots \\ a'_{.m} \end{bmatrix}_n \tag{9.1.3}$$

The transpose of Eq. (9.1.2) is

$$a' = [a_{1.} \quad a_{2.} \quad \ldots \quad a_{n.}]_m \tag{9.1.4}$$

142

The b matrix as a type II supervector

Suppose now that we have another matrix b with t rows and s columns. We can write b in row supervector form as

$$b = [b_{.1} \quad b_{.2} \quad \ldots \quad b_{.s}]_t \qquad (9.1.5)$$

We write it as a column supervector as

$$b = \begin{bmatrix} b'_{1.} \\ b'_{2.} \\ \ldots \\ b'_{t.} \end{bmatrix}_s \qquad (9.1.6)$$

The transpose of Eq. (9.1.5) is

$$b' = \begin{bmatrix} b'_{.1} \\ b'_{.2} \\ \ldots \\ b'_{.s} \end{bmatrix}_t \qquad (9.1.7)$$

The transpose of Eq. (9.1.6) is

$$b' = [b_{1.} \quad b_{2.} \quad \ldots \quad b_{t.}]_s \qquad (9.1.8)$$

9.2   The Product of Two Matrices Expressed as the Minor Product of Type II Supervectors

The product of matrix a and matrix b can now be considered to be the minor product of type II supervectors. This means, of course, that we must have a row vector on the left and a column vector on the right. We therefore use the right side of Eq. (9.1.1) for the a matrix and the right side of Eq. (9.1.6) for the b matrix. This gives us Eq. (9.2.1).

$$ab = [a_{.1} \quad a_{.2} \quad \ldots \quad a_{.m}]_n \begin{bmatrix} b'_{1.} \\ b'_{2.} \\ \ldots \\ b'_{t.} \end{bmatrix}_s \qquad (9.2.1)$$

To get the product of the vectors on the right side of Eq. (9.2.1), we use the same rule as in multiplying simple vectors. We take the sum of the product of corresponding elements from the two vectors. This gives us

$$ab = [a_{.1}b'_{1.} + a_{.2}b'_{2.} + \ldots + a_{.m}b'_{t.}]_{ns} \qquad (9.2.2)$$

Now notice carefully three things about the order of the symbols in Eqs. (9.2.1) and (9.2.2). First, on the left of both equations, a precedes b. The prefactor is a and the postfactor is b. Second, on the right of Eq. (9.2.1) the row vector for a precedes the column vector for b; that is, the row vector for a is the prefactor and column vector for b is the postfactor. Third, on the right in Eq. (9.2.2), $a_{.i}$ always precedes $b'_{i.}$. Once we start out with a given order, we must be very careful to preserve this order consistently. Keep this in mind until you learn certain exceptions to this rule and how to use them.

Notice now that since each term on the right side of Eq. (9.2.2) is the product of two vectors, the first of which represents a column vector from a; the second, a row vector from b; this product, expanded, or in number form, is a matrix, and thus a major product. We see, therefore, that the product of two matrices may be expressed as a sum of major products of vectors. A numerical example of this rule should make the point clear. Let us represent the two matrices a and b as in Fig. 9.2.1.

$$a = \begin{matrix} (a_{.1})(a_{.2}) \\ \begin{bmatrix} 2 & 1 \\ 3 & 5 \\ 6 & 1 \end{bmatrix} \end{matrix} \qquad b = \begin{bmatrix} 1 & 2 \\ 3 & 1 \end{bmatrix} \begin{matrix} (b'_{1.}) \\ (b'_{2.}) \end{matrix}$$

Fig. 9.2.1

In Fig. 9.2.2, we show the product, $a \times b$, as the sum of two major products of vectors. In Fig. 9.2.3, we actually multiply out the product of the two pairs of vectors.

$$ab = \begin{matrix} (a_{.1} \times b'_{1.}) \\ \begin{bmatrix} 2 \\ 3 \\ 6 \end{bmatrix} \begin{bmatrix} 1 & 2 \end{bmatrix} \end{matrix} + \begin{matrix} (a_{.2} \times b'_{2.}) \\ \begin{bmatrix} 1 \\ 5 \\ 1 \end{bmatrix} \begin{bmatrix} 3 & 1 \end{bmatrix} \end{matrix}$$

Fig. 9.2.2

$$ab = \begin{bmatrix} 2 & 4 \\ 3 & 6 \\ 6 & 12 \end{bmatrix} + \begin{bmatrix} 3 & 1 \\ 15 & 5 \\ 3 & 1 \end{bmatrix} = \begin{bmatrix} 5 & 5 \\ 18 & 11 \\ 9 & 13 \end{bmatrix}$$

Fig. 9.2.3

The product of two matrices defined as the sum of major products of vectors is generally not given in textbooks on matrix algebra. However, this definition has been found very useful in the analysis of data from the behavioral sciences. It is particularly important in high-speed computer programming.

9.3 The Product of Two Matrices Expressed as the Major Product of Type II Supervectors

A much more common way to express the product of two matrices than the one given above is as the major product of type II supervectors. This means that the product $a \times b$ would be expressed as a column supervector postmultiplied by a row supervector. We therefore take the column supervector form of a given in Eq. (9.1.2) and the row supervector form of b given in Eq. (9.1.5) and express the product as in Eq. (9.3.1)

$$ab = \begin{bmatrix} a'_{1.} \\ a'_{2.} \\ \cdots \\ a'_{n.} \end{bmatrix}_m [b_{.1} \quad b_{.2} \quad \cdots \quad b_{.s}]_t \tag{9.3.1}$$

To get the product of the two supervectors on the right of Eq. (9.3.1), we use the same rule for getting the major product of simple vectors. This gives us

$$ab = \begin{bmatrix} a'_{1.}b_{.1} & a'_{1.}b_{.2} & \cdots & a'_{1.}b_{.s} \\ a'_{2.}b_{.1} & a'_{2.}b_{.2} & \cdots & a'_{2.}b_{.s} \\ \cdots & \cdots & \cdots & \cdots \\ a'_{n.}b_{.1} & a'_{n.}b_{.2} & \cdots & a'_{n.}b_{.s} \end{bmatrix} \tag{9.3.2}$$

Remember that each of the elements in the product matrix represents a row vector postmultiplied by a column vector. Therefore the right side of Eq. (9.3.2) is a matrix whose elements are the scalar or minor products of vectors. Notice also that a comes before b on the left side of Eq. (9.3.2). So also the $a'_{i.}$ comes before $b_{.j}$ on the right side of Eq. (9.3.2).

We may take the same example given in Fig. 9.2.1 to show numerically how the product of two matrices may be expressed as a matrix whose elements are the minor products of vectors. Figure 9.3.1 gives the product matrix before the calculation of minor products of the vectors.

$$ab = \begin{bmatrix} [2 \quad 1]\begin{bmatrix} 1 \\ 3 \end{bmatrix} & [2 \quad 1]\begin{bmatrix} 2 \\ 1 \end{bmatrix} \\ [3 \quad 5]\begin{bmatrix} 1 \\ 3 \end{bmatrix} & [3 \quad 5]\begin{bmatrix} 2 \\ 1 \end{bmatrix} \\ [6 \quad 1]\begin{bmatrix} 1 \\ 3 \end{bmatrix} & [6 \quad 1]\begin{bmatrix} 2 \\ 1 \end{bmatrix} \end{bmatrix}$$

Fig. 9.3.1

In Fig. 9.3.2, the multiplication for each of the elements is carried out.

$$ab = \begin{bmatrix} 2 \times 1 + 1 \times 3 & 2 \times 2 + 1 \times 1 \\ 3 \times 1 + 5 \times 3 & 3 \times 2 + 5 \times 1 \\ 6 \times 1 + 1 \times 3 & 6 \times 2 + 1 \times 1 \end{bmatrix} = \begin{bmatrix} 5 & 5 \\ 18 & 11 \\ 9 & 13 \end{bmatrix}$$

Fig. 9.3.2

You see that Fig. 9.3.2 gives exactly the same answer as Fig. 9.2.3, but we arrive at this result by two distinctly different computational procedures.

The minor product form for the product of two matrices is by far the most common among mathematicians. We shall, however, have occasion to use both the minor and the major products of type II supervectors for the product of matrices.

9.4   Reversing the Order of the Factors

The minor product form

In the previous section, we considered the product of the matrices a and b where a was the prefactor and b the postfactor. Let us now consider the case where b is the prefactor and a the postfactor. First let us express this product in reverse order as the minor product of supervectors. For the b matrix, we use the right side of Eq. (9.1.5) and for the a matrix the right side of Eq. (9.1.2). This gives Eq. (9.4.1).

$$ba = [b_{.1} \quad b_{.2} \quad \ldots \quad b_{.s}]_t \begin{bmatrix} a'_{1.} \\ a'_{2.} \\ \ldots \\ a'_{n.} \end{bmatrix}_m \tag{9.4.1}$$

If we multiply out the right side of Eq. (9.4.1), we get

$$ba = [b_{.1}a'_{1.} + b_{.2}a'_{2.} + \ldots + b_{.s}a'_{n.}]_{tm} \tag{9.4.2}$$

The major product form

Next we express the product $b \times a$ as the major product of supervectors. We use the right side of Eq. (9.1.6) for the b matrix and the right side of Eq. (9.1.1) for the a matrix. This gives us Eq. (9.4.3).

$$ba = \begin{bmatrix} b'_{1.} \\ b'_{2.} \\ \ldots \\ b'_{t.} \end{bmatrix}_s [a_{.1} \quad a_{.2} \quad \ldots \quad a_{.m}]_n \tag{9.4.3}$$

If we multiply out the right side of Eq. (9.4.3) we get

$$ba = \begin{bmatrix} b'_{1.}a_{.1} & b'_{1.}a_{.2} & \ldots & b'_{1.}a_{.m} \\ b'_{2.}a_{.1} & b'_{2.}a_{.2} & \ldots & b'_{2.}a_{.m} \\ \ldots & \ldots & \ldots & \ldots \\ b'_{t.}a_{.1} & b'_{t.}a_{.2} & \ldots & b'_{t.}a_{.m} \end{bmatrix} \tag{9.4.4}$$

9.5   Dimensions of Factors and Their Product

We are now ready to draw some conclusions about restrictions on the orders of two matrices that are to be multiplied together, and also the order of their product.

The common order

If we look at the right side of Eq. (9.2.2), the minor product, we see that each column vector of a is paired with a row vector of b. Therefore, where a is the prefactor and b is the postfactor, in order for Eq. (9.2.2) to make any sense, we must have the same number of columns in a as we have rows in b. Then if $a \times b$ exists, Eq. (9.2.2) shows that m equals t. The same point can be made from the right side of Eq. (9.3.2), the major product. From Eq. (9.3.1) you can see that each of the row vectors from a has m elements in it, and each of the column vectors from b has t elements in it. But in Eq. (9.3.2) each element is a row vector from a postmultiplied by a column vector from b. Therefore the rows and columns must each have the same number of elements. Thus Eq. (9.3.2) also proves that m equals t if it is possible to postmultiply a by b.

Now look at Eq. (9.4.2). Here the prefactor is b and the postfactor is a. In order for Eq. (9.4.2) to make any sense, the number of columns in b must be equal to the number of rows in a; therefore, in this case, s must equal n. The same rule can also be deduced from Eq. (9.4.4). Here each element is a row vector from b postmultiplied by a column vector from a, but Eq. (9.4.3) shows that the row vectors from b have s elements and the column vectors from a each have n elements. If it is possible to postmultiply a row vector from b by a column vector from a, then they must both have the same number of elements. Therefore s must be equal to n, if b times a exists.

We are now ready to state the general rule that two matrices can be multiplied together if, and only if, the number of columns of the prefactor is equal to the number of rows of the postfactor. To put it another way, the width of the prefactor must be the height of the postfactor. When these two dimensions are equal they may be called the *common dimension* or *common order* of the matrices. Therefore we have the rule that the common order of the two matrices of a product is the width (columns) of the left-hand matrix (prefactor) and the height (rows) of the right-hand matrix (postfactor).

If two matrices do not have a common order, we say that they are *nonconformable*. If they have a common order, we say that they are *conformable*.

The order of the product

It is easy to see now what the order of a product is in terms of the orders of the factors. Equation (9.3.2) shows that the product ab has n rows and s columns; or that it is of order $n \times s$. Equation (9.4.4) shows that the product ba has t rows and m columns or that it is of the order t by m. But in Eq. (9.3.2), n is the number of rows in the prefactor and s is the number of columns in the postfactor. Similarly, in Eq. (9.4.4), t is the number of rows in the prefactor and m is the number of columns in the postfactor. Therefore, the number of rows in the product of two matrices is the number of rows in the prefactor, and the number of columns in the product is the number of columns in the postfactor. These numbers we call the *distinct dimensions* or *distinct orders* of the factors. We therefore conclude that any two matrices may be multiplied together provided they are conformable, no matter how many rows the left-hand matrix has or how many columns the right-hand matrix has.

We also see that the order of a product of two matrices does not indicate what the common dimension of the two factors is. All we know about the order of two factors from their product is their distinct orders and the fact that they do have a common dimension.

The rule for the dimensions of factors and the dimensions of products is summarized by Fig. 9.5.1.

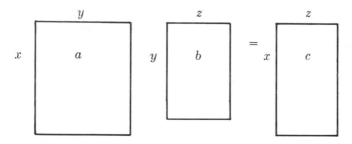

FIG. 9.5.1

Here you see that the height of the matrix a is x; therefore, the height of the product c is x. The width of the matrix b is z; therefore, the width of the product c is z. The width of the matrix a is y; therefore, the height of the matrix b is y.

Another way to remember the dimension rules for multiplication is to use the subscript notation to indicate the order of the matrices. Thus, we can indicate the product of Fig. 9.5.1 by the equation in Fig. 9.5.2.

$$a_{xy}b_{yz} = c_{xz}$$

FIG. 9.5.2

Here you see that the subscript y, which appears twice on the left, is a common dimension and disappears on the right side of the equation. Also, on the left, x is the first of the four subscripts and on the right, x is the first of the two subscripts, whereas, on the left, z is the last of the four subscripts, and on the right, z is the last of the two subscripts. When subscripts are used to indicate the order of the matrices, the last subscript of the prefactor must always be the same as the first subscript of the postfactor. Also, the first subscript of the prefactor must always be the first subscript of the product, and the last subscript of the postfactor must always be the last subscript of the product.

9.6 The Commutative Law of Matrix Multiplication

You have seen that the commutative law for the addition of matrices holds. You also know that the commutative law for the multiplication of scalar quantities and numbers hold. Now let us consider matrix multiplication. Does the commutative law still hold? In other words, is $a \times b$ equal to $b \times a$? You have seen this is not true for vector multiplication. To answer the question, let us compare the ijth element of ab with the ijth element of ba. From Eq. (9.3.2), we see that the ijth element of ab is $a'_{i.}b_{.j}$. From Eq. (9.4.4), we see that the ijth element of ba is $b'_{i.}a_{.j}$. If we indicate the product ab by c,

$$ab = c \tag{9.6.1}$$

and the product ba by f,

$$ba = f \tag{9.6.2}$$

then, the ijth element of c in Eq. (9.6.1) is given by

$$c_{ij} = a'_{i.}b_{.j} \tag{9.6.3}$$

Similarly, the ijth element of f in (9.6.2) is given by

$$f_{ij} = b'_{i.}a_{.j} \tag{9.6.4}$$

Now the right side of Eq. (9.6.2) is the scalar product of the ith row of a and the jth column of b, whereas the right side of (9.6.4) is the scalar product of the ith row of b and the jth column of a. These two would not in general be equal, since they consist of entirely different elements. Therefore, the right sides of Eqs. (9.6.3) and (9.6.4) are not in general equal, and consequently the left sides would not in general be equal. We can, therefore, write Eq. (9.6.5) to show that the two are not equal.

$$c_{ij} \neq f_{ij} \tag{9.6.5}$$

Equation (9.6.5) shows that corresponding elements of the matrices c and f in Eqs. (9.6.1) and (9.6.2) are not in general equal; therefore, c and f are not in general equal, and we can write the inequality

$$c \neq f \tag{9.6.6}$$

Substituting the left side of Eqs. (9.6.1) and (9.6.2) into the left and right sides respectively of Eq. (9.6.6), we have the inequality

$$ab \neq ba \tag{9.6.7}$$

We must conclude, therefore, that the commutative law for multiplication does not hold, in general, for matrix algebra as it does for scalar algebra. In other words, we say that *matrix multiplication is noncommutative*. There are certain special cases when the commutative law of multiplication does hold in matrix algebra, some of which we shall consider later.

A numerical example confirms that matrix multiplication could not in general be commutative. We may let the matrices a and b be given as in Fig. 9.6.1.

$$a = \begin{bmatrix} 3 & 2 \\ 7 & 5 \end{bmatrix} \qquad b = \begin{bmatrix} 2 & 4 \\ 3 & 1 \end{bmatrix}$$

FIG. 9.6.1

Then the product ab is given by Fig. 9.6.2.

$$ab = \begin{bmatrix} 3 & 2 \\ 7 & 5 \end{bmatrix} \begin{bmatrix} 2 & 4 \\ 3 & 1 \end{bmatrix} = \begin{bmatrix} 12 & 14 \\ 29 & 33 \end{bmatrix}$$

FIG. 9.6.2

On the other hand, the product ba is given by

$$ba = \begin{bmatrix} 2 & 4 \\ 3 & 1 \end{bmatrix} \begin{bmatrix} 3 & 2 \\ 7 & 5 \end{bmatrix} = \begin{bmatrix} 34 & 24 \\ 16 & 11 \end{bmatrix}$$

FIG. 9.6.3

The right sides of Figs. 9.6.2 and 9.6.3 are obviously not equal, and if you actually multiply out the appropriate rows from the prefactor by the appropriate columns from the postfactor, you can see why they could not be equal.

9.7 The Associative Law of Multiplication

We saw in Sections 7.2 and 7.3 how the associative law holds for the addition and subtraction of matrices. Now let us consider the law as applied to multiplication of more than two matrices. First, note that in arithmetic and scalar algebra, of course, any number of quantities may be multiplied. In matrix algebra, any number of matrices may be multiplied, provided adjacent pairs of factors have a common order.

Therefore, as you might suppose, the number of rows in the first, or left-hand, factor of a series of factors determines the number of rows in the product. The number of columns in the last, or right-hand, factor of the series determines the number of columns in the product. The orders of the factors between the first and last factor in the series have nothing whatever to do with the order of the product.

Figure 9.7.1 illustrates diagrammatically how the order of the product is determined only by the first and last factors and how adjacent pairs of factors must be conformable.

FIG. 9.7.1

In Fig. 9.7.1, you see that the width of a preceding matrix is the same as the height of the matrix immediately following. For example, the second matrix v has a width of c and the third matrix w has a height of c. You also see that the first matrix u in the series has a height of a and the product matrix y has a height of a. Similarly, the last matrix x has a width of e and the product matrix y has a width of e.

We can illustrate the same rule for order or dimensionality by using the subscript notation to indicate the order of the matrices as in

$$u_{ab}v_{bc}w_{cd}x_{de} = y_{ae}$$

FIG. 9.7.2

Here you see that the last subscript of a matrix is the same as the first subscript of the following matrix. You also see that the first subscript of the first matrix in the series is the same as the first subscript of the product matrix, and the last subscript of the last matrix in the series is the same as the last subscript of the product matrix.

Let us take a numerical example, Fig. 9.7.3, to show how we can get the product of more than two matrices.

$$\begin{bmatrix} 2 & 3 \\ 1 & 2 \end{bmatrix} \begin{bmatrix} 3 & 4 \\ 1 & 2 \end{bmatrix} \begin{bmatrix} 5 & 2 \\ 1 & 3 \end{bmatrix} = x$$
$$\quad a \qquad\quad b \qquad\quad c$$

FIG. 9.7.3

Now there are two ways that we can get the product here. First, let us get the product ab and call this r so that we can have Fig. 9.7.4.

$$ab = r = \begin{bmatrix} 2 & 3 \\ 1 & 2 \end{bmatrix} \begin{bmatrix} 3 & 4 \\ 1 & 2 \end{bmatrix} - \begin{bmatrix} 9 & 14 \\ 5 & 8 \end{bmatrix}$$

FIG. 9.7.4

Then let us multiply the matrix r in Fig. 9.7.4 by the postfactor c. This gives us the product of all three matrices, as shown in Fig. 9.7.5.

$$rc = \begin{bmatrix} 9 & 14 \\ 5 & 8 \end{bmatrix} \begin{bmatrix} 5 & 2 \\ 1 & 3 \end{bmatrix} = \begin{bmatrix} 59 & 60 \\ 33 & 34 \end{bmatrix} = x$$

Fig. 9.7.5

We can obtain this triple product a second way. First, we can get the product bc, which we may call s as in Fig. 9.7.6.

$$bc = s = \begin{bmatrix} 3 & 4 \\ 1 & 2 \end{bmatrix} \begin{bmatrix} 5 & 2 \\ 1 & 3 \end{bmatrix} = \begin{bmatrix} 19 & 18 \\ 7 & 8 \end{bmatrix}$$

Fig. 9.7.6

Then we can premultiply s in Fig. 9.7.6 by the matrix a to get the triple product, as shown in Fig. 9.7.7.

$$as = \begin{bmatrix} 2 & 3 \\ 1 & 2 \end{bmatrix} \begin{bmatrix} 19 & 18 \\ 7 & 8 \end{bmatrix} = \begin{bmatrix} 59 & 60 \\ 33 & 34 \end{bmatrix} = x$$

Fig. 9.7.7

Now you see that Figs. 9.7.5 and 9.7.7 give exactly the same answer. It does not matter whether we first premultiply b by a and then postmultiply the product by c, or whether we first premultiply c by b and then premultiply the products by a. So long as we do not change the order of the matrices in the series, the product is the same. We can state this principle in algebraic form for the case of three factors as

$$(ab)c = a(bc) \tag{9.7.1}$$

If we have four factors in a product, the various combinations shown in Fig. 9.7.8 can be used.

(1) $abcd = a \ b \ c \ d$

(2) $abcd = a \ b \ c \ d$

(3) $abcd = a \ b \ c \ d$

(4) $abcd = a \ b \ c \ d$

(5) $abcd = a \ b \ c \ d$

Fig. 9.7.8

The first equation in Fig. 9.7.8 means that first the product ab is formed, second the product cd is formed, and third, the product of these two products is formed. The second equation in Fig. 9.7.8 means that first the product ab is formed, second the product of this product with c is formed, and third, the product of this triple product with d is formed. The other equations in Fig. 9.7.8 are interpreted in the same way. You see then that it does not matter how we form the multiple product for more than two factors, as long as we continue to multiply adjacent pairs of matrices and do not change the order of any matrix in the original series. Therefore, the rule is clear that matrix multiplication is associative for any number of factors in the product.

9.8 The Order of Multiplication for Multiple Products

Since matrix multiplication is associative for any number of factors, we may get the product by any combination method, so long as we do not change the order of the factors. Nevertheless, one method may be preferable because it requires least time and work.

To determine the most economical order of multiplication, we first need a rule to determine the amount of work required in getting the product of any two matrices. It is easy to show that the number of scalar multiplications involved in multiplying two matrices together is the triple product of their common dimension and their distinct dimensions. For example, let us indicate the orders of the matrices by subscript notation and let the product of two matrices a and b be given as c as in Fig. 9.8.1.

$$a_{xy}b_{yz} = c_{xz}$$

Fig. 9.8.1

The matrix a has x rows; therefore, the matrix c has x rows. The matrix b has z columns; therefore, the matrix c has z columns. The total number of elements in the product c is xz. But to get each element of c requires the multiplication of y different pairs of scalar quantities since y is the common dimension of a and b. Therefore, the total number of multiplications required to get the elements in c is xyz. But this, as we have seen, is the triple product of the distinct dimensions and the common dimensions of the factors. This principle is illustrated in Fig. 9.8.2.

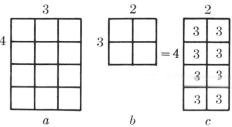

Fig. 9.8.2

Here we see that a is a 4 by 3 matrix and b is a 3 by 2 matrix; therefore, the product c is a 4×2 matrix. But since the common dimension of a and b is 3, each element in the product matrix c requires the multiplication of three pairs of elements. Since there are 4×2 of these elements, the total number of scalar multiplications is $4 \times 3 \times 2$ or 24 scalar multiplications.

In determining the most economical sequence of multiplications for a series of factors we will not go far wrong if we consider only the total number of scalar multiplications for any particular sequence. It is true that scalar additions are also involved but these take much less time than do the scalar multiplications. This is true whether desk calculators or high-speed electronic computers are used.

9.9 The Associative Law and Algebraic Checks

Numerical computations may be checked by two general methods. You may simply do the work over again to see if you get the same results, or you may use algebraic checks. The associative law can be applied to the use of algebraic checks in computational work. Let us first take a very simple example of what is commonly known as *cross footing*. Suppose we have a simple table of numbers as in Fig. 9.9.1.

$$
\begin{array}{ccc|c}
3 & 4 & 5 & 12 \\
2 & 1 & 6 & 9 \\
4 & 3 & 2 & 9 \\
5 & 7 & 1 & 13 \\
\hline
14 & 15 & 14 & 43
\end{array}
$$

FIG. 9.9.1

Here the last or fourth column on the right of the table is the sum of the numbers in the rows, and the bottom row is the sum for each column. Notice that the sum for the fourth column is 43 and the sum for the last row is also 43. This number 43 therefore checks both the row sums and the column sums. Let us see now how the associative law of multiplication applies to a check of this sort. Suppose we let x be the original table of numbers and we postmultiply it by a column unit vector as in Fig. 9.9.2.

$$
x1 = \begin{bmatrix} 3 & 4 & 5 \\ 2 & 1 & 6 \\ 4 & 3 & 2 \\ 5 & 7 & 1 \end{bmatrix} \begin{bmatrix} 1 \\ 1 \\ 1 \end{bmatrix} = \begin{bmatrix} 12 \\ 9 \\ 9 \\ 13 \end{bmatrix}
$$

FIG. 9.9.2

Next let us premultiply x by a row unit vector, as in Fig. 9.9.3.

$$1'x = \begin{bmatrix} 1 & 1 & 1 & 1 \end{bmatrix} \begin{bmatrix} 3 & 4 & 5 \\ 2 & 1 & 6 \\ 4 & 3 & 2 \\ 5 & 7 & 1 \end{bmatrix} = \begin{bmatrix} 14 & 15 & 14 \end{bmatrix}$$

FIG. 9.9.3

You see that Fig. 9.9.2 gives us a column vector of row sums and Fig. 9.9.3 gives a row vector of column sums.

Now let us premultiply the column vector in Fig. 9.9.2 by a row unit vector as in Fig. 9.9.4.

$$1'(x1) = \begin{bmatrix} 1 & 1 & 1 & 1 \end{bmatrix} \begin{bmatrix} 12 \\ 9 \\ 9 \\ 13 \end{bmatrix} = 43$$

FIG. 9.9.4

Let us also postmultiply the row vector in Fig. 9.9.3 by a column unit vector as in Fig. 9.9.5.

$$(1'x)1 = \begin{bmatrix} 14 & 15 & 14 \end{bmatrix} \begin{bmatrix} 1 \\ 1 \\ 1 \end{bmatrix} = 43$$

FIG. 9.9.5

You see that Figs. 9.9.4 and 9.9.5 give the same results. We therefore can write in matrix notation

$$1'(x1) = (1'x)1 \tag{9.9.1}$$

Obviously, then, both the sum of the totals column and the sum of the totals row is a triple product of the row unit vector, the matrix, and the column unit vector in that order. Since the associative law holds for matrix multiplication, it does not matter whether we multiply the last two factors together first and then premultiply by the first factor or whether we multiply the first two factors and then postmultiply by the last factor. This is one of the simplest examples of the application of the associative law to the checking of computations.

Next let us consider a somewhat more complex application of the associative law in checking matrix multiplication. We take as an example the matrices a and b given in Fig. 9.2.1. The product of these matrices is given in Fig. 9.9.6.

$$ab = \begin{bmatrix} 5 & 5 \\ 18 & 11 \\ 9 & 13 \end{bmatrix}$$

Fig. 9.9.6

Let us now postmultiply Fig. 9.9.6 by a column unit vector, as in Fig. 9.9.7,

$$(ab)1 = \begin{bmatrix} 5 & 5 \\ 18 & 11 \\ 9 & 13 \end{bmatrix} \begin{bmatrix} 1 \\ 1 \end{bmatrix} = \begin{bmatrix} 10 \\ 29 \\ 22 \end{bmatrix}$$

Fig. 9.9.7

and next postmultiply the matrix b by a column unit vector, as in Fig. 9.9.8.

$$b1 = \begin{bmatrix} 1 & 2 \\ 3 & 1 \end{bmatrix} \begin{bmatrix} 1 \\ 1 \end{bmatrix} = \begin{bmatrix} 3 \\ 4 \end{bmatrix}$$

Fig. 9.9.8

Then we may postmultiply the matrix a by the vector obtained in Fig. 9.9.8, as in Fig. 9.9.9.

$$a(b1) = \begin{bmatrix} 2 & 1 \\ 3 & 5 \\ 6 & 1 \end{bmatrix} \begin{bmatrix} 3 \\ 4 \end{bmatrix} = \begin{bmatrix} 10 \\ 29 \\ 22 \end{bmatrix}$$

Fig. 9.9.9

If now you compare the vectors on the right side in Figs. 9.9.7 and 9.9.9, you will see that they are exactly the same. From the associative law you know, of course, they must be the same because the left sides must be the same as indicated in

$$(ab)1 = a(b1) \tag{9.9.2}$$

Equation (9.9.2), together with Figs. 9.9.6 through 9.9.9, shows how we use the associative law of multiplication to check the multiplication of the product, ab. After getting the product of the two matrices, we sum the rows of the product to get a column vector as on the right of Fig. 9.9.7. Then we postmultiply the matrix a by a vector of row sums of b as in Fig. 9.9.9. The two resulting vectors must be the same if our computations have been correct.

Another check on the product of the two matrices is

$$1'(ab) = (1'a)b \tag{9.9.3}$$

Let us interpret this equation in terms of actual computations. According to the left side of Eq. (9.9.3), we take the column sums of the product matrix, ab. This is illustrated by Fig. 9.9.10.

$$1'(ab) = [1 \quad 1 \quad 1] \begin{bmatrix} 5 & 5 \\ 18 & 11 \\ 9 & 13 \end{bmatrix} = [32 \quad 29]$$

FIG. 9.9.10

Next we get the row vector of column sums for the a matrix as in Fig. 9.9.11.

$$1'a = [1 \quad 1 \quad 1] \begin{bmatrix} 2 & 1 \\ 3 & 5 \\ 6 & 1 \end{bmatrix} = [11 \quad 7]$$

FIG. 9.9.11

Finally, we postmultiply the row vector of column sums of a in Fig. 9.9.11 by the matrix b as in

$$(1'a)b = [11 \quad 7] \begin{bmatrix} 1 & 2 \\ 3 & 7 \end{bmatrix} = [32 \quad 29]$$

FIG. 9.9.12

Figure 9.9.12 gives the right side of Eq. (9.9.3). If you compare the right sides of Figs. 9.9.10 and 9.9.12 you will see that they are exactly the same, as they should be according to the associative law expressed in Eq. (9.9.3).

9.10 The Distributive Law and Matrix Multiplication

If we take the sum of two or more numbers and then multiply this sum by another number, we get the same results as if we had first multiplied each of the numbers in the sum by this multiplier and then taken the sum of the products. This principle is illustrated by Fig. 9.10.1.

$$2(3 + 4 + 5) = 6 + 8 + 10 = 24$$
$$2(3 + 4 + 5) = 2 \times 12 = 24$$

FIG. 9.10.1

This principle is known as the *distributive law*. The distributive law of scalar algebra or arithmetic also applies to matrix algebra. We can state the distributive law of matrix algebra as

$$a(b + c) = ab + ac \tag{9.10.1}$$

Equation (9.10.1) can be generalized to include the sum of more than two matrices multiplied by another. This is shown in Eq. (9.10.2), where the subscripts inside the parentheses refer to different matrices.

$$a(b_1 + b_2 + \ldots + b_n) = ab_1 + ab_2 + \ldots + ab_n \qquad (9.10.2)$$

Assuming that the orders of the matrices are conformable, we could just as well have written the matrix a as a postfactor as in

$$(b_1 + b_2 + \ldots + b_n)a = b_1a + b_2a + \ldots + b_na \qquad (9.10.3)$$

You notice that on the right of Eq. (9.10.2), the matrix a precedes each of the b matrices while in Eq. (9.10.3), the matrix a is a postfactor for each of the b matrices. The reason for this you will recall is that matrix multiplication is not in general commutative.

As a numerical example of the distributive law for matrix algebra, in Fig. 9.10.2 we have three matrices, a, b, and c.

$$a = \begin{bmatrix} 1 & 2 \\ 2 & 1 \end{bmatrix} \qquad b = \begin{bmatrix} 3 & 1 \\ 4 & 2 \end{bmatrix} \qquad c = \begin{bmatrix} 4 & 1 \\ 2 & 1 \end{bmatrix}$$

Fig. 9.10.2

Substituting from Fig. 9.10.2 into the left side of Eq. (9.10.1) we have Fig. 9.10.3.

$$\underset{a}{\begin{bmatrix} 1 & 2 \\ 2 & 1 \end{bmatrix}} \underset{b+c}{\begin{bmatrix} 7 & 2 \\ 6 & 3 \end{bmatrix}} = \begin{bmatrix} 19 & 8 \\ 20 & 7 \end{bmatrix}$$

Fig. 9.10.3

Substituting from Fig. 9.10.2 into the right side of Eq. (9.10.1), we have Fig. 9.10.4.

$$\underset{ab}{\begin{bmatrix} 11 & 5 \\ 10 & 4 \end{bmatrix}} + \underset{ac}{\begin{bmatrix} 8 & 3 \\ 10 & 3 \end{bmatrix}} = \begin{bmatrix} 19 & 8 \\ 20 & 7 \end{bmatrix}$$

Fig. 9.10.4

We see that the right sides of Figs. 9.10.3 and 9.10.4 are exactly the same, although we used two different methods to get the answer.

In scalar algebra, if we desire the product of a sum of quantities, we ordinarily take the sum first and then multiply by the common multiplier. In terms of matrix algebra, this is the method illustrated in Fig. 9.10.3. It is preferred to the method illustrated in Fig. 9.10.4 because fewer operations are required. That is, ordinarily, we get the sums of the matrices first and then multiply them by the common multiplying matrix, because it is much easier to add matrices than it is to multiply them. In Fig. 9.10.3 we have only one matrix product to compute, whereas in Fig. 9.10.4 we have two.

The rule does not always hold, however. It is possible in special cases that computational economy would result if the multiplications were carried out before the additions.

Thus far, Eqs. (9.10.1), (9.10.2), and (9.10.3) have shown how the distributive law works in matrix algebra if we have the sum of matrices either pre- or postmultiplied by another matrix. We can also have a sum of matrices premultiplied by one matrix and postmultiplied by another. Here, too, the distributive law would hold. This situation would be written as

$$a(b_1 + b_2 + \ldots + b_n) = ab_1c + ab_2c + \ldots + ab_nc \qquad (9.10.4)$$

Notice in Eq. (9.10.4) that because in matrix algebra multiplication is noncommutative, on the right side of the equation the b matrix is always in the middle while the a matrix is on the left and the c matrix on the right. The order of these matrices must never be changed except in special cases which we shall consider later on.

9.11 Algebraic Checks and the Distributive Law

We have seen how the associative law of algebra can be used for checking computations. The distributive law can also be used for checking certain kinds of computations. Suppose we wish to add the matrices a and b given in Fig. 9.11.1.

$$a = \begin{bmatrix} 3 & 2 \\ 1 & 7 \\ 4 & 1 \end{bmatrix} \qquad b = \begin{bmatrix} 4 & 1 \\ 1 & 3 \\ 2 & 5 \end{bmatrix}$$

FIG. 9.11.1

The sum of the two matrices is given in Fig. 9.11.2.

$$a + b = \begin{bmatrix} 3 & 2 \\ 1 & 7 \\ 4 & 1 \end{bmatrix} + \begin{bmatrix} 4 & 1 \\ 1 & 3 \\ 2 & 5 \end{bmatrix} = \begin{bmatrix} 7 & 3 \\ 2 & 10 \\ 6 & 6 \end{bmatrix}$$

FIG. 9.11.2

Now let us postmultiply both a and b by a unit vector to get the row sums for each of the matrices, as in Fig. 9.11.3.

$$a1 = \begin{bmatrix} 3 & 2 \\ 1 & 7 \\ 4 & 1 \end{bmatrix} \begin{bmatrix} 1 \\ 1 \end{bmatrix} = \begin{bmatrix} 5 \\ 8 \\ 5 \end{bmatrix} \qquad b1 = \begin{bmatrix} 4 & 1 \\ 1 & 3 \\ 2 & 5 \end{bmatrix} \begin{bmatrix} 1 \\ 1 \end{bmatrix} = \begin{bmatrix} 5 \\ 4 \\ 7 \end{bmatrix}$$

FIG. 9.11.3

Next let us postmultiply the right side of Fig. 9.11.2 by the unit vector to get a column vector of row sums. This is given in Fig. 9.11.4.

$$(a + b)1 = \begin{bmatrix} 7 & 3 \\ 2 & 10 \\ 6 & 6 \end{bmatrix} \begin{bmatrix} 1 \\ 1 \end{bmatrix} = \begin{bmatrix} 10 \\ 12 \\ 12 \end{bmatrix}$$

FIG. 9.11.4

We also add together the right sides of both equations in Fig. 9.11.3 to get Fig. 9.11.5.

$$a1 + b1 = \begin{bmatrix} 5 \\ 8 \\ 5 \end{bmatrix} + \begin{bmatrix} 5 \\ 4 \\ 7 \end{bmatrix} = \begin{bmatrix} 10 \\ 12 \\ 12 \end{bmatrix}$$

FIG. 9.11.5

You see that the right sides of Figs. 9.11.4 and 9.11.5 are exactly the same. Their left sides are also the same, then, and we can write

$$(a + b)1 = a1 + b1 \tag{9.11.1}$$

But Eq. (9.11.1) must be true, for in general it is simply a special case of the distributive law where the sum of two matrices has been postmultiplied by the unit vector.

9.12   The Transpose of a Product of Matrices

The transpose of a product of two matrices

Let us now consider the transpose of a product of two matrices. If we go back to Eq. (9.3.2) and take the transpose of both sides, we get Eq. (9.12.1).

$$(ab)' = \begin{bmatrix} a'_{1.}b_{.1} & a'_{1.}b_{.2} & \dots & a'_{1.}b_{.s} \\ a'_{2.}b_{.1} & a^{i}_{2.}b_{.2} & \dots & a'_{2.}b_{.s} \\ \dots & \dots & \dots & \dots \\ a'_{n.}b_{.1} & a'_{n.}b_{.2} & \dots & a'_{n.}b_{.s} \end{bmatrix}' \tag{9.12.1}$$

But we know that to get the transpose of a matrix we write rows as columns and take the transpose of each element. Therefore, we can rewrite Eq. (9.12.1) as Eq. (9.12.2).

$$(ab)' = \begin{bmatrix} b'_{.1}a_{1.} & b'_{.1}a_{2.} & \dots & b'_{.1}a_{n.} \\ b'_{.2}a_{1.} & b'_{.2}a_{2.} & \dots & b'_{.2}a_{n.} \\ \dots & \dots & \dots & \dots \\ b'_{.s}a_{1.} & b'_{.s}a_{2.} & \dots & b'_{.s}a_{n.} \end{bmatrix} \tag{9.12.2}$$

Let us next see what happens if we use Eqs. (9.1.8) and (9.1.4) to write Eq. (9.12.3).

$$b'a' = \begin{bmatrix} b'_{.1} \\ b'_{.2} \\ \cdots \\ b'_{.s} \end{bmatrix}_t [a_1. \quad a_2. \quad \cdots \quad a_n.]_m \qquad (9.12.3)$$

We multiply out the right sides of Eq. (9.12.3) to get Eq. (9.12.4).

$$b'a' = \begin{bmatrix} b'_{.1}a_1. & b'_{.1}a_2. & \cdots & b'_{.1}a_n. \\ b'_{.2}a_1. & b'_{.2}a_2. & \cdots & b'_{.2}a_n. \\ \cdots & \cdots & \cdots & \cdots \\ b'_{.s}a_1. & b'_{.s}a_2. & \cdots & b'_{.s}a_n. \end{bmatrix} \qquad (9.12.4)$$

But now compare the right sides of Eqs. (9.12.2) and (9.12.4), and you will see that corresponding elements are identically the same. Therefore, the left-hand sides of the equations must be equal, and we write

$$(ab)' = b'a' \qquad (9.12.5)$$

We can therefore state the rule that the transpose of a product of two matrices is equal to the product of their transposes in reverse order. You recall that this is the same as the rule for the product of vectors. A numerical example of the rule is given in Figs. 9.12.1 and 9.12.2.

$$(ab)' = \left(\begin{bmatrix} 3 & 2 \\ 7 & 5 \end{bmatrix} \begin{bmatrix} 2 & 4 \\ 3 & 1 \end{bmatrix} \right)' = \begin{bmatrix} 12 & 14 \\ 29 & 33 \end{bmatrix}' = \begin{bmatrix} 12 & 29 \\ 14 & 33 \end{bmatrix}$$
$$\quad\quad\quad\; a \quad\quad\;\; b$$

FIG. 9.12.1

$$b'a' = \begin{bmatrix} 2 & 3 \\ 4 & 1 \end{bmatrix} \begin{bmatrix} 3 & 7 \\ 2 & 5 \end{bmatrix} = \begin{bmatrix} 12 & 29 \\ 14 & 33 \end{bmatrix}$$
$$\quad\quad\;\; b' \quad\quad\; a'$$

FIG. 9.12.2

Notice that in Fig. 9.12.1 we first postmultiplied the matrix a by b and then took the transpose of the product, whereas, in Fig. 9.12.2 we first took the transpose of b and postmultiplied it by the transpose of a. As you see, both methods give the same result.

The transpose of a product of any number of matrices

Let us see now whether, knowing the rule for the transpose of the product of two matrices, we can deduce the rule for the transpose of the product of any number of matrices. If we indicate the product ab by k,

$$ab = k \qquad (9.12.6)$$

and the product kc by f,

$$kc = f \tag{9.12.7}$$

from the rule for two matrices, we can indicate the transpose of f in Eq. (9.12.7) as

$$f' = (kc)' = c'k' \tag{9.12.8}$$

From Eq. (9.12.6), we get the transpose of k as

$$k' = (ab)' = b'a' \tag{9.12.9}$$

If we substitute the right side of Eq. (9.12.9) for k in the right side of Eq. (9.12.8), we can write

$$f' = c'b'a' \tag{9.12.10}$$

But if we substitute the left side of Eq. (9.12.6) in the left side of Eq. (9.12.7), we have

$$abc = f \quad \text{or} \quad f = abc \tag{9.12.11}$$

Equations (9.12.10) and (9.12.11) show that the transpose of the product of three matrices is equal to the product of their transposes in reverse order. By this same line of reasoning, if we have a matrix a that is the product of any number of matrices $a_1 a_2$, to a_n, as in

$$a = a_1 a_2 \ldots a_n \tag{9.12.12}$$

then by the same methods we have just used, we can express the product of a' as

$$a' = a'_n \ldots a'_2 a'_1 \tag{9.12.13}$$

We therefore have the very important rule that the transpose of the product of any number of matrices is equal to the product of their transposes in reverse order.

SUMMARY

1. Matrices expressed as type II supervectors:
   a. The matrix a

$$a = [a_{.1} \quad a_{.2} \quad \ldots \quad a_{.m}]_n = \begin{bmatrix} a'_{1.} \\ a'_{2.} \\ \ldots \\ a'_{n.} \end{bmatrix}_m$$

$$a' = \begin{bmatrix} a'_{.1} \\ a'_{.2} \\ \ldots \\ a'_{.m} \end{bmatrix}_n = [a_{1.} \quad a_{2.} \quad \ldots \quad a_{n.}]_m$$

b. The matrix b

$$b = [b_{.1} \quad b_{.2} \quad \ldots \quad b_{.s}]_t = \begin{bmatrix} b'_{1.} \\ b'_{2.} \\ \ldots \\ b'_{t.} \end{bmatrix}_s$$

2. The product ab as the minor product of type II supervectors:

   a. The equation in minor product form

$$ab = [a_{.1} \quad a_{.2} \quad \ldots \quad a_{.m}]_n \begin{bmatrix} b'_{1.} \\ b'_{2.} \\ \ldots \\ b'_{t.} \end{bmatrix}_s = [a_{.1}b'_{1.} + a_{.2}b'_{2.} + a_{.m}b'_{t.}]_{ns}$$

   b. The product of two matrices expressed as the minor product of type II supervectors is a sum of major products of simple vectors—the first factor of a given term coming from the corresponding column vector of the prefactor and the second from the corresponding row of the postfactor.

3. The product ab as the major product of type II vectors:

   a. The equation

$$ab = \begin{bmatrix} a'_{1.} \\ a'_{2.} \\ \ldots \\ a'_{n.} \end{bmatrix}_m [b_{.1} \quad b_{.2} \quad \ldots \quad b_{.s}]_t = \begin{bmatrix} a'_{1.}b_{.1} & a'_{1.}b_{.2} & \ldots & a'_{1.}b_{.s} \\ a'_{2.}b_{.1} & a'_{2.}b_{.2} & \ldots & a'_{2.}b_{.s} \\ \ldots & \ldots & \ldots & \ldots \\ a'_{n.}b_{.1} & a'_{n.}b_{.2} & \ldots & a'_{n.}b_{.s} \end{bmatrix}$$

   b. The product of two matrices expressed as the major product of type II vectors is a matrix of minor products of vectors where the ijth element is the minor product of the ith row of the prefactor postmultiplied by the jth column of the postfactor; that is, if

$$ab = c \qquad c_{ij} = a'_{i.}b_{.j}$$

4. Reversing the order of the factors:

   a. The minor product

$$ba = [b_{.1} \quad b_{.2} \quad \ldots \quad b_{.s}]_t \begin{bmatrix} a'_{1.} \\ a'_{2.} \\ \ldots \\ a'_{n.} \end{bmatrix}_m = [b_{.1}a'_{2.} + b_{.2}a'_{2.} + \ldots + b_{.s}a'_{n.}]_{tm}$$

   b. The major product

$$ba = \begin{bmatrix} b'_{1.} \\ b'_{2.} \\ \ldots \\ b'_{t.} \end{bmatrix} [a_{.1} \quad a_{.2} \quad \ldots \quad a_{.m}]_n = \begin{bmatrix} b'_{1.}a_{.1} & b'_{1.}a_{.2} & \ldots & b'_{1.}a_{.m} \\ b'_{2.}a_{.2} & b'_{2.}a_{.2} & \ldots & b'_{2.}a_{.m} \\ \ldots & \ldots & \ldots & \ldots \\ b'_{t.}a_{.1} & b'_{t.}a_{.2} & \ldots & b'_{t.}a_{.m} \end{bmatrix}$$

5. Dimensions of factors and their products:
    a. *Restrictions on the orders of the factors.*
        (1) If ab exists, m and t must be equal; that is, the number of columns of a and rows of b must be the same
        (2) If ba exists, s and n must be equal; that is, the number of columns of b and rows of a must be the same
    b. *Order of the product*—The order of a product is the *height* of the prefactor and the *width* of the postfactor.
    c. *Terminology for dimensions of the factors.*
        (1) The height of the prefactor and the width of the postfactor are called the distinct orders or dimensions of the factors
        (2) The width of the prefactor and height of the postfactor are called the common order or common dimension of the factors
    d. The rules of order for matrix multiplication may be indicated with subscript notation by

$$a_{xy}b_{yz} = c_{xz}$$

6. The commutative law and matrix multiplication:
    a. ab and ba exist only if the orders of a and b' are the same.
    b. Even if ab and ba exist, they are not in general equal; therefore, the commutative law does not in general hold for matrix multiplication.

7. The transpose of a product of two matrices is equal to the product of their transposes in reverse order, that is,

$$(ab)' = b'a'$$

8. The associative law of matrix multiplication:
    a. The product of more than two matrices may be formed by finding successively the products of any two adjacent factors, for example,

$$(ab)c = a(bc)$$

    b. The order restrictions on more than two factors and their product may be indicated in subscript notation as follows:

$$U_{ab}V_{bc}W_{cd}X_{de} = Y_{ae}$$

9. The associative law and algebraic checks: The computed product of two matrices may be checked by the associative law as follows:

$$[a \quad (b \quad b1)] = [ab \quad a(b1)] \qquad (ab)1 = a(b1)$$

or alternately

$$\begin{bmatrix} a \\ 1'a \end{bmatrix} b = \begin{bmatrix} ab \\ (1'a)b \end{bmatrix} \qquad 1'[ab] = [1'a]b$$

10. The distributive law of matrix multiplication is given by

$$a(b_1 + b_2 + \ldots + b_n) = ab_1 + ab_2 + \ldots + ab_n$$

or

$$(b_1 + b_2 + \ldots + b_n)a = b_1a + b_2a + \ldots + b_na$$

11. The distributive law may be used in checking the addition of matrices as follows:

$$[a \quad a1] + [b \quad b1] = [(a + b), \quad (a1 + b1)]$$
$$a1 + b1 = (a + b)1$$

or alternately

$$\begin{bmatrix} a \\ 1'a \end{bmatrix} + \begin{bmatrix} b \\ 1'b \end{bmatrix} = \begin{bmatrix} a + b \\ 1'a + 1'b \end{bmatrix}$$
$$1'a + 1'b = 1'[a + b]$$

12. The transpose of the product of any number of matrices is the product of their transposes in reverse order, that is,

$$(a_1 a_2 \ldots a_n')' = a_n' \ldots a_2' a_1'$$

EXERCISES

Assume that the following products all exist: (a) xy (b) yx (c) $x'y$ (d) yx'

1. If these products have all been expressed as the major product of type II supervectors, what is the klth element for the transpose of each of the products?

2. If these products have all been expressed as the minor product of type II supervectors what is the kth term in the sum of vector products for the transpose of each of the products?

ANSWERS

1. (a) $y'_{.k}x_{l.}$,   (b) $x'_{.k}y_{l.}$,   (c) $y'_{.k}x_{.l}$,   (d) $x'_{k.}y_{l.}$

2. (a) $y_{k.}x'_{.k}$,   (b) $x_{k.}y'_{.k}$,   (c) $y_{k.}x'_{k.}$,   (d) $x_{.k}y'_{.k}$

Chapter 10

Special Matrix Products

10.1   The Multiplication of a Matrix by Its Transpose

In the analysis of behavioral science data, one of the most common types of products is that of a matrix and its transpose. The matrix is usually a data matrix. As you know, there may be two kinds of products of a matrix and its transpose. The matrix can be premultiplied by its transpose or it can be postmultiplied by its transpose. The product of a matrix by its transpose is called the *product moment matrix* of the original matrix. Clearly, since the natural order of the matrix is vertical, the order of the product moment matrix is smaller when the matrix is premultiplied by its transpose than when the matrix is postmultiplied by its transpose. We shall therefore refer to the product of a matrix premultiplied by its transpose as the *minor product moment* of the matrix. The product of a matrix postmultiplied by its transpose will be referred to as the *major product moment* of the matrix. This use of the terms *minor* and *major* follows their use as applied to the products of vectors.

The minor product moment

Let us now prove that the minor product moment of a matrix is a symmetric matrix. You recall that we can represent the matrix a as a type II row supervector, thus:

$$a = [a_{.1} \quad a_{.2} \quad \ldots \quad a_{.m}]_n \tag{10.1.1}$$

Then the transpose of a can be written from Eq. (10.1.1) as Eq. (10.1.2).

$$a' = \begin{bmatrix} a'_{.1} \\ a'_{.2} \\ \ldots \\ a'_{.m} \end{bmatrix}_n \tag{10.1.2}$$

166

From Eqs. (10.1.1) and (10.1.2), the minor product moment matrix of a can be written as Eq. (10.1.3).

$$a'a = \begin{bmatrix} a'_{.1} \\ a'_{.2} \\ \cdots \\ a'_{.m} \end{bmatrix}_n [a_{.1} \quad a_{.2} \quad \cdots \quad a_{.m}]_n \qquad (10.1.3)$$

In multiplying out the right side of Eq. (10.1.3), we have Eq. (10.1.4).

$$a'a = \begin{bmatrix} a'_{.1}a_{.1} & a'_{.1}a_{.2} & \cdots & a'_{.1}a_{.m} \\ a'_{.2}a_{.1} & a'_{.2}a_{.2} & \cdots & a'_{.2}a_{.m} \\ \cdots & \cdots & \cdots & \cdots \\ a'_{.m}a_{.1} & a'_{.m}a_{.2} & \cdots & a'_{.m}a_{.m} \end{bmatrix} = c \qquad (10.1.4)$$

Now notice that the ijth element on the right of Eq. (10.1.4) is $a'_{.i}a_{.j}$. Notice that the corresponding jith element is $a'_{.j}a_{.i}$. But these two elements are equal since they are minor products of vectors and one is the transpose of the other; therefore we can write

$$a'_{.i}a_{.j} = a'_{.j}a_{.i} \qquad (10.1.5)$$

But a symmetric matrix is one whose ijth element is equal to its jith element. Therefore, the minor product moment matrix of a is symmetric.

A numerical example of a minor product moment matrix is given in Fig. 10.1.1.

$$\begin{bmatrix} 3 & 1 & 2 \\ 1 & 4 & 2 \end{bmatrix} \begin{bmatrix} 3 & 1 \\ 1 & 4 \\ 2 & 2 \end{bmatrix} = \begin{bmatrix} 14 & 11 \\ 11 & 21 \end{bmatrix}$$

$$\quad a' \qquad\qquad a \qquad\qquad a'a$$

FIG. 10.1.1

We have shown by the type II supervector notation and also by a numerical example that the minor product moment of a matrix is symmetrical. Actually, however, you have learned enough by this time to give a much simpler proof of this rule. You know that one definition of a symmetric matrix is that it is equal to its transpose. Therefore, if you prove that the transpose of a minor product moment is equal to the minor product moment, you have proved that it is a symmetric matrix. This you can do very simply as follows: use the rule that the transpose of a product is equal to the product of the transpose in reverse order. From this you get

$$(a'a)' = a'(a')' \qquad (10.1.6)$$

But you know that the transpose of a transpose is the original matrix, or that two primes cancel each other; therefore you can write the right side of Eq. (10.1.6) as

$$(a'a)' = a'a \qquad (10.1.7)$$

Thus we have proved that the minor product moment of a matrix is symmetric.

The major product moment

We can prove just as simply now that the major product moment of a matrix is a symmetric matrix. We do this in exactly the same fashion by writing

$$(aa')' = (a')'a' \qquad (10.1.8)$$

You see that on the right side of Eq. (10.1.8) we have reversed the order of the factors and taken their transposes. From Eq. (10.1.8) we get

$$(aa')' = aa' \qquad (10.1.9)$$

Thus we have proved that the major product moment of a matrix is equal to its transpose, or that it is a symmetric matrix. Using the matrix from Fig. 10.1.1 again as a numerical illustration, we can get the major product moment as in Fig. 10.1.2.

$$\underbrace{\begin{bmatrix} 3 & 1 \\ 1 & 4 \\ 2 & 2 \end{bmatrix}}_{a} \underbrace{\begin{bmatrix} 3 & 1 & 2 \\ 1 & 4 & 2 \end{bmatrix}}_{a'} = \underbrace{\begin{bmatrix} 10 & 7 & 8 \\ 7 & 17 & 10 \\ 8 & 10 & 8 \end{bmatrix}}_{aa'}$$

FIG. 10.1.2

The order of product moments

You can readily deduce for yourself the simple rules for the order of product moment matrices. The order of the minor product moment is the width of the original matrix. The order of the major product moment is the height of the original matrix. The rule for the order of product moment matrices may be demonstrated by means of subscripts. For example, let the original matrix be vertical, with n rows and m columns. Then we can write the original matrix as in Eq. (10.1.10) and the transpose as in Eq. (10.1.11).

$$a = a_{nm} \qquad (10.1.10)$$

$$a' = a_{mn} \qquad (10.1.11)$$

We would then have for the minor product moment

$$a'a = a_{mn}a_{nm} \qquad (10.1.12)$$

But the rule for the order of a product tells us that the first subscript of the first factor is the number of rows in the product, the last subscript in the last factor is the number of columns. We see from Eq. (10.1.12) that the minor product moment matrix is of order m, or the width of the original matrix.

From Eqs. (10.1.10) and (10.1.11), we can write the major product moment as

$$aa' = a_{nm}a_{mn} \tag{10.1.13}$$

Equation (10.1.13) shows that the major product moment is of order $n \times n$, or the same as the height of the original matrix.

10.2 The Inequality of Major and Minor Product Moments

We should make clear that the minor and major product moments of a matrix are, in general, not the same. It is obvious that the two cannot be equal if the original matrix is not square. In such a case, as you have already seen, the minor and major product moments are not of the same order; therefore they cannot possibly be equal. But even if the matrices are square, two product moments are not in general equal. The general rule is given in Fig. 10.2.1.

$$aa' \neq a'a$$

FIG. 10.2.1

It is easy to prove the inequality of the two product moments even for square matrices. You recall, of course, that the natural order of a square matrix is not automatically defined. Suppose, however, that we let n equal m so that we arbitrarily express the minor product moment of the matrix in Eq. (10.1.4). To get the major product moment, we go back to the type II supervector representation of a matrix and define a in terms of a column vector as in Eq. (10.2.1).

$$a = \begin{bmatrix} a'_{1.} \\ a'_{2.} \\ \cdots \\ a'_{n.} \end{bmatrix}_m \tag{10.2.1}$$

Then from this the transpose of a would be given as

$$a' = [a_{1.} \quad a_{2.} \quad \cdots \quad a_{n.}]_m \tag{10.2.2}$$

From Eqs. (10.2.1) and (10.2.2) we get the major product as in Eq. 10.2.3.

$$aa' = \begin{bmatrix} a'_{1.} \\ a'_{2.} \\ \cdots \\ a'_{n.} \end{bmatrix}_m [a_{1.} \quad a_{2.} \quad \cdots \quad a_{n.}]_m \tag{10.2.3}$$

Multiplying out the right side of Eq. (10.2.3), we get Eq. (10.2.4).

$$aa' = \begin{bmatrix} a'_{1.}a_{1.} & a'_{1.}a_{2.} & \ldots & a'_{1.}a_{n.} \\ a'_{2.}a_{1.} & a'_{2.}a_{2.} & \ldots & a'_{2.}a_{n.} \\ \ldots & \ldots & \ldots & \ldots \\ a'_{n.}a_{1.} & a'_{n.}a_{2.} & \ldots & a'_{n.}a_{n.} \end{bmatrix} \qquad (10.2.4)$$

Now going back to Eq. (10.1.4), you see that the ijth element of the minor product matrix c is given by

$$c_{ij} = a'_{.i}a_{.j} \qquad (10.2.5)$$

You also see that the ijth element of the major product matrix in Eq. (10.2.4) is given by

$$f_{ij} = a'_{i.}a_{j.} \qquad (10.2.6)$$

But the right-hand sides of Eqs. (10.2.5) and (10.2.6) cannot in general be the same, for in Eq. (10.2.5) we have the minor product of column vectors, and in Eq. (10.2.6) we have the minor product of row vectors. Therefore, since the elements of the minor and major product moment matrices are not equal, the matrices themselves cannot in general be equal, as stated in the rule, Fig. 10.2.1. It is important that this distinction be kept in mind, for in using matrix equations to analyze large masses of data, both of these products are frequently required, and it is very easy to confuse them. A numerical example will emphasize the difference between them when the matrices are square; the difference is clearly brought out by Figs. 10.1.1 and 10.1.2 for a rectangular matrix. Even for a square matrix, when we premultiply a by its transpose to get the c matrix in Fig. 10.2.2, the result is quite different than when we postmultiply it by the transpose to get the f matrix in Fig. 10.2.3.

$$\begin{bmatrix} 1 & 2 & 1 \\ 3 & 1 & 4 \\ 2 & 2 & 3 \end{bmatrix} \begin{bmatrix} 1 & 3 & 2 \\ 2 & 1 & 2 \\ 1 & 4 & 3 \end{bmatrix} = \begin{bmatrix} 6 & 9 & 9 \\ 9 & 26 & 20 \\ 9 & 20 & 17 \end{bmatrix}$$
$$\quad a' \qquad\qquad a \qquad\qquad c$$

FIG. 10.2.2

$$\begin{bmatrix} 1 & 3 & 2 \\ 2 & 1 & 2 \\ 1 & 4 & 3 \end{bmatrix} \begin{bmatrix} 1 & 2 & 1 \\ 3 & 1 & 4 \\ 2 & 2 & 3 \end{bmatrix} = \begin{bmatrix} 14 & 9 & 19 \\ 9 & 9 & 12 \\ 19 & 12 & 26 \end{bmatrix}$$
$$\quad a \qquad\qquad a' \qquad\qquad f$$

FIG. 10.2.3

10.3   A Symmetric Matrix Pre- and Postmultiplied by a Matrix and Its Transpose

Any symmetric matrix that is premultiplied by a matrix and postmultiplied by the transpose of the matrix is a symmetric matrix. For example, if we let the symmetric matrix be s, then we can write

$$(asa')' = (a')'s'a' \tag{10.3.1}$$

But remembering that two primes cancel each other and the transpose of a symmetric matrix is the matrix itself, we can write the right side of Eq. (10.3.1) as

$$(asa')' = asa' \tag{10.3.2}$$

In the same way, we can prove that if a symmetric matrix is postmultiplied by a matrix and premultiplied by its transpose, the triple product is a symmetric matrix. First we write

$$(a'sa)' = a's'(a')' \tag{10.3.3}$$

But from this we get

$$(a'sa)' = a'sa \tag{10.3.4}$$

Equations (10.3.2) and (10.3.4) state that if a symmetric matrix is multiplied on either side by another matrix and its transpose, the triple product is a symmetric matrix.

10.4 Square Products of Matrices

The factors of a square product

You have seen that the product moment of a matrix is a square matrix whether it is the major product or the minor product. Obviously, of course, a matrix and its transpose are not the only factors that yield a square product, but if the product of a matrix and another matrix *is* square, we can draw some interesting conclusions. Since the order of the product of two matrices is determined by the distinct dimensions of the factors, we know that a square product comes from two factors whose distinct dimensions are equal. We also know that, in order for the product of two matrices to exist, they must have a common dimension. We know, therefore, that if the product of two factors is a square matrix, the height of the prefactor is equal to the width of the postfactor, and the width of the prefactor is equal to the height of the postfactor. Suppose then that x and y are two different matrices. If now xy' is a square matrix, c, let us say, then and only then does the product $x'y$ exist; and the product $x'y$ is also a square matrix. We can put it another way by saying that if both the products xy' and $x'y$ exist, then each of these products is square, irrespective of the order of x. We also know that if xy' and $x'y$ exist, the orders of x and y must be the same. This principle in diagrammatic form is shown in Fig. 10.4.1.

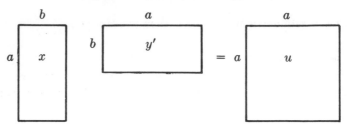

FIG. 10.4.1

Figure 10.4.1 shows clearly that if the product matrix u is of order a, then the height of x must be a and the width of y' must be a. Now in Fig. 10.4.2, you see that if the height of x is a and the width of y' is a we can take the transpose of each of these in the same order and they will have the common dimensions a and distinct dimension b.

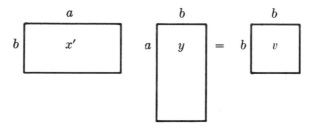

FIG. 10.4.2

Major and minor products

It is sometimes convenient to talk about the major and minor products of two matrices when both distinct orders are greater or less than their common order. But if the major product xy' exists, the transpose of this major product, which would be yx', also exists. Both of these are major products. Similarly, if the minor product $x'y$ exists, its transpose, $y'x$, exists. We see, therefore, that if two matrices x and y have both a major and minor product, the matrices are of the same order and have two minor and two major products. Conversely, if two matrices are of the same order, they have two major and two minor products. For examples of these various types of products, we can let x and y be the matrices given in Fig. 10.4.3.

$$x = \begin{bmatrix} 1 & 2 \\ 3 & 1 \\ 2 & 1 \end{bmatrix} \qquad y = \begin{bmatrix} 2 & 3 \\ 1 & 2 \\ 3 & 1 \end{bmatrix}$$

FIG. 10.4.3

Then one of the major products will be c, let us say, as in Fig. 10.4.4.

$$xy' = \begin{bmatrix} 1 & 2 \\ 3 & 1 \\ 2 & 1 \end{bmatrix} \begin{bmatrix} 2 & 1 & 3 \\ 3 & 2 & 1 \end{bmatrix} = \begin{bmatrix} 8 & 5 & 5 \\ 9 & 5 & 10 \\ 7 & 4 & 7 \end{bmatrix}$$

$$ \quad x \qquad\quad y' \qquad = \qquad c$$

FIG. 10.4.4

The transpose of this major product is shown in Fig. 10.4.5.

$$yx' = \begin{bmatrix} 2 & 3 \\ 1 & 2 \\ 3 & 1 \end{bmatrix} \begin{bmatrix} 1 & 3 & 2 \\ 2 & 1 & 1 \end{bmatrix} = \begin{bmatrix} 8 & 9 & 7 \\ 5 & 5 & 4 \\ 5 & 10 & 7 \end{bmatrix}$$

$$ \quad y \qquad\quad x' \qquad = \qquad c'$$

FIG. 10.4.5

We also have two minor products. One of these is given in Fig. 10.4.6.

$$x'y = \begin{bmatrix} 1 & 3 & 2 \\ 2 & 1 & 1 \end{bmatrix} \begin{bmatrix} 2 & 3 \\ 1 & 2 \\ 3 & 1 \end{bmatrix} = \begin{bmatrix} 11 & 11 \\ 8 & 9 \end{bmatrix}$$

$$ \quad x' \qquad\quad y \qquad = \qquad F$$

FIG. 10.4.6

The transpose of this minor product is shown in Fig. 10.4.7.

$$y'x = \begin{bmatrix} 2 & 1 & 3 \\ 3 & 2 & 1 \end{bmatrix} \begin{bmatrix} 1 & 2 \\ 3 & 1 \\ 2 & 1 \end{bmatrix} = \begin{bmatrix} 11 & 8 \\ 11 & 9 \end{bmatrix}$$

$$ \quad y' \qquad\quad x \qquad = \qquad F'$$

FIG. 10.4.7

10.5 The Trace of a Matrix

An important feature of the square matrix is commonly called by mathematicians the *trace* of a matrix. By the trace of a matrix, we mean no more than the sum of its principal diagonal elements. Since, properly speaking, we talk about diagonal elements only with reference to square matrices, we conclude that only square matrices have traces. The method used by mathematicians to indicate the trace of a matrix a is *tr a*. This is a convenient shorthand notation; however, we shall examine in detail traces of special kinds of products and how these can be expressed in terms of matrix operations.

The diagonal of a matrix

First, as you learned in Section 3.4, it is convenient to use a symbol to indicate the diagonal matrix made up from the corresponding principal diagonal elements of a square matrix. This we frequently do by using D with the matrix symbol as a subscript, as in Fig. 10.5.1.

$$a = \begin{bmatrix} a_{11} & a_{12} & \cdots & a_{1m} \\ a_{21} & a_{22} & \cdots & a_{2m} \\ \cdots & \cdots & \cdots & \cdots \\ a_{m1} & a_{m2} & \cdots & a_{mm} \end{bmatrix} \qquad D_a = \begin{bmatrix} a_{11} & & & \\ & a_{22} & & \\ & & \cdots & \\ & & & a_{mm} \end{bmatrix}$$

FIG. 10.5.1

A numerical example of a square matrix and its diagonal matrix is Fig. 10.5.2.

$$b = \begin{bmatrix} 2 & 4 & 3 \\ 1 & 4 & 2 \\ 2 & 3 & 1 \end{bmatrix} \qquad D_b = \begin{bmatrix} 2 & & \\ & 4 & \\ & & 1 \end{bmatrix}$$

FIG. 10.5.2

The diagonal of a product moment

An example of the diagonal matrix for the minor product moment given in Eq. (10.1.4) would be Fig. 10.5.3.

$$D_{a'a} = \begin{bmatrix} a'_{.1}a_{.1} & & & \\ & a'_{.2}a_{.2} & & \\ & & \cdots & \\ & & & a'_{.m}a_{.m} \end{bmatrix}$$

FIG. 10.5.3

An example of the diagonal matrix for the major product moment given in Eq. (10.2.4) is Fig. 10.5.4.

$$D_{aa'} = \begin{bmatrix} a'_{1.}a_{1.} & & & \\ & a'_{2.}a_{2.} & & \\ & & \cdots & \\ & & & a'_{n.}a_{n.} \end{bmatrix}$$

FIG. 10.5.4

The diagonal of a major product

Let us now consider the diagonal matrix for the major product of two matrices a and b. The first major product we shall consider is ab'. Let us represent a and b' by type II supervectors as in Fig. 10.5.5.

$$a = \begin{bmatrix} a'_{1.} \\ a'_{2.} \\ \cdots \\ a'_{n.} \end{bmatrix}_m \qquad b' = [b_{1.} \quad b_{2.} \quad \cdots \quad b_{n.}]_m$$

FIG. 10.5.5

Then the major product is given in Eq. (10.5.1).

$$ab' = \begin{bmatrix} a'_{1.}b_{1.} & a'_{1.}b_{2.} & \cdots & a'_{1.}b_{n.} \\ a'_{2.}b_{1.} & a'_{2.}b_{2.} & \cdots & a'_{2.}b_{n.} \\ \cdots & \cdots & \cdots & \cdots \\ a'_{n.}b_{1.} & a'_{n.}b_{2.} & \cdots & a'_{n.}b_{n.} \end{bmatrix} \qquad (10.5.1)$$

The diagonal matrix is then given in Eq. (10.5.1).

$$D_{ab'} = \begin{bmatrix} a'_{1.}b_{1.} & & \\ & a'_{2.}b_{2.} & \\ & & \cdots \\ & & & a'_{n.}b_{n.} \end{bmatrix} \qquad (10.5.2)$$

Let us now take the transpose of Eq. (10.5.1). Here the diagonal elements are still the same except that the order of the factors is reversed. Equation (10.5.3) is the transpose of Eq. (10.5.1).

$$ba' = \begin{bmatrix} b'_{1.}a_{1.} & b'_{1.}a_{2.} & \cdots & b'_{1.}a_{n.} \\ b'_{2.}a_{1.} & b'_{2.}a_{2.} & \cdots & b'_{2.}a_{n.} \\ \cdots & \cdots & \cdots & \cdots \\ b'_{n.}a_{1.} & b'_{n.}a_{2.} & \cdots & b'_{n.}a_{n.} \end{bmatrix} \qquad (10.5.3)$$

Then the diagonal of the matrix in Eq. (10.5.3) is given in Eq. (10.5.4).

$$D_{ba'} = \begin{bmatrix} b'_{1.}a_{1.} & & \\ & b'_{2.}a_{2.} & \\ & & \cdots \\ & & & b'_{n.}a_{n.} \end{bmatrix} \qquad (10.5.4)$$

But corresponding elements in the right sides of Eqs. (10.5.2) and (10.5.4) are the same because one is the transpose of the other. They are scalar products of vectors, and the transpose of a scalar is the same scalar. Therefore, we can write

$$D_{ba'} = D_{ab'} \qquad (10.5.5)$$

Equation (10.5.5) says that the diagonal matrix of a square product is equal to the diagonal matrix of the transpose of the square product. This, as you can well see, is true in general of all square matrices. The diagonal of any square matrix is equal to the diagonal of its transpose.

The diagonal of a minor product

We may now consider the minor product of two matrices. Instead of using the notation in Fig. 10.5.5, let us express a as a type II row supervector and b' as a type II column supervector. We then have the matrices expressed as in Fig. 10.5.6.

$$b' = \begin{bmatrix} b'_{.1} \\ b'_{.2} \\ \cdots \\ b'_{.m} \end{bmatrix}_n \qquad a = [a_{.1} \quad a_{.2} \quad \cdots \quad a_{.m}]_n$$

Fig. 10.5.6

We can express one of the minor products of b and a by Eq. (10.5.6).

$$b'a = \begin{bmatrix} b'_{.1}a_{.1} & b'_{.1}a_{.2} & \cdots & b'_{.1}a_{.m} \\ b'_{.2}a_{.1} & b'_{.2}a_{.2} & \cdots & b'_{.2}a_{.m} \\ \cdots & \cdots & \cdots & \cdots \\ b'_{.m}a_{.1} & b'_{.m}a_{.2} & \cdots & b'_{.m}a_{.m} \end{bmatrix} \tag{10.5.6}$$

Then the diagonal matrix of the product is as given in Eq. (10.5.7).

$$D_{b'a} = \begin{bmatrix} b'_{.1}a_{.1} & & & \\ & b'_{.2}a_{.2} & & \\ & & \cdots & \\ & & & b'_{.m}a_{.m} \end{bmatrix} \tag{10.5.7}$$

Thus, corresponding to the case of the major product, the diagonal of the minor product of two matrices is equal to the diagonal of the transpose of the minor product of those two matrices. This is expressed as

$$D_{b'a} = D_{a'b} \tag{10.5.8}$$

We may take a numerical example of Eq. (10.5.5) from Figs. (10.4.4) and (10.4.5). From these two figures we get Fig. 10.5.7.

$$D_{xy'} = \begin{bmatrix} 8 & & \\ & 5 & \\ & & 7 \end{bmatrix} \qquad D_{yx'} = \begin{bmatrix} 8 & & \\ & 5 & \\ & & 7 \end{bmatrix}$$

Fig. 10.5.7

An example of Eq. (10.5.8) can be obtained from Figs. (10.4.6) and (10.4.7). From these we get Fig. 10.5.8.

$$D_{x'y} = \begin{bmatrix} 11 & \\ & 9 \end{bmatrix} \qquad D_{y'x} = \begin{bmatrix} 11 & \\ & 9 \end{bmatrix}$$

Fig. 10.5.8

Computational notation for trace

We have defined the trace of a square matrix as the sum of its diagonal elements. Therefore, if we premultiply the diagonal of a square matrix by a row unit vector and then postmultiply by a column unit vector, we have the sum of its diagonal elements. This is illustrated in Fig. 10.5.9, where we use the diagonal matrix D in Fig. 10.5.2.

$$[1 \quad 1 \quad 1] \begin{bmatrix} 2 & 0 & 0 \\ 0 & 4 & 0 \\ 0 & 0 & 1 \end{bmatrix} \begin{bmatrix} 1 \\ 1 \\ 1 \end{bmatrix} = [1 \quad 1 \quad 1] \begin{bmatrix} 2 \\ 4 \\ 1 \end{bmatrix} = 7$$

FIG. 10.5.9

It is easy to see in Fig. 10.5.9 that the answer 7 is actually the sum of the diagonal elements of the matrix.

Now let us go back to Eq. (10.5.2). We can express the sum of the diagonal elements as

$$1'D_{ab'}1 = a'_{1.}b_{1.} + a'_{2.}b_{2.} + \ldots + a'_{n.}b_{n.} \tag{10.5.9}$$

In Eq. (10.5.9) we have on the right side merely the sum of the diagonal elements of the matrix in Eq. (10.5.1). In the same way we can give the sum of the elements on the right side of Eq. (10.5.7) as

$$1'D_{b'a}1 = b'_{.1}a_{.1} + b'_{.2}a_{.2} + \ldots + b'_{.m}a_{.m} \tag{10.5.10}$$

Now let us look at the right side of Eq. (10.5.9) and see just what it is made of. The first term on the right is the minor or scalar product of the first row from a by the first row from b, which means that it is the sum of products of corresponding elements from the first row of each matrix. The second term on the right side of Eq. (10.5.9) is the scalar product of the second row of a and the second row of b. This means that it is the sum of the products of corresponding element from the second rows of a and b. So for each of the remaining terms in Eq. (10.5.9), we have the scalar product of corresponding rows of a and b, or the sum of products of corresponding elements from the two rows. Therefore, the sum of all of these scalar products on the right side of Eq. (10.5.9) would be simply the sum of the products of all corresponding elements in the matrices a and b. Let us go back to Figs. 10.4.4 and 10.5.7 to see how these diagonal elements are actually made up in a numerical example. The first diagonal element of the product in Fig. 10.4.4 is 8. This is the scalar product of the first row of x and the first row of y (the first column of y'). In the same way, 5 comes from the second row of x and the second row of y, while the third element, 7, comes from the third row of x and the third row of y. We can illustrate the composition of these elements by Fig. 10.5.10.

$$8 = (1 \times 2) + (2 \times 3)$$
$$5 = (3 \times 1) + (1 \times 2)$$
$$7 = (2 \times 3) + (1 \times 1)$$
$$\overline{}$$
$$20$$

FIG. 10.5.10

But now if we add together all of the diagonal elements represented in Fig. 10.5.10, we simply have the sum of the product of all corresponding elements in the two matrices. For this particular case, of course, the sum is 20.

Now let us see what the right side of Eq. (10.5.10) consists of. We see that the first term is the scalar product of the first column of b and the first column of a. In general, the ith term is the product of the ith column of b and the ith column of a; therefore, since each of these scalar products is the sum of the product of corresponding elements, then the sum of all of the scalar products would be the sum of the product of all corresponding elements in the two matrices a and b. We can take a numerical example from Figs. 10.4.6 and 10.5.8. Figure 10.5.11 illustrates the composition of the two diagonal elements.

$$11 = (1 \times 2) + (3 \times 1) + (2 \times 3)$$
$$9 = (2 \times 3) + (1 \times 2) + (1 \times 1)$$
$$\overline{}$$
$$20$$

FIG. 10.5.11

If we add together the two diagonal elements 11 and 9, we get 20. This is the sum of the products of all corresponding elements of the two matrices. We see, therefore, that this is the same as Fig. 10.5.10. We also see by now that since the right side of Eq. (10.5.9) is the sum of the product of all corresponding elements of the matrices a and b, and the right side of (10.5.10) is also the sum of the products of all such corresponding elements, then the right sides of Eqs. (10.5.9) and (10.5.10) are equal. Therefore, the left sides of these equations must also be equal, and we can write

$$1'D_{ab'}1 = 1'D_{a'b}1 \tag{10.5.12}$$

Equation (10.5.12) says that the trace of the major product of two matrices is equal to the trace of the minor product of those two matrices. To summarize: the trace of a square product of matrices is the sum of products of their corresponding elements.

The major and minor product moments of a matrix are special cases of the major and minor square products of two matrices. We can therefore write Eq. (10.5.13) as a special case of (10.5.12).

$$1'D_{aa'}1 = 1'D_{a'a}1 \tag{10.5.13}$$

This says that the trace of the minor product moment matrix is equal to the trace of the major product moment matrix. A numerical example of Eq. (10.5.13) may be taken from the matrices in Figs. 10.1.1 and 10.1.2. The diagonal matrices for their minor and major product moments are given in Fig. 10.5.12.

$$D_{a'a} = \begin{bmatrix} 14 & \\ & 21 \end{bmatrix} \qquad D_{aa'} = \begin{bmatrix} 10 & & \\ & 17 & \\ & & 8 \end{bmatrix}$$

FIG. 10.5.12

The trace of the minor product moment is given in Fig. 10.5.13.

$$1'D_{a'a}1 = \begin{bmatrix} 1 & 1 \end{bmatrix} \begin{bmatrix} 14 & \\ & 21 \end{bmatrix} \begin{bmatrix} 1 \\ 1 \end{bmatrix} = 35$$

FIG. 10.5.13

The trace of the major product moment is given in Fig. 10.5.14.

$$1'D_{aa'}1 = \begin{bmatrix} 1 & 1 & 1 \end{bmatrix} \begin{bmatrix} 10 & & \\ & 17 & \\ & & 8 \end{bmatrix} \begin{bmatrix} 1 \\ 1 \\ 1 \end{bmatrix} = 35$$

FIG. 10.5.14

You see that in both these figures the trace is 35. To summarize: the trace of a product moment of a matrix is the sum of squares of the matrix elements.

10.6   Products Involving Vectors

Products in which both matrices and vectors are factors have special characteristics, which you should learn to recognize at once.

Premultiplication by a vector

First, any vector that premultiplies a matrix must be a row vector. Second, if a matrix is premultiplied by a vector, the product is a row vector. Third, the order of the product vector depends only upon the width of the matrix, and not at all upon the order of the vector by which it is premultiplied. These three rules follow from what we already know about order and the products of matrices. They are illustrated by the diagram in Fig. 10.6.1.

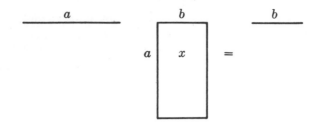

Fig. 10.6.1

Here we have an $a \times b$ matrix x premultiplied by a row vector of order a. Notice that it must be of length a to be conformable with x. The product yields a vector in row form of length b, or of length equal to the width of the matrix x. A numerical example is given in Fig. 10.6.2.

$$[2 \quad 1 \quad 3] \begin{bmatrix} 4 & 2 \\ 1 & 3 \\ 2 & 3 \end{bmatrix} = [15 \quad 16]$$

Fig. 10.6.2

Notice that here the prefactor is a row vector of length equal to the height of the matrix, and the product is a row vector equal to the width of the matrix.

Postmultiplication by a vector

The rules for postmultiplying a matrix by a row vector can also be deduced from the rules of multiplication you already know. First, any vector that postmultiplies a matrix must always be a column vector. Second, if a matrix is postmultiplied by a column vector, the product is a column vector. Third, the order of the product vector is determined by the height of the matrix and has nothing to do with the order of the multiplying vector. The diagram in Fig. 10.6.3 illustrates these rules.

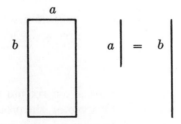

Fig. 10.6.3

A numerical example is given in Fig. 10.6.4.

$$\begin{bmatrix} 4 & 2 \\ 1 & 3 \\ 2 & 3 \end{bmatrix} \begin{bmatrix} 2 \\ 3 \end{bmatrix} = \begin{bmatrix} 14 \\ 11 \\ 13 \end{bmatrix}$$

Fig. 10.6.4

Here we have a 3×2 matrix postmultiplied by a second-order column vector, which gives a third-order column vector.

Pre- and postmultiplication by a vector

If we both pre- and postmultiply a matrix by a vector, we always get a scalar quantity. Because of the rules of order or dimensionality for multiplication, the first vector must always be a row vector, and the last vector, or postfactor, must always be a column vector. We can illustrate the general principle in Fig. 10.6.5, where V_a is one vector and V_b another vector, and X is a matrix.

$$V_a' X V_b = \text{a scalar quantity}$$

Fig. 10.6.5

You will notice that the vector on the left of X in Fig. 10.6.5 is primed to show that it is a row, while the vector V_b on the right is unprimed to show that it is a column.

A numerical example of pre- and postmultiplication of a matrix by vectors can be constructed from Figs. 10.6.2 and 10.6.4. Let us use the same matrix and also the same vectors for pre- and postmultiplication. We have then Fig. 10.6.6.

$$[2 \quad 1 \quad 3] \begin{bmatrix} 4 & 2 \\ 1 & 3 \\ 2 & 3 \end{bmatrix} \begin{bmatrix} 2 \\ 3 \end{bmatrix}$$

Fig. 10.6.6

We could first get the product of the first two factors and then multiply by the last factor. But the product of the first two factors is already given in Fig. 10.6.2, so we may write the product as in Fig. 10.6.7.

$$[15 \quad 16] \begin{bmatrix} 2 \\ 3 \end{bmatrix} = 30 + 48 = 78$$

Fig. 10.6.7

We could also get the product of the last two factors in Fig. 10.6.6 as in Fig. 10.6.4, and then premultiply this column vector by the first factor, or row vector. This we do in Fig. 10.6.8.

$$[2 \quad 1 \quad 3] \begin{bmatrix} 14 \\ 11 \\ 13 \end{bmatrix} = 28 + 11 + 39 = 78$$

Fig. 10.6.8

In either case, the result is the same, as we knew it would have to be, because of the associative law of multiplication.

10.7 Multiplication of a Matrix by a Unit Vector

The unit vector plays a much more important role in the analysis of behavioral science data than in the more purely mathematical treatment of matrix algebra. We shall, therefore, consider the most important cases of multiplication of matrices by unit vectors.

Premultiplication by a unit vector

The premultiplication of a matrix by a unit vector yields a row vector whose elements are the sums of the elements of corresponding columns. A numerical example is given in

$$[1 \quad 1 \quad 1] \begin{bmatrix} 2 & 4 \\ 3 & 2 \\ 1 & 3 \end{bmatrix} = [6 \quad 9]$$

Fig. 10.7.1

You see that on the right side of Fig. 10.7.1, the first element in the row vector is 6, and it is the sum of the elements in the first column of the matrix. The second element is 9, and it is the sum of the elements in the second column of the matrix. Therefore, if we want to sum the columns of a matrix we premultiply it by a row unit vector.

Postmultiplication by a unit vector

The postmultiplication of a matrix by a unit vector gives a column vector whose elements are the sums of the elements in the corresponding rows of the matrix. A numerical example is given in Fig. 10.7.2.

$$\begin{bmatrix} 2 & 4 \\ 3 & 2 \\ 1 & 3 \end{bmatrix} \begin{bmatrix} 1 \\ 1 \end{bmatrix} = \begin{bmatrix} 6 \\ 5 \\ 4 \end{bmatrix}$$

Fig. 10.7.2

You see that the first element in the column vector on the right side in Fig. 10.7.2 is 6. This is the sum of the elements in the first row of the matrix. The second element, 5, is the sum of the elements in the second row of the matrix, namely 3 and 2; and the third element, 4, is the sum of the elements in the last row of the matrix. Therefore, to sum the elements in the rows of a matrix, we postmultiply the matrix by a column unit vector.

Pre- and postmultiplication by a unit vector

If we pre- and postmultiply a matrix by unit vectors we get a scalar quantity that is the sum of all of the elements in the matrix. Using the same matrix as in Figs. 10.7.1 and 10.7.2, we pre- and postmultiply it by unit vectors as in Fig. 10.7.3.

$$[1 \quad 1 \quad 1] \begin{bmatrix} 2 & 4 \\ 3 & 2 \\ 1 & 3 \end{bmatrix} \begin{bmatrix} 1 \\ 1 \end{bmatrix}$$

Fig. 10.7.3

Here we can first multiply together the first two factors and then multiply by the last, or we can first multiply the last two factors and then premultiply by the first. We shall use the former method. Already in Fig. 10.7.1 the first two factors have been multiplied together. We use this product and postmultiply it by the unit vector as in Fig. 10.7.4.

$$[6 \quad 9] \begin{bmatrix} 1 \\ 1 \end{bmatrix} = 15$$

Fig. 10.7.4

You can verify, as shown in Fig. 10.7.4, that the sum of the elements in the matrix of Figs. 10.7.1 and 10.7.2 is 15.

10.8 Multiplication of a Matrix by an e_i Vector

In the application of matrix algebra to behavioral science data, we frequently have occasion to multiply a matrix by an e_i vector. The rules for such multiplication are very simple and useful.

Premultiplication by an e_i vector

A matrix premultiplied by an e_i vector gives a vector which is the ith row of the matrix. Thus

$$e_i'X = X_{i.}' \tag{10.8.1}$$

A numerical example for Eq. (10.8.1) is Fig. 10.8.1.

$$[0 \quad 1 \quad 0 \quad 0] \begin{bmatrix} 1 & 3 & 5 \\ 2 & 6 & 4 \\ 5 & 7 & 6 \\ 4 & 3 & 2 \end{bmatrix} = [2 \quad 6 \quad 4]$$

Fig. 10.8.1

You see that in Fig. 10.8.1, e_i' is e_2', and therefore the product on the right is the second row vector of the matrix.

Postmultiplication by an e_i vector

Postmultiplication of a matrix by an e_i vector gives the ith column vector of the matrix. Thus

$$Xe_i = X_{.i} \tag{10.8.2}$$

A numerical example of Eq. (10.8.2) is Fig. 10.8.2.

$$\begin{bmatrix} 1 & 3 & 5 \\ 2 & 6 & 4 \\ 5 & 7 & 6 \\ 4 & 3 & 2 \end{bmatrix} \begin{bmatrix} 0 \\ 0 \\ 1 \end{bmatrix} = \begin{bmatrix} 5 \\ 4 \\ 6 \\ 2 \end{bmatrix}$$

FIG. 10.8.2

Here e_i is an e_3 column vector; therefore, the product is the third column vector of the matrix.

Pre- and postmultiplication by an e_i vector

Pre- and postmultiplication of a matrix by an e_i' vector and an e_j vector respectively gives a scalar quantity which is the ijth element of the matrix. Thus

$$e_i' X e_j = X_{ij} \tag{10.8.3}$$

A numerical example of Eq. (10.8.3) is Fig. 10.8.3.

$$\begin{bmatrix} 0 & 1 & 0 & 0 \end{bmatrix} \begin{bmatrix} 1 & 2 & 3 \\ 2 & 6 & 4 \\ 5 & 7 & 6 \\ 4 & 3 & 2 \end{bmatrix} \begin{bmatrix} 0 \\ 0 \\ 1 \end{bmatrix} = 4$$

FIG. 10.8.3

10.9   Products Involving Diagonal Matrices

Diagonal matrices play an important role in many applications of matrix algebra. It is, therefore, well to know the most important rules governing their products with other matrices and with one another.

Premultiplication by a diagonal matrix

If we premultiply any matrix by a diagonal matrix, the product is a matrix such that each element in a given row is the product of the corresponding element in the original matrix and the diagonal element corresponding to that row from the diagonal matrix. A numerical example of premultiplication by a diagonal matrix will clarify this rule. In Fig. 10.9.1, we show a matrix premultiplied by a diagonal matrix.

$$\begin{bmatrix} 3 & 0 & 0 \\ 0 & 2 & 0 \\ 0 & 0 & 4 \end{bmatrix}\begin{bmatrix} 2 & 4 & 3 & 1 \\ 6 & 2 & 5 & 2 \\ 2 & 3 & 4 & 1 \end{bmatrix} = c$$

$$Da$$

FIG. 10.9.1

Now let us express each of the elements of c as the scalar products of the corresponding row vectors from the prefactor in Fig. 10.9.1 and the appropriate column vector from the postfactor. Taking each of the elements of c in turn, we get the results in Fig. 10.9.2 for the first row of c.

$$c_{11} = (3 \times 2) + (0 \times 6) + (0 \times 2) = 3 \times 2$$
$$c_{12} = (3 \times 4) + (0 \times 2) + (0 \times 3) = 3 \times 4$$
$$c_{13} = (3 \times 3) + (0 \times 5) + (0 \times 4) = 3 \times 3$$
$$c_{14} = (3 \times 1) + (0 \times 2) + (0 \times 1) = 3 \times 1$$

FIG. 10.9.2

For the second row of the elements of c, we can show in the same way that they are given by Fig. 10.9.3, as are also the elements in the third row.

$$c_{21} = 2 \times 6 \quad c_{22} = 2 \times 2 \quad c_{23} = 2 \times 5 \quad c_{24} = 2 \times 2$$
$$c_{31} = 4 \times 2 \quad c_{32} = 4 \times 3 \quad c_{33} = 4 \times 4 \quad c_{34} = 4 \times 1$$

FIG. 10.9.3

Now we put together the elements of Figs. 10.9.2 and 10.9.3 to constitute the entire c matrix in Fig. 10.9.4.

$$c = \begin{bmatrix} 3 \times 2 & 3 \times 4 & 3 \times 3 & 3 \times 1 \\ 2 \times 6 & 2 \times 2 & 2 \times 5 & 2 \times 2 \\ 4 \times 2 & 4 \times 3 & 4 \times 4 & 4 \times 1 \end{bmatrix}$$

FIG. 10.9.4

But we see here that in the first row the first factor of each element is always 3, or the first element in the diagonal matrix in Fig. 10.9.1. The second factor of each element in the first row is always the corresponding element from the right-hand matrix of Fig. 10.9.1. The same is true of rows 2 and 3 except that the constant multipliers now are the corresponding elements 2 and 4 from the diagonal matrix.

The general rule for the premultiplication of a matrix by a diagonal matrix is given in Eq. (10.9.1).

$$\begin{bmatrix} D_1 & 0 & \cdots & 0 \\ 0 & D_2 & \cdots & 0 \\ \cdots & \cdots & \cdots & \cdots \\ 0 & 0 & \cdots & D_n \end{bmatrix}\begin{bmatrix} a_{11} & a_{12} & \cdots & a_{1m} \\ a_{21} & a_{22} & \cdots & a_{2m} \\ \cdots & \cdots & \cdots & \cdots \\ a_{n1} & a_{n2} & \cdots & a_{nm} \end{bmatrix} =$$

$$\begin{bmatrix} D_1 a_{11} & D_1 a_{12} & \cdots & D_1 a_{1m} \\ D_2 a_{21} & D_2 a_{22} & \cdots & D_2 a_{2m} \\ \cdots & \cdots & \cdots & \cdots \\ D_n a_{n1} & D_n a_{n2} & \cdots & D_n a_{nm} \end{bmatrix} \quad (10.9.1)$$

The rule for premultiplication by a diagonal matrix is further emphasized by the diagram in Fig. 10.9.4.

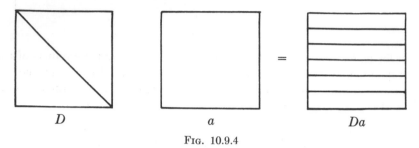

$$D \qquad\qquad a \qquad\qquad Da$$

FIG. 10.9.4

Postmultiplication by a diagonal matrix

If a matrix is postmultiplied by a diagonal matrix, the product is the original matrix with each element in a given column multiplied by the corresponding element in the diagonal matrix. Figure 10.9.5 illustrates this rule.

$$\begin{bmatrix} 2 & 3 & 6 \\ 1 & 5 & 2 \\ 4 & 2 & 3 \end{bmatrix} \begin{bmatrix} 2 & 0 & 0 \\ 0 & 4 & 0 \\ 0 & 0 & 3 \end{bmatrix} = \begin{bmatrix} 4 & 12 & 18 \\ 2 & 20 & 6 \\ 8 & 8 & 9 \end{bmatrix}$$
$$a \qquad\qquad D \qquad\qquad c$$

FIG. 10.9.5

Here you see that the first column of c on the right is obtained by multiplying each element in the first column of a on the left by the first element in D, which is 2. So for the second column in c, each element in the second column of a has been multiplied by 4, the second element in the diagonal matrix. The third column is similarly obtained.

Equation (10.9.2) illustrates the general rule for postmultiplying a matrix by a diagonal matrix.

$$\begin{bmatrix} a_{11} & a_{12} & \ldots & a_{1m} \\ a_{21} & a_{22} & \ldots & a_{2m} \\ \ldots & \ldots & \ldots & \ldots \\ a_{n1} & a_{n2} & \ldots & a_{nm} \end{bmatrix} \begin{bmatrix} D_1 & 0 & \ldots & 0 \\ 0 & D_2 & \ldots & 0 \\ \ldots & \ldots & \ldots & \ldots \\ 0 & 0 & \ldots & D_m \end{bmatrix} =$$

$$\begin{bmatrix} a_{11}D_1 & a_{12}D_2 & \ldots & a_{1m}D_m \\ a_{21}D_1 & a_{22}D_2 & \ldots & a_{2m}D_m \\ \ldots & \ldots & \ldots & \ldots \\ a_{n1}D_1 & a_{n2}D_2 & \ldots & a_{nm}D_m \end{bmatrix} \qquad (10.9.2)$$

This rule for postmultiplication by a diagonal matrix is emphasized further by Fig. 10.9.6.

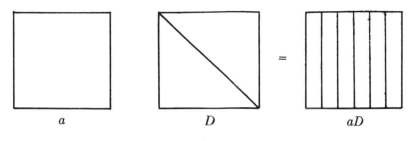

FIG. 10.9.6

Pre- and postmultiplication by a diagonal matrix

If a matrix is pre- and postmultiplied by a diagonal matrix, the product is the matrix whose ijth element is the triple product of the ith element from the left diagonal multiplier, the ijth element of the matrix, and the jth element of the right diagonal multiplier. Figure 10.9.7 illustrates this rule.

$$\begin{bmatrix} 2 & 0 & 0 \\ 0 & 3 & 0 \\ 0 & 0 & 4 \end{bmatrix} \begin{bmatrix} 2 & 3 \\ 1 & 5 \\ 4 & 2 \end{bmatrix} \begin{bmatrix} 1 & 0 \\ 0 & 5 \end{bmatrix} = \begin{bmatrix} 2 \times 2 \times 1 & 2 \times 3 \times 5 \\ 3 \times 1 \times 1 & 3 \times 5 \times 5 \\ 4 \times 4 \times 1 & 4 \times 2 \times 5 \end{bmatrix} = \begin{bmatrix} 4 & 30 \\ 3 & 75 \\ 16 & 40 \end{bmatrix}$$

FIG. 10.9.7

Equation (10.9.3) illustrates the general rule for pre- and postmultiplying a matrix by diagonal matrices.

$$\begin{bmatrix} D_1 & 0 & \cdots & 0 \\ 0 & D_2 & \cdots & 0 \\ \cdots & \cdots & \cdots & \cdots \\ 0 & 0 & \cdots & D_n \end{bmatrix} \begin{bmatrix} a_{11} & a_{12} & \cdots & a_{1m} \\ a_{21} & a_{22} & \cdots & a_{2m} \\ \cdots & \cdots & \cdots & \cdots \\ a_{n1} & a_{n2} & \cdots & a_{nm} \end{bmatrix} \begin{bmatrix} d_1 & 0 & \cdots & 0 \\ 0 & d_2 & \cdots & 0 \\ \cdots & \cdots & \cdots & \cdots \\ 0 & 0 & \cdots & d_m \end{bmatrix} =$$

$$\begin{bmatrix} D_1 d_1 a_{11} & D_1 d_2 a_{12} & \cdots & D_1 d_m a_{1m} \\ D_2 d_1 a_{21} & D_2 d_2 a_{22} & \cdots & D_2 d_m a_{2m} \\ \cdots & \cdots & \cdots & \cdots \\ D_n d_1 a_{n1} & D_n d_2 a_{n2} & \cdots & D_n d_m a_{nm} \end{bmatrix} \quad (10.9.3)$$

Note that in the elements on the right side of Eq. (10.9.3) we have placed the d_i's before the a_{ij}'s. This we can do, obviously, because scalar multiplication is commutative.

Multiplication of a unit vector and a diagonal matrix

A vector may always be expressed as a product of a diagonal matrix and a unit vector. For example, any row vector can always be expressed as a row unit vector postmultiplied by a diagonal matrix. This is illustrated in Eq. (10.9.4).

$$[a_1 \quad a_2 \quad \ldots \quad a_n] = [1 \quad 1 \quad \ldots \quad 1] \begin{bmatrix} a_1 & 0 & \ldots & 0 \\ 0 & a_2 & \ldots & 0 \\ \ldots & \ldots & \ldots & \ldots \\ 0 & 0 & \ldots & a_n \end{bmatrix} \quad (10.9.4)$$

A more concise notation for Eq. (10.9.4) would be

$$V'_a = 1'D_a \quad (10.9.5)$$

This means that a row vector may be expressed as the product of a unit row vector postmultiplied by a diagonal matrix whose diagonal elements are the same as the elements in the row vector.

A numerical example of Eqs. (10.9.4) and (10.9.5) is given in Fig. 10.9.8.

$$[2 \quad 3 \quad 7] = [1 \quad 1 \quad 1] \begin{bmatrix} 2 & 0 & 0 \\ 0 & 3 & 0 \\ 0 & 0 & 7 \end{bmatrix}$$

Fɪɢ. 10.9.8

In the same way, any column vector may be expressed as the product of a unit column vector premultiplied by a diagonal matrix whose elements are the same as the corresponding elements in the vector. This rule may be illustrated by Eq. (10.9.6).

$$\begin{bmatrix} b_1 \\ b_2 \\ \ldots \\ b_n \end{bmatrix} = \begin{bmatrix} b_1 & 0 & \ldots & 0 \\ 0 & b_2 & \ldots & 0 \\ \ldots & \ldots & \ldots & \ldots \\ 0 & 0 & \ldots & b_n \end{bmatrix} \begin{bmatrix} 1 \\ 1 \\ \ldots \\ 1 \end{bmatrix} \quad (10.9.6)$$

It may be expressed more compactly as

$$V_b = D_b 1 \quad (10.9.7)$$

A numerical illustration of this rule is Fig. 10.9.9.

$$\begin{bmatrix} 4 \\ 2 \\ 1 \end{bmatrix} = \begin{bmatrix} 4 & 0 & 0 \\ 0 & 2 & 0 \\ 0 & 0 & 1 \end{bmatrix} \begin{bmatrix} 1 \\ 1 \\ 1 \end{bmatrix}$$

Fɪɢ. 10.9.9

10.10   Products in Which All the Factors Are Diagonal Matrices

The product of two diagonal matrices

The product of two diagonal matrices is a diagonal matrix each of whose diagonal elements is the product of the corresponding elements of the factors. This rule is illustrated in Fig. 10.10.1.

$$\begin{bmatrix} 3 & 0 & 0 \\ 0 & 1 & 0 \\ 0 & 0 & 2 \end{bmatrix} \begin{bmatrix} 2 & 0 & 0 \\ 0 & 3 & 0 \\ 0 & 0 & 4 \end{bmatrix} = \begin{bmatrix} 3 \times 2 & 0 & 0 \\ 0 & 1 \times 3 & 0 \\ 0 & 0 & 2 \times 4 \end{bmatrix} = \begin{bmatrix} 6 & 0 & 0 \\ 0 & 3 & 0 \\ 0 & 0 & 8 \end{bmatrix}$$

Fig. 10.10.1

It is evident that this rule must hold. We have seen that to premultiply a matrix by a diagonal matrix amounts to multiplying each element in the given row of the matrix by the corresponding diagonal element in the diagonal matrix. But since in this particular matrix we have a diagonal matrix premultiplied by a diagonal matrix, then all elements in a given row would be 0 except the diagonal ones, and these, in turn, would be multiplied by the corresponding element of the other diagonal matrix. Using the rule of postmultiplication by a diagonal matrix in which the columns of the matrix are multiplied by the corresponding elements of the diagonal matrix, we again get the same results, namely, that all 0 elements in the prefactor still remain 0. The rule for the product of two diagonal matrices is given by Eq. (10.10.1).

$$\begin{bmatrix} a_1 & 0 & \cdots & 0 \\ 0 & a_2 & \cdots & 0 \\ \cdots & \cdots & \cdots & \cdots \\ 0 & 0 & \cdots & a_n \end{bmatrix} \begin{bmatrix} b_1 & 0 & \cdots & 0 \\ 0 & b_2 & \cdots & 0 \\ \cdots & \cdots & \cdots & \cdots \\ 0 & 0 & \cdots & b_n \end{bmatrix} = \begin{bmatrix} a_1b_1 & 0 & \cdots & 0 \\ 0 & a_2b_2 & \cdots & 0 \\ \cdots & \cdots & \cdots & \cdots \\ 0 & 0 & \cdots & a_nb_n \end{bmatrix}$$

$$(10.10.1)$$

This can be indicated in compact notation as

$$D_a D_b = D_c \tag{10.10.2}$$

We understand, of course, that in Eq. (10.10.2) all three matrices are diagonal. Then any particular diagonal element of the matrix D_c, say c_i, can be expressed as the product of the corresponding diagonal elements from the matrix D_a and D_b. Thus

$$c_i = a_i b_i \tag{10.10.3}$$

The product of any number of diagonal matrices

Knowing the rule for the multiplication of two diagonal matrices, we can easily state the rule for the product of any number of diagonal matrices. This rule is that the product of any number of diagonal matrices is a diagonal matrix each of whose diagonal elements is the product of the corresponding diagonal elements of the factors. This rule must follow because no matter how we apply the associative law in forming the product of more than two diagonal matrices, we shall always have at each step the product of two diagonal matrices to compute. An example of the product of three diagonal matrices is given in

$$\begin{bmatrix} 3 & 0 & 0 \\ 0 & 1 & 0 \\ 0 & 0 & 2 \end{bmatrix} \begin{bmatrix} 4 & 0 & 0 \\ 0 & 3 & 0 \\ 0 & 0 & 2 \end{bmatrix} \begin{bmatrix} 1 & 0 & 0 \\ 0 & 2 & 0 \\ 0 & 0 & 4 \end{bmatrix} = \begin{bmatrix} 12 & 0 & 0 \\ 0 & 6 & 0 \\ 0 & 0 & 16 \end{bmatrix}$$

Fig. 10.10.2

Now let us indicate the product of any number of diagonal matrices by Eq. (10.10.4), where again each of the matrices in the equation is diagonal.

$$D_a D_b \ldots D_z = D_c \tag{10.10.4}$$

Then the ith diagonal element of the product matrix, or the element c_i, may be expressed as the product of the corresponding elements of the matrices on the left of Eq. (10.10.4).

$$c_i = a_i b_i \ldots z_i \tag{10.10.5}$$

The commutative law and diagonal matrices

Although matrix multiplication is in general noncommutative, clearly all diagonal matrices of the same order are commutative one with another. This must be true because any element in the product matrix is the scalar product of the corresponding elements of the factors. The only effect of changing the order of the factors is to change the order of the elements in the scalar products that form the elements of the product matrix. But since all scalar quantities are commutative with respect to multiplication, the change in the order of the factors has no effect on the product matrix. For the product of three diagonal matrices, we could therefore write Fig. 10.10.3.

$$D_a D_b D_c = D_b D_a D_c = D_c D_b D_a = D_a D_c D_b = D_b D_c D_a = D_c D_a D_b$$

Fig. 10.10.3

10.11   The Powers of a Diagonal Matrix

The general case

Just as scalar quantities can be raised to various powers, so also can diagonal matrices. To raise any diagonal matrix to a given power we simply raise each of its diagonal elements to that power. This principle is stated in Eq. (10.11.1).

$$\begin{bmatrix} a_1 & 0 & \ldots & 0 \\ 0 & a_2 & \ldots & 0 \\ \ldots & \ldots & \ldots & \ldots \\ 0 & 0 & \ldots & a_n \end{bmatrix}^k = \begin{bmatrix} a_1^k & 0 & \ldots & 0 \\ 0 & a_2^k & \ldots & 0 \\ \ldots & \ldots & \ldots & \ldots \\ 0 & 0 & \ldots & a_n^k \end{bmatrix} \tag{10.11.1}$$

Let us now consider several special cases of this rule. Suppose we wish to raise a matrix to the second power. This would be expressed by Eq. (10.11.2).

$$D_a^2 = \begin{bmatrix} a_1^2 & 0 & \cdots & 0 \\ 0 & a_2^2 & \cdots & 0 \\ \cdots & \cdots & \cdots & \cdots \\ 0 & 0 & \cdots & a_n^2 \end{bmatrix} \tag{10.11.2}$$

A numerical example of this equation is Fig. 10.11.4.

$$\begin{bmatrix} 4 & 0 & 0 \\ 0 & 3 & 0 \\ 0 & 0 & 1 \end{bmatrix}^2 = \begin{bmatrix} 16 & 0 & 0 \\ 0 & 9 & 0 \\ 0 & 0 & 1 \end{bmatrix}$$

FIG. 10.11.4

The square root

But we must also remember that in scalar algebra we can raise quantities to fractional powers. For example, we can raise the number 4 to the 1/2 power, which means that we take the square root of 4, which is 2. In the same way, we can also take the square root and other roots of a diagonal matrix. Equation (10.11.3) shows how we can indicate the square root of a diagonal matrix.

$$D_a^{1/2} = \begin{bmatrix} \sqrt{a_1} & 0 & \cdots & 0 \\ 0 & \sqrt{a_2} & \cdots & 0 \\ \cdots & \cdots & \cdots & \cdots \\ 0 & 0 & \cdots & \sqrt{a_n} \end{bmatrix} \tag{10.11.3}$$

A numerical example of the square root of a diagonal matrix is given in Fig. 10.11.5.

$$\begin{bmatrix} 9 & 0 & 0 \\ 0 & 25 & 0 \\ 0 & 0 & 36 \end{bmatrix}^{1/2} = \begin{bmatrix} 3 & 0 & 0 \\ 0 & 5 & 0 \\ 0 & 0 & 6 \end{bmatrix}$$

FIG. 10.11.5

It should be noted that if any of the diagonal elements are negative, their square roots are imaginary. In this book we do not make use of such quantities.

The inverse

We also recall from scalar algebra that a number may be raised to a negative power. A number raised to a negative power is equal to 1 divided by that number raised to the same positive power. For example, 4 raised to the −1 power is 1/4. We can also raise a diagonal matrix to the −1 power simply by letting k be -1 in Eq. (10.11.1). Equation (10.11.4) shows how we indicate the -1 power, or, as it is called, the *inverse* of the diagonal matrix.

$$D_a^{-1} = \begin{bmatrix} a_1^{-1} & 0 & \ldots & 0 \\ 0 & a_2^{-1} & \ldots & 0 \\ \ldots & \ldots & \ldots & \ldots \\ 0 & 0 & \ldots & a_n^{-1} \end{bmatrix} = \begin{bmatrix} \dfrac{1}{a_1} & 0 & \ldots & 0 \\ 0 & \dfrac{1}{a_2} & \ldots & 0 \\ \ldots & \ldots & \ldots & \ldots \\ 0 & 0 & \ldots & \dfrac{1}{a_n} \end{bmatrix} \quad (10.11.4)$$

Obviously, the a_i in Eq. (10.11.4) are all assumed to be different from 0. Figure 10.11.6 gives a numerical example of raising a matrix to the -1 power.

$$\begin{bmatrix} 2 & 0 & 0 \\ 0 & 4 & 0 \\ 0 & 0 & 1 \end{bmatrix}^{-1} = \begin{bmatrix} .5 & 0 & 0 \\ 0 & .25 & 0 \\ 0 & 0 & 1 \end{bmatrix}$$

FIG. 10.11.6

But we can also have a negative fraction for the exponent of a diagonal matrix. The most common exponent of this kind which we have for diagonal matrices in the analysis of data matrices is $-1/2$. Equation (10.11.5) shows how we raise a diagonal matrix to the $-1/2$ power.

$$D_a^{-1/2} = \begin{bmatrix} a_1^{-1/2} & 0 & \ldots & 0 \\ 0 & a_2^{-1/2} & \ldots & 0 \\ \ldots & \ldots & \ldots & \ldots \\ 0 & 0 & \ldots & a_n^{-1/2} \end{bmatrix} = \begin{bmatrix} \dfrac{1}{\sqrt{a_1}} & 0 & \ldots & 0 \\ 0 & \dfrac{1}{\sqrt{a_2}} & \ldots & 0 \\ \ldots & \ldots & \ldots & \ldots \\ 0 & 0 & \ldots & \dfrac{1}{\sqrt{a_n}} \end{bmatrix}$$
$$(10.11.5)$$

A numerical example of a matrix raised to the $-1/2$ power is Fig. 10.11.7.

$$\begin{bmatrix} 4 & 0 & 0 \\ 0 & 16 & 0 \\ 0 & 0 & 9 \end{bmatrix}^{-1/2} = \begin{bmatrix} .5 & 0 & 0 \\ 0 & .25 & 0 \\ 0 & 0 & .33 \end{bmatrix}$$

FIG. 10.11.7

10.12   Matrix Products Involving Scalar Matrices

We can multiply any matrix by a scalar matrix simply by multiplying each of its elements by the scalar quantity in the diagonal element of the scalar matrix. Figure 10.12.1 shows how we premultiply a matrix by a scalar matrix.

$$\begin{bmatrix} 2 & 0 & 0 & 0 \\ 0 & 2 & 0 & 0 \\ 0 & 0 & 2 & 0 \\ 0 & 0 & 0 & 2 \end{bmatrix} \begin{bmatrix} 4 & 2 & 1 \\ 3 & 1 & 6 \\ 2 & 4 & 7 \\ 1 & 3 & 1 \end{bmatrix} = \begin{bmatrix} 8 & 4 & 2 \\ 6 & 2 & 12 \\ 4 & 8 & 14 \\ 2 & 6 & 2 \end{bmatrix}$$

FIG. 10.12.1

This rule follows directly from the multiplication of a matrix by a diagonal matrix. It is simply a special case of such multiplication in which the elements in the diagonal matrix are all equal. Let us now postmultiply the matrix in Fig. 10.12.1 by a scalar matrix of diagonal elements 2, as in Fig. 10.12.2.

$$\begin{bmatrix} 4 & 2 & 1 \\ 3 & 1 & 6 \\ 2 & 4 & 7 \\ 1 & 3 & 1 \end{bmatrix} \begin{bmatrix} 2 & 0 & 0 \\ 0 & 2 & 0 \\ 0 & 0 & 2 \end{bmatrix} = \begin{bmatrix} 8 & 4 & 2 \\ 6 & 2 & 12 \\ 4 & 8 & 14 \\ 2 & 6 & 2 \end{bmatrix}$$

FIG. 10.12.2

Now you will notice that the right-hand sides of the equations in Figs. 10.12.1 and 10.12.2 are identical. The only difference between the left-hand sides of the two equations is that in 10.12.1 we have premultiplied by a scalar matrix of elements 2, and in 10.12.2 we have postmultiplied by a scalar matrix of elements 2. Notice, however, that the two scalar matrices are not equal. They cannot be equal since they are not of the same order.

10.13 Products Involving Scalar Quantities

We shall now consider a somewhat simpler concept than that of scalar matrices, which will serve essentially the same purpose. Instead of talking about scalar matrices and the multiplication of other matrices by them, we can talk about multiplying matrices by scalar quantities. When we say that we multiply a matrix by a scalar quantity, we mean that we multiply each of the elements by that quantity. An example of a matrix multiplied by the scalar quantity 2 is given in Figs. 10.13.1 and 10.13.2.

$$2 \begin{bmatrix} 4 & 3 & 7 \\ 4 & 2 & 1 \\ 3 & 1 & 4 \\ 4 & 3 & 2 \end{bmatrix} = \begin{bmatrix} 8 & 6 & 14 \\ 8 & 4 & 2 \\ 6 & 2 & 8 \\ 8 & 6 & 4 \end{bmatrix}$$

FIG. 10.13.1

$$\begin{bmatrix} 4 & 3 & 7 \\ 4 & 2 & 1 \\ 3 & 1 & 4 \\ 4 & 3 & 2 \end{bmatrix} 2 = \begin{bmatrix} 8 & 6 & 14 \\ 8 & 4 & 2 \\ 6 & 2 & 8 \\ 8 & 6 & 4 \end{bmatrix}$$

FIG. 10.13.2

Now in Fig. 10.13.1 we have premultiplied the matrix by the scalar quantity 2. In Fig. 10.13.2 we have postmultiplied the matrix by the scalar quantity 2. You see now that we get the same results whether we pre- or postmultiply the matrix by 2. This result leads us to the rule that in matrix multiplication a scalar quantity is commutative with a matrix. As you recall, this statement is not in general true for scalar matrices, because unless the matrix is square, it cannot be premultiplied by the same scalar matrix used for postmultiplication, since the two are not of the same order.

Commutativity and scalar factors

Given the very simple rule for multiplying a matrix by a scalar quantity, it should be easy to see that all scalar quantities in the factors of a product of matrices and scalar quantities are commutative with one another and with all of the matrix factors. For example, let us suppose two scalar quantities a and b and two matrices x and y, and consider all possible products of these four factors in which x precedes y. All such products would be equal. Some of the products are given in Fig. 10.13.3.

$$abxy = baxy = axby = bxay = bxya = xbya = xayb = xyab$$

FIG. 10.13.3

Notice that in each of the products x always precedes y.

Ordinarily in the application of matrix algebra to the analysis of data, it is convenient to write all of the scalar quantities to the left in the product, and in actual computation, to form the product of the matrices and the product of the scalar quantities separately.

You should be aware of a paradox arising out of our definition of matrix multiplication by a scalar quantity. Suppose, for example, we have a product of matrices and vectors as shown in Fig. 10.13.4.

$$\begin{bmatrix} 4 \\ 3 \end{bmatrix} \left(\begin{bmatrix} 2 & 3 \end{bmatrix} \begin{bmatrix} 1 & 2 \\ 2 & 3 \end{bmatrix} \begin{bmatrix} 2 \\ 3 \end{bmatrix} \right)$$

FIG. 10.13.4

Here we have the product of four factors, three of which are vectors and one of which is a matrix. Suppose now we indicate each of the vectors by symbols as in Fig. 10.13.5.

$$V_1 = \begin{bmatrix} 4 \\ 3 \end{bmatrix} \qquad V_2' = [2 \quad 3]$$

$$M = \begin{bmatrix} 1 & 2 \\ 2 & 3 \end{bmatrix} \qquad V_3 = \begin{bmatrix} 2 \\ 3 \end{bmatrix}$$

Fig. 10.13.5

We could then write the product in Fig. 10.13.4 as

$$V_1(V_2'M V_3)$$

Fig. 10.13.6

Notice now that the product in parentheses here constitutes a matrix premultiplied by a row vector and postmultiplied by a column vector. You will recall that such a product is always a scalar quantity. But we have said that scalar quantities are commutative with matrices. Since a vector is a special case of a matrix, a scalar would also be commutative with vectors. Therefore, we can write Fig. 10.13.6 as

$$(V_2'M V_3)V_1 = V_1(V_2'M V_3)$$

Fig. 10.13.7

Or removing the parentheses, we would have

$$V_2'M V_3 V_1 = V_1 V_2'M V_3$$

Fig. 10.13.8

But on the left side of Fig. 10.13.8 the last two factors are both column vectors. This is a paradox because we cannot premultiply a column vector by a column vector. We can, of course, multiply the first three factors together and get a scalar quantity. Then we can premultiply the column vector V_1 by the scalar quantity. But clearly the associative law of multiplication does not hold on the left side of Fig. 10.13.8, because we cannot multiply together the last two adjacent factors. This exception to the associative law of multiplication comes as a consequence of the definition of multiplication of a matrix by a scalar quantity. It is for this reason that you should learn to recognize a scalar product of factors immediately in any matrix equation. Whenever you see a product in which a row vector is followed by a column vector, whether or not one or more matrices separate the two, you should know at once that you have a scalar quantity that can be commuted with other factors in the series, and that such cases are an exception to the associative law of multiplication. Another simple example of the exception to the associative law of matrix multiplication is given in Fig. 10.13.9.

$$\begin{bmatrix} 2 \\ 1 \end{bmatrix} \begin{bmatrix} 3 & 1 \end{bmatrix} \begin{bmatrix} 2 \\ 3 \end{bmatrix} = \begin{bmatrix} 3 & 1 \end{bmatrix} \begin{bmatrix} 2 \\ 3 \end{bmatrix} \Big] \begin{bmatrix} 2 \\ 1 \end{bmatrix}$$

or $\qquad \begin{bmatrix} 2 \\ 1 \end{bmatrix} [3 \times 2 + 1 \times 3] = [3 \times 2 + 1 \times 3] \begin{bmatrix} 2 \\ 1 \end{bmatrix}$

or $\qquad\qquad\qquad \begin{bmatrix} 2 \\ 1 \end{bmatrix} 9 = 9 \begin{bmatrix} 2 \\ 1 \end{bmatrix}$

or $\qquad\qquad\qquad\quad \begin{bmatrix} 18 \\ 9 \end{bmatrix} = \begin{bmatrix} 18 \\ 9 \end{bmatrix}$

Fig. 10.13.9

It is particularly important that you be aware of the nonassociative nature of matrix multiplication involving scalar products of vectors, because this exception is not usually pointed out by writers on the subject of matrix algebra.

The scalar matrix as the product of a scalar and the identity matrix

Although the scalar matrix as such is used extensively in the theory of matrices, we find little use for it in the application of matrix algebra to the solution of problems involving primary matrices. In multiplication, as you have seen, we usually substitute a scalar quantity for a scalar matrix. In addition, of course, we cannot use this device. If it is necessary, however, to add a scalar matrix to other square matrices of the same order, we usually indicate the scalar matrix as the identity matrix multiplied by a scalar quantity. Suppose, for example, that the elements of a third-order scalar matrix are 3. We can indicate the scalar matrix as in Fig. 10.13.10.

$$\begin{bmatrix} 3 & 0 & 0 \\ 0 & 3 & 0 \\ 0 & 0 & 3 \end{bmatrix} = 3 \begin{bmatrix} 1 & 0 & 0 \\ 0 & 1 & 0 \\ 0 & 0 & 1 \end{bmatrix}$$

Fig. 10.13.10

Or in general if the elements of the scalar matrix are simply some specified scalar quantity, say a, we can indicate this matrix as in Fig. 10.13.11.

$$\begin{bmatrix} a & 0 & \ldots & 0 \\ 0 & a & \ldots & 0 \\ \ldots & \ldots & \ldots & \ldots \\ 0 & 0 & \ldots & a \end{bmatrix} = aI$$

Fig. 10.13.11

10.14   Multiplication by the Identity Matrix

Product of the identity matrix with a matrix

We shall now see why the identity matrix serves the same purpose in matrix algebra as the number 1 does in arithmetic and scalar algebra. In scalar algebra, if we multiply a number by 1, we get the same number. For example, $1 \times 3 = 3$. In the same way, if we multiply a matrix by the identity matrix, we get the same matrix for the product. This must be true because the identity matrix is a special case of the diagonal matrix, so that if we premultiply a matrix by the identity matrix we multiply each of its rows by 1, and if we postmultiply by the identity matrix, we multiply each of the columns by 1. Obviously, this leaves the matrix unchanged. These rules we illustrate by Eq. (10.14.1), where A is the matrix and I is the identity matrix.

$$IA = AI = A \qquad (10.14.1)$$

It is important to remember, however, that the identity matrix on the left of Eq. (10.14.1) and the identity matrix in the middle term are not necessarily equal. The reason for this is, of course, that if the matrix A is not square, these two identity matrices will not be of the same order. Figure 10.14.1 is a numerical example of the rule we have just stated.

$$\begin{bmatrix} 1 & 0 & 0 & 0 \\ 0 & 1 & 0 & 0 \\ 0 & 0 & 1 & 0 \\ 0 & 0 & 0 & 1 \end{bmatrix} \begin{bmatrix} 2 & 4 & 2 \\ 4 & 1 & 3 \\ 1 & 1 & 2 \\ 2 & 1 & 4 \end{bmatrix} = \begin{bmatrix} 2 & 4 & 2 \\ 4 & 1 & 3 \\ 1 & 1 & 2 \\ 2 & 1 & 4 \end{bmatrix} \begin{bmatrix} 1 & 0 & 0 \\ 0 & 1 & 0 \\ 0 & 0 & 1 \end{bmatrix} = \begin{bmatrix} 2 & 4 & 2 \\ 4 & 1 & 3 \\ 1 & 1 & 2 \\ 2 & 1 & 4 \end{bmatrix}$$

Fig. 10.14.1

Here you see that the value of the matrix is unchanged when pre- or postmultiplied by the identity matrix. You also see that in this particular case the identity matrix as a prefactor is not equal to the identity matrix as a postfactor. The prefactor is a fourth-order identity matrix, whereas the postfactor is a third-order identity matrix.

Powers of the identity matrix

Another way in which the identity matrix behaves as the number 1 is that it is unchanged by being raised to any power, say k. This is illustrated in Eq. (10.14.2).

$$\begin{bmatrix} 1 & 0 & \dots & 0 \\ 0 & 1 & \dots & 0 \\ \dots & \dots & \dots & \dots \\ 0 & 0 & \dots & 1 \end{bmatrix}^k = \begin{bmatrix} 1^k & 0 & \dots & 0 \\ 0 & 1^k & \dots & 0 \\ \dots & \dots & \dots & \dots \\ 0 & 0 & \dots & 1^k \end{bmatrix} = \begin{bmatrix} 1 & 0 & \dots & 0 \\ 0 & 1 & \dots & 0 \\ \dots & \dots & \dots & \dots \\ 0 & 0 & \dots & 1 \end{bmatrix}$$

$$(10.14.2)$$

The rule can also be stated more compactly by

$$I^k = I \qquad (10.14.3)$$

Equations (10.14.2) and (10.14.3) are true no matter what the value of k. Some of the most common examples are

$$I^{-1} = I \qquad I^2 = I \qquad I^{1/2} = I \qquad (10.14.4)$$

We can therefore take the inverse of I; we can raise it to the second power; we can take the square root or raise it to the 1/2 power, and in no case do we change its value. In expanded notation, Fig. 10.14.2 shows how by raising I to the -1 power, we do not change the matrix.

$$\begin{bmatrix} 1 & 0 & \cdots & 0 \\ 0 & 1 & \cdots & 0 \\ \cdots & \cdots & \cdots & \cdots \\ 0 & 0 & \cdots & 1 \end{bmatrix}^{-1} = \begin{bmatrix} 1^{-1} & 0 & \cdots & 0 \\ 0 & 1^{-1} & \cdots & 0 \\ \cdots & \cdots & \cdots & \cdots \\ 0 & 0 & \cdots & 1^{-1} \end{bmatrix} =$$

$$\begin{bmatrix} \frac{1}{1} & 0 & \cdots & 0 \\ 0 & \frac{1}{1} & \cdots & 0 \\ 0 & 0 & \cdots & \cdots \\ 0 & 0 & \cdots & \frac{1}{1} \end{bmatrix} = \begin{bmatrix} 1 & 0 & \cdots & 0 \\ 0 & 1 & \cdots & 0 \\ \cdots & \cdots & \cdots & \cdots \\ 0 & 0 & \cdots & 1 \end{bmatrix}$$

Fig. 10.14.2

10.15 Multiplication by the Sign Matrix

Premultiplication of matrix by a sign matrix

Premultiplication of a matrix by a sign matrix changes the signs of all elements in rows corresponding to the -1 in the sign matrix. An example is given in Fig. 10.15.1.

$$\begin{bmatrix} 1 & 0 & 0 \\ 0 & -1 & 0 \\ 0 & 0 & 1 \end{bmatrix} \begin{bmatrix} 4 & 2 \\ 2 & 1 \\ 3 & 4 \end{bmatrix} = \begin{bmatrix} 4 & 2 \\ -2 & -1 \\ 3 & 4 \end{bmatrix}$$

Fig. 10.15.1

Here the second diagonal element in the sign matrix was -1; therefore the sign of the elements in the second row of the matrix was changed from plus to minus.

Postmultiplication of a matrix by a sign matrix

Postmultiplication of a matrix by a sign matrix changes the sign of the elements in the column corresponding to the -1 in the sign matrix. An example is given in Fig. 10.15.2.

$$\begin{bmatrix} 4 & 3 & 1 \\ 3 & 5 & 2 \\ 3 & 1 & 6 \end{bmatrix} \begin{bmatrix} 1 & 0 & 0 \\ 0 & -1 & 0 \\ 0 & 0 & -1 \end{bmatrix} = \begin{bmatrix} 4 & -3 & -1 \\ 3 & -5 & -2 \\ 3 & -1 & -6 \end{bmatrix}$$

Fig. 10.15.2

Since the last two diagonal elements of the sign matrix are -1, the sign of the elements in the last two columns of the matrix was changed from plus to minus.

Pre- and postmultiplication of a square matrix by the same sign matrix

If a square matrix is pre- and postmultiplied by the same sign matrix, the signs of the diagonal elements are left unchanged. To illustrate this we let i be a particular sign matrix, M a square matrix of the same order, and H the product of M pre- and postmultiplied by i. Then

$$iMi = H \tag{10.15.1}$$

The rule we have stated then says that the diagonal matrix of M is the same as the diagonal matrix of H. Thus

$$D_M = D_H \tag{10.15.2}$$

Figure 10.15.3 is a numerical example and shows why this rule must hold.

$$\begin{bmatrix} 1 & 0 & 0 & 0 \\ 0 & -1 & 0 & 0 \\ 0 & 0 & 1 & 0 \\ 0 & 0 & 0 & -1 \end{bmatrix} \begin{bmatrix} 4 & 2 & 3 & 4 \\ 2 & 3 & 7 & 1 \\ 3 & 7 & 4 & 2 \\ 4 & 1 & 2 & 2 \end{bmatrix} \begin{bmatrix} 1 & 0 & 0 & 0 \\ 0 & -1 & 0 & 0 \\ 0 & 0 & 1 & 0 \\ 0 & 0 & 0 & -1 \end{bmatrix} =$$

$$\begin{bmatrix} 4 & 2 & 3 & 4 \\ -2 & -3 & -7 & -1 \\ 3 & 7 & 4 & 2 \\ -4 & -1 & -2 & -2 \end{bmatrix} \begin{bmatrix} 1 & 0 & 0 & 0 \\ 0 & -1 & 0 & 0 \\ 0 & 0 & 1 & 0 \\ 0 & 0 & 0 & -1 \end{bmatrix} = \begin{bmatrix} 4 & -2 & 3 & -4 \\ -2 & 3 & -7 & 1 \\ 3 & -7 & 4 & -2 \\ -4 & 1 & -2 & 2 \end{bmatrix}$$

Fig. 10.15.3

When you premultiply by the sign matrix you change the signs of elements in certain rows. When you postmultiply, you change the signs of the elements in the corresponding columns, but the element that is in a row and also in the corresponding column is a diagonal element. Therefore, the sign of each diagonal element is changed by premultiplication and then changed again by postmultiplication. Since a minus times a minus is a plus, this means that the net result is no change in sign.

Powers of a sign matrix

While it is possible mathematically to talk about powers of a sign matrix in which the exponent is a fraction, in this book we shall consider only powers in which the exponent is a whole number. We notice first, then, that if the exponent is a whole number k, the sign matrix to the kth power is equal to the sign matrix to the $-k$th power. If we let i be the sign matrix, this relationship is expressed as

$$i^k = i^{-k} \tag{10.15.3}$$

An example of Eq. (10.15.3) where k is equal to 2 is given in Fig. 10.15.4.

$$
\begin{bmatrix}
1 & 0 & 0 & 0 \\
0 & -1 & 0 & 0 \\
0 & 0 & 1 & 0 \\
0 & 0 & 0 & -1
\end{bmatrix}^2
=
\begin{bmatrix}
1 & 0 & 0 & 0 \\
0 & -1 & 0 & 0 \\
0 & 0 & 1 & 0 \\
0 & 0 & 0 & -1
\end{bmatrix}^{-2}
$$

or

$$
\begin{bmatrix}
1^2 & 0 & 0 & 0 \\
0 & (-1)^2 & 0 & 0 \\
0 & 0 & 1^2 & 0 \\
0 & 0 & 0 & (-1)^2
\end{bmatrix}
=
\begin{bmatrix}
\dfrac{1}{1^2} & 0 & 0 & 0 \\
0 & \dfrac{1}{(-1)^2} & 0 & 0 \\
0 & 0 & \dfrac{1}{1^2} & 0 \\
0 & 0 & 0 & \dfrac{1}{(-1)^2}
\end{bmatrix}
$$

or

$$
\begin{bmatrix}
1 & 0 & 0 & 0 \\
0 & 1 & 0 & 0 \\
0 & 0 & 1 & 0 \\
0 & 0 & 0 & 1
\end{bmatrix}
=
\begin{bmatrix}
\dfrac{1}{1} & 0 & 0 & 0 \\
0 & \dfrac{1}{1} & 0 & 0 \\
0 & 0 & \dfrac{1}{1} & 0 \\
0 & 0 & 0 & \dfrac{1}{1}
\end{bmatrix}
$$

or

$$
\begin{bmatrix}
1 & 0 & 0 & 0 \\
0 & 1 & 0 & 0 \\
0 & 0 & 1 & 0 \\
0 & 0 & 0 & 1
\end{bmatrix}
=
\begin{bmatrix}
1 & 0 & 0 & 0 \\
0 & 1 & 0 & 0 \\
0 & 0 & 1 & 0 \\
0 & 0 & 0 & 1
\end{bmatrix}
$$

Fig. 10.15.4

You will notice that in Fig. 10.15.4, we took the square on the left side of the sign matrix. Taking the square of a sign matrix means raising each of the diagonal elements to the second power. But the square of -1 is $+1$; furthermore, any even power of -1 is $+1$. Therefore, any sign matrix raised to an even power is the identity matrix. In particular, the sign matrix times itself is the identity matrix. This relationship is expressed as

$$i^2 = I \tag{10.15.4}$$

or in general, for k an even number, Eq. (10.15.5) summarizes both Eq. (10.15.2) and the general rule for even powers of i.

$$i^k = i^{-k} = I \qquad (10.15.5)$$

A numerical example of Eq. (10.15.4) is Fig. 10.15.5.

$$\begin{bmatrix} 1 & 0 & 0 \\ 0 & -1 & 0 \\ 0 & 0 & -1 \end{bmatrix} \begin{bmatrix} 1 & 0 & 0 \\ 0 & -1 & 0 \\ 0 & 0 & -1 \end{bmatrix} = \begin{bmatrix} 1 & 0 & 0 \\ 0 & 1 & 0 \\ 0 & 0 & 1 \end{bmatrix}$$

FIG. 10.15.5

If the exponent k of the sign matrix is an odd number, then the result is the sign matrix itself. This relationship can be expressed as

$$i^k = i^{-k} = i \qquad (10.15.6)$$

It is easy to see why this rule must be true if we take a numerical example in which the exponent is 3, as in Fig. 10.5.6.

$$\begin{bmatrix} 1 & 0 & 0 \\ 0 & -1 & 0 \\ 0 & 0 & -1 \end{bmatrix}^3 = \begin{bmatrix} 1 & 0 & 0 \\ 0 & -1 & 0 \\ 0 & 0 & -1 \end{bmatrix} \begin{bmatrix} 1 & 0 & 0 \\ 0 & -1 & 0 \\ 0 & 0 & -1 \end{bmatrix} \begin{bmatrix} 1 & 0 & 0 \\ 0 & -1 & 0 \\ 0 & 0 & -1 \end{bmatrix}$$

or

$$\begin{bmatrix} 1 & 0 & 0 \\ 0 & -1 & 0 \\ 0 & 0 & -1 \end{bmatrix}^3 = \begin{bmatrix} 1 & 0 & 0 \\ 0 & 1 & 0 \\ 0 & 0 & 1 \end{bmatrix} \begin{bmatrix} 1 & 0 & 0 \\ 0 & -1 & 0 \\ 0 & 0 & -1 \end{bmatrix} = \begin{bmatrix} 1 & 0 & 0 \\ 0 & -1 & 0 \\ 0 & 0 & -1 \end{bmatrix}$$

FIG. 10.15.6

You see that a matrix raised to an odd-number power can always be expressed as that sign matrix raised to the next smallest even number power and then multiplied by the sign matrix itself. Therefore, any sign matrix raised to an odd-numbered power is the sign matrix itself.

The sign matrix is used extensively in the analysis of data matrices where the attributes are personality traits that may be expressed positively or negatively such as ascendance-submission, generosity-miserliness, introversion-extroversion, and so on. If certain attributes in a personality data matrix are to be reversed, this is done by postmultiplication with an appropriate sign matrix.

10.16 Multiplication by a Permutation Matrix

You will recall that a permutation matrix is obtained from an identity matrix by interchanging certain rows or certain columns. Therefore, in a permutation matrix each row and each column has only a single 1 in it and all the other elements are 0. As we shall see, multiplication by a permutation matrix has two distinct results, depending on whether we pre- or postmultiply the matrix by the permutation matrix.

Any permutation matrix may always be regarded in one of two ways, that is, as having been obtained from the identity matrix by interchanging either certain columns or certain rows of the identity matrix.

Postmultiplication by a permutation matrix

If we postmultiply a matrix by a permutation matrix, we look upon the permutation matrix as having been obtained by interchanging certain columns of the identity matrix. If then we postmultiply a matrix by a permutation matrix, we get a matrix that is the same as the original except that the columns have been interchanged in exactly the same way as the columns of the identity matrix were interchanged to get the permutation matrix. This is illustrated by Fig. 10.16.1.

$$\begin{bmatrix} 2 & 4 & 1 \\ 3 & 2 & 3 \\ 1 & 4 & 5 \\ 6 & 5 & 2 \end{bmatrix} \begin{bmatrix} 0 & 1 & 0 \\ 0 & 0 & 1 \\ 1 & 0 & 0 \end{bmatrix} = \begin{bmatrix} 1 & 2 & 4 \\ 3 & 3 & 2 \\ 5 & 1 & 4 \\ 2 & 6 & 5 \end{bmatrix}$$
$$\quad\quad a \quad\quad\quad\quad \pi \quad\quad\quad\quad b$$

Fig. 10.16.1

Here you see that to get the permutation matrix, the last column of an identity matrix was put as the first column. Therefore, in the product b, the last column of the matrix a becomes the first. The first column of the identity matrix was put second; therefore, the first column of the matrix a becomes the second in the matrix b. The second column of the identity matrix was put last in the permutation matrix; therefore, the second column of the matrix a becomes the last in the product matrix b.

Premultiplication by a permutation matrix

If we premultiply a matrix by a permutation matrix, we look upon the permutation matrix as having been obtained by interchanging certain rows of an identity matrix. Then the product of a matrix premultiplied by a permutation matrix gives the same matrix with rows interchanged in exactly the same manner as the rows of an identity matrix were interchanged to get the permutation matrix. An example of premultiplication by a permutation matrix is given in Fig. 10.6.2.

$$\begin{bmatrix} 0 & 0 & 1 & 0 \\ 0 & 1 & 0 & 0 \\ 0 & 0 & 0 & 1 \\ 1 & 0 & 0 & 0 \end{bmatrix} \begin{bmatrix} 2 & 4 & 1 \\ 3 & 2 & 3 \\ 1 & 4 & 5 \\ 6 & 5 & 2 \end{bmatrix} = \begin{bmatrix} 1 & 4 & 5 \\ 3 & 2 & 3 \\ 6 & 5 & 2 \\ 2 & 4 & 1 \end{bmatrix}$$

Fig. 10.16.2

Here you see that the third row of an identity matrix was put first in the permutation matrix; therefore the third row of the matrix a becomes the first in the product matrix b. The second row of the identity matrix remains unchanged in the permutation matrix; therefore the second row of the matrix a remains unchanged in the second row of the matrix b. The fourth row of the identity matrix was put third in the permutation matrix; therefore, the fourth row of the matrix a was put third in the product matrix. The first row of the identity matrix was put last in the permutation matrix; therefore the first row of the matrix a was put last in the product matrix.

Pre- and postmultiplication by a permutation matrix

As you would expect, if we postmultiply a matrix by a permutation matrix, and premultiply it by another permutation matrix, we get a product in which we interchange the columns according to the right permutation matrix and rows according to the left permutation matrix. This is illustrated in Fig. 10.16.3.

$$\begin{bmatrix} 0 & 0 & 1 & 0 \\ 0 & 1 & 0 & 0 \\ 0 & 0 & 0 & 1 \\ 1 & 0 & 0 & 0 \end{bmatrix} \begin{bmatrix} 2 & 4 & 1 \\ 3 & 2 & 3 \\ 1 & 4 & 5 \\ 6 & 5 & 2 \end{bmatrix} \begin{bmatrix} 0 & 1 & 0 \\ 0 & 0 & 1 \\ 1 & 0 & 0 \end{bmatrix} = \begin{bmatrix} 5 & 1 & 4 \\ 3 & 3 & 2 \\ 2 & 6 & 5 \\ 1 & 2 & 4 \end{bmatrix}$$

Fig. 10.16.3

The product moment of a permutation matrix

The product of a permutation matrix by its transpose is the identity matrix. This is indicated as

$$\pi\pi' = I \tag{10.16.1}$$

The rule is illustrated by Fig. 10.16.4.

$$\underbrace{\begin{bmatrix} 0 & 0 & 1 & 0 \\ 0 & 1 & 0 & 0 \\ 0 & 0 & 0 & 1 \\ 1 & 0 & 0 & 0 \end{bmatrix}}_{\pi} \underbrace{\begin{bmatrix} 0 & 0 & 0 & 1 \\ 0 & 1 & 0 & 0 \\ 1 & 0 & 0 & 0 \\ 0 & 0 & 1 & 0 \end{bmatrix}}_{\pi'} = \underbrace{\begin{bmatrix} 1 & 0 & 0 & 0 \\ 0 & 1 & 0 & 0 \\ 0 & 0 & 1 & 0 \\ 0 & 0 & 0 & 1 \end{bmatrix}}_{I}$$

Fig. 10.16.4

It does not matter whether the permutation matrix is pre- or postmultiplied by its transpose; we still get the identity matrix as indicated in

$$\pi'\pi = I \tag{10.16.2}$$

Using the same example as in Fig. 10.16.6, we get Fig. 10.16.5.

$$\begin{bmatrix} 0 & 0 & 0 & 1 \\ 0 & 1 & 0 & 0 \\ 1 & 0 & 0 & 0 \\ 0 & 0 & 1 & 0 \end{bmatrix} \begin{bmatrix} 0 & 0 & 1 & 0 \\ 0 & 1 & 0 & 0 \\ 0 & 0 & 0 & 1 \\ 1 & 0 & 0 & 0 \end{bmatrix} = \begin{bmatrix} 1 & 0 & 0 & 0 \\ 0 & 1 & 0 & 0 \\ 0 & 0 & 1 & 0 \\ 0 & 0 & 0 & 1 \end{bmatrix}$$

Fig. 10.16.5

Figure 10.16.5 is the same as Fig. 10.16.4 except that the transpose precedes the matrix. Although from Figs. 10.16.4 and 10.16.5 you can probably see why the permutation matrix pre- or postmultiplied by its transpose must equal the identity matrix, the reason for the rule may be clearer if we consider each row or column vector of the permutation matrix as an e_i vector. The product moment of a matrix is simply one whose diagonal elements are the scalar products of a vector by itself, and whose non-diagonal elements are scalar products of one vector by another. The scalar product of an e_i vector by itself, as you know, is 1, whereas the scalar product of an e_i vector by another e_j vector is equal to 0.

It should be clear then that when we rearrange the order of the entities or the attributes in a data matrix, we are pre- or postmultiplying the matrix by a permutation matrix. You should also be able to determine under what conditions multiplication of a matrix by a permutation matrix does not alter one of its product moments.

Multiplication of the unit vector by a permutation matrix

If a permutation matrix is pre- or postmultiplied by a unit vector, the result is the unit vector itself. This is stated in Eq. (10.16.3) for a column vector and Eq. (10.16.4) for a row vector.

$$\pi 1 = 1 \qquad\qquad (10.16.3)$$

$$1'\pi = 1' \qquad\qquad (10.16.4)$$

While these rules should be obvious because multiplication by the permutation matrix does no more than interchange the elements of the unit vector, they are important for some kinds of statistical analysis. Therefore, we shall give a numerical example for both row and column vectors in Fig. 10.16.6.

$$\begin{bmatrix} 0 & 1 & 0 & 0 \\ 0 & 0 & 1 & 0 \\ 1 & 0 & 0 & 0 \\ 0 & 0 & 0 & 1 \end{bmatrix} \begin{bmatrix} 1 \\ 1 \\ 1 \\ 1 \end{bmatrix} = \begin{bmatrix} 1 \\ 1 \\ 1 \\ 1 \end{bmatrix} \quad [1 \quad 1 \quad 1 \quad 1] \begin{bmatrix} 0 & 1 & 0 & 0 \\ 0 & 0 & 1 & 0 \\ 1 & 0 & 0 & 0 \\ 0 & 0 & 0 & 1 \end{bmatrix} = [1 \quad 1 \quad 1 \quad 1]$$

Fig. 10.16.6

10.17 Multiplication by an e_{ij} Matrix

Any product involving one or more e_{ij} matrices may always be simplified by expressing the e_{ij} matrix as the product of an e_i column vector postmultiplied by an e_j row vector.

Postmultiplication by an e_{ij} matrix

The product of a matrix a postmultiplied by an e_{ij} matrix may be written as

$$ae_{ij} = ae_ie_j' \tag{10.17.1}$$

But we know that a matrix postmultiplied by an e_i vector is simply the ith column vector of the matrix, as in

$$ae_i = a_{.i} \tag{10.17.2}$$

Substituting the right side of Eq. (10.17.2) in the right side of Eq. (10.17.1) gives

$$ae_{ij} = a_{.i}e_j \tag{10.17.3}$$

From this we see that a matrix a postmultiplied by an e_{ij} matrix gives the ith column vector of a postmultiplied by an e_j row vector. Another way of stating this rule is that a matrix a postmultiplied by an e_{ij} matrix results in a matrix all of whose columns are 0 except the jth which is the ith column of a. A numerical example is given in Fig. 10.17.1.

$$\begin{bmatrix} 3 & 4 & 2 \\ 4 & 2 & 1 \\ 6 & 4 & 3 \\ 7 & 2 & 2 \end{bmatrix}\begin{bmatrix} 0 & 0 & 0 & 0 \\ 0 & 0 & 1 & 0 \\ 0 & 0 & 0 & 0 \end{bmatrix} = \begin{bmatrix} 4 \\ 2 \\ 4 \\ 2 \end{bmatrix}\begin{bmatrix} 0 & 0 & 1 & 0 \end{bmatrix} = \begin{bmatrix} 0 & 0 & 4 & 0 \\ 0 & 0 & 2 & 0 \\ 0 & 0 & 4 & 0 \\ 0 & 0 & 2 & 0 \end{bmatrix}$$

$$a \qquad e_{23} \qquad a_{.2} \qquad e_3' \qquad ae_{23}$$

Fig. 10.17.1

Premultiplication by an e_{ij} matrix

We can also write the product of a matrix a premultiplied by an e_{ij} matrix as

$$e_{ij}a = e_ie_j'a \tag{10.17.4}$$

But we know that a matrix premultiplied by an e_j row vector gives the jth row vector of the matrix, so that

$$e_j'a = a_{j.}' \tag{10.17.5}$$

Substituting the right side of Eq. (10.17.5) in the right side of Eq. (10.17.4) gives

$$e_{ij}a = e_ia_{j.}' \tag{10.17.6}$$

From Eq. (10.17.6) we see that a matrix a premultiplied by an e_{ij} matrix gives the jth row of a premultiplied by a column e_i vector. To put it another way, a matrix a premultiplied by an e_{ij} matrix gives a matrix all of whose rows are 0 except the ith which is the jth row of a. A numerical example of this rule is given in Fig. 10.17.2.

$$
\underset{e_{23}}{\begin{bmatrix} 0 & 0 & 0 & 0 \\ 0 & 0 & 1 & 0 \\ 0 & 0 & 0 & 0 \end{bmatrix}}
\underset{a}{\begin{bmatrix} 3 & 4 & 2 \\ 4 & 2 & 1 \\ 6 & 4 & 3 \\ 7 & 2 & 2 \end{bmatrix}}
= \underset{e_2}{\begin{bmatrix} 0 \\ 1 \\ 0 \end{bmatrix}}
\underset{a'_3.}{\begin{bmatrix} 6 & 4 & 3 \end{bmatrix}}
= \underset{e_{23}a}{\begin{bmatrix} 0 & 0 & 0 \\ 6 & 4 & 3 \\ 0 & 0 & 0 \end{bmatrix}}
$$

Fig. 10.17.2

Pre- and postmultiplication by an e_{ij} matrix

A matrix may be premultiplied by an e_{ij} matrix and postmultiplied by an e_{pq} matrix, that is, one in which the element in the pth row and jth column is 1. The product may be written as

$$e_{ij}ae_{pq} = e_ie'_jae_pe'_q \tag{10.17.7}$$

But we know that a matrix premultiplied by an e_j vector and postmultiplied by an e_p vector gives simply the jpth element of a, so that

$$e'_jae_p = a_{jp} \tag{10.17.8}$$

Substituting the right side of Eq. (10.17.8) into the right side of Eq. (10.17.7) gives

$$e_{ij}ae_{pq} = e_ia_{jp}e'_q \tag{10.17.9}$$

But the right side of Eq. (10.17.9) can be written as

$$e_ia_{jp}e'_q = a_{jp}e_{iq} \tag{10.17.10}$$

If we substitute the right side of Eq. (10.17.10) into the right side of Eq. (10.17.9), we get

$$e_{ij}ae_{pq} = a_{jp}e_{iq} \tag{10.17.11}$$

We see therefore that if a matrix a is premultiplied by an e_{ij} matrix and postmultiplied by an e_{pq} matrix the product is an e_{iq} matrix multiplied by the jpth element of a. A numerical example of this rule is given in Fig. 10.17.3.

$$
\underset{e_{23}}{\begin{bmatrix} 0 & 0 & 0 & 0 \\ 0 & 0 & 0 & 0 \\ 0 & 0 & 1 & 0 \\ 0 & 0 & 0 & 0 \end{bmatrix}}
\underset{a}{\begin{bmatrix} 3 & 4 & 2 \\ 4 & 2 & 1 \\ 6 & 4 & 3 \\ 7 & 2 & 2 \end{bmatrix}}
\underset{e_{31}}{\begin{bmatrix} 0 & 0 & 0 & 0 \\ 0 & 0 & 0 & 0 \\ 1 & 0 & 0 & 0 \end{bmatrix}}
= \underset{a_{33}}{3}
\underset{e_{21}}{\begin{bmatrix} 0 & 0 & 0 & 0 \\ 1 & 0 & 0 & 0 \\ 0 & 0 & 0 & 0 \end{bmatrix}}
$$

Fig. 10.17.3

The product of two e_{ij} matrices

If the matrix a in Eq. (10.17.11) is a unit matrix, and if j is equal to p, the product is e_{iq}. This we can show as, first,

$$e_{ij}Ie_{jq} = a_{jp}e_{iq} \qquad (10.17.12)$$

Then since the element from the jth row and jth column of the identity matrix is simply 1, Eq. (10.17.12) may be written as

$$e_{ij}e_{jq} = e_{iq} \qquad (10.17.13)$$

Now suppose that j is different from p. The element in the jth row and pth column of the identity matrix is 0 unless j and p are equal; therefore, for j and p not equal, we have

$$e_{ij}e_{pq} = 0 \qquad (10.17.14)$$

In other words, an e_{ij} matrix multiplied by an e_{pq} matrix is 0 unless j and p are equal. An example of Eq. (10.17.13) is given in Fig. 10.17.4.

$$\underbrace{\begin{bmatrix} 0 & 0 & 0 \\ 0 & 0 & 1 \\ 0 & 0 & 0 \end{bmatrix}}_{e_{23}} \underbrace{\begin{bmatrix} 0 & 0 & 0 \\ 0 & 0 & 0 \\ 0 & 1 & 0 \end{bmatrix}}_{e_{32}} = \underbrace{\begin{bmatrix} 0 & 0 & 0 \\ 0 & 1 & 0 \\ 0 & 0 & 0 \end{bmatrix}}_{e_{22}}$$

Fig. 10.17.4

An example of Eq. (10.17.14) is given in Fig. 10.17.5.

$$\underbrace{\begin{bmatrix} 0 & 0 & 0 \\ 0 & 0 & 1 \\ 0 & 0 & 0 \end{bmatrix}}_{e_{23}} \underbrace{\begin{bmatrix} 0 & 0 & 0 \\ 0 & 0 & 1 \\ 0 & 0 & 0 \end{bmatrix}}_{e_{23}} = \underbrace{\begin{bmatrix} 0 & 0 & 0 \\ 0 & 0 & 0 \\ 0 & 0 & 0 \end{bmatrix}}_{0}$$

Fig. 10.17.5

10.18   Powers of Matrices

To raise matrices to powers, we simply multiply the matrix by itself a given number of times. This is the same as in scalar algebra. We shall here consider raising matrices only to those powers that are positive whole numbers.

Only square matrices can be raised to powers. To raise a matrix to the second power we multiply it by itself once as in Fig. 10.18.1.

$$\underbrace{\begin{bmatrix} 3 & 2 \\ 1 & 5 \end{bmatrix}^2}_{a^2} = \underbrace{\begin{bmatrix} 3 & 2 \\ 1 & 5 \end{bmatrix}}_{a} \underbrace{\begin{bmatrix} 3 & 2 \\ 1 & 5 \end{bmatrix}}_{a} = \underbrace{\begin{bmatrix} 11 & 16 \\ 8 & 27 \end{bmatrix}}_{b}$$

Fig. 10.18.1

Figure 10.18.1 shows why a matrix must be square if we want to raise it to a power. The prefactor must have the same number of columns as the postfactor has rows; therefore, since it is the same matrix, the matrix must have the same number of columns as it has rows. The rules of exponents hold in the case of matrix algebra just as they do in scalar algebra. These rules we review in

$$a^x a^y = a^{x+y} \qquad\qquad (10.18.1)$$

$$[a^x]^y = a^{xy} \qquad\qquad (10.18.2)$$

Equation (10.18.1) says that if we multiply a matrix a raised to the power x by the same matrix raised to the power y, we get a matrix raised to the power $(x + y)$. Equation (10.18.2) states that if we raise a matrix a to the power x and then raise the resulting matrix to the power y, we get the same matrix as if we raised a to the power xy. Examples using numbers for exponents are given in Fig. 10.18.2.

$$a^2 a^3 = a^5$$
$$[a^2]^3 = a^6$$

Fig. 10.18.2

SUMMARY

1. The product of a matrix and its transpose:
   a. When the natural order of a matrix is premultiplied by its transpose, the product is called the *minor product moment* of the matrix.
   b. The minor product moment of a matrix is symmetrical; that is,

   $$(a'a)' = a'a$$

   c. When the natural order of a matrix is postmultiplied by its transpose, the product is called the major product moment of the matrix.
   d. The major product moment of a matrix is symmetrical; that is,

   $$(aa')' = aa'$$

2. The major and minor product moments of matrices are not in general equal, that is,

   $$aa' \neq a'a$$

3. The product of a symmetric matrix pre- and postmultiplied by a matrix and its transpose is a symmetric matrix. If

   $$s = s'$$

   then

   $$(a'sa)' = a'sa$$

   and

   $$(asa')' = asa'$$

4. Square products of matrices:

    a. If x and y are vertical matrices, and $x'y$ exists and is a square matrix, then xy' also exists and is square.

    b. If $x'y$ and xy' exist, x and y are of the same order.

    c. The products $x'y$ and $y'x$ are called *minor products* and the products xy' and yx' are called major products.

5. The trace of a matrix:

    a. The sum of the diagonal elements of a matrix is called the trace of a matrix and indicated as

$$tr\ a = \Sigma\ a_{ii}$$

    b. The trace of a matrix is equal to the trace of its transpose:

$$tr\ a = tr\ a'$$

    c. The trace of the sum of matrices is the sum of their traces:

$$tr\ [a + b] = tr\ a + tr\ b$$

    d. The trace of the minor product of two matrices is equal to the trace of their major product moment:

$$tr\ x'y = tr\ xy' = tr\ y'x = tr\ yx'$$

    e. The trace of a major product moment is equal to the trace of a minor product moment:

$$tr\ aa' = tr\ a'a$$

    f. The trace of a product of two matrices is the sum of products of corresponding elements of the two matrices.

    g. The trace of the product moment of a matrix is the sum of squares of its elements.

6. Products involving vectors:

    a. *Premultiplication of a matrix by a vector.*

        (1) The vector must be a row whose order is the height of the matrix.

        (2) The product is a row vector whose order is the width of the matrix.

    b. *Postmultiplication of a matrix by a vector.*

        (1) The vector must be a column whose order is the width of the matrix.

        (2) The product is a column whose order is the height of the matrix.

    c. *Pre- and postmultiplication by a vector.*

        (1) The prefactor must be a row and the postfactor a column.

        (2) The triple product is a scalar quantity.

7. Multiplication of a matrix by a unit vector:
   a. Premultiplication of a matrix by a unit vector gives a row vector whose elements are the sums of corresponding column elements of the matrix.
   b. Postmultiplication of a matrix by a unit vector gives a column vector whose elements are the sums of corresponding row elements of the matrix.
   c. Pre- and postmultiplication of a matrix by a unit vector is a scalar that is the sum of all elements in the matrix.

8. Multiplication of a matrix by an e_i vector:
   a. Premultiplication of a matrix by an e_i vector gives the ith row vector of the matrix:

$$e_i'a = a_{i.}'$$

   b. Postmultiplication of a matrix by an e_i vector gives the ith column vector of the matrix:

$$ae_i = a_{.i}$$

   c. Pre- and postmultiplication of a matrix by an e_i and e_j vector, respectively, gives the ijth elements of the matrix:

$$e_i'ae_j = a_{ij}$$

9. Products involving diagonal matrices:
   a. If a matrix is premultiplied by a diagonal matrix, each element in the ith row of the product is the corresponding element from the matrix multiplied by the ith element of the diagonal matrix. If

$$Da = c$$

then

$$c_{ij} = D_i a_{ij}$$

   b. If a matrix is postmultiplied by a diagonal matrix each element in the jth column of the product is the corresponding element from the matrix multiplied by the jth element of the diagonal matrix. If

$$ad = G$$

then

$$G_{ij} = a_{ij}d_j$$

   c. If a matrix is pre- and postmultiplied by a diagonal matrix, the ijth element of the product is the corresponding element from the matrix multiplied by the ith diagonal element of the prefactor and the jth diagonal element of the postfactor. If

$$Dad = M$$

then

$$M_{ij} = D_i M_{ij} d_j$$

10. Products in which all factors are diagonal:
    a. The product of any number of diagonal matrices is a diagonal matrix whose ith element is the product of the ith elements of the matrices.
    If

$$D_a D_b D_c = D_e$$

then

$$e_i = a_i b_i c_i$$

    b. All diagonal matrices of the same order are commutative one with another:

$$D_a D_b D_c = D_c D_b D_a = D_c D_a D_b \text{ etc.}$$

11. Powers of diagonal matrices:
    a. Any power of a diagonal matrix with no vanishing diagonal elements is obtained by raising each diagonal element to the power.
    If

$$D^k = \delta$$

then

$$\delta_i = D_i^k$$

    b. The inverse of a diagonal matrix with no vanishing diagonal elements is obtained by taking the reciprocal of each diagonal element.
    If

$$D^{-1} = d$$

then

$$d_i = \frac{1}{D_i}$$

    c. The square root of any diagonal matrix is obtained by taking the square root of each diagonal element.
    If

$$D^{1/2} = d$$

then

$$d_i = \sqrt{D_i}$$

12. Matrix products involving scalar quantities:
    a. If a matrix is pre- or postmultiplied by a scalar matrix each element of the matrix is multiplied by the scalar diagonal constant.
    b. A scalar matrix is commutative only with square matrices.

13. Products involving scalar quantities:

   a. Multiplication of any matrix by a scalar quantity consists of multiplying each element of the matrix by the scalar. If g is a scalar and b a matrix, and if

$$gb = f$$

   then

$$f_{ij} = gb_{ij}$$

   b. All scalar quantities commute with all matrices and with one another. If g and k are scalars and x, y, and z matrices conformable for multiplication, then

$$gkxyz = xgykz = xykgz \text{ etc.}$$

14. Multiplication by the identity matrix:

   a. Multiplication of a matrix by an identity matrix leaves the matrix unchanged.

$$Ia = aI = a$$

   b. Identity matrices commute only with square matrices.

   c. Any power of an identity matrix is the identity matrix, that is,

$$I^k = I$$

15. Multiplication by the sign matrix:

   a. Premultiplication of a matrix by a sign matrix changes the signs of the elements in the rows corresponding to the -1's.

   b. Postmultiplication by a sign matrix changes the signs of the elements in the columns corresponding to the -1's.

   c. Pre- and postmultiplication of a square matrix by the same sign matrix leaves the principal diagonal unchanged.

   d. Powers of a sign matrix:

   (1) A sign matrix raised to a positive or negative odd power yields the sign matrix. If k is odd,

$$i^k = i^{-k} = i$$

   (2) A sign matrix raised to a positive or negative even power is the identity matrix:

$$i^{2k} = i^{-2k} = I$$

16. Multiplication by a permutation matrix:

   a. If a matrix is postmultiplied by a permutation matrix its columns are interchanged as the columns of an identity matrix were interchanged to obtain the permutation matrix.

   b. If a matrix is premultiplied by a permutation matrix its rows are interchanged as the rows of an identity matrix were interchanged to obtain the permutation matrix.

c. Any permutation matrix pre- or postmultiplied by its transpose gives the identity matrix, that is,

$$\pi'\pi = \pi\pi' = I$$

d. Any permutation matrix pre- or postmultiplied by a unit vector yields the same unit vector, that is,

$$1'\pi = 1' \quad \text{and} \quad \pi 1 = 1$$

17. Multiplication by an e_{ij} matrix:
   a. A matrix postmultiplied by an e_{ij} matrix gives the ith column vector of the matrix postmultiplied by an e_i vector:

$$ae_{ij} = a_{\cdot i}e'_j$$

   b. A matrix premultiplied by an e_{ij} matrix gives the jth row of the matrix premultiplied by an e_i vector:

$$e_{ij}a = e_i a'_{j\cdot}$$

   c. A matrix pre- and postmultiplied by e_{ij} and e_{pq} matrices, respectively, yield the product of the ipth element of the matrix and the e_{iq} matrix:

$$e_{ij}ae_{pq} = a_{jp}e_{iq}$$

   d. The product of an e_{ij} by an e_{jq} matrix is an e_{iq} matrix:

$$e_{ij}e_{jq} = e_{iq}$$

   e. The product of an e_{ij} by an e_{pq} matrix is 0 for $i \neq p$:

$$e_{ij}e_{pq} = 0$$

19. The powers of a matrix:
   a. A square matrix may be raised to any power that is a positive whole number by multiplying it by itself the number of times indicated by the exponent.
   b. The laws of exponents for scalars hold for square matrices:

$$a^x a^y = a^{x+y}$$

$$[a^x]^y = a^{xy}$$

EXERCISES

1. Given

$$\lambda = a_{11}x_1^2 + a_{22}x_2^2 + \dots a_{nn}x_n^2 + 2a_{12}x_1x_2 + 2a_{13}x_1x_3 + \dots + 2a_{1n}x_1x_n +$$
$$2a_{23}x_2x_3 + \dots 2a_{2n}x_2x_n + \dots + 2a_{[n-1]n}x_{[n-1]}x_n$$

(a) Define matrix and vector factors in terms of which k can be expressed.

(b) Express k in terms of these factors.

2. Given the n by m matrix x. In the following summations, $\sum_{i=1}^{n} x_{ij}$ means "the sum of the elements from 1 to n in the jth column of x; $\sum_{j=1}^{m} x_{ij}$ means the sum of the elements from 1 to m in the ith row of x. Show how each of the summations can be expressed in terms of x, 1, and e_i or e_j vectors. Be sure to indicate the order of computation.

(a) $\sum_{i=1}^{n} x_{ij}$ (b) $\sum_{j=1}^{m} x_{ij}$ (c) $\sum_{j=1}^{m} \sum_{i=1}^{n} x_{ij}$ (d) $\sum_{i=1}^{n} \sum_{j=1}^{m} x_{ij}$

For each of the following exercises, two matrices, x and y, are given. Try to find as simply as possible two other factors, a and b, such that $a \times b = y$. If one of these is the identity, indicate this merely by I with the appropriate subscript.

3.
$$x = \begin{bmatrix} 2 & 4 & 1 \\ 3 & 2 & 7 \\ 5 & 7 & 8 \\ 1 & 2 & 4 \end{bmatrix} \qquad y = \begin{bmatrix} 3 & 2 & 7 \end{bmatrix}$$

4.
$$x = \begin{bmatrix} 2 & 4 & 1 \\ 3 & 2 & 7 \\ 5 & 7 & 8 \\ 1 & 2 & 4 \end{bmatrix} \qquad y = \begin{bmatrix} 4 \\ 2 \\ 7 \\ 2 \end{bmatrix}$$

5.
$$x = \begin{bmatrix} 1 & 2 & 3 \\ 5 & 4 & 6 \\ 8 & 7 & 9 \\ 2 & 4 & 1 \end{bmatrix} \qquad y = 5$$

6.
$$x = \begin{bmatrix} 3 & 2 & 1 & 1 \\ 4 & 3 & 2 & 3 \\ 2 & 5 & 2 & 4 \end{bmatrix} \qquad y = \begin{bmatrix} 6 & 4 & 2 & 2 \\ 4 & 3 & 2 & 3 \\ 6 & 15 & 6 & 12 \end{bmatrix}$$

7.
$$x = \begin{bmatrix} 3 & 2 & 1 & 1 \\ 4 & 3 & 2 & 3 \\ 2 & 5 & 2 & 4 \end{bmatrix} \qquad y = \begin{bmatrix} 9 & 4 & 1 & 2 \\ 12 & 6 & 2 & 6 \\ 6 & 10 & 2 & 8 \end{bmatrix}$$

8. Indicate solutions in each of which a and b are both matrices.
$$x = \begin{bmatrix} 3 & 4 & 2 \\ 4 & 5 & 6 \\ 2 & 1 & 3 \\ 4 & 2 & 1 \end{bmatrix} \qquad y = \begin{bmatrix} 6 & 8 & 4 \\ 8 & 10 & 12 \\ 4 & 2 & 6 \\ 8 & 4 & 2 \end{bmatrix}$$

9.
$$x = \begin{bmatrix} 4 & 2 & 6 \\ 3 & 5 & 1 \\ 2 & 4 & 3 \\ 5 & 2 & 7 \end{bmatrix} \qquad y = \begin{bmatrix} 0 & 0 & 0 \\ 2 & 4 & 3 \\ 0 & 0 & 0 \end{bmatrix}$$

10.
$$x = \begin{bmatrix} 5 & 3 & 2 \\ 2 & 1 & 4 \\ 4 & 6 & 7 \\ 5 & 3 & 2 \end{bmatrix} \qquad y = \begin{bmatrix} 0 & 0 & 0 & 3 \\ 0 & 0 & 0 & 1 \\ 0 & 0 & 0 & 6 \\ 0 & 0 & 0 & 3 \end{bmatrix}$$

11.
$$x = \begin{bmatrix} 4 & 3 & 5 \\ 2 & 6 & 4 \\ 4 & 3 & 2 \\ 5 & 1 & 0 \end{bmatrix} \quad y = \begin{bmatrix} 4 & -3 & 5 \\ 2 & -6 & 4 \\ 4 & -3 & 2 \\ 5 & -1 & 0 \end{bmatrix}$$

12.
$$x = \begin{bmatrix} 2 & 5 & 1 \\ 3 & 6 & 4 \\ 2 & 4 & 8 \\ 5 & 7 & 2 \end{bmatrix} \quad y = \begin{bmatrix} -2 & -5 & -1 \\ 3 & 6 & 4 \\ -2 & -4 & -8 \\ 5 & 7 & 2 \end{bmatrix}$$

13.
$$x = \begin{bmatrix} 2 & 3 & 5 \\ 4 & 6 & 2 \\ 7 & 1 & 7 \\ 3 & 2 & 8 \end{bmatrix} \quad y = \begin{bmatrix} 5 & 2 & 3 \\ 2 & 4 & 6 \\ 7 & 7 & 1 \\ 8 & 3 & 2 \end{bmatrix}$$

14.
$$x = \begin{bmatrix} 4 & 7 & 3 \\ 5 & 1 & 4 \\ 2 & 6 & 5 \\ 3 & 1 & 2 \end{bmatrix} \quad y = \begin{bmatrix} 2 & 6 & 5 \\ 4 & 7 & 3 \\ 3 & 1 & 2 \\ 5 & 1 & 4 \end{bmatrix}$$

15. What is the relationship between
    (a) the major product moments of x and y in Ex. (11) and (13)?
    (b) the minor product moments of x and y in Ex. (12) and (14)?

16. Let

$$x = \begin{bmatrix} 4 & 0 & 0 \\ 0 & 9 & 0 \\ 0 & 0 & 25 \end{bmatrix} \quad x^k = \begin{bmatrix} \dfrac{1}{2} & 0 & 0 \\ 0 & \dfrac{1}{3} & 0 \\ 0 & 0 & \dfrac{1}{5} \end{bmatrix}$$

What is k?

17. Let

$$x = \begin{bmatrix} 3 & 2 & 4 \\ 2 & 3 & 3 \\ 4 & 3 & 5 \\ 3 & 1 & 2 \end{bmatrix} \quad l = \begin{bmatrix} 1 \\ -1 \\ 1 \end{bmatrix}$$

What is xl?

18. Given a matrix x, let $a = x11'$. Assume a square. Express very simply (a) in symbols, (b) in words — the trace of a as a function of the elements of x

ANSWERS

1. (a) a is an nth-order symmetric matrix; x is an nth-order vector
   (b) $k = x'ax$

2. (a) $1'[xe_j]$ (b) $[e_i'x]1$ (c) $[1'x]1$ (d) $1'[x1]$

3. $a = [0 \quad 1 \quad 0 \quad 0] \qquad b = I_3$

4. $a = I_4 \qquad b = [0 \quad 1 \quad 0]$

5.
$$a = [0 \quad 1 \quad 0 \quad 0] \qquad b = \begin{bmatrix} 1 \\ 0 \\ 0 \end{bmatrix} \qquad \textbf{6.} \ a = \begin{bmatrix} 2 & 0 & 0 \\ 0 & 1 & 0 \\ 0 & 0 & 3 \end{bmatrix} \qquad b = I_4$$

7.
$$a = I_3 \qquad b = \begin{bmatrix} 3 & 0 & 0 & 0 \\ 0 & 2 & 0 & 0 \\ 0 & 0 & 1 & 0 \\ 0 & 0 & 0 & 2 \end{bmatrix}$$

8.
$$a = \begin{bmatrix} 2 & 0 & 0 & 0 \\ 0 & 2 & 0 & 0 \\ 0 & 0 & 2 & 0 \\ 0 & 0 & 0 & 2 \end{bmatrix} \qquad b = I_3 \qquad a = I_4 \qquad b = \begin{bmatrix} 2 & 0 & 0 \\ 0 & 2 & 0 \\ 0 & 0 & 2 \end{bmatrix}$$

9.
$$a = \begin{bmatrix} 0 \\ 1 \\ 0 \end{bmatrix} [0 \quad 0 \quad 1 \quad 0] \qquad b = I_4$$

10.
$$a = I_4 \qquad b = \begin{bmatrix} 0 \\ 1 \\ 0 \end{bmatrix} [0 \quad 0 \quad 0 \quad 1]$$

11.
$$a = I_4 \qquad b = \begin{bmatrix} 1 & 0 & 0 \\ 0 & -1 & 0 \\ 0 & 0 & 1 \end{bmatrix}$$

12.
$$a = \begin{bmatrix} -1 & 0 & 0 & 0 \\ 0 & 1 & 0 & 0 \\ 0 & 0 & -1 & 0 \\ 0 & 0 & 0 & 1 \end{bmatrix} \qquad b = I_3$$

13.
$$a = I_4 \qquad b = \begin{bmatrix} 0 & 1 & 0 \\ 0 & 0 & 1 \\ 1 & 0 & 0 \end{bmatrix}$$

14.
$$a = \begin{bmatrix} 0 & 0 & 1 & 0 \\ 1 & 0 & 0 & 0 \\ 0 & 0 & 0 & 1 \\ 0 & 1 & 0 & 0 \end{bmatrix} \qquad b = I_3$$

15. (a) $xx' = yy'$ (b) $x'x = y'y$

16. $k = -1/2$

17.
$$xl = \begin{bmatrix} 5 \\ 2 \\ 6 \\ 4 \end{bmatrix}$$

18. (a) $tr\ a = 1'x1$
    (b) The trace of a is the sum of the elements in x

Chapter II

Multiplication of Supermatrices

In the behavioral sciences it is often desirable to partition data matrices by entities or attributes or both. A particular study may logically call for subsets of entities and attributes. For example, in a matrix of test scores, grades, ratings, and physiological and other measurements on a large group of military service personnel, it may be helpful to partition the entities according to service schools attended, and the attributes according to type of measure, for example, test scores and grades. It is also convenient sometimes to partition entities or attributes or both for computational convenience. It may be that the computational facilities available will not handle the complete data matrix as a whole or matrices derived from it. It is therefore desirable to have a systematic procedure for handling the data in subsets or appropriate combinations of submatrices.

The addition and subtraction of supermatrices cause little difficulty either in principle or application. The multiplication of supermatrices is simple in principle, but in practical applications even competent mathematicians are easily confused. In typical textbooks on matrices, the principles of supermatrix multiplication are briefly explained but no adequate notation is provided to help the student or research worker avoid confusion.

11.1 A Notation for the Multiplication of Supermatrices

Partitioning vectors

To understand how supermatrices are multiplied we must first develop a notation. You recall that a supermatrix may be obtained by merely partitioning a simple matrix between certain rows and columns. Let us begin with a simple matrix X of P rows and Q columns. But, instead of writing the matrix as X_{PQ} to show that it has P rows and Q columns, let us write it as $_PX_Q$. Here we call P a *prescript* and Q a *postscript*. The prescript tells us how many rows the matrix has and the postscript tells us how many columns it has.

Next let us define what we shall call *partitioning vectors*. A *prepartitioning*

217

vector or, as we shall sometimes call it, a *prepartitioner*, indicates how we shall partition the matrix by rows, and a *postpartitioner* indicates how we shall partition it by columns. A prepartitioner is a column vector whose order is equal to the number of matric rows in the partitioned matrix. Each element in the prepartitioner is a scalar indicating the number of simple rows of the matric elements in the corresponding matric row. A postpartitioner is a row vector whose order is equal to the number of matric columns of the partitioned matrix. Each element in the postpartitioner is a scalar indicating the number of simple columns of the matric elements in the corresponding matric column.

You see, therefore, that a prepartitioner is always a column vector and a postpartitioner is always a row vector. You see also that the sum of the elements in the prepartitioner must always be equal to the number of simple rows in the matrix, and the sum of the elements in the postpartitioner must be equal to the number of simple columns in the matrix.

Prepartitioning

Let us consider now the prepartitioning vector in Eq. (11.1.1).

$$[a_r] = \begin{bmatrix} a_1 \\ a_2 \\ \cdots \\ a_r \end{bmatrix} \tag{11.1.1}$$

Notice that a_r on the left of Eq. (11.1.1) is enclosed in brackets. This is to show that it is a vector rather than the rth element of a vector. The subscript r in the symbol $[a_r]$ means that the vector has r elements. If now we want to show that the matrix $_PX_Q$ has been prepartitioned by the vector $[a_r]$ we would write it as

$$_{[a_r]}[_PX_Q] = {}_{[a_r]}X_Q \tag{11.1.2}$$

The column vector prescript $[a_r]$ on the right side of this equation means that $_PX_Q$ has been prepartitioned into r matric rows. We now have a type III column vector. You will recall that a type III vector is a vector whose elements are matrices. According to Eq. (11.1.1), the first matric element has a_1 simple rows, the second matric element has a_2 simple rows, and so on. Since the sum of the elements in $[a_r]$ must be equal to P, we can write

$$[1_r]'[a_r] = P \tag{11.1.3}$$

where, of course, $[1_r]'$ is a row unit vector of order r.

To illustrate Eq. (11.1.2), in $_PX_Q$ we let $P = 5$ and $Q = 4$. We then take the numerical example in Fig. 11.1.1.

$$_PX_Q = {_5}X_4 = \begin{bmatrix} 1 & 2 & 3 & 5 \\ 4 & 3 & 7 & 6 \\ 4 & 2 & 1 & 5 \\ 8 & 6 & 9 & 5 \\ 4 & 3 & 2 & 7 \end{bmatrix}$$

FIG. 11.1.1

Now suppose we wish to prepartition $_5X_4$ into two matric rows so that the first matric element has 3 simple rows and the second matric element has 2 simple rows. The prepartitioner would then be as given in Fig. 11.1.2.

$$[a_r] = \begin{bmatrix} 3_1 \\ 2_2 \end{bmatrix}$$

FIG. 11.1.2

As indicated here, when we write out the numerical elements of a partitioning vector we shall frequently attach a subscript to each element to indicate its position in the vector. Using the example in Fig. 11.1.1, we can then illustrate Eq. (11.1.2) by Fig. 11.1.3.

$$_{[a_r]}X_Q = {_{\begin{bmatrix} 3_1 \\ 2_2 \end{bmatrix}}}X_4 = \left[\begin{array}{cccc} 1 & 2 & 3 & 5 \\ 4 & 3 & 7 & 6 \\ 4 & 2 & 1 & 5 \\ \hline 8 & 6 & 9 & 5 \\ 4 & 3 & 2 & 7 \end{array} \right]$$

FIG. 11.1.3

Postpartitioning

Let us next consider the postpartitioning vector in the equation

$$[b_s]' = [b_1 \quad b_2 \quad \ldots \quad b_s] \tag{11.1.4}$$

Notice that the left side of Eq. (11.1.4) is primed to show that we have a row vector, just as in ordinary vector notation. The subscript s in the symbol $[b_s]'$ means that the vector has s elements. If now we want to show that the matrix $_PX_Q$ has been postpartitioned by the vector $[b_s]'$ we write it as

$$[_PX_Q]_{[b_s]'} = {_P}X_{[b_s]'} \tag{11.1.5}$$

The row vector postscript on the right side of Eq. (11.1.5) means that $_PX_Q$ has been postpartitioned into s matric columns. This then gives us a type III row vector. According to Eq. (11.1.4), the first matric element has b_1 simple columns, the second matric element has b_2 simple columns, and so on.

Since the sum of the elements in $[b_s]'$ must be equal to Q, we can write

$$[b_s]' \, [1_s] = Q \qquad\qquad (11.1.6)$$

where 1_s is a unit vector of order s.

Let us assume now that we wish to postpartition the matrix $_5X_4$ in Fig. 11.1.1 into two matric columns so that both the first and second matric elements have two simple columns. The postpartitioner would then be given as

$$[b_s]' = [2_1 2_2]$$

Fig. 11.1.4

Then using the numerical example of Fig. 11.1.1 to illustrate Eq. (11.1.5), we have Fig. 11.1.5.

$$_PX_{[b_s]'} = {}_5X_{[2_1 2_2]} = \begin{bmatrix} 1 & 2 & 3 & 5 \\ 4 & 3 & 7 & 6 \\ 4 & 2 & 1 & 5 \\ 8 & 6 & 9 & 5 \\ 4 & 3 & 2 & 7 \end{bmatrix}$$

Fig. 11.1.5

Pre- and postpartitioning

Suppose next we wish to show that the matrix $_PX_Q$ has been both prepartitioned by $[a_r]$ and postpartitioned by $[b_s]'$. This we would indicate by

$$_{[a_r]} [_PX_Q]_{[b_s]'} = {}_{[a_r]}X_{[b_s]'} \qquad\qquad (11.1.7)$$

To illustrate this let us again take $_PX_Q$ as given in Fig. 11.1.1 and assume that the prepartitioner is as given in Fig. 11.1.2 and the postpartitioner as in Fig. 11.1.4. This would result in the supermatrix given in Fig. 11.1.6.

$$_{[a_r]}X_{[b_s]'} = {}_{[{}^{3_1}_{2_2}]}X_{[2_1 2_2]} = \left[\begin{array}{cc|cc} 1 & 2 & 3 & 5 \\ 4 & 3 & 7 & 6 \\ 4 & 2 & 1 & 5 \\ \hline 8 & 6 & 9 & 5 \\ 4 & 3 & 2 & 7 \end{array} \right]$$

Fig. 11.1.6

As indicated by the numerical examples in Figs. 11.1.3, 11.1.5, and 11.1.6, we can actually write out the elements of the vector subscripts when their numerical values are specified. However, we can also write these vector subscripts in expanded notation, even when the numerical values are not given. For example, we can indicate a prepartitioning of $_PX_Q$ as in Eq. (11.1.8).

$$_{[a_r]}X_Q = \begin{bmatrix} a_1 \\ a_2 \\ \cdots \\ a_r \end{bmatrix} X_Q \qquad (11.1.8)$$

We can also indicate a postpartitioning as

$$_PX_{[b_s]'} = {_PX}_{[b_1 \quad b_2 \quad \cdots \quad b_s]} \qquad (11.1.9)$$

To show both a pre- and postpartitioning in expanded subvector notation we would have Eq. (11.1.10).

$$_{[a_r]}X_{[b_s]'} = \begin{bmatrix} a_1 \\ a_2 \\ \cdots \\ a_r \end{bmatrix} X_{[b_1 \quad b_2 \quad \cdots \quad b_s]} \qquad (11.1.10)$$

Expanded notation for the type III vector

Remembering now that Eq. (11.1.8) is a type III column vector, let us see how the right side may be rewritten to show the individual matric elements of the vector. This we do in Eq. (11.1.11).

$$_{[a_r]}X_Q = \begin{bmatrix} a_1 X_Q \\ a_2 X_Q \\ \cdots \\ a_r X_Q \end{bmatrix} \qquad (11.1.11)$$

On the right side of Eq. (11.1.11) we now have given the matric elements of the type III column vector. The prescript gives the number of simple rows of the element and the postscript the number of simple columns. Since there is only one matric column and since all matric elements within a column must have the same number of simple columns, we have the same postscript, Q, for each element. In an actual case, each element on the right of Eq. (11.1.11) might be a matrix of all measures available on the military service personnel in a particular technical training school.

To express the type III row vector in terms of its matric elements, we write the right side of Eq. (11.1.9) as

$$_PX_{[b_s]'} = [\,_PX_{b_1} \quad _PX_{b_2} \quad \cdots \quad _PX_{b_s}\,] \qquad (11.1.12)$$

We now see that the prescript P in the right side of this equation indicates the number of simple rows of the matric elements, and the postscripts indicate the number of simple columns. Again, in an actual case, each element on the right of Eq. (11.1.12) might be a matrix of particular measures, such as test scores on all service personnel in all technical training schools.

Expanded notation for the supermatrix

To express the supermatrix in terms of its matric elements, we write the right side of Eq. (11.1.10) as in Eq. (11.1.13).

$$_{[a_r]}X_{[b_s]'} = \begin{bmatrix} a_1X_{b_1} & a_1X_{b_2} & \cdots & a_1X_{b_s} \\ a_2X_{b_1} & a_2X_{b_2} & \cdots & a_2X_{b_s} \\ \cdots & \cdots & \cdots & \cdots \\ a_rX_{b_1} & a_rX_{b_2} & \cdots & a_rX_{b_s} \end{bmatrix} \tag{11.1.13}$$

Again the prescripts in the right side of Eq. (11.1.13) indicate the number of simple rows of the submatrix, and the postscripts indicate the number of simple columns. You will note that all submatrices within a given matric row have the same prescript, and all submatrices within a given matric column have the same postscript. This must be so, of course, because of the way supermatrices are formed from simple matrices. Here each element on the right of Eq. (11.1.13) might be a matrix of a particular type of measures on personnel in a particular school.

We may use the numerical example of Fig. 11.1.3 to illustrate the submatrices on the right of Eq. (11.1.11). This we do in Fig. 11.1.7.

$$_3X_4 = \begin{bmatrix} 1 & 2 & 3 & 5 \\ 4 & 3 & 7 & 6 \\ 4 & 2 & 1 & 5 \end{bmatrix}$$

$$_2X_4 = \begin{bmatrix} 8 & 6 & 9 & 5 \\ 4 & 3 & 2 & 7 \end{bmatrix}$$

Fig. 11.1.7

To illustrate the submatrices on the right of Eq. (11.1.12), we use the numerical example in Fig. 11.1.5. This gives us Fig. 11.1.8.

$$_5X_{2_1} = \begin{bmatrix} 1 & 2 \\ 4 & 3 \\ 4 & 2 \\ 8 & 6 \\ 4 & 3 \end{bmatrix} \qquad _5X_{2_2} = \begin{bmatrix} 3 & 5 \\ 7 & 6 \\ 1 & 5 \\ 9 & 5 \\ 2 & 7 \end{bmatrix}$$

Fig. 11.1.8

Figure 11.1.9 uses the numerical example in Fig. 11.1.6 to illustrate the submatrices on the right of Eq. (11.1.13).

$$_{3_1}X_{2_1} = \begin{bmatrix} 1 & 2 \\ 4 & 3 \\ 4 & 2 \end{bmatrix} \qquad _{3_1}X_{2_2} = \begin{bmatrix} 3 & 5 \\ 7 & 6 \\ 1 & 5 \end{bmatrix}$$

$$_{2_2}X_{2_1} = \begin{bmatrix} 8 & 6 \\ 4 & 3 \end{bmatrix} \qquad _{2_2}X_{2_2} = \begin{bmatrix} 9 & 5 \\ 2 & 7 \end{bmatrix}$$

Fig. 11.1.9

Simplified notation for type III vectors

Next let us see how we may introduce a slight simplification into the notation of Eq. (11.1.11). We can express the right side as in Eq. (11.1.14).

$$
_{[a_r]}X_Q = \begin{bmatrix} a_1X \\ a_2X \\ \dots \\ a_rX \end{bmatrix}_Q
\tag{11.1.14}
$$

Notice that on the right of Eq. (11.1.14) we have dropped the postscript Q from each of the matric elements and written it as a postscript outside the brackets. This means that Q belongs as a postscript to each of the elements inside the brackets.

We can also simplify the right side of Eq. (11.1.12) by writing it as

$$
_PX_{[b_s]'} = {}_P[X_{b_1} \quad X_{b_2} \quad \dots \quad X_{b_s}]
\tag{11.1.15}
$$

On the right of Eq. (11.1.15) the P as a prescript to the brackets means that it belongs as a prescript to each of the matric elements.

In Eq. (11.1.14) the scalar subscript Q is a postscript; therefore we know from the left side of Eq. (11.1.14) alone that we have a type III column vector. If the scalar subscript is a prescript, we have a type III row vector. In Eq. (11.1.15) the scalar subscript P is a prescript, therefore we know from the left side of Eq. (11.1.15) alone that we have a type III row vector.

11.2   The Type IV Supervector

The supermatrix as a column supervector with type III row vector elements

We shall now see how we may express a supermatrix as a type of supervector. You recall that we expressed a simple matrix as a type II supervector whose elements were simple vectors. We begin by writing the matric row vectors on the right of Eq. (11.1.13) as in Eq. (11.2.1).

$$
\begin{aligned}
_{a_1}X_{[b_s]'} &= [_{a_1}X_{b_1} \quad _{a_1}X_{b_2} \quad \dots \quad _{a_1}X_{b_s}] \\
_{a_2}X_{[b_s]'} &= [_{a_2}X_{b_1} \quad _{a_2}X_{b_2} \quad \dots \quad _{a_2}X_{b_s}] \\
&\dots \quad\quad \dots \quad\quad \dots \quad \dots \\
_{a_r}X_{[b_s]'} &= [_{a_r}X_{b_1} \quad _{a_r}X_{b_2} \quad \dots \quad _{a_r}X_{b_s}]
\end{aligned}
\tag{11.2.1}
$$

But we have already seen that the prescripts for the elements of a type III row vector may be taken as a prescript outside the brackets, so that Eq. (11.2.1) may be written as in Eq. (11.2.2).

$$
\begin{aligned}
_{a_1}X_{[b_s]'} &= {}_{a_1}[X_{b_1} \quad X_{b_2} \quad \dots \quad X_{b_s}] \\
_{a_2}X_{[b_s]'} &= {}_{a_2}[X_{b_1} \quad X_{b_2} \quad \dots \quad X_{b_s}] \\
&\dots \quad\quad\quad \dots \\
_{a_r}X_{[b_s]'} &= {}_{a_r}[X_{b_1} \quad X_{b_2} \quad \dots \quad X_{b_s}]
\end{aligned}
\tag{11.2.2}
$$

Then because of Eqs. (11.2.1) and (11.2.2) we may express the supermatrix $_{[a_r]}X_{[b_s]'}$ as a column vector whose elements are type III row vectors. This we do in Eq. (11.2.3).

$$[_{[a_r]}X]_{[b_s]'} = \begin{bmatrix} a_1 X_{[b_s]'} \\ a_2 X_{[b_s]'} \\ \cdots \\ a_r X_{[b_s]'} \end{bmatrix} \tag{11.2.3}$$

But since the vector postscript is the same for each element in the right of Eq. (11.2.3) we may write

$$[_{[a_r]}X]_{[b_s]'} = \begin{bmatrix} a_1 X \\ a_2 X \\ \cdots \\ a_r X \end{bmatrix}_{[b_s]'} \tag{11.2.4}$$

As you know by this time, the vector postscript $[b_s]'$ outside the brackets on the right of Eq. (11.2.4) means that it belongs as a postscript to each element inside the brackets.

The supermatrix as a row supervector with type III column vector elements

We may also express the type III column vectors of the supermatrix on the right of Eq. (11.1.13) by means of Eq. (11.2.5).

$$_{[a_r]}X_{b_1} = \begin{bmatrix} a_1 X \\ a_2 X \\ \cdots \\ a_r X \end{bmatrix}_{b_1} \quad _{[a_r]}X_{b_2} = \begin{bmatrix} a_1 X \\ a_2 X \\ \cdots \\ a_r X \end{bmatrix}_{b_2} \quad \cdots \quad _{[a_r]}X_{b_s} = \begin{bmatrix} a_1 X \\ a_2 X \\ \cdots \\ a_r X \end{bmatrix}_{b_s}$$
$$\tag{11.2.5}$$

Because of Eq. (11.2.5), therefore, we may express the supermatrix $_{[a_r]}X_{[b_s]'}$ as a row supervector whose elements are type III column vectors. This we do in Eq. (11.2.6).

$$_{[a_r]}[X_{[b_s]'}] = {}_{[a_r]}[X_{b_1} \quad X_{b_2} \quad \cdots \quad X_{b_s}] \tag{11.2.6}$$

The vector prescript $[a_r]$ outside the brackets on the right of Eq. (11.2.6) means, of course, that it belongs as a prescript to each element within the brackets.

We shall now designate supervectors of the type given by Eqs. (11.2.4) and (11.2.6) as type IV supervectors. A type IV column vector is a column supervector whose elements are type III row vectors. A type IV row vector is a row supervector whose elements are type III column vectors.

Notice that we can tell from the left-hand sides of Eqs. (11.2.4) and (11.2.6) whether we have a column or a row type IV vector. If the post-partitioner is outside the brackets, we have a column type IV vector as in

Eq. (11.2.4). If the prepartitioner is outside the brackets, we have a type IV row vector, as in Eq. (11.2.6).

We may illustrate the type IV column vector of Eq. (11.2.4) by using the numerical example of Fig. 11.1.1 in Fig. 11.2.1.

$$[_{[a_r]}X]_{[b_s]'} = \left[_{\left[\begin{smallmatrix}3_1\\2_2\end{smallmatrix}\right]}X\right]_{[2_1 \quad 2_2]} = \begin{bmatrix} \begin{pmatrix} 1 & 2 \\ 4 & 3 \\ 4 & 2 \end{pmatrix} & \begin{pmatrix} 3 & 5 \\ 7 & 6 \\ 1 & 5 \end{pmatrix} \\ \begin{pmatrix} 8 & 6 \\ 4 & 3 \end{pmatrix} & \begin{pmatrix} 9 & 5 \\ 2 & 7 \end{pmatrix} \end{bmatrix}$$

FIG. 11.2.1

To illustrate the type IV row vector of Eq. (11.2.6), we use the same numerical example in Fig. 11.2.2.

$$_{[a_r]}[X_{[b_s]'}] = {}_{\left[\begin{smallmatrix}3_1\\2_2\end{smallmatrix}\right]}[X_{[2_1 \quad 2_2]}] = \begin{bmatrix} \begin{pmatrix} 1 & 2 \\ 4 & 3 \\ 4 & 2 \\ \hline 8 & 6 \\ 4 & 3 \end{pmatrix} & \begin{pmatrix} 3 & 5 \\ 7 & 6 \\ 1 & 5 \\ \hline 9 & 5 \\ 2 & 7 \end{pmatrix} \end{bmatrix}$$

FIG. 11.2.2

11.3 The Transpose Notation for the Supermatrix

Transpose of the type III column vector

Next let us see how we indicate the transpose of a supermatrix by means of the notation we have just developed. We shall begin with the type III column vector given by Eq. (11.1.11). To take the transpose of the left side of Eq. (11.1.11) we use a rule very similar to that for the transpose of a product of simple matrices. Here you recall we placed the symbols for the matrices in reverse order and took their transposes. Similarly we place the three symbols on the left of Eq. (11.1.11) in reverse order and take the transpose of each symbol, thus

$$[_{[a_r]}X_Q]' = {}_Q X'_{[a_r]'} \tag{11.3.1}$$

You will notice on the right of Eq. (11.3.1) that we have not primed the prescript Q since it is a scalar quantity.

To take the transpose of the right side of Eq. (11.1.11), we first write the column as a row, and indicate the transpose of each element just as we learned to do in taking the transpose of a supermatrix. This gives us Eq. (11.3.2).

$$\begin{bmatrix} {}_{a_1}X_Q \\ {}_{a_2}X_Q \\ \cdots \\ {}_{a_r}X_Q \end{bmatrix}' = [({}_{a_1}X_Q)' \quad ({}_{a_2}X_Q)' \quad \cdots \quad ({}_{a_r}X_Q)'] \tag{11.3.2}$$

To take the transposes of the elements in the right of Eq. (11.3.2) we must put the symbols in reverse order and take their transposes. But since the subscripts are all scalars, we prime only the X symbol. Therefore we have Eq. (11.3.3).

$$\left.\begin{array}{c} [_{a_1}X_Q]' = {}_Q X'_{a_1} \\ [_{a_2}X_Q]' = {}_Q X'_{a_2} \\ \cdots \qquad \cdots \\ [_{a_r}X_Q]' = {}_Q X'_{a_r} \end{array}\right\} \qquad (11.3.3)$$

Substituting the right side of Eq. (11.3.3) in the right of Eq. (11.3.2) gives Eq. (11.3.4).

$$\begin{bmatrix} _{a_1}X_Q \\ _{a_2}X_Q \\ \cdots \\ _{a_r}X_Q \end{bmatrix}' = [{}_Q X'_{a_1} \quad {}_Q X'_{a_2} \quad \cdots \quad {}_Q X'_{a_r}] \qquad (11.3.4)$$

But we can take the subscript Q in the right of Eq. (11.3.4) as a prescript to the brackets, as in Eq. (11.3.5).

$$\begin{bmatrix} _{a_1}X_Q \\ _{a_2}X_Q \\ \cdots \\ _{a_r}X_Q \end{bmatrix}' = {}_Q[X'_{a_1} \quad X'_{a_2} \quad \cdots \quad X'_{a_r}] \qquad (11.3.5)$$

From Eqs. (11.1.11), (11.3.1), and (11.3.5), therefore, we may write

$$({}_{[a_r]}X_Q)' = {}_Q X'_{[a_r]'} = {}_Q[X'_{a_1} \quad X'_{a_2} \quad \cdots \quad X'_{a_r}] \qquad (11.3.6)$$

Remember that on the right of Eq. (11.3.6) the Q belongs as a prescript to each of the elements.

Let us now compare the right side of Eq. (11.3.6) with the right side of Eq. (11.1.14) to see what rules we have used to take the transpose of the column type III vector. There are four distinct rules: (1) the column is written as a row, (2) the symbol X for the submatrices is primed, (3) the prescripts of the submatrices are written as postscripts, and (4) the postscript Q to the brackets is written as a prescript.

Transpose of a type III row vector

We should now be able to write the transpose of a type III row vector as given in Eq. (11.1.15). To take the transpose of the left side of Eq. (11.1.15) we reverse the order of the symbols and take their transposes. To take the transpose of the right side of Eq. (11.1.15) we use the following four rules: (1) The row is written as a column. (2) The symbol X for the submatrices is primed. (3) The postscripts of the submatrices are written as prescripts. (4) The prescript P to the brackets is written as a postscript.

We therefore get Eq. (11.3.7) from Eq. (11.1.15).

$$(_P X_{[b_s]'})' = {}_{[b_s]} X'_P = \begin{bmatrix} b_1 X' \\ b_2 X' \\ \cdots \\ b_s X' \end{bmatrix}_P \tag{11.3.7}$$

Examples of transposes of type III vectors

Let us see now what the transpose form of a type III column vector given by Eq. (11.3.6) means in terms of a concrete example. We have the column form given in Fig. 11.1.3. We can illustrate the transposed row form in Fig. 11.3.1.

$$_Q X'_{[a_r]'} = {}_4 X'_{[3_1 \quad 2_2]} = \begin{bmatrix} 1 & 4 & 4 & | & 8 & 4 \\ 2 & 3 & 2 & | & 6 & 3 \\ 3 & 7 & 1 & | & 9 & 2 \\ 5 & 6 & 5 & | & 5 & 7 \end{bmatrix}$$

Fig. 11.3.1

The submatrices of the row form are therefore as in Fig. 11.3.2.

$$_4 X'_{3_1} = \begin{bmatrix} 1 & 4 & 4 \\ 2 & 3 & 2 \\ 3 & 7 & 1 \\ 5 & 6 & 5 \end{bmatrix} \qquad _4 X'_{3_2} = \begin{bmatrix} 8 & 4 \\ 6 & 3 \\ 9 & 2 \\ 5 & 7 \end{bmatrix}$$

Fig. 11.3.2

It is clear that in simple form the right-hand side of Fig. 11.3.2 is simply the transpose of the right-hand side of Fig. 11.1.3. It is also clear that the left and right examples in Fig. 11.3.2 are simply the transposes of the upper and lower examples of Fig. 11.1.7. This is, of course, as it must be, and merely serves to illustrate concretely the notation we use in indicating the transpose of a type III column vector.

Next we shall see what the transpose form of a type III row vector given in Eq. (11.3.7) means in terms of a concrete example. We have the row form given in Fig. 11.1.5. We illustrate the transposed column form by Fig. 11.3.3.

$$_{[b_s]} X'_P = {}_{[2_1]}_{[2_2]} X'_5 = \begin{bmatrix} 1 & 4 & 4 & 8 & 4 \\ 2 & 3 & 2 & 6 & 3 \\ \hline 3 & 7 & 1 & 9 & 2 \\ 5 & 6 & 5 & 5 & 7 \end{bmatrix}$$

Fig. 11.3.3

The submatrices of the column form would be the transposes of the submatrices in Fig. 11.1.8 of the row form. These we show in Fig. 11.3.4.

$$_{21}X'_5 = \begin{bmatrix} 1 & 4 & 4 & 8 & 4 \\ 2 & 3 & 2 & 6 & 3 \end{bmatrix}$$

$$_{22}X_5 = \begin{bmatrix} 3 & 7 & 1 & 9 & 2 \\ 5 & 6 & 5 & 5 & 7 \end{bmatrix}$$

Fig. 11.3.4

The transpose of a type IV supervector

We can now show how we indicate the transpose of a type IV vector. First we shall consider the transpose of the type IV column vector in Eq. (11.2.4). The rules are very much the same as for indicating the transpose of a type III column vector.

To take the transpose of the expression on the left of Eq. (11.2.4), we write the symbols in reverse order and take the transpose of each symbol. Since both subscripts are vectors, this means that the transpose of each must be indicated. We therefore write the transpose of the left side of Eq. (11.2.4) as Eq. 11.3.8.

$$([_{[a_r]}X]_{[b_s]'}) = {}_{[b_s]}[X'_{[a_r]'}] \tag{11.3.8}$$

The transpose of the right side of Eq. (11.2.4) is taken exactly as it is for a type III vector except that the row vector postscript $[b_s]'$ must be unprimed when it becomes a prescript. We have therefore Eq. (11.3.9) as the transpose of Eq. (11.2.4).

$$([_{[a_r]}X]_{[b_s]'})' = {}_{[b_s]}[X'_{[a_r]'}] = {}_{[b_s]}[X'_{a_1} \quad X'_{a_2} \quad \ldots \quad X'_{a_r}] \tag{11.3.9}$$

Using the numerical example in Fig. 11.2.1, we may illustrate Eq. (11.3.9) by Fig. 11.3.5.

$$_{[b_s]}[X_{[a_r]'}] = {}_{\begin{bmatrix}2_1\\2_2\end{bmatrix}}[X'_{[3_1} \quad {}_{2_2]}] = \left[\left(\begin{array}{ccc} 1 & 4 & 4 \\ 2 & 3 & 2 \\ \hline 3 & 7 & 1 \\ 5 & 6 & 5 \end{array} \right) \left(\begin{array}{cc} 8 & 4 \\ 6 & 3 \\ \hline 9 & 2 \\ 5 & 7 \end{array} \right) \right]$$

Fig. 11.3.5

The type III column vector elements on the right side of Eq. (11.3.9) are illustrated by Fig. 11.3.6.

$$_{\begin{bmatrix}2_1\\2_2\end{bmatrix}}X'_{3_1} = \begin{bmatrix} 1 & 4 & 4 \\ 2 & 3 & 2 \\ \hline 3 & 7 & 1 \\ 5 & 6 & 5 \end{bmatrix} \qquad _{\begin{bmatrix}2_1\\2_2\end{bmatrix}}X'_{2_2} = \begin{bmatrix} 8 & 4 \\ 6 & 3 \\ \hline 9 & 2 \\ 5 & 7 \end{bmatrix}$$

Fig. 11.3.6

You see then that aside from the fact that the transposes of vector subscripts must be indicated, the rules for taking the transpose of the type IV column vector given by Eq. (11.2.4) are the same as those for taking the transpose of the type III column vector given by Eq. (11.1.14). This is also true for the transpose of the type IV row vector. We therefore write the transpose of the type IV row vector directly from Eq. (11.2.6) as in Eq. (11.3.10).

$$(_{[a_r]}[X_{[b_s]'}])' = [_{[b_s]}X']_{[a_r]'} = \begin{bmatrix} _{b_1}X' \\ _{b_2}X' \\ \cdots \\ _{b_s}X' \end{bmatrix}_{[a_r]'} \tag{11.3.10}$$

Using the numerical example in Fig. 11.2.2, Eq. (11.3.10) may be illustrated by Fig. 11.3.7.

$$[_{[b_s]}X']_{[a_r]'} = \begin{bmatrix} _{[2_1 \atop 2_2]}X' \end{bmatrix}_{[3_1 \quad 2_2]} = \begin{bmatrix} \begin{pmatrix} 1 & 4 & 4 & | & 8 & 4 \\ 2 & 3 & 2 & | & 6 & 3 \\ 3 & 7 & 1 & | & 9 & 2 \\ 5 & 6 & 5 & | & 5 & 7 \end{pmatrix} \end{bmatrix}$$

Fig. 11.3.7

The type III row vector elements on the right of Eq. (11.3.10) are illustrated by Fig. 11.3.8.

$$_{2_1}X'_{[3_1 \quad 2_2]} = \begin{bmatrix} 1 & 4 & 4 & | & 8 & 4 \\ 2 & 3 & 2 & | & 6 & 3 \end{bmatrix}$$

$$_{2_2}X'_{[3_1 \quad 2_2]} = \begin{bmatrix} 3 & 7 & 1 & | & 9 & 2 \\ 5 & 6 & 5 & | & 5 & 7 \end{bmatrix}$$

Fig. 11.3.8

11.4 The Minor Product of Type III Vectors

Notation for the minor product

Before considering how the products of type IV vectors or supermatrices are formed we shall first see how we multiply type III vectors. As in multiplication of simple vectors, we have both a minor and major product. First we shall consider the minor product of two type III vectors. As you would suppose, the minor product of two type III vectors consists of a row vector postmultiplied by a column vector, just as in the case of simple vectors. Here the minor product of two type III vectors is the sum of products of corresponding elements. Therefore, it must be clear that for the minor product of two type III vectors to exist the vectors must have the same number of elements. Furthermore, since the products are formed from corresponding elements of the two vectors, the number of simple columns in a given element of the prefactor must be the same as the number of simple rows in the corresponding element of the postfactor. A concise way of combining these rules is as follows: for the minor product of two type III vectors to exist, the postpartitioner of the prefactor must be the transpose of the prepartitioner of the postfactor. This rule can be easily demonstrated. Suppose we have a column type III vector given by Eq. (11.4.1).

$$_{[b_s]}Y_M = \begin{bmatrix} _{b_1}Y \\ _{b_2}Y \\ \cdots \\ _{b_s}Y \end{bmatrix}_M \tag{11.4.1}$$

We wish to premultiply both sides of Eq. (11.4.1) by the corresponding sides of Eq. (11.1.15). This gives us the minor product indicated by Eq. (11.4.2).

$$_PX_{[b_s]'}\,_{[b_s]}Y_M = _P[X_{b_1} \quad X_{b_2} \quad \cdots \quad X_{b_s}]\begin{bmatrix} _{b_1}Y \\ _{b_2}Y \\ \cdots \\ _{b_s}Y \end{bmatrix}_M \tag{11.4.2}$$

Multiplying out the right side of Eq. (11.4.2) and attaching the external scalar subscripts P and M to the product terms, we have

$$_PX_{[b_s]'}\,_{[b_s]}Y_M = [_PX_{b_1}\,_{b_1}Y_M + _PX_{b_2}\,_{b_2}Y_M + \ldots + _PX_{b_s}\,_{b_s}Y_M] \tag{11.4.3}$$

Or if we sum the products on the right of Eq. (11.4.3) to get a single matrix Z, we have

$$_PX_{[b_s]'}\,_{[b_s]}Y_M = _PZ_M \tag{11.4.4}$$

Notice now on the left of Eq. (11.4.4) that the vector postscript $[b_s]'$ of X is the transpose of the vector prescript $[b_s]$ of Y. Notice also that the right side of Eq. (11.4.4) is a simple matrix with the same number of rows, P, as the prefactor X and the same number of columns, M, as the postfactor Y. Furthermore, there is nothing to indicate on the right of Eq. (11.4.4) what were the adjacent subscripts or common superorder of the factors on the left. These rules of dimensionality are analogous to those for simple matrices.

Numerical illustration

We can illustrate Eqs. (11.4.3), (11.4.4), and (11.4.5) by Fig. 11.4.1.

$$_{3_1}X_{[2_1 \quad 1_2 \quad 3_3]}\begin{bmatrix} 2_1 \\ 1_2 \\ 3_3 \end{bmatrix}Y_{2_1} = \begin{bmatrix} 2 & 3 & | & 4 & | & 3 & 2 & 3 \\ 1 & 4 & | & 1 & | & 1 & 1 & 2 \\ 2 & 1 & | & 2 & | & 4 & 3 & 2 \end{bmatrix}\begin{bmatrix} 2 & 1 \\ 3 & 2 \\ \hline 4 & 1 \\ \hline 5 & 3 \\ 1 & 2 \\ 4 & 3 \end{bmatrix} =$$

Fig. 11.4.1 (Cont. on next page)

$$
\begin{bmatrix} 2 & 3 \\ 1 & 4 \\ 2 & 1 \end{bmatrix}\begin{bmatrix} 2 & 1 \\ 3 & 2 \end{bmatrix} + \begin{bmatrix} 4 \\ 1 \\ 2 \end{bmatrix}\begin{bmatrix} 4 & 1 \end{bmatrix} + \begin{bmatrix} 3 & 2 & 3 \\ 1 & 1 & 2 \\ 4 & 3 & 2 \end{bmatrix}\begin{bmatrix} 5 & 3 \\ 1 & 2 \\ 4 & 3 \end{bmatrix} =
$$

$$
\begin{bmatrix} 13 & 8 \\ 14 & 9 \\ 7 & 4 \end{bmatrix} + \begin{bmatrix} 16 & 4 \\ 4 & 1 \\ 8 & 2 \end{bmatrix} + \begin{bmatrix} 29 & 22 \\ 14 & 11 \\ 31 & 24 \end{bmatrix} = \begin{bmatrix} 58 & 34 \\ 32 & 21 \\ 46 & 30 \end{bmatrix}
$$

Fig. 11.4.1 (Cont.)

Figure 11.4.2 is a diagrammatic representation of the right-hand side of the equation in Fig. 11.4.1.

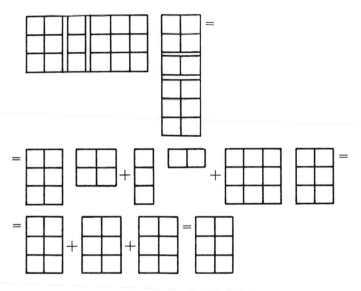

Fig. 11.4.2

Transpose of the minor product

To take the transpose of the minor product of type III vectors involves no new principles. On the left side of Eq. (11.4.3) we take the product of the transposes in reverse order. Equations (11.3.5) and (11.3.6) show how we take the transpose of a type III column vector, and Eq. (11.3.7) shows how we take the transpose of a type III row vector. We therefore get Eq. (11.4.5) from (11.4.2).

$$
[_P X_{[b_s]'} \ _{[b_s]} Y_M]' = \ _M Y'_{[b_s]'} \ _{[b_s]} X'_P = \ _M[Y'_{b_1} \ \ Y'_{b_2} \ \ \cdots \ \ Y'_{b_s}]\begin{bmatrix} _{b_1}X' \\ _{b_2}X' \\ \cdots \\ _{b_s}X' \end{bmatrix}_P
$$

$$(11.4.5)$$

Multiplying out the right side of Eq. (11.4.5) and attaching the external scalar subscripts M and P to the product terms, we get

$$_M Y'_{[b_s]'}{}_{[b_s]} X'_P = [_M Y'_{b_1}{}_{b_1} X'_P + _M Y'_{b_2}{}_{b_2} X'_P + \ldots + _M Y'_{b_s}{}_{b_s} X'_P]$$

$$(11.4.6)$$

Notice that we could have obtained the right side of Eq. (11.4.6) directly by taking the transposes of each of the product terms on the right of Eq. (11.4.3). If we sum the products on the right of Eq. (11.4.6), we may write

$$_M Y'_{[b_s]'}{}_{[b_s]} X'_P = _M Z'_P \qquad (11.4.7)$$

Using the same numerical example as in Fig. 11.4.1, Eqs. (11.4.6) and (11.4.7) are illustrated in Fig. 11.4.3.

$$_{2_1} Y'_{[2_1 \ 1_2 \ 3_3]} {}_{\begin{bmatrix}2_1\\1_2\\3_3\end{bmatrix}} X'_{3_1} = \begin{bmatrix}2 & 3\\1 & 2\end{bmatrix}\begin{bmatrix}2 & 1 & 2\\3 & 4 & 1\end{bmatrix} + \begin{bmatrix}4\\1\end{bmatrix}[4 \ 1 \ 2] + \begin{bmatrix}5 & 1 & 4\\3 & 2 & 3\end{bmatrix}\begin{bmatrix}3 & 1 & 4\\2 & 1 & 3\\3 & 2 & 2\end{bmatrix} =$$

$$\begin{bmatrix}13 & 14 & 7\\8 & 9 & 4\end{bmatrix} + \begin{bmatrix}16 & 4 & 8\\4 & 1 & 2\end{bmatrix} + \begin{bmatrix}29 & 14 & 31\\22 & 11 & 24\end{bmatrix} = \begin{bmatrix}58 & 32 & 46\\34 & 21 & 30\end{bmatrix}$$

FIG. 11.4.3

11.5   The Minor Product Moment of a Type III Vector

We define the minor product moment of a type III vector in the same way as for a simple vector. The minor product moment of a type III vector is a type III column vector premultiplied by its transpose. We must therefore distinguish two kinds of minor products of type III vectors depending on whether the type III vector is a column vector or row vector. Remember that a matrix may be partitioned either into a type III column or row vector, whether the simple form is vertical or horizontal. Ordinarily we would think of the natural order of a simple data matrix as being vertical. However, a vertical simple matrix, as you know, may be partitioned into a row type III vector. Therefore, whether a type III vector is a row or column tells us nothing about whether the simple form is vertical or horizontal. Therefore, even though with simple matrices and vectors the use of the prime ordinarily indicates a horizontal form, with supermatrices and their elements it indicates only a transpose.

Minor product moment of type III row vector

First, then, let us consider the minor product moment of a type III row vector. Using Eqs. (11.1.15) and (11.3.7), we have

$$_P X_{[b_s]'}{}_{[b_s]} X'_P = _P[X_{b_1} \ \ X_{b_2} \ \ \ldots \ \ X_{b_s}]\begin{bmatrix}_{b_1}X'\\_{b_2}X'\\\ldots\\_{b_s}X'\end{bmatrix}_P \qquad (11.5.1)$$

Multiplying out the right side of Eq. (11.5.1) gives

$$_PX_{[b_s]'\ [b_s]}X'_P = [_PX_{b_1\ b_1}X'_P + _PX_{b_2\ b_2}X'_P + \ldots + _PX_{b_s\ b_s}X'_P]$$
(11.5.2)

Notice now that although Eq. (11.5.2) is a minor product moment, the primed factor is the postfactor on the left side of the equation and in each term on the right side also. This is different from the practice you are used to in the case of minor products of simple matrices and vectors, where the prefactor is customarily the primed factor. Note that each term on the right of Eq. (11.5.2) is a product moment of a simple matrix. It may, however, be either a major or minor product, depending on whether its common order b_i is greater or less than the distinct order P. To illustrate Eq. (11.5.1) and (11.5.2), we can use the numerical example for X in Fig. 11.4.1 and get Fig. 11.5.1.

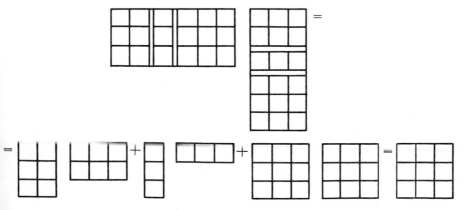

$$_{3_1}X_{[?_1\ \ 1_2\ \ 3_2]}\begin{bmatrix}2_1\\1_2\\3_2\end{bmatrix}X'_{3_1} = \left[\begin{array}{cc|c|ccc}2&3&4&3&2&3\\1&4&1&1&1&2\\2&1&2&4&3&2\end{array}\right]\left[\begin{array}{ccc}2&1&2\\3&4&1\\\hline4&1&2\\\hline3&1&4\\2&1&3\\3&2&2\end{array}\right] =$$

$$\begin{bmatrix}2&3\\1&4\\2&1\end{bmatrix}\begin{bmatrix}2&1&2\\3&4&1\end{bmatrix} + \begin{bmatrix}4\\1\\2\end{bmatrix}[4\ 1\ 2] + \begin{bmatrix}3&2&3\\1&1&2\\4&3&2\end{bmatrix}\begin{bmatrix}3&1&4\\2&1&3\\3&2&2\end{bmatrix} =$$

$$\begin{bmatrix}13&14&7\\14&17&6\\7&6&5\end{bmatrix} + \begin{bmatrix}16&4&8\\4&1&2\\8&2&4\end{bmatrix} + \begin{bmatrix}22&11&24\\11&6&11\\24&11&29\end{bmatrix} = \begin{bmatrix}51&29&39\\29&24&19\\39&19&38\end{bmatrix}$$

FIG. 11.5.1

Figure 11.5.2 shows diagrammatically how the minor product moment of a type III vector is formed.

FIG. 11.5.2

You will note that the product terms in Figs. 11.5.1 and 11.5.2 are all major product moments except the last.

Minor product moment of type III column vector

Next let us consider the minor product moment of a type III column vector. Suppose we take the minor product moment of the type III column vector given by Eq. (11.4.1). This would give us Eq. (11.5.3).

$$
{}_M Y'_{[b_s]'}\,{}_{[b_s]} Y_M = {}_M [Y'_{b_1} \quad Y'_{b_2} \quad \ldots \quad Y'_{b_s}]
\begin{bmatrix}
{}_{b_1} Y \\
{}_{b_2} Y \\
\ldots \\
{}_{b_s} Y
\end{bmatrix}_M
\tag{11.5.3}
$$

Multiplying out the right side of Eq. (11.5.3) gives

$$
{}_M Y'_{[b_s]'}\,{}_{[b_s]} Y_M = [{}_M Y'_{b_1}\,{}_{b_1} Y_M + {}_M Y'_{b_2}\,{}_{b_2} Y_M + \ldots + {}_M Y'_{b_s}\,{}_{b_s} Y_M]
\tag{11.5.4}
$$

Using the numerical example for Y in Fig. 11.4.1, we illustrate Eq. (11.5.3) with Fig. 11.5.3.

$$
{}_{2_1} Y'_{[2_1 \quad 1_2 \quad 3_3]}
\begin{bmatrix} {}_{2_1} \\ {}_{1_2} \\ {}_{3_3} \end{bmatrix} Y_{2_1}
=
\begin{bmatrix} 2 & 3 & | & 4 & | & 5 & 1 & 4 \\ 1 & 2 & | & 1 & | & 3 & 2 & 3 \end{bmatrix}
\begin{bmatrix}
2 & 1 \\
3 & 2 \\
\hline
4 & 1 \\
\hline
5 & 3 \\
1 & 2 \\
4 & 3
\end{bmatrix}
=
$$

$$
\begin{bmatrix} 2 & 3 \\ 1 & 2 \end{bmatrix}\begin{bmatrix} 2 & 1 \\ 3 & 2 \end{bmatrix}
+
\begin{bmatrix} 4 \\ 1 \end{bmatrix}[4 \quad 1]
+
\begin{bmatrix} 5 & 1 & 4 \\ 3 & 2 & 3 \end{bmatrix}\begin{bmatrix} 5 & 3 \\ 1 & 2 \\ 4 & 3 \end{bmatrix}
=
$$

$$
\begin{bmatrix} 13 & 8 \\ 8 & 5 \end{bmatrix}
+
\begin{bmatrix} 16 & 4 \\ 4 & 1 \end{bmatrix}
+
\begin{bmatrix} 42 & 29 \\ 29 & 22 \end{bmatrix}
=
\begin{bmatrix} 71 & 41 \\ 41 & 28 \end{bmatrix}
$$

FIG. 11.5.3

If now you look at the right-hand sides of the equations in Figs. 11.5.1 and 11.5.3, you have no way of knowing which is the minor product moment of a type III row vector and which is the minor product moment of a type III column vector. However, you can readily tell from the left-hand sides of the equations. In the former case the primed factor is the post-factor, while in the latter case it is the prefactor. In the same way you can tell from both sides of the equation that Eq. (11.5.2) is the minor product moment of a row vector and (11.5.4) is the minor product moment of a column vector. You can always tell that a product moment is a minor product moment because adjacent subscripts are vectors (enclosed in brackets) rather than scalars.

11.6 The Major Product of Type III Vectors

Notation for the major product

As in the case of simple vectors, the major product of type III vectors has a column vector for the prefactor and a row vector for the postfactor. The element in the ith row and jth column is the ith element of the prefactor postmultiplied by the jth element of the postfactor. Equation (11.6.1) shows how we indicate the major product of two type III vectors.

$$
_{[a_r]}X_Q \, _QY_{[c_t]'} = \begin{bmatrix} a_1X \\ a_2X \\ \cdots \\ a_rX \end{bmatrix}_Q \, _Q[Y_{c_1} \quad Y_{c_1} \quad \cdots \quad Y_{c_t}] \tag{11.6.1}
$$

Expanding the right side of Eq. (11.6.1) gives Eq. (11.6.2).

$$
_{[a_r]}X_Q \, _QY_{[c_t]'} = \begin{bmatrix} a_1X_Q \, _QY_{c_1} & a_1X_Q \, _QY_{c_2} & \cdots & a_1X_Q \, _QY_{c_t} \\ a_2X_Q \, _QY_{c_1} & a_2X_Q \, _QY_{c_2} & \cdots & a_2X_Q \, _QY_{c_t} \\ \cdots & \cdots & \cdots & \cdots \\ a_rX_Q \, _QY_{c_1} & a_rX_Q \, _QY_{c_2} & \cdots & a_rX_Q \, _QY_{c_t} \end{bmatrix}
$$

$$\tag{11.6.2}$$

The right side of Eq. (11.6.2) may be written more compactly as

$$
_{[a_r]}X_Q \, _QY_{[c_t]'} = \,_{[a_r]}Z_{[c_t]'} \tag{11.6.3}
$$

Now let us summarize the rules for the major product of type III vectors as indicated in Eq. (11.6.3). First, we see from the left side of this equation that the scalar postscript Q of the prefactor must be the same as the scalar prescript of the postfactor. This merely means that the factors must be conformable, as they must be for simple matrices. Second, we see that the product on the right of the equation is a supermatrix that is prepartitioned as the prefactor and postpartitioned as the postfactor. Finally, we see that the product tells us nothing about the common dimension of the factors.

Numerical illustration

A numerical example of the major product of type III vectors is given in Fig. 11.6.1.

$$
_{\begin{bmatrix} 2_1 \\ 1_2 \\ 3_2 \end{bmatrix}}X_2 \, _2Y_{[1_1 \ 2_2]} = \begin{bmatrix} 2 & 1 \\ 3 & 2 \\ \hline 1 & 4 \\ \hline 2 & 3 \\ 1 & 1 \\ 4 & 2 \end{bmatrix} \begin{bmatrix} 4 & 2 & 3 \\ 1 & 1 & 2 \end{bmatrix} =
$$

Fig. 11.6.1 (Cont. on next page)

$$\left[\begin{array}{c|c} \begin{pmatrix} 2 & 1 \\ 3 & 2 \end{pmatrix}\begin{pmatrix} 4 \\ 1 \end{pmatrix} & \begin{pmatrix} 2 & 1 \\ 3 & 2 \end{pmatrix}\begin{pmatrix} 2 & 3 \\ 1 & 2 \end{pmatrix} \\ \hline (1\ \ 4)\begin{pmatrix} 4 \\ 1 \end{pmatrix} & (1\ \ 4)\begin{pmatrix} 2 & 3 \\ 1 & 2 \end{pmatrix} \\ \hline \begin{pmatrix} 2 & 3 \\ 1 & 1 \\ 4 & 2 \end{pmatrix}\begin{pmatrix} 4 \\ 1 \end{pmatrix} & \begin{pmatrix} 2 & 3 \\ 1 & 1 \\ 4 & 2 \end{pmatrix}\begin{pmatrix} 2 & 3 \\ 1 & 2 \end{pmatrix} \end{array} \right] = \left[\begin{array}{c|cc} 9 & 5 & 8 \\ 14 & 8 & 13 \\ \hline 8 & 6 & 11 \\ \hline 11 & 7 & 12 \\ 5 & 3 & 5 \\ 18 & 10 & 16 \end{array} \right]$$

Fig. 11.6.1 (Cont.)

Using the same orders as in Fig. 11.6.1, we illustrate the major product diagrammatically in Fig. 11.6.2.

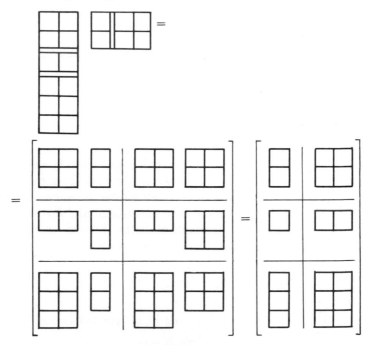

Fig. 11.6.2

Transpose of the major product

The transpose of the major product of type III vectors involves no new rules. We proceed as follows with Eq. (11.6.2). On the left we write the symbols in reverse order and take their transposes. On the right we write rows as columns and take the transpose of each element. We have then Eq. (11.6.4).

$$
_{[c_t]}Y'_Q {}_QX'_{[a_r]'} = \begin{bmatrix} c_1Y'_Q {}_QX'_{a_1} & c_1Y'_Q {}_QX'_{a_2} & \cdots & c_1Y'_Q {}_QX'_{a_r} \\ c_2Y'_Q {}_QX'_{a_1} & c_2Y'_Q {}_QX'_{a_2} & \cdots & c_2Y'_Q {}_QX'_{a_r} \\ \cdots & \cdots & \cdots & \cdots \\ c_tY'_Q {}_QX'_{a_1} & c_tY'_Q {}_QX'_{a_2} & \cdots & c_tY'_Q {}_QX'_{a_r} \end{bmatrix}
$$

$$(11.6.4)$$

More compactly we could take the transpose of both sides of Eq. (11.6.3) to get

$$
_{[c_t]}Y'_Q {}_QX'_{[a_r]'} = {}_{[c_t]}Z'_{[a_r]}
$$

$$(11.6.5)$$

Figure 11.6.3 shows the transpose of the numerical example given in Fig. 11.6.1.

$$
{}_{\begin{bmatrix}1_1\\2_2\end{bmatrix}}Y'_2 {}_2X'_{[2_1 \quad 1_2 \quad 3_3]} = \begin{bmatrix} \dfrac{\begin{matrix}4 & 1\end{matrix}}{\begin{matrix}2 & 1\\3 & 2\end{matrix}} \end{bmatrix} \begin{bmatrix} 2 & 3 & | & 1 & | & 2 & 1 & 4 \\ 1 & 2 & | & 4 & | & 3 & 1 & 2 \end{bmatrix} =
$$

$$
\begin{bmatrix} (4 \quad 1)\begin{pmatrix}2 & 3\\1 & 2\end{pmatrix} & (4 \quad 1)\begin{pmatrix}1\\4\end{pmatrix} & (4 \quad 1)\begin{pmatrix}2 & 1 & 4\\3 & 1 & 2\end{pmatrix} \\ \hline \begin{pmatrix}2 & 1\\3 & 2\end{pmatrix}\begin{pmatrix}2 & 3\\1 & 2\end{pmatrix} & \begin{pmatrix}2 & 1\\3 & 2\end{pmatrix}\begin{pmatrix}1\\4\end{pmatrix} & \begin{pmatrix}2 & 1\\3 & 2\end{pmatrix}\begin{pmatrix}2 & 1 & 4\\3 & 1 & 2\end{pmatrix} \end{bmatrix} =
$$

$$
\begin{bmatrix} 9 & 14 & | & 8 & | & 11 & 5 & 18 \\ \hline 5 & 8 & | & 6 & | & 7 & 3 & 10 \\ 8 & 13 & | & 11 & | & 12 & 5 & 16 \end{bmatrix}
$$

FIG. 11.6.3

11.7 The Major Product Moment of Type III Vectors

The major product moment of a type III column vector

As in the case of minor product moments of type III vectors, we also have two kinds of major product moments, namely, the major product moment from a column vector and the major product moment from a row vector. Equation (11.7.1) indicates the major product moment of a column vector.

$$
_{[a_r]}X_Q {}_QX'_{[a_r]'} = \begin{bmatrix} a_1X \\ a_2X \\ \cdots \\ a_rX \end{bmatrix}_Q {}_Q[X'_{a_1} \quad X'_{a_2} \quad \cdots \quad X'_{a_r}]
$$

$$(11.7.1)$$

Expanding the right side of Eq. (11.7.1) gives Eq. (11.7.2).

$$
_{[a_r]}X_Q {}_QX'_{[a_r]'} = \begin{bmatrix} a_1X_Q {}_QX'_{a_1} & a_1X_Q {}_QX'_{a_2} & \cdots & a_1X_Q {}_QX'_{a_r} \\ a_2X_Q {}_QX'_{a_1} & a_2X_Q {}_QX'_{a_2} & \cdots & a_2X_Q {}_QX'_{a_r} \\ \vdots & \vdots & & \vdots \\ a_rX_Q {}_QX'_{a_1} & a_rX_Q {}_QX'_{a_2} & \cdots & a_rX_Q {}_QX'_{a_r} \end{bmatrix}
$$

$$(11.7.2)$$

A numerical example of a major product moment of a type III column vector is given by Fig. 11.7.1.

$$\begin{bmatrix} 2_1 \\ 1_2 \\ 3_3 \end{bmatrix} X_{2\,2} X'_{[2_1 \quad 1_2 \quad 3_3]} = \begin{bmatrix} 2 & 1 \\ 3 & 2 \\ \hline 1 & 4 \\ \hline 2 & 3 \\ 1 & 1 \\ 4 & 2 \end{bmatrix} \begin{bmatrix} 2 & 3 & 1 & 2 & 1 & 4 \\ 1 & 2 & 4 & 3 & 1 & 2 \end{bmatrix} =$$

$$\begin{bmatrix} \begin{pmatrix} 2 & 1 \\ 3 & 2 \end{pmatrix}\begin{pmatrix} 2 & 3 \\ 1 & 2 \end{pmatrix} & \begin{pmatrix} 2 & 1 \\ 3 & 2 \end{pmatrix}\begin{pmatrix} 1 \\ 4 \end{pmatrix} & \begin{pmatrix} 2 & 1 \\ 3 & 2 \end{pmatrix}\begin{pmatrix} 2 & 1 & 4 \\ 3 & 1 & 2 \end{pmatrix} \\ \hline (1 \quad 4)\begin{pmatrix} 2 & 3 \\ 1 & 2 \end{pmatrix} & (1 \quad 4)\begin{pmatrix} 1 \\ 4 \end{pmatrix} & (1 \quad 4)\begin{pmatrix} 2 & 1 & 4 \\ 3 & 1 & 2 \end{pmatrix} \\ \hline \begin{pmatrix} 2 & 3 \\ 1 & 1 \\ 4 & 2 \end{pmatrix}\begin{pmatrix} 2 & 3 \\ 1 & 2 \end{pmatrix} & \begin{pmatrix} 2 & 3 \\ 1 & 1 \\ 4 & 2 \end{pmatrix}\begin{pmatrix} 1 \\ 4 \end{pmatrix} & \begin{pmatrix} 2 & 3 \\ 1 & 1 \\ 4 & 2 \end{pmatrix}\begin{pmatrix} 2 & 1 & 4 \\ 3 & 1 & 2 \end{pmatrix} \end{bmatrix} =$$

$$\begin{bmatrix} 5 & 8 & 6 & 7 & 3 & 10 \\ 8 & 13 & 11 & 12 & 5 & 16 \\ \hline 6 & 11 & 17 & 14 & 5 & 12 \\ \hline 7 & 12 & 14 & 13 & 5 & 14 \\ 3 & 5 & 5 & 5 & 2 & 6 \\ 10 & 16 & 12 & 14 & 6 & 20 \end{bmatrix}$$

Fig. 11.7.1

It should be clear from Fig. 11.7.1 that the major product moment of a type III vector is a symmetric simple matrix symmetrically partitioned.

The major product moment is illustrated diagrammatically by Fig. 11.7.2.

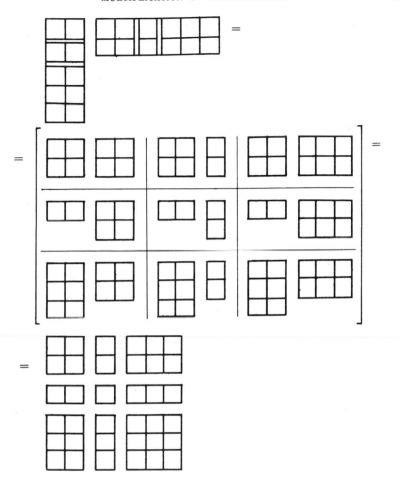

Fig. 11.7.2

The major product moment of a type III row vector

To indicate the major product moment of a type III row vector we would simply premultiply the row vector by its transpose. The major product moment of a type III row vector is given by Eq. (11.7.3).

$$
_{[b_s]}X'_P{}_PX_{[b_s]'} = \begin{bmatrix} _{b_1}X'_P{}_PX_{b_1} & _{b_1}X'_P{}_PX_{b_2} & \cdots & _{b_1}X'_P{}_PX_{b_s} \\ _{b_2}X'_P{}_PX_{b_1} & _{b_2}X'_P{}_PX_{b_2} & \cdots & _{b_2}X'_P{}_PX_{b_s} \\ \cdots & \cdots & \cdots & \cdots \\ _{b_s}X'_P{}_PX_{b_1} & _{b_s}X'_P{}_PX_{b_2} & \cdots & _{b_s}X'_P{}_PX_{b_s} \end{bmatrix}
$$

$$(11.7.0)$$

In Eq. (11.7.2), which represents the major product moment of a type 3 column vector, the primed matrices are the postfactors on both sides of the equation. In Eq. (11.7.3), which represents the major product moment of a type III row vector, the primed matrices are the prefactors on both sides of the equations.

11.8 Minor Products of Type IV Vectors

We have seen in Section 11.2 how a supermatrix may be expressed as a row vector whose elements are type III column vectors, or a column vector whose elements are type III row vectors. We have also seen how the major and minor products of type III vectors are formed. We shall now consider how the products of supermatrices may be expressed in terms of the products of the type III vector elements from type IV vectors.

Notation for the minor product of type IV vectors

First we shall consider the minor product of two type IV vectors. The method corresponds to that used for simple vectors and lower type supervectors; the minor product is formed by premultiplying a type IV column vector by a type IV row vector.

Suppose we have given the supermatrix Y as in Eq. (11.8.1).

$$
_{[b_s]}Y_{[c_t]'} = \begin{bmatrix} b_1 Y_{c_1} & b_1 Y_{c_2} & \ldots & b_1 Y_{c_t} \\ b_2 Y_{c_1} & b_2 Y_{c_2} & \ldots & b_2 Y_{c_t} \\ \ldots & \ldots & \ldots & \ldots \\ b_s Y_{c_1} & b_s Y_{c_2} & \ldots & b_s Y_{c_t} \end{bmatrix} \tag{11.8.1}
$$

We know from Section 11.2 that Y can be expressed as a type IV column vector by Eq. (11.8.2).

$$
[_{[b_s]}Y]_{[c_t]'} = \begin{bmatrix} b_1 Y \\ b_2 Y \\ \ldots \\ b_s Y \end{bmatrix}_{[c_t]'} \tag{11.8.2}
$$

Now let us premultiply both sides of Eq. (11.8.2) by the corresponding sides of Eq. (11.2.6). This gives us Eq. (11.8.3).

$$
_{[a_r]}[X_{[b_s]'}][_{[b_s]}Y]_{[c_t]'} = {}_{[a_r]}[X_{b_1} \quad X_{b_2} \quad \ldots \quad X_{b_s}] \begin{bmatrix} b_1 Y \\ b_2 Y \\ \ldots \\ b_s Y \end{bmatrix}_{[c_t]'} \tag{11.8.3}
$$

Multiplying out the right side of Eq. (11.8.3) gives

$$
_{[a_r]}[X_{[b_s]'}][_{[b_s]}Y]_{[c_t]} = {}_{[a_r]}X_{b_1}\, b_1 Y_{[c_t]} + {}_{[a_r]}X_{b_2}\, b_2 Y_{[c_t]} + \ldots + {}_{[a_r]}X_{b_s}\, b_s Y_{[c_t]'} \tag{11.8.4}
$$

If we sum the products on the right side of Eq. (11.8.4) to get a single matrix Z, we have

$$
_{[a_r]}[X_{[b_s]'}][_{[b_s]}Y]_{[c_t]'} = {}_{[a_r]}Z_{[c_t]'} \tag{11.8.5}
$$

The right side of Eq. (11.8.4) shows how the product of two supermatrices may be expressed as the sum of major products of type III vectors, just as we learned in Section 9.2 that the product of two simple matrices can be expressed as the sum of major products of simple vectors.

The left sides of Eqs. (11.8.3), (11.8.4), and (11.8.5) show that for the products of two supermatrices to exist, the postpartitioner of the prefactor must be the transpose of the prepartitioner of the postfactor. This rule is the same as for the minor product of type III vectors. The right side of Eq. (11.8.5) shows three important characteristics about the product of two supermatrices: (1) The prepartitioner of the product is the same as the prepartitioner of the prefactor. (2) The postpartitioner of the product is the same as the postpartitioner of the postfactor. (3) The product tells us nothing about the common superorder of the factors. These rules, you will recall, are analogous to those for the product of simple matrices.

Diagrammatic representation of minor product

Figure 11.8.1 illustrates diagrammatically how the product of two supermatrices may be expressed as the minor product of type IV vectors.

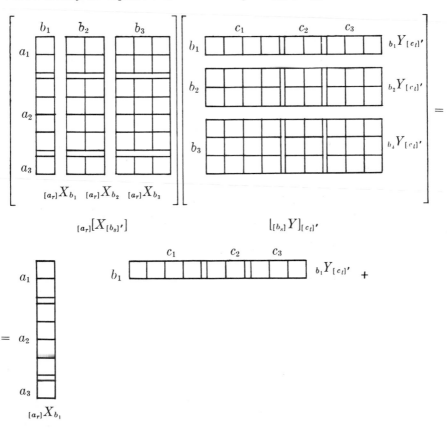

FIG. 11.8.1 (Cont. on next page)

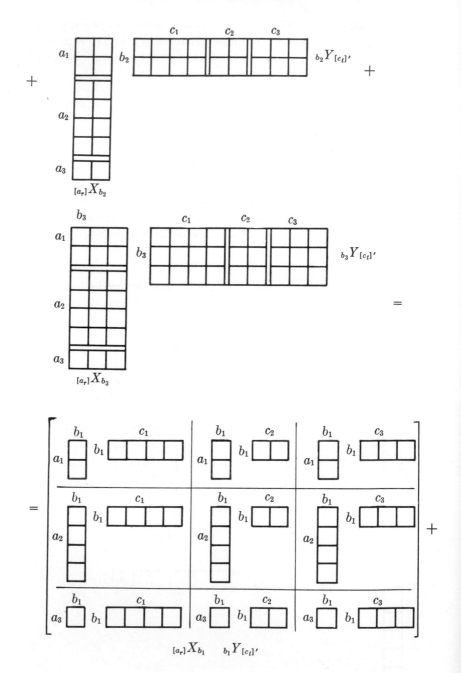

FIG. 11.8.1 (Cont. on next page)

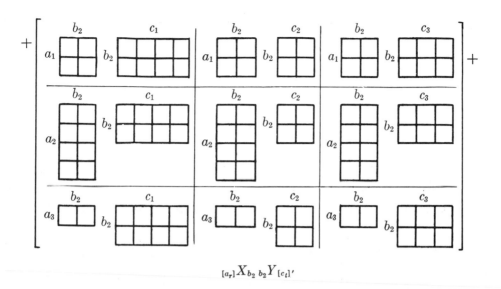

$$[a_r] X_{b_2} {}_{b_2} Y_{[c_i]}'$$

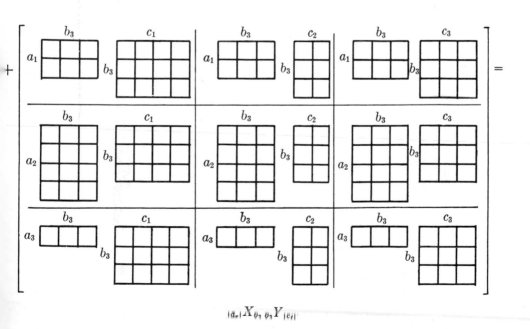

$$[a_r] X_{b_3} {}_{b_3} Y_{[c_i]}'$$

FIG. 11.8.1 (Cont. on next page)

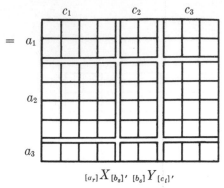

$$[a_r]X_{[b_s]'} \; [b_s]Y_{[c_t]'}$$

Fig. 11.8.1 (Cont.)

Numerical example of minor product

A numerical example using the same partitioning vectors as Fig. 11.8.1 is given in Fig. 11.8.2.

$$
\begin{bmatrix}
1 & 2 & 1 & 3 & 2 & 1 \\
2 & 3 & 1 & 2 & 1 & 2 \\
\hline
1 & 4 & 2 & 3 & 2 & 2 \\
4 & 1 & 3 & 2 & 1 & 1 \\
2 & 3 & 2 & 3 & 2 & 3 \\
3 & 4 & 1 & 1 & 4 & 2 \\
\hline
2 & 1 & 2 & 2 & 1 & 3
\end{bmatrix}
\begin{bmatrix}
1 & 3 & 2 & 1 & 2 & 4 & 4 & 2 & 1 \\
\hline
3 & 4 & 1 & 2 & 1 & 3 & 4 & 1 & 2 \\
1 & 1 & 2 & 3 & 4 & 2 & 3 & 4 & 1 \\
\hline
3 & 1 & 3 & 4 & 2 & 1 & 4 & 2 & 1 \\
3 & 2 & 2 & 3 & 4 & 1 & 1 & 5 & 1 \\
1 & 2 & 1 & 2 & 1 & 3 & 2 & 1 & 1
\end{bmatrix} =
$$

$$\qquad\quad {}_{[a_r]}[X_{[b_s]'}] \qquad\qquad\qquad\qquad [{}_{[b_s]}Y]_{[c_t]'}$$

$$
\begin{bmatrix}
1 \\
2 \\
- \\
1 \\
4 \\
2 \\
3 \\
- \\
2
\end{bmatrix}
[1 \quad 3 \quad 2 \quad 1 \,|\, 2 \quad 4 \,|\, 4 \quad 2 \quad 1] +
$$

$$
\begin{bmatrix}
2 & 1 \\
3 & 1 \\
\hline
4 & 2 \\
1 & 3 \\
3 & 2 \\
4 & 1 \\
\hline
1 & 2
\end{bmatrix}
\begin{bmatrix}
3 & 4 & 1 & 2 & 1 & 3 & 4 & 1 & 2 \\
1 & 1 & 2 & 3 & 4 & 2 & 3 & 4 & 1
\end{bmatrix} +
$$

Fig. 11.8.2 (Cont. on next page)

$$\begin{bmatrix} 3 & 2 & 1 \\ 2 & 1 & 2 \\ \hline 3 & 2 & 2 \\ 2 & 1 & 1 \\ 3 & 2 & 3 \\ 1 & 4 & 2 \\ \hline 2 & 1 & 3 \end{bmatrix} \begin{bmatrix} 3 & 1 & 3 & 4 & 2 & 1 & 4 & 2 & 1 \\ 3 & 2 & 2 & 3 & 4 & 1 & 1 & 5 & 1 \\ 1 & 2 & 1 & 2 & 1 & 3 & 2 & 1 & 1 \end{bmatrix} =$$

$$\begin{bmatrix} \begin{pmatrix} 1 \\ 2 \end{pmatrix}(1 \ 3 \ 2 \ 1) & \begin{pmatrix} 1 \\ 2 \end{pmatrix}(2 \ 4) & \begin{pmatrix} 1 \\ 2 \end{pmatrix}(4 \ 2 \ 1) \\ \hline \begin{pmatrix} 1 \\ 4 \\ 2 \\ 3 \end{pmatrix}(1 \ 3 \ 2 \ 1) & \begin{pmatrix} 1 \\ 4 \\ 2 \\ 3 \end{pmatrix}(2 \ 4) & \begin{pmatrix} 1 \\ 4 \\ 2 \\ 3 \end{pmatrix}(4 \ 2 \ 1) \\ \hline (2)(1 \ 3 \ 2 \ 1) & (2)(2 \ 4) & (2)(4 \ 2 \ 1) \end{bmatrix} +$$

$$\begin{bmatrix} \begin{pmatrix} 2 & 1 \\ 3 & 1 \end{pmatrix}\begin{pmatrix} 3 & 4 & 1 & 2 \\ 1 & 1 & 2 & 3 \end{pmatrix} & \begin{pmatrix} 2 & 1 \\ 3 & 1 \end{pmatrix}\begin{pmatrix} 1 & 3 \\ 4 & 2 \end{pmatrix} & \begin{pmatrix} 2 & 1 \\ 3 & 1 \end{pmatrix}\begin{pmatrix} 4 & 1 & 2 \\ 3 & 4 & 1 \end{pmatrix} \\ \hline \begin{pmatrix} 4 & 2 \\ 1 & 3 \\ 3 & 2 \\ 4 & 1 \end{pmatrix}\begin{pmatrix} 3 & 4 & 1 & 2 \\ 1 & 1 & 2 & 3 \end{pmatrix} & \begin{pmatrix} 4 & 2 \\ 1 & 3 \\ 3 & 2 \\ 4 & 1 \end{pmatrix}\begin{pmatrix} 1 & 3 \\ 4 & 2 \end{pmatrix} & \begin{pmatrix} 4 & 2 \\ 1 & 3 \\ 3 & 2 \\ 4 & 1 \end{pmatrix}\begin{pmatrix} 4 & 1 & 2 \\ 3 & 4 & 1 \end{pmatrix} \\ \hline (1 \ 2)\begin{pmatrix} 3 & 4 & 1 & 2 \\ 1 & 1 & 2 & 3 \end{pmatrix} & (1 \ 2)\begin{pmatrix} 1 & 3 \\ 4 & 2 \end{pmatrix} & (1 \ 2)\begin{pmatrix} 4 & 1 & 2 \\ 3 & 4 & 1 \end{pmatrix} \end{bmatrix} +$$

$$\begin{bmatrix} \begin{pmatrix} 3 & 2 & 1 \\ 2 & 1 & 2 \end{pmatrix}\begin{pmatrix} 3 & 1 & 3 & 4 \\ 3 & 2 & 2 & 3 \\ 1 & 2 & 1 & 2 \end{pmatrix} & \begin{pmatrix} 3 & 2 & 1 \\ 2 & 1 & 2 \end{pmatrix}\begin{pmatrix} 2 & 1 \\ 4 & 1 \\ 1 & 3 \end{pmatrix} & \begin{pmatrix} 3 & 2 & 1 \\ 2 & 1 & 2 \end{pmatrix}\begin{pmatrix} 4 & 2 & 1 \\ 1 & 5 & 1 \\ 2 & 1 & 1 \end{pmatrix} \\ \hline \begin{pmatrix} 3 & 2 & 2 \\ 2 & 1 & 1 \\ 3 & 2 & 3 \\ 1 & 4 & 2 \end{pmatrix}\begin{pmatrix} 3 & 1 & 3 & 4 \\ 3 & 2 & 2 & 3 \\ 1 & 2 & 1 & 2 \end{pmatrix} & \begin{pmatrix} 3 & 2 & 2 \\ 2 & 1 & 1 \\ 3 & 2 & 3 \\ 1 & 4 & 2 \end{pmatrix}\begin{pmatrix} 2 & 1 \\ 4 & 1 \\ 1 & 3 \end{pmatrix} & \begin{pmatrix} 3 & 2 & 2 \\ 2 & 1 & 1 \\ 3 & 2 & 3 \\ 1 & 4 & 2 \end{pmatrix}\begin{pmatrix} 4 & 2 & 1 \\ 1 & 5 & 1 \\ 2 & 1 & 1 \end{pmatrix} \\ \hline (2 \ 1 \ 3)\begin{pmatrix} 3 & 1 & 3 & 4 \\ 3 & 2 & 2 & 3 \\ 1 & 2 & 1 & 2 \end{pmatrix} & (2 \ 1 \ 3)\begin{pmatrix} 2 & 1 \\ 4 & 1 \\ 1 & 3 \end{pmatrix} & (2 \ 1 \ 3)\begin{pmatrix} 4 & 2 & 1 \\ 1 & 5 & 1 \\ 2 & 1 & 1 \end{pmatrix} \end{bmatrix} =$$

FIG. 11.8.2 (Cont. on next page)

1	3	2	1	2	4	4	2	1
2	6	4	2	4	8	8	4	2
1	3	2	1	2	4	4	2	1
4	12	8	4	8	16	16	8	4
2	6	4	2	4	8	8	4	2
3	9	6	3	6	12	12	6	3
2	6	4	2	4	8	8	4	2

$+$

7	9	4	7	6	8	11	6	5
10	13	5	9	7	11	15	7	7
14	18	8	14	12	16	22	12	10
6	7	7	11	13	9	13	13	5
11	14	7	12	11	13	18	11	8
13	17	6	11	8	14	19	8	9
5	6	5	8	9	7	10	9	4

$+$

16	9	14	20	15	8	16	17	6
11	8	10	15	10	9	13	11	5
17	11	15	22	16	11	18	18	7
10	6	9	13	9	6	11	10	4
18	13	16	24	17	14	20	19	8
17	13	13	20	20	11	12	24	7
12	10	11	17	11	12	15	12	6

$=$

24	21	20	28	23	20	31	25	12
23	27	19	26	21	28	36	22	14
32	32	25	37	30	31	44	32	18
20	25	24	28	30	31	40	31	13
31	33	27	38	32	35	46	34	18
33	39	25	34	34	37	43	38	19
19	22	20	27	24	27	33	25	12

Fig. 11.8.2 (Cont.)

Transpose of minor product

The transpose of the minor product of type IV vectors may be indicated by taking the transpose of both sides of Eq. (11.8.4). On the left we arrange the symbols in reverse order, and take the transpose of each symbol. On the right we take the transpose of each product term in the same way. This gives us Eq. (11.8.6).

$$({}_{[a_r]}[X_{[b_s]'}][{}_{[b_s]}Y]_{[c_t]'})' = {}_{[c_t]}[Y'_{[b_s]'}][{}_{[b_s]}X']_{[a_r]'}$$

$$= {}_{[c_t]}Y'_{b_1 \, b_1}X'_{[a_r]'} + {}_{[c_t]}Y'_{b_2 \, b_2}X'_{[a_r]'} + \ldots + {}_{[c_t]}Y'_{b_s \, b_s}X'_{[a_r]'} \qquad (11.8.6)$$

Equation (11.8.6) can be expressed more compactly if we take the transposes of both sides of Eq. (11.8.5) to get

$$_{[c_t]}[Y'_{[b_s]'}][{}_{[b_s]}X']_{[a_r]'} = {}_{[c_t]}Z'_{[a_r]'} \qquad (11.8.7)$$

We may illustrate the transpose of the minor product of type IV vectors by using the numerical example in Fig. 11.8.2 to get Fig. 11.8.3.

FIG. 11.8.3 (Cont. on next page)

$$
\left[
\begin{array}{c|c|c}
\begin{pmatrix}3&3&1\\1&2&2\\3&2&1\\4&3&2\end{pmatrix}\begin{pmatrix}3&2\\2&1\\1&2\end{pmatrix} &
\begin{pmatrix}3&3&1\\1&2&2\\3&2&1\\4&3&2\end{pmatrix}\begin{pmatrix}3&2&3&1\\2&1&2&4\\2&1&3&2\end{pmatrix} &
\begin{pmatrix}3&3&1\\1&2&2\\3&2&1\\4&3&2\end{pmatrix}\begin{pmatrix}2\\1\\3\end{pmatrix} \\
\hline
\begin{pmatrix}2&4&1\\1&1&3\end{pmatrix}\begin{pmatrix}3&2\\2&1\\1&2\end{pmatrix} &
\begin{pmatrix}2&4&1\\1&1&3\end{pmatrix}\begin{pmatrix}3&2&3&1\\2&1&2&4\\2&1&3&2\end{pmatrix} &
\begin{pmatrix}2&4&1\\1&1&3\end{pmatrix}\begin{pmatrix}2\\1\\3\end{pmatrix} \\
\hline
\begin{pmatrix}4&1&2\\2&5&1\\1&1&1\end{pmatrix}\begin{pmatrix}3&2\\2&1\\1&2\end{pmatrix} &
\begin{pmatrix}4&1&2\\2&5&1\\1&1&1\end{pmatrix}\begin{pmatrix}3&2&3&1\\2&1&2&4\\2&1&3&2\end{pmatrix} &
\begin{pmatrix}4&1&2\\2&5&1\\1&1&1\end{pmatrix}\begin{pmatrix}2\\1\\3\end{pmatrix}
\end{array}
\right] =
$$

$$
\left[
\begin{array}{cc|cccc|c}
1&2&1&4&2&3&2\\
3&6&3&12&6&9&6\\
2&4&2&8&4&6&4\\
1&2&1&4&2&3&2\\
\hline
2&4&2&8&4&6&4\\
4&8&4&16&8&12&8\\
\hline
4&8&4&16&8&12&8\\
2&4&2&8&4&6&4\\
1&2&1&4&2&3&2
\end{array}
\right]
+
\left[
\begin{array}{cc|cccc|c}
7&10&14&6&11&13&5\\
9&13&18&7&14&17&6\\
4&5&8&7&7&6&5\\
7&9&14&11&12&11&8\\
\hline
6&7&12&13&11&8&9\\
8&11&16&9&13&14&7\\
\hline
11&15&22&13&18&19&10\\
6&7&12&13&11&8&9\\
5&7&10&5&8&9&4
\end{array}
\right]
+
$$

$$
\left[
\begin{array}{cc|cccc|c}
16&11&17&10&18&17&12\\
9&8&11&6&13&13&10\\
14&10&15&9&16&13&11\\
20&15&22&13&24&20&17\\
\hline
15&10&16&9&17&20&11\\
8&9&11&6&14&11&12\\
\hline
16&13&18&11&20&12&15\\
17&11&18&10&19&24&12\\
6&5&7&4&8&7&6
\end{array}
\right]
=
\left[
\begin{array}{cc|cccc|c}
24&23&32&20&31&33&19\\
21&27&32&25&33&39&22\\
20&19&25&24&27&25&20\\
28&26&37&28&38&34&27\\
\hline
23&21&30&30&32&34&24\\
20&28&31&31&35&37&27\\
\hline
31&36&44&40&46&43&33\\
25&22&32&31&34&38&25\\
12&14&18&13&18&19&12
\end{array}
\right]
$$

Fig. 11.8.3 (Cont.)

11.9   The Minor Product Moment of a Type IV Vector

For type IV vectors, as for type III vectors, we must distinguish between the minor product moment of a row vector and a column vector. Correspondingly also, either a row or column type IV vector may, in simple form, be either a vertical or horizontal matrix.

Minor product moment of type IV row vector

First, let us consider the minor product moment of a type IV row vector. Here we must postmultiply the row vector by its transpose. Taking the row form given by Eq. (11.2.6) and its transpose given by Eq. (11.3.10), we get for the minor product moment Eq. (11.9.1).

$$_{[a_r]}[X_{[b_s]'}]\,[_{[b_s]}X']_{[a_r]'} = {}_{[a_r]}[X_{b_1}\quad X_{b_2}\quad \ldots \quad X_{b_s}]\begin{bmatrix} b_1 X' \\ b_2 X' \\ \ldots \\ b_s X' \end{bmatrix}_{[a_r]'}$$

$$(11.9.1)$$

Expanding the right side of Eq. (11.9.1) gives us Eq. (11.9.2).

$$_{[a_r]}[X_{[b_s]'}]\,[_{[b_s]}X']_{[a_r]'} = {}_{[a_r]}X_{b_1}\,_{b_1}X'_{[a_r]'} + {}_{[a_r]}X_{b_2}\,_{b_2}X'_{[a_r]'} + \ldots + {}_{[a_r]}X_{b_s}\,_{b_s}X'_{[a_r]'} \quad (11.9.2)$$

Equation (11.9.2) shows how the minor product moment of a type IV row vector may be expressed as the sum of the major product moments of its type III vector elements.

We may take the example for $_{[a_r]}X_{[b_s]'}$ in Fig. 11.8.2 to illustrate Eq. (11.9.2). This we do in Fig. 11.9.1.

$$\begin{bmatrix} \begin{pmatrix} 1 \\ 2 \\ \hline 1 \\ 4 \\ 2 \\ 3 \\ \hline 2 \end{pmatrix} & \begin{pmatrix} 2 & 1 \\ 3 & 1 \\ \hline 4 & 2 \\ 1 & 3 \\ 3 & 2 \\ 4 & 1 \\ \hline 1 & 2 \end{pmatrix} & \begin{pmatrix} 3 & 2 & 1 \\ 2 & 1 & 2 \\ \hline 3 & 2 & 2 \\ 2 & 1 & 1 \\ 3 & 2 & 3 \\ 1 & 4 & 2 \\ \hline 2 & 1 & 3 \end{pmatrix} \end{bmatrix}\begin{bmatrix} \begin{pmatrix} 1 & 2 & | & 1 & 4 & 2 & 3 & | & 2 \\ 2 & 3 & | & 4 & 1 & 3 & 4 & | & 1 \\ 1 & 1 & | & 2 & 3 & 2 & 1 & | & 2 \\ 3 & 2 & | & 3 & 2 & 3 & 1 & | & 2 \\ 2 & 1 & | & 2 & 1 & 2 & 4 & | & 1 \\ 1 & 2 & | & 2 & 1 & 3 & 2 & | & 3 \end{pmatrix} \end{bmatrix} =$$

$$_{[a_r]}[X_{[b_s]'}] \qquad\qquad\qquad\qquad [_{[b_s]}X'_{[a_r]'}$$

$$\begin{bmatrix} 1 \\ 2 \\ \hline 1 \\ 4 \\ 2 \\ 3 \\ \hline 2 \end{bmatrix}\begin{bmatrix} 1 & 2 & | & 1 & 4 & 2 & 3 & | & 2 \end{bmatrix} +$$

Fig. 11.9.1 (Cont. on next page)

$$\begin{bmatrix} 2 & 1 \\ 3 & 1 \\ \hline 4 & 2 \\ 1 & 3 \\ 3 & 2 \\ 4 & 1 \\ \hline 1 & 2 \end{bmatrix} \left[\begin{array}{cc|cccc|c} 2 & 3 & 4 & 1 & 3 & 4 & 1 \\ 1 & 1 & 2 & 3 & 2 & 1 & 2 \end{array}\right] +$$

$$\begin{bmatrix} 3 & 2 & 1 \\ 2 & 1 & 2 \\ \hline 3 & 2 & 2 \\ 2 & 1 & 1 \\ 3 & 2 & 3 \\ 1 & 4 & 2 \\ \hline 2 & 1 & 3 \end{bmatrix} \left[\begin{array}{cc|cccc|c} 3 & 2 & 3 & 2 & 3 & 1 & 2 \\ 2 & 1 & 2 & 1 & 2 & 4 & 1 \\ 1 & 2 & 2 & 1 & 3 & 2 & 3 \end{array}\right] =$$

$$\left[\begin{array}{c|c|c} \begin{pmatrix}1\\2\end{pmatrix}\begin{pmatrix}1 & 2\end{pmatrix} & \begin{pmatrix}1\\2\end{pmatrix}\begin{pmatrix}1 & 4 & 2 & 3\end{pmatrix} & \begin{pmatrix}1\\2\end{pmatrix}\begin{pmatrix}2\end{pmatrix} \\ \hline \begin{pmatrix}1\\4\\2\\3\end{pmatrix}\begin{pmatrix}1 & 2\end{pmatrix} & \begin{pmatrix}1\\4\\2\\3\end{pmatrix}\begin{pmatrix}1 & 4 & 2 & 3\end{pmatrix} & \begin{pmatrix}1\\4\\2\\3\end{pmatrix}\begin{pmatrix}2\end{pmatrix} \\ \hline \begin{pmatrix}2\end{pmatrix}\begin{pmatrix}1 & 2\end{pmatrix} & \begin{pmatrix}2\end{pmatrix}\begin{pmatrix}1 & 4 & 2 & 3\end{pmatrix} & \begin{pmatrix}2\end{pmatrix}\begin{pmatrix}2\end{pmatrix} \end{array}\right] +$$

$$\left[\begin{array}{c|c|c} \begin{pmatrix}2 & 1\\3 & 1\end{pmatrix}\begin{pmatrix}2 & 3\\1 & 1\end{pmatrix} & \begin{pmatrix}2 & 1\\3 & 1\end{pmatrix}\begin{pmatrix}4 & 1 & 3 & 4\\2 & 3 & 2 & 1\end{pmatrix} & \begin{pmatrix}2 & 1\\3 & 1\end{pmatrix}\begin{pmatrix}1\\2\end{pmatrix} \\ \hline \begin{pmatrix}4 & 2\\1 & 3\\3 & 2\\4 & 1\end{pmatrix}\begin{pmatrix}2 & 3\\1 & 1\end{pmatrix} & \begin{pmatrix}4 & 2\\1 & 3\\3 & 2\\4 & 1\end{pmatrix}\begin{pmatrix}4 & 1 & 3 & 4\\2 & 3 & 2 & 1\end{pmatrix} & \begin{pmatrix}4 & 2\\1 & 3\\3 & 2\\4 & 1\end{pmatrix}\begin{pmatrix}1\\2\end{pmatrix} \\ \hline \begin{pmatrix}1 & 2\end{pmatrix}\begin{pmatrix}2 & 3\\1 & 1\end{pmatrix} & \begin{pmatrix}1 & 2\end{pmatrix}\begin{pmatrix}4 & 1 & 3 & 4\\2 & 3 & 2 & 1\end{pmatrix} & \begin{pmatrix}1 & 2\end{pmatrix}\begin{pmatrix}1\\2\end{pmatrix} \end{array}\right] +$$

Fig. 11.9.1 (Cont. on next page)

$$
\left[
\begin{array}{c|c|c}
\begin{pmatrix} 3 & 2 & 1 \\ 2 & 1 & 2 \end{pmatrix}\begin{pmatrix} 3 & 2 \\ 2 & 1 \\ 1 & 2 \end{pmatrix} &
\begin{pmatrix} 3 & 2 & 1 \\ 2 & 1 & 2 \end{pmatrix}\begin{pmatrix} 3 & 2 & 3 & 1 \\ 2 & 1 & 2 & 4 \\ 2 & 1 & 3 & 2 \end{pmatrix} &
\begin{pmatrix} 3 & 2 & 1 \\ 2 & 1 & 2 \end{pmatrix}\begin{pmatrix} 2 \\ 1 \\ 3 \end{pmatrix} \\
\hline
\begin{pmatrix} 3 & 2 & 2 \\ 2 & 1 & 1 \\ 3 & 2 & 3 \\ 1 & 4 & 2 \end{pmatrix}\begin{pmatrix} 3 & 2 \\ 2 & 1 \\ 1 & 2 \end{pmatrix} &
\begin{pmatrix} 3 & 2 & 2 \\ 2 & 1 & 1 \\ 3 & 2 & 3 \\ 1 & 4 & 2 \end{pmatrix}\begin{pmatrix} 3 & 2 & 3 & 1 \\ 2 & 1 & 2 & 4 \\ 2 & 1 & 3 & 2 \end{pmatrix} &
\begin{pmatrix} 3 & 2 & 2 \\ 2 & 1 & 1 \\ 3 & 2 & 3 \\ 1 & 4 & 2 \end{pmatrix}\begin{pmatrix} 2 \\ 1 \\ 3 \end{pmatrix} \\
\hline
(2\ 1\ 3)\begin{pmatrix} 3 & 2 \\ 2 & 1 \\ 1 & 2 \end{pmatrix} &
(2\ 1\ 3)\begin{pmatrix} 3 & 2 & 3 & 1 \\ 2 & 1 & 2 & 4 \\ 2 & 1 & 3 & 2 \end{pmatrix} &
(2\ 1\ 3)\begin{pmatrix} 2 \\ 1 \\ 3 \end{pmatrix}
\end{array}
\right] =
$$

$$
\left[
\begin{array}{cc|cccc|c}
1 & 2 & 1 & 4 & 2 & 3 & 2 \\
2 & 4 & 2 & 8 & 4 & 6 & 4 \\
\hline
1 & 2 & 1 & 4 & 2 & 3 & 2 \\
4 & 8 & 4 & 16 & 8 & 12 & 8 \\
2 & 4 & 2 & 8 & 4 & 6 & 4 \\
3 & 6 & 3 & 12 & 6 & 9 & 6 \\
\hline
2 & 4 & 2 & 8 & 4 & 6 & 4
\end{array}
\right] +
\left[
\begin{array}{cc|cccc|c}
5 & 7 & 10 & 5 & 8 & 9 & 4 \\
7 & 10 & 14 & 6 & 11 & 13 & 5 \\
\hline
10 & 14 & 20 & 10 & 16 & 18 & 8 \\
5 & 6 & 10 & 10 & 9 & 7 & 7 \\
8 & 11 & 16 & 9 & 13 & 14 & 7 \\
9 & 13 & 18 & 7 & 14 & 17 & 6 \\
\hline
4 & 5 & 8 & 7 & 7 & 6 & 5
\end{array}
\right] +
$$

$$
\left[
\begin{array}{cc|cccc|c}
14 & 10 & 15 & 9 & 16 & 13 & 11 \\
10 & 9 & 12 & 7 & 14 & 10 & 11 \\
\hline
15 & 12 & 17 & 10 & 19 & 15 & 14 \\
9 & 7 & 10 & 6 & 11 & 8 & 8 \\
16 & 14 & 19 & 11 & 22 & 17 & 17 \\
13 & 10 & 15 & 8 & 17 & 21 & 12 \\
\hline
11 & 11 & 14 & 8 & 17 & 12 & 14
\end{array}
\right] =
\left[
\begin{array}{cc|cccc|c}
20 & 19 & 26 & 18 & 26 & 25 & 17 \\
19 & 23 & 28 & 21 & 29 & 29 & 20 \\
\hline
26 & 28 & 38 & 24 & 37 & 36 & 24 \\
18 & 21 & 24 & 32 & 28 & 27 & 23 \\
26 & 29 & 37 & 28 & 39 & 37 & 28 \\
25 & 29 & 36 & 27 & 37 & 47 & 24 \\
\hline
17 & 20 & 24 & 23 & 28 & 24 & 23
\end{array}
\right]
$$

Fig. 11.9.1 (Cont.)

It is clear from the final matrix in Fig. 11.9.1 that the minor product moment of a type IV vector is a symmetrically partitioned symmetric simple matrix.

Minor product moment of type IV column vector

To get the minor product moment of a type IV column vector, we must premultiply the column vector by its transpose. Taking the column form given by Eq. (11.2.4), and its transpose given by Eq. (11.3.9), we get Eq. (11.9.3).

$$_{[b_s]}[X'_{[a_r]'}]\,[_{[a_r]}X]_{[b_s]'} = {}_{[b_s]}[X'_{a_1} \quad X'_{a_2} \quad \dots \quad X'_{a_r}] \begin{bmatrix} a_1 X \\ a_2 X \\ \dots \\ a_r X \end{bmatrix}_{[b_s]'} \tag{11.9.3}$$

Multiplying out the right side of Eq. (11.9.3) gives Eq. (11.9.4).

$$_{[b_s]}[X'_{[a_r]'}]\,[_{[a_r]}X]_{[b_s]'} = {}_{[b_s]}[X'_{a_1}\,{}_{a_1}X_{[b_s]'} + {}_{[b_s]}X'_{a_2}\,{}_{a_2}X_{[b_s]'} + \dots$$
$$+ {}_{[b_s]}X'_{a_r}\,{}_{a_r}X_{[b_s]'} \tag{11.9.4}$$

The numerical procedure for getting the minor product moment of a type IV column vector as indicated in Eq. (11.9.4), is the same as for the minor product moment of the type IV row vector, as indicated in Eq. (11.9.2) and illustrated in Fig. 11.9.1. Notice, however, that in Eq. (11.9.2) we have the product of the supermatrix $_{[a_r]}X_{[b_s]'}$ postmultiplied by its transpose and expressed as the sum of major product moments of its type III column vectors. In Eq. (11.9.4) we have the product of the supermatrix premultiplied by its transpose and expressed as the sum of major product moments of its type III row vectors.

11.10   Major Products of Type IV Vectors

Notation for major product of type IV vectors

To get the major product of type IV vectors we must premultiply a row vector by a column vector. From Eq. (11.8.1) we can express Y as a type IV row vector as in Eq. (11.10.1).

$$_{[b_s]}[Y_{[c_t]'}] = {}_{[b_s]}[Y_{c_1} \quad Y_{c_2} \quad \dots \quad Y_{c_t}] \tag{11.10.1}$$

We shall then have a major product of type IV vectors if we premultiply both sides of Eq. (11.10.1) by the corresponding sides of Eq. (11.2.4) as in Eq. (11.10.2).

$$[_{[a_r]}X]_{[b_s]'}\,{}_{[b_s]}[Y_{[c_t]'}] = \begin{bmatrix} a_1 X \\ a_2 X \\ \dots \\ a_r X \end{bmatrix}_{[b_s]'} {}_{[b_s]}[Y_{c_1} \quad Y_{c_2} \quad \dots \quad Y_{c_t}]$$

$$\tag{11.10.2}$$

Multiplying out the right side of Eq. (11.10.2), we get Eq. (11.10.3).

$$[_{[a_r]}X]_{[b_s]'}\,{}_{[b_s]}[Y_{[c_t]'}] =$$

$$\begin{bmatrix} a_1 X_{[b_s]'}\,{}_{[b_s]}Y_{c_1} & a_1 X_{[b_s]'}\,{}_{[b_s]}Y_{c_2} & \dots & a_1 X_{[b_s]'}\,{}_{[b_s]}Y_{c_t} \\ a_2 X_{[b_s]'}\,{}_{[b_s]}Y_{c_1} & a_2 Y_{[b_s]'}\,{}_{[b_s]}X_{c_2} & \dots & a_2 X_{[b_s]'}\,{}_{[b_s]}Y_{c_t} \\ \dots & \dots & \dots & \dots \\ a_r X_{[b_s]'}\,{}_{[b_s]}Y_{c_1} & a_r X_{[b_s]'}\,{}_{[b_s]}Y_{c_2} & \dots & a_r X_{[b_s]'}\,{}_{[b_s]}Y_{c_t} \end{bmatrix} \tag{11.10.3}$$

The right side of Eq. (11.10.3) shows how the product of two super-matrices may be expressed as a matrix of minor products of type III vectors, just as the product of two simple matrices can be expressed as a matrix of minor products of simple vectors (Section 9.2).

The right side of Eq. (11.10.3) may be written more compactly as Eq. (11.10.4).

$$[_{[a_r]}X]_{[b_s]'} \,_{[b_s]}[Y_{[c_t]'}] = \begin{bmatrix} a_1 Z_{c_1} & a_1 Z_{c_2} & \cdots & a_1 Z_{c_t} \\ a_2 Z_{c_1} & a_2 Z_{c_2} & \cdots & a_2 Z_{c_t} \\ \cdots & \cdots & \cdots & \cdots \\ a_r Z_{c_1} & a_r Z_{c_2} & \cdots & a_r Z_{c_t} \end{bmatrix} \qquad (11.10.4)$$

Numerical example of major product of type IV vectors

We can illustrate Eqs. (11.10.3) and (11.10.4) by starting with the column form of X and the row form of Y of the supermatrices in Fig. 11.8.2. These we show in Fig. 11.10.1.

$$[_{[a_r]}X]_{[b_s]'} = \begin{bmatrix} \begin{pmatrix} 1 & 2 & 1 & 3 & 2 & 1 \\ 2 & 3 & 1 & 2 & 1 & 2 \end{pmatrix} & {}_{a_1}X_{[b_s]'} \\ \begin{pmatrix} 1 & 4 & 2 & 3 & 2 & 2 \\ 4 & 1 & 3 & 2 & 1 & 1 \\ 2 & 3 & 2 & 3 & 2 & 3 \\ 3 & 4 & 1 & 1 & 4 & 2 \end{pmatrix} & {}_{a_2}X_{[b_s]'} \\ \begin{pmatrix} 2 & 1 & 2 & 2 & 1 & 3 \end{pmatrix} & {}_{a_3}X_{[b_s]'} \end{bmatrix}$$

$$_{[b_s]}[Y_{[c_t]'}] = \begin{bmatrix} \begin{pmatrix} 1 & 3 & 2 & 1 \\ 3 & 4 & 1 & 2 \\ 1 & 1 & 2 & 3 \end{pmatrix} & \begin{pmatrix} 2 & 4 \\ 1 & 3 \\ 4 & 2 \end{pmatrix} & \begin{pmatrix} 4 & 2 & 1 \\ 4 & 1 & 2 \\ 3 & 4 & 1 \end{pmatrix} \\ \begin{pmatrix} 3 & 1 & 3 & 4 \\ 3 & 2 & 2 & 3 \\ 1 & 2 & 1 & 2 \end{pmatrix} & \begin{pmatrix} 2 & 1 \\ 4 & 1 \\ 1 & 3 \end{pmatrix} & \begin{pmatrix} 4 & 2 & 1 \\ 1 & 5 & 1 \\ 2 & 1 & 1 \end{pmatrix} \end{bmatrix}$$

$$\qquad\qquad _{[b_s]}Y_{c_1} \qquad\qquad _{[b_s]}Y_{c_2} \qquad _{[b_s]}Y_{c_3}$$

FIG. 11.10.1

To illustrate how the minor products are formed from the type III row vectors of X and the type III column vectors of Y to give the matrix elements on the right of Eqs. (11.10.3) and (11.10.4), it will suffice to calculate the first two elements of the first two rows. This we do in Fig. 11.10.2.

$$_{a_1}Z_{c_1} = {}_{a_1}X_{[b_s]'} {}_{[b_s]}Y_{c_1} = \begin{bmatrix} 1 & 2 & 1 & 3 & 2 & 1 \\ 2 & 3 & 1 & 2 & 1 & 2 \end{bmatrix} \begin{bmatrix} 1 & 3 & 2 & 1 \\ \hline 3 & 4 & 1 & 2 \\ 1 & 1 & 2 & 3 \\ \hline 3 & 1 & 3 & 4 \\ 3 & 2 & 2 & 3 \\ 1 & 2 & 1 & 2 \end{bmatrix} =$$

$$\begin{bmatrix} 1 \\ 2 \end{bmatrix} \begin{bmatrix} 1 & 3 & 2 & 1 \end{bmatrix} + \begin{bmatrix} 2 & 1 \\ 3 & 1 \end{bmatrix} \begin{bmatrix} 3 & 4 & 1 & 2 \\ 1 & 1 & 2 & 3 \end{bmatrix} +$$

$$\begin{bmatrix} 3 & 2 & 1 \\ 2 & 1 & 2 \end{bmatrix} \begin{bmatrix} 3 & 1 & 3 & 4 \\ 3 & 2 & 2 & 3 \\ 1 & 2 & 1 & 2 \end{bmatrix} = \begin{bmatrix} 1 & 3 & 2 & 1 \\ 2 & 6 & 4 & 2 \end{bmatrix} + \begin{bmatrix} 7 & 9 & 4 & 7 \\ 10 & 13 & 5 & 9 \end{bmatrix} +$$

$$\begin{bmatrix} 16 & 9 & 14 & 20 \\ 11 & 8 & 10 & 15 \end{bmatrix} = \begin{bmatrix} 24 & 21 & 20 & 28 \\ 23 & 27 & 19 & 26 \end{bmatrix}$$

$$_{a_1}Z_{c_2} = {}_{a_1}X_{[b_s]'} {}_{[b_s]}Y_{c_2} = \begin{bmatrix} 1 & 2 & 1 & 3 & 2 & 1 \\ 2 & 3 & 1 & 2 & 1 & 2 \end{bmatrix} \begin{bmatrix} 2 & 4 \\ \hline 1 & 3 \\ 4 & 2 \\ \hline 2 & 1 \\ 4 & 1 \\ 1 & 3 \end{bmatrix} =$$

$$\begin{bmatrix} 1 \\ 2 \end{bmatrix} \begin{bmatrix} 2 & 4 \end{bmatrix} + \begin{bmatrix} 2 & 1 \\ 3 & 1 \end{bmatrix} \begin{bmatrix} 1 & 3 \\ 4 & 2 \end{bmatrix} + \begin{bmatrix} 3 & 2 & 1 \\ 2 & 1 & 2 \end{bmatrix} \begin{bmatrix} 2 & 1 \\ 4 & 1 \\ 1 & 3 \end{bmatrix} =$$

$$\begin{bmatrix} 2 & 4 \\ 4 & 8 \end{bmatrix} + \begin{bmatrix} 6 & 8 \\ 7 & 11 \end{bmatrix} + \begin{bmatrix} 15 & 8 \\ 10 & 9 \end{bmatrix} = \begin{bmatrix} 23 & 20 \\ 21 & 28 \end{bmatrix}$$

$$_{a_2}Z_{c_1} = {}_{a_2}X_{[b_s]'} {}_{[b_s]}Y_{c_1} = \begin{bmatrix} 1 & 4 & 3 & 3 & 2 & 2 \\ 4 & 1 & 3 & 2 & 1 & 1 \\ 2 & 3 & 2 & 3 & 2 & 3 \\ 3 & 4 & 1 & 1 & 4 & 2 \end{bmatrix} \begin{bmatrix} 1 & 3 & 2 & 1 \\ \hline 3 & 4 & 1 & 2 \\ 1 & 1 & 2 & 3 \\ \hline 3 & 1 & 3 & 4 \\ 3 & 2 & 2 & 3 \\ 1 & 2 & 1 & 2 \end{bmatrix} =$$

$$\begin{bmatrix} 1 \\ 4 \\ 2 \\ 3 \end{bmatrix} \begin{bmatrix} 1 & 3 & 2 & 1 \end{bmatrix} + \begin{bmatrix} 4 & 2 \\ 1 & 3 \\ 3 & 2 \\ 4 & 1 \end{bmatrix} \begin{bmatrix} 3 & 4 & 1 & 2 \\ 1 & 1 & 2 & 3 \end{bmatrix} +$$

Fig. 11.10.2 (Cont. on next page)

$$
\begin{bmatrix} 3 & 2 & 2 \\ 2 & 1 & 1 \\ 3 & 2 & 3 \\ 1 & 4 & 2 \end{bmatrix}
\begin{bmatrix} 3 & 1 & 3 & 4 \\ 3 & 2 & 2 & 3 \\ 1 & 2 & 1 & 2 \end{bmatrix}
=
\begin{bmatrix} 1 & 3 & 2 & 1 \\ 4 & 12 & 8 & 4 \\ 2 & 6 & 4 & 2 \\ 3 & 9 & 6 & 3 \end{bmatrix} +
$$

$$
\begin{bmatrix} 14 & 18 & 8 & 14 \\ 6 & 7 & 7 & 11 \\ 11 & 14 & 7 & 12 \\ 13 & 17 & 6 & 11 \end{bmatrix} +
\begin{bmatrix} 17 & 11 & 15 & 22 \\ 10 & 6 & 9 & 13 \\ 18 & 13 & 16 & 24 \\ 17 & 13 & 13 & 20 \end{bmatrix}
=
\begin{bmatrix} 32 & 32 & 25 & 37 \\ 20 & 25 & 24 & 28 \\ 31 & 33 & 27 & 38 \\ 33 & 39 & 25 & 34 \end{bmatrix}
$$

$$
_{a_2}Z_{c_2} = {}_{a_2}X_{[b_s]}{}' {}_{[b_s]}X_{c_1} =
\left[\begin{array}{c|cc|ccc} 1 & 4 & 2 & 3 & 2 & 2 \\ 4 & 1 & 3 & 2 & 1 & 1 \\ 2 & 3 & 2 & 3 & 2 & 3 \\ 3 & 4 & 1 & 1 & 4 & 2 \end{array}\right]
\left[\begin{array}{cc} 2 & 4 \\ 1 & 3 \\ 4 & 2 \\ \hline 2 & 1 \\ 4 & 1 \\ 1 & 3 \end{array}\right]
=
$$

$$
\begin{bmatrix} 1 \\ 4 \\ 2 \\ 3 \end{bmatrix} \begin{bmatrix} 2 & 4 \end{bmatrix} +
\begin{bmatrix} 4 & 2 \\ 1 & 3 \\ 3 & 2 \\ 4 & 1 \end{bmatrix} \begin{bmatrix} 1 & 3 \\ 4 & 2 \end{bmatrix} +
\begin{bmatrix} 3 & 2 & 2 \\ 2 & 1 & 1 \\ 3 & 2 & 3 \\ 1 & 4 & 2 \end{bmatrix} \begin{bmatrix} 2 & 1 \\ 4 & 1 \\ 1 & 3 \end{bmatrix} =
$$

$$
\begin{bmatrix} 2 & 4 \\ 8 & 16 \\ 4 & 8 \\ 6 & 12 \end{bmatrix} +
\begin{bmatrix} 12 & 16 \\ 13 & 9 \\ 11 & 13 \\ 8 & 14 \end{bmatrix} +
\begin{bmatrix} 16 & 11 \\ 9 & 6 \\ 17 & 14 \\ 20 & 11 \end{bmatrix} =
\begin{bmatrix} 30 & 31 \\ 30 & 31 \\ 32 & 35 \\ 34 & 37 \end{bmatrix}
$$

Fɪɢ. 11.10.2 (Cont.)

You will notice that the matric elements calculated in Fig. 11.10.2 are the same as the corresponding matric elements in the final matrix of Fig. 11.8.2. This must of course be so, since we have here merely two different methods for getting the product of two supermatrices.

Transpose of major product of type IV vectors

The transpose of the major product of type IV vectors can be indicated by taking the transpose of both sides of Eq. (11.10.3). This gives Eq. (11.10.5).

$$
\{[_{[a_r]}X]_{[b_s]}{}' {}_{[b_s]}[Y_{[c_t]'}]\}' = [_{[c_t]}Y']_{[b_s]}{}' {}_{[b_s]}[X_{[a_r]'}]' =
$$

$$
\begin{bmatrix}
c_1 Y'_{[b_s]}{}' {}_{[b_s]}X'_{a_1} & c_1 Y'_{[b_s]}{}' {}_{[b_s]}Y'_{a_2} & \cdots & c_1 Y'_{[b_s]}{}' {}_{[b_s]}X'_{a_r} \\
c_2 Y'_{[b_s]}{}' {}_{[b_s]}X'_{a_1} & c_2 Y'_{[b_s]}{}' {}_{[b_s]}X'_{a_2} & \cdots & c_2 Y'_{[b_s]}{}' {}_{[b_s]}X'_{a_r} \\
\cdots & \cdots & \cdots & \cdots \\
c_t Y'_{[b_s]}{}' {}_{[b_s]}X'_{a_1} & c_t Y'_{[b_s]}{}' {}_{[b_s]}X'_{a_2} & \cdots & c_t Y'_{[b_s]}{}' {}_{[b_s]}X'_{a_r}
\end{bmatrix}
\quad (11.10.5)
$$

More simply we can write this as Eq. (11.10.6).

$$[_{[c_t]}Y']_{[b_s]'\ [b_s]}[X_{[a_r]'}]' = \begin{bmatrix} _{c_1}Z'_{a_1} & _{c_1}Z'_{a_2} & \cdots & _{c_1}Z'_{a_r} \\ _{c_2}Z'_{a_1} & _{c_2}Z'_{a_2} & \cdots & _{c_2}Z'_{a_r} \\ \cdots & \cdots & \cdots & \cdots \\ _{c_t}Z'_{a_1} & _{c_t}Z'_{a_2} & \cdots & _{c_t}Z'_{a_r} \end{bmatrix} \tag{11.10.6}$$

If, then, we use the examples in Fig. 11.10.1, the matric element in the second row, first column, of Eq. (11.10.5) would be as given in Fig. 11.10.3.

$$_{c_2}Z'_{a_1} = _{c_2}Y'_{[b_s]'\ [b_s]}X'_{a_1} = \begin{bmatrix} 2 & 1 & 4 & 2 & 4 & 1 \\ 4 & 3 & 2 & 1 & 1 & 3 \end{bmatrix} \begin{bmatrix} 1 & 2 \\ \hline 2 & 3 \\ 1 & 1 \\ \hline 3 & 2 \\ 2 & 1 \\ 1 & 2 \end{bmatrix} =$$

$$\begin{bmatrix} 2 \\ 4 \end{bmatrix} \begin{bmatrix} 1 & 2 \end{bmatrix} + \begin{bmatrix} 1 & 4 \\ 3 & 2 \end{bmatrix} \begin{bmatrix} 2 & 3 \\ 1 & 1 \end{bmatrix} + \begin{bmatrix} 2 & 4 & 1 \\ 1 & 1 & 3 \end{bmatrix} \begin{bmatrix} 3 & 2 \\ 2 & 1 \\ 1 & 2 \end{bmatrix} =$$

$$\begin{bmatrix} 2 & 4 \\ 4 & 8 \end{bmatrix} + \begin{bmatrix} 6 & 7 \\ 8 & 11 \end{bmatrix} + \begin{bmatrix} 15 & 10 \\ 8 & 9 \end{bmatrix} = \begin{bmatrix} 23 & 21 \\ 20 & 28 \end{bmatrix}$$

Fig. 11.10.3

As you would expect, the matric element we computed in Fig. 11.10.3 for the second row, first column, is the transpose of the element we computed for the first row, second column, in Fig. 11.10.2.

11.11　The Major Product Moment of a Type IV Vector

In view of what we know of type 3 vectors and the minor product moment of a type IV vector, we may expect to have two kinds of major product moments of type IV vectors. These are, of course, the major product moment of a type IV row vector and that of a type IV column vector.

Major product moment of type IV row vector

First, let us consider the major product moment of a type IV row vector. This implies that we premultiply the row vector by its tranpose. Taking the row form given by Eq. (11.2.6) and its transpose given by Eq. (11.3.10), we get Eq. (11.11.1) for the major product moment.

$$[_{[b_s]}X']_{[a_r]'}\ _{[a_r]}[X_{[b_s]'}] = \begin{bmatrix} _{b_1}X' \\ _{b_2}X' \\ \cdots \\ _{b_s}X' \end{bmatrix}_{[a_r]'}\ _{[a_r]}[X_{b_1}\quad X_{b_2}\quad \cdots \quad X_{b_s}]$$

(11.11.1)

Expanding the right side of Eq. (11.11.1) gives us Eq. (11.11.2).

$[_{[b_s]}X']_{[a_r]'}\ _{[a_r]}[X_{[b_s]'}] =$

$$\begin{bmatrix} _{b_1}X'_{[a_r]'}\ _{[a_r]}X_{b_1} & _{b_1}X'_{[a_r]'}\ _{[a_r]}X_{b_2} & \cdots & _{b_1}X'_{[a_r]'}\ _{[a_r]}X_{b_s} \\ _{b_2}X'_{[a_r]'}\ _{[a_r]}X_{b_1} & _{b_2}X'_{[a_r]'}\ _{[a_r]}X_{b_2} & \cdots & _{b_2}X'_{[a_r]'}\ _{[a_r]}X_{b_s} \\ \cdots & \cdots & \cdots & \cdots \\ _{b_s}X'_{[a_r]'}\ _{[a_r]}X_{b_1} & _{b_s}X'_{[a_r]'}\ _{[a_r]}X_{b_2} & \cdots & _{b_s}X'_{[a_r]'}\ _{[a_r]}X_{b_s} \end{bmatrix}$$

(11.11.2)

We can illustrate a typical matric element on the right side of Eq. (11.11.2) by using the numerical example for the row form of X given in Fig. 11.3.10. The element in the first row, second column, of Eq. (11.11.2) would then be as indicated in Fig. 11.11.1.

$$_{b_1}X'_{[a_r]'}\ _{[a_r]}X_{b_2} = [1\quad 2\ |\ 1\quad 4\quad 2\quad 3\ |\ 2]\begin{bmatrix} 2 & 1 \\ 3 & 1 \\ \hline 4 & 2 \\ 1 & 3 \\ 3 & 2 \\ 4 & 1 \\ \hline 1 & 2 \end{bmatrix} =$$

$$[1\quad 2]\begin{bmatrix} 2 & 1 \\ 3 & 1 \end{bmatrix} + [1\quad 4\quad 2\quad 3]\begin{bmatrix} 4 & 2 \\ 1 & 3 \\ 3 & 2 \\ 4 & 1 \end{bmatrix} + [2][1\quad 2] =$$

$$[8\quad 3] + [26\quad 21] + [2\quad 4] = [36\quad 28]$$

Fig. 11.11.1

You can readily show also that the element in the second row, first column, is the transpose of the element computed in Fig. 11.11.1.

Major product moment of type IV column vector

To show the major product moment of a type IV column vector, we must postmultiply the column vector by its transpose. Taking the column form given by Eq. (11.2.4) and its transpose given by Eq. (11.3.9), we get Eq. (11.11.3).

$$[_{[a_r]}X]_{[b_s]'} \; _{[b_s]}[X'_{[a_r]'}] = \begin{bmatrix} a_1 X \\ a_2 X \\ \cdots \\ a_r X \end{bmatrix}_{[b_s]'} \; _{[b_s]}[X_{a_1} \quad X_{a_2} \quad \cdots \quad X_{a_r}]$$

(11.11.3)

Expanding the right side of Eq. (11.11.3) gives

$$[_{[a_r]}X]_{[b_s]'} \; _{[b_s]}[X_{[a_r]'}] =$$

$$\begin{bmatrix} a_1 X_{[b_s]'} \; _{[b_s]}X'_{a_1} & a_1 X_{[b_s]'} \; _{[b_s]}X'_{a_2} & \cdots & a_1 X_{[b_s]'} \; _{[b_s]}X'_{a_r} \\ a_2 X_{[b_s]'} \; _{[b_s]}X'_{a_1} & a_2 X_{[b_s]'} \; _{[b_s]}X'_{a_2} & \cdots & a_2 X_{[b_s]'} \; _{[b_s]}X'_{a_r} \\ \cdots & \cdots & \cdots & \cdots \\ a_r X_{[b_s]'} \; _{[b_s]}X'_{a_1} & a_r X_{[b_s]'} \; _{[b_s]}X'_{a_2} & \cdots & a_r X_{[b_s]'} \; _{[b_s]}X'_{a_r} \end{bmatrix}$$

(11.11.4)

To illustrate a typical matric element on the right side of Eq. (11.11.4), we may use the numerical example for the column form of X given in Fig. 11.10.1. Then the element in the first row, second column, of Eq. (11.11.4) would be as shown in Fig. 11.11.3.

$$a_1 X_{[b_s]} \; _{[b_s]}X'_{a_2} = \begin{bmatrix} 1 & 2 & 1 & 3 & 2 & 1 \\ 2 & 3 & 1 & 2 & 1 & 2 \end{bmatrix} \begin{bmatrix} 1 & 4 & 2 & 3 \\ \hline 4 & 1 & 3 & 4 \\ 2 & 3 & 2 & 1 \\ \hline 3 & 2 & 3 & 1 \\ 2 & 1 & 2 & 4 \\ 2 & 1 & 3 & 2 \end{bmatrix} =$$

$$\begin{bmatrix} 1 \\ 2 \end{bmatrix} [1 \quad 4 \quad 2 \quad 3] + \begin{bmatrix} 2 & 1 \\ 3 & 1 \end{bmatrix} \begin{bmatrix} 4 & 1 & 3 & 4 \\ 2 & 3 & 2 & 1 \end{bmatrix} +$$

$$\begin{bmatrix} 3 & 2 & 1 \\ 2 & 1 & 2 \end{bmatrix} \begin{bmatrix} 3 & 2 & 3 & 1 \\ 2 & 1 & 2 & 4 \\ 2 & 1 & 3 & 2 \end{bmatrix} = \begin{bmatrix} 1 & 4 & 2 & 3 \\ 2 & 8 & 4 & 6 \end{bmatrix} +$$

$$\begin{bmatrix} 10 & 5 & 8 & 9 \\ 14 & 6 & 11 & 13 \end{bmatrix} + \begin{bmatrix} 15 & 9 & 16 & 13 \\ 12 & 7 & 14 & 10 \end{bmatrix} = \begin{bmatrix} 26 & 18 & 26 & 25 \\ 28 & 21 & 29 & 29 \end{bmatrix}$$

FIG. 11.11.3

It should be clear from Eqs. (11.11.2) and (11.11.4) that the major product moment of a type IV vector is a symmetrically partitioned simple matrix. The diagonal elements on the right side of both equations are symmetric matrices since they are the minor product moments of type III vectors. Corresponding elements above and below the diagonals are transposes of each other.

Note that in the practical applications of supermatrix multiplication in the social sciences, product moments of type IV vectors, either row or column, and major or minor, are more typical than the major or minor products of two distinct type IV vectors. Where the products of distinct type IV vectors are required, the factors may frequently be regarded as submatrices of a larger supermatrix. It is also true that product moments of type III supervectors, either row or column, major or minor, are also more typical than products of distinct type III vectors. Where products of distinct type III vectors occur they are usually minor products and constitute the elements of the major product moment of a type IV vector.

11.12   The Product of Any Number of Supermatrices

We have seen in Section 11.8 that for the product of two supermatrices to exist, the postpartitioner of the prefactor must be the transpose of the prepartitioner of the postfactor. We have also learned that the prepartitioner of the product is the same as the prepartitioner of the prefactor, and the postpartitioner of the product is the same as the postpartitioner of the postfactor. Since these rules must hold for the product of any two supermatrices, they must hold also if we multiply the product of two matrices by another matrix. Therefore, we can indicate the subscripts for the product of three supermatrices as

$$_{[a_r]}W_{[b_s]'}\;_{[b_s]}X_{[c_t]'}\;_{[c_t]}Y_{[d_u]'} = {}_{[a_r]}Z_{[d_u]'} \qquad (11.12.1)$$

We can readily generalize from Eq. (11.12.1) to the product of any number of supermatrices: (1) For the product of any number of supermatrices to exist, it is only necessary that for each pair of adjacent factors, the postscript of the prefactor be the transpose of the prescript of the postfactor. (2) The product of any number of supermatrices has the same prescript as the left-hand factor and the same postscript as the right-hand factor.

The transpose of the product given in Eq. (11.12.1) is given in Eq. (11.12.2).

$$(_{[a_r]}Z_{[d_u]'})' = {}_{[d_u]}Z'_{[a_r]'} = {}_{[d_u]}Y'_{[c_t]'}\;_{[c_t]}X'_{[b_s]'}\;_{[b_s]}W'_{[a_r]'} \qquad (11.12.2)$$

We can easily generalize from Eq. (11.12.2) to the transpose of the product of any number of supermatrices. Obviously, the rule is the same as for simple matrices. The transpose of the product of any number of supermatrices is the product of their transposes in reverse order.

SUMMARY

1. Notation for supermatrices:

   a. *Prepartitioning.*

     (1) The simple matrix $_PX_Q$ is partitioned into a type III column vector by the prepartitioning column vector prescript $[a_r]$, thus

$$_{[a_r]}[_PX_Q] = {}_{[a_r]}X_Q$$

     where

$$[a_r] = \begin{bmatrix} a_1 \\ a_2 \\ \ldots \\ a_r \end{bmatrix}$$

     (2) The number of matric elements in the type III column vector $_{[a_r]}X_Q$ is r, and the number of simple rows in the ith matric element is a_i.

     (3) The sum of the elements in $[a_r]$ is

$$1'[a_r] = P$$

   b. *Postpartitioning.*

     (1) The simple matrix $_PX_Q$ is partitioned into a type III row vector by the postpartitioning row postscript $[b_s]'$, thus

$$[_PX_Q]_{[b_s]'} = {}_PX_{[b_s]'}$$

     where

$$[b_s]' = [b_1 \quad b_2 \quad \ldots \quad b_s]$$

     (2) The number of matric elements in the type III row vector $_PX_{[b_s]'}$ is s and the number of simple columns in the ith matric element is b_i.

     (3) The sum of the elements in $[b_s]'$ is

$$[b_s]'1 = Q$$

   c. The matrix $_PX_Q$ is pre- and postpartitioned into r matric rows and s matric columns by

$$_{[a_r]}[_PX_Q]_{[b_s]'} = {}_{[a_r]}X_{[b_s]'}$$

   d. The ijth matric element of the supermatrix $_{[a_r]}X_{[b_s]'}$ is $_{a_i}X_{b_j}$.

2. The type IV supervector:

   a. The supermatrix $_{[a_r]}X_{[b_s]'}$ may be expressed as a type IV column vector whose elements are type III row vectors by

$$[_{[a_r]}X]_{[b_s]'} = \begin{bmatrix} a_1X \\ a_2X \\ \cdots \\ a_rX \end{bmatrix}_{[b_s]'}$$

b. The supermatrix $_{[a_r]}X_{[b_s]'}$ may be expressed as a type IV row vector whose elements are type III column vectors by

$$_{[a_r]}[X_{[b_s]'}] = {}_{[a_r]}[X_{b_1} \quad X_{b_2} \quad \cdots \quad X_{b_s}]$$

3. Transpose notation for the supermatrix:

a. The transpose of the type III column vector is indicated by writing the transposes of the symbols in reverse order, thus

$$[_{[a_r]}X_Q]' = {}_QX'_{[a_r]'}$$

b. The transpose of the type III row vector is similarly indicated by

$$[_PX_{[b_s]'}]' = {}_{[b_s]}X'_P$$

c. The transpose of the type IV column vector is

$$([_{[a_r]}X]_{[b_s]'})' = {}_{[b_s]}[X_{[a_r]'}]$$

d. The transpose of the type IV row vector is

$$(_{[a_r]}[X_{[b_s]'}])' = [_{[b_s]}X']_{[a_r]'}$$

4. The minor product of type III vectors:

a. For the minor product of type III vectors to exist, the postpartitioner of the prefactor must be the transpose of the prepartitioner of the postfactor.

b. The minor product of type III vectors is a sum of products of corresponding matric elements from the two vectors, and hence a simple matrix, thus

$$_PX_{[b_s]'} {}_{[b_s]}Y_M = {}_PX_{b_1} {}_{b_1}Y_M + \ldots + {}_PX_{b_s} {}_{b_s}Y_M = {}_PZ_M$$

c. The transpose of the minor product of type III vectors is the product of the transposes in reverse order, thus

$$[_PX_{[b_s]'} {}_{[b_s]}Y_M]' = {}_MY'_{[b_s]'} {}_{b_s}X'_P = {}_MY'_{b_1} {}_{b_1}X'_P + \ldots + {}_MY'_{b_s} {}_{b_s}X'_P = {}_MZ'_P$$

5. The minor product moment of a type III vector

a. The minor product moment of a natural order type III row vector is a sum of product moments of the matric elements of the vector in which the *postfactors* are primed, that is,

$$_PX_{[b_s]'} {}_{[b_s]}X'_P = {}_PX_{b_1} {}_{b_1}X'_P + \ldots + {}_PX_{b_s} {}_{b_s}X'_P$$

It is a simple symmetric matrix.

b. The minor product moment of a natural order type III column vector is a sum of product moments of the matric elements of the vector in which the prefactor is primed, that is,

$$_M Y'_{[b_s]'\ [b_s]} Y_M = {}_M Y'_{b_1\ b_1} Y_M + \ldots + {}_M Y'_{b_s\ b_s} Y_M$$

6. The major product of type III vectors

    a. The major product of type III vectors is a supermatrix partitioned as the factors, thus

$$_{[a_r]} X_Q {}_Q Y_{[c_t]'} = {}_{[a_r]} Z_{[c_t]'}$$

    b. The element in the ith row and jth column of the products is the product of the ith matric element from the prefactor postmultiplied by the jth matric element of the postfactor, that is,

$$_{a_i} X_Q {}_Q Y_{c_j} = {}_{a_i} Z_{c_j}$$

    c. The transpose of the major product of type III vectors is given by

$$[_{[a_r]} X_Q {}_Q Y_{[c_t]'}]' = {}_{[c_t]} Y'_Q {}_Q X_{[a_r]'} = {}_{c_t} Z'_{[a_r]'}$$

7. The major product moment of type III vectors:

    a. The major product moment of a type III column vector is a symmetrically partitioned symmetric simple matrix given by

$$_{[a_r]} X_Q {}_Q X'_{[a_r]'}$$

in which the ijth element is

$$_{a_i} X_Q {}_Q X'_{a_j}. \quad \text{(The postfactor is primed)}$$

    b. The major product moment of a type III row vector is a symmetrically partitioned symmetric simple matrix given by

$$_{[b_s]} X'_P {}_P X_{[b_s]'}$$

in which the ijth element is

$$_{b_i} X'_P {}_P X_{b_j} \quad \text{(the prefactor is primed)}$$

8. Minor product of type IV vectors:

    a. The product of two supermatrices may be expressed as the minor product of type IV supervectors, thus

$$_{[a_r]}[X_{[b_s]'}]\,[_{[b_s]} Y]_{[c_t]'} = {}_{[a_r]} X_{b_1\ b_1} Y_{[c_t]'} + \ldots + {}_{[a_r]} X_{b_s\ b_s} Y_{[c_t]'} = {}_{[a_r]} Z_{[c_t]'}$$

    b. For the minor product of type IV supervectors to exist, the postpartitioner of the prefactor must be the transpose of the prepartitioner of the postfactor.

    c. The minor product of type IV vectors is a supermatrix prepartitioned as the prefactor and postpartitioned as the postfactor.

    d. The minor product of type IV supervectors is the sum of major products of type III vectors in which the ith term is the ith element of the prefactor postmultiplied by the ith element of the postfactor.

e. The transpose of the minor product of type IV vectors is indicated by the transposes of the symbols in reverse order, thus

$$(_{[a_r]}[X_{[b_s]'}]\,[_{[b_s]}Y]_{[c_t]'})' = {}_{[c_t]}[Y'_{[b_s]'}]\,[_{[b_s]}X']_{[a_r]'} = {}_{[c_t]}Z'_{[a_r]'}$$

9. The minor product moment of a type IV vector:

a. The minor product moment of a natural order type IV row vector is a sum of major product moments of its type III column vector elements given by

$$_{[a_r]}[X_{[b_s]'}]\,[_{[b_s]}X']_{[a_r]'} = {}_{[a_r]}X_{b_1}\,{}_{b_1}X'_{[a_r]'} + \ldots + {}_{[a_r]}X_{b_s}\,{}_{b_s}X'_{[a_r]'}$$

b. The minor product moment of a natural order type IV column vector is a sum of major product moments of its type III row vector elements given by

$$_{[b_s]}[X'_{[a_r]'}]\,[_{[a_r]}X]_{[b_s]'} = {}_{[b_s]}X'_{a_1}\,{}_{a_1}X_{[b_s]'} + \ldots + {}_{[b_s]}X_{a_r}\,{}_{a_r}X_{[b_s]'}$$

10. The major product of type IV vectors:

a. The major product of type IV vectors is expressed as a matrix of minor products of type 3 vectors given by

$$[_{[a_r]}X]_{[b_s]'}\,{}_{[b_s]}[Y_{[c_t]'}] = \begin{bmatrix} a_1 X \\ a_2 X \\ \cdots \\ a_r X \end{bmatrix}_{[b_s]'} \qquad _{[b_s]}[Y_{c_1}\quad Y_{c_2}\quad \cdots \quad Y_{c_t}] = {}_{a_r}Z_{[c_t]'}$$

b. The ijth element of a major product of type IV vectors is given by

$$_{a_i}Z_{c_j} = {}_{a_i}X_{[b_s]'}\,{}_{[b_s]}Y_{c_j}$$

c. The transpose of the major product of type IV vectors is written by taking the transposes of the symbols in reverse order, thus

$$\{[_{[a_r]}X]_{[b_s]'}\,{}_{[b_s]}[Y_{[c_t]'}]\}' = [_{[c_t]}Y']_{[b_s]'}\,{}_{[b_s]}[X'_{[a_r]'}] = {}_{[c_t]}Z'_{[a_r]'}$$

11. The major product moment of a type IV vector:

a. The major product moment of a natural order type IV row vector is a matrix of minor product moments of its type III column vector elements. It is indicated by

$$[_{[b_s]}X']_{[a_r]'}\,{}_{[a_r]}[X_{[b_s]'}]$$

and its ijth element is

$$_{b_i}X'_{[a_r]'}\,{}_{[a_r]}X_{b_j}$$

b. The major product of a natural order type IV column vector is a matrix of minor product moments of its type III row vector elements. It is indicated by

$$[_{[a_r]}X]_{[b_s]'}\ {}_{[b_s]}[X'_{[a_r]'}]$$

and its ijth element is

$$_{a_i}X_{[b_s]'}\ {}_{[b_s]}X'_{a_j}$$

EXERCISES

1. Which of the following are type III row vectors and which type III column vectors?

    (a) $_{[b_s]}W_T$ (b) $_TW_{[a_s]'}$ (c) $_{[a_r]}W_P$ (d) $_QW_{[a_s]'}$

2. What is the simple height of each of the matrices in Ex. 1?

3. What is the simple width of each of the matrices in Ex. 1?

4. What is the superheight of each of the matrices in Ex. 1?

5. What is the superwidth of each of the matrices in Ex. 1?

6. Which of the following are incorrectly written?

    (a) $_{[b_s]}W_{]a_r]}$ (b) $_{[b_s]'}W_{[a_r]}$ (c) $_{[b_s]}W_{[a_r]'}$ (d) $_{[a_r]}W_{[b_s]'}$

    (e) $_{[b_s]'}W_{[a_r]'}$ (f) $_{[a_r]'}W_{[b_s]}$ (g) $_{[a_r]}W_{[b_s]}$

7. Which of the following are square matrices in simple form?

    (a) $_{[a_r]}W_{[b_s]'}$ (b) $_{[a_r]}W'_{[b_s]'}$ (c) $_{[b_s]}W_{[a_r]'}$ (d) $_{[a_r]}W_{[a_r]'}$

    (e) $_{[b_s]}W_{[b_s]'}$ (f) $_{[a_r]}W'_{[a_r]'}$ (g) $_{[b_s]}W'_{[b_s]'}$

8. Assuming all the following exist, which are type IV row vectors, and which are type IV column vectors?

    (a) $[_{[a_r]}X]_{[b_s]'}$ (b) $_{[a_r]}[X_{[b_s]'}]$ (c) $[_{[b_s]}X]_{[a_r]'}$ (d) $_{[b_s]}[X'_{[a_r]'}]$

    (e) $_{[a_r]}[X'_{[b_s]'}]$ (f) $[_{[b_s]}X']_{[a_r]'}$

9. Remove the parentheses from each of the following:

    (a) $(_{[b_r]}Y_P)'$ (b) $(_QY_{[a_r]'})'$ (c) $(_QW'_{[b_r]'})'$ (d) $(_{[a_s]}M'_L)'$

10. Remove the parentheses from each of the following:

    (a) $([_{[b_r]}Y]_{[a_s]'})'$ (b) $(_{[b_s]}[X_{[c_t]'}])'$ (c) $(_{[c_t]}[W'_{[b_r]'}])'$ (d) $([_{[a_r]}M']_{[b_s]'})'$

11. (a) Indicate the kind of product and the kind of vectors represented by

$$_QW_{[c_t]'}\ {}_{[c_t]}M_L$$

    (b) What is the ith term in the sum indicated by this product?
    (c) Is the product a simple or supermatrix?
    (d) What is its order?

12. Without the use of parentheses write the first term in the sum indicated by

$$({}_Q X_{[a_r]'} {}_{[a_r]} Y_P)'$$

13. If the ith term of a sum representing a product of supervectors of superorder r is

$$_Q X_{a_i} {}_{a_i} X'_Q$$

(a) What type of supervectors are involved?
(b) What kind of product is it?
(c) Is the natural order supervector, a row or column?
(d) Indicate the notation for this vector product.

14. The ith term of a sum representing a product of supervectors of superorder t is

$$_P Y'_{c_i} {}_{c_i} Y_P$$

(a) What type of supervectors are involved?
(b) What kind of product is it?
(c) Is the natural order supervector, a row or column?
(d) Indicate the notation for this vector product.

15. (a) Indicate the kind of product and kind of vectors represented by

$$_{[a_s]} Y_P {}_P X_{[b_r]'}$$

(b) What is the ijth element in the matrix indicated by this product?
(c) Is the product a simple or supermatrix?
(d) What do you know about its order?

16. The ijth element in a product of supervectors of superorder r is

$$_{a_i} X_Q {}_Q X'_{a_j}$$

(a) What type of supervectors are involved?
(b) What kind of product is it?
(c) Is the natural order supervector a row or column?
(d) Indicate the notation for this vector product.

17. The ijth element in a product of supervectors of superorder t is

$$_{c_i} Y'_L {}_L Y_{c_j}$$

(a) What type of supervectors are involved?
(b) What kind of product is it?
(c) Is the natural order supervector, a row or column?
(d) Indicate the notation for this vector product.

18. (a) What type of supervectors are involved in the product

$$_{[b_s]} [X_{[a_r]'}] {}_{[a_r]} Y]_{[c_t]'}$$

(b) What kind of product is it?
(c) What is the first term in the sum representing this product?
(d) What type of vectors are involved in this term?
(e) What kind of product is this term?
(f) How many terms are there in the sum?
(g) What is the order of the product?

19. Remove the parentheses from

$$\left({}_{[b_s]}[X_{[a_r]'}]\, [{}_{[a_r]}Y]_{[c_t]'} \right)'$$

20. (a) Write the product moment of a type IV vector if the ith term representing the product is

$$_{[a_r]}X_{b_i\ b_i}X'_{[a_r]'}$$

    and there are s terms
    (b) What kind of product moment is this?
    (c) Is the natural order of the vector a row or column?

21. (a) Write the product moment of a type IV vector if the ith term representing the product is

$$_{[b_s]}X'_{a_i\ a_i}X_{[b_s]'}$$

    and there are r terms.
    (b) What kind of product moment is this?
    (c) Is the natural order of the vector a row or column?

22. The ijth element in a product of supervectors of superorder $r \times t$ is

$$_{a_i}X_{[b_s]'\ [b_s]}Y_{c_j}$$

    (a) Indicate the notation for this product.
    (b) What type of supervectors is involved?
    (c) What kind of product is it?

23. Remove the parentheses from

$$\left([{}_{[c_t]}Y']_{[b_s]'\ [b_s]}[X'_{[a_r]'}] \right)'$$

24. The ijth element of a product of supervectors of order s is

$$_{b_i}X'_{[a_r]\ [a_r]'}X_{b_j}$$

    (a) What type of supervectors are they?
    (b) What kind of product is it?

25. Given the matrices

$$[{}_{[b_s]}X']_{[a_r]'\ [a_r]}[X_{[b_s]'}] = A$$

    and

$$[{}_{[a_r]}X]_{[b_s]'\ [b_s]}[X'_{[a_r]'}] = B$$

    (a) What type of vectors is involved in A and B?
    (b) What kinds of products?
    (c) What is the difference between A and B?

26. Given the product

$$[\ \ ]X_{[b_s]}\,[\ \ ]Y_{[\ \ ]\ [c_t]}Z_{[\ \ ]} = {}_{[a_r]}W_{[d_u]}$$

    Fill in the brackets and put primes where needed.

ANSWERS

1. (a) column   (b) row   (c) column   (d) row

2. (a) $1'[b_s]$ (b) T (c) $1'[a_r]$ (d) Q

3. (a) T (b) $[a_s]'1$ (c) P (d) $[a_s]'1$

4. (a) s (b) 1   (c) r (d) 1

5. (a) 1   (b) s (c) 1   (d) s

6. (a), (b), (e), (f), and (g) are incorrect

7. (d), (e), (f), and (g) are square

8. (b), (d), and (e) are rows; (a), (c), and (f) are columns

9. (a) $_P Y'_{[b_r]'}$ (b) $_{[a_r]} Y'_Q$ (c) $_{[b_r]} W_Q$ (d) $_L M_{[a_s]'}$

10. (a) $_{[a_s]} [Y'_{[b_r]'}]$ (b) $[_{[c_t]} X']_{[b_s]'}$ (c) $[_{[b_r]} W]_{[c_t]'}$ (d) $_{[b_s]} [M_{[a_r]'}]$

11. (a) Minor product of type III vectors   (b) $_Q W_{c_i\ c_i} M_L$ (c) Simple
    (d) $Q \times L$

12. $_P X'_{a_1\ a_1} X'_Q$

13. (a) Type III   (b) Minor product moment   (c) row   (d) $_Q X_{[a_r]'\ [a_r]} X'_Q$

14. (a) Type III   (b) Minor product moment   (c) column   (d) $_P Y'_{[c_t]'\ [c_t]} Y_P$

15. (a) Major product of type III vectors   (b) $_{a_i} Y_P\ _P X_{b_j}$ (c) Supermatrix
    (d) Superorder is $s \times r$, simple order is $1'[a_s] \times [b_r]'1$

16. (a) Type III   (b) Major product moment   (c) column vector
    (d) $_{[a_i]} X_Q\ _Q X'_{[a_r]'}$

17. (a) Type III   (b) Major product moment   (c) row vector
    (d) $_{[c_t]} Y'_L\ _L Y_{[c_t]'}$

18. (a) Type IV   (b) Minor   (c) $_{[b_s]} X_{a_1\ a_1} Y_{[c_t]'}$ (d) Type III
    (e) Major   (f) r (g) $_{[b_s]} X_{[c_t]'}$

19. $_{[c_t]} [Y'_{[a_r]'}]\ [_{[a_r]} X']_{[b_s]'}$

20. (a) $_{[a_r]} [X_{[b_s]'}]\ [_{[b_s]} X']_{[a_r]'}$ (b) Minor   (c) Row

21. (a) $_{[b_s]} [X'_{[a_r]'}]\ [_{[a_r]} X]_{[b_s]'}$ (b) Minor   (c) Column

22. (a) $[_{[a_r]} X]_{[b_s]'}\ _{[b_s]} [Y_{[c_t]'}]$ (b) Type IV   (c) Major

23. $[_{[a_r]} X]_{[b_s]'}\ _{[b_s]} [Y_{[c_t]'}]$

24. (a) Type IV   (b) Major product moment

25. (a) Type IV   (b) Major product moment   (c) In A the natural order is a
    row, in B a column

26. $_{[a_r]} X_{[b_s]'}\ _{[b_s]} Y_{[c_t]'}\ _{[c_t]} Z_{[d_u]'} = {}_{[a_r]} W_{[d_u]'}$

Chapter 12

Simple Statistical Equations

12.1 The Mean

We shall now see how matrix notation enables us to express simple statistical concepts in convenient form. Perhaps one of the simplest concepts in statistics is the *mean*, or arithmetic average of a set of numbers. Suppose we have a set of N measures. We get the mean simply by adding them together and dividing by the number of measures. As we have already seen, the mean of a number of measures can be expressed in another way; if we let the measures be the elements in a vector, the minor product of this vector and the unit vector is the sum of the measures. To express the number of measures we take the minor product moment of a unit vector whose order is equal to N. We then multiply the sum of the measures by the inverse of the number of measures. We can let Eq. (12.1.1) represent the relationship between the scalar notation for the sum of measures and the matrix notation.

$$\sum_1^N X = X'1_N = 1'_N X \tag{12.1.1}$$

We notice that it does not matter which of the two vectors is used as the prefactor in getting the scalar or minor product, since a scalar is equal to its transpose.

To indicate the number of measures, we have

$$N = 1'_N 1_N \tag{12.1.2}$$

Ordinarily, the mean is expressed as

$$M_X = \frac{\sum_1^N X}{N} \tag{12.1.3}$$

Substituting from the right sides of Eqs. (12.1.1) and (12.1.2), we have

$$M_X = \frac{X'1_N}{1'_N 1_N} \tag{12.1.4}$$

We could also express this as

$$M_X = (X'1_N)(1'_N1_N)^{-1} = (1'_N1_N)^{-1}X'1_N \qquad (12.1.5)$$

You will notice that since the minor product moment of the unit vector to the -1 power is a scalar quantity, it is commutative with all other factors. Therefore, we can write it either before or after the minor product of the X vector by the unit vector. Since the primary purpose of matrix algebra is to simplify the analysis of data, we customarily do not indicate the number of measures by a minor product moment of the unit vector. We simply use the letter N to indicate the number of cases. Furthermore, since the unit vector must be conformable with the X vector, we do not need to use the subscript N to indicate the order of the unit vector. Ordinarily, then, in matrix notation we would indicate the mean of a number of measures as

$$M_X = \frac{X'1}{N} \qquad (12.1.6)$$

We could also write it as

$$M_X = \frac{1'X}{N} \qquad (12.1.7)$$

12.2 The Variance

If we let X be a vector of raw measures and M_X the mean, then the standard formula for the variance, or the square of the standard deviation σ of a set of measures is given as

$$\sigma^2 = \frac{\Sigma X^2}{N} - M_X^2 \qquad (12.2.1)$$

Now we have already seen in Section 8.5 that the sum of squares of a number of measures is the minor product moment of the vector whose elements are those measures. Therefore, we can write

$$\Sigma X^2 = X'X \qquad (12.2.2)$$

If now we substitute from the right sides of Eqs. (12.1.6) and (12.2.2) into the right side of Eq. (12.2.1), we have

$$\sigma^2 = \frac{X'X}{N} - \left(\frac{X'1}{N}\right)^2 \qquad (12.2.3)$$

But the second term on the right of Eq. (12.2.3) can be written as

$$\left(\frac{X'1}{N}\right)^2 = \left(\frac{X'1}{N}\right)\left(\frac{X'1}{N}\right) \qquad (12.2.4)$$

And since the transpose of a scalar product of vectors is equal to the scalar product, we have

$$\frac{X'1}{N} = \frac{1'X}{N} \qquad (12.2.5)$$

If we substitute the right side of Eq. (12.2.5) for the second factor on the right of Eq. (12.2.4), we have

$$\left(\frac{X'1}{N}\right)^2 = \left(\frac{X'1}{N}\right)\left(\frac{1'X}{N}\right) \tag{12.2.6}$$

Since matrix multiplication is associative, we can write the right side of Eq. (12.2.6) as

$$\left(\frac{X'1}{N}\right)^2 = X'\left(\frac{11'}{N^2}\right)X \tag{12.2.7}$$

Substituting from the right side of Eq. (12.2.7) into the second term of the right side of Eq. (12.2.3), we get

$$\sigma^2 = \frac{X'X}{N} - X'\left(\frac{11'}{N^2}\right)X \tag{12.2.8}$$

Now the first term on the right of Eq. (12.2.8) can be written as

$$\frac{X'X}{N} = X'\left(\frac{I}{N}\right)X \tag{12.2.9}$$

You notice that interposing the identity matrix between X' and X does not alter the value of the product. Next let us substitute from Eq. (12.2.9) for the first term on the right side of Eq. (12.2.8), and we get

$$\sigma^2 = X'\left(\frac{I}{N}\right)X - X'\left(\frac{11'}{N^2}\right)X \tag{12.2.10}$$

Since matrix multiplication is distributive, we know that the right side of Eq. (12.2.10) can be written as

$$X'\left(\frac{I}{N}\right)X - X'\left(\frac{11'}{N^2}\right)X = X'\left[\left(\frac{I}{N}\right)X - \left(\frac{11'}{N^2}\right)X\right] \tag{12.2.11}$$

Again, since matrix multiplication is distributive, the factor in brackets on the right side of Eq. (12.2.11) can be written as

$$\left(\frac{I}{N}\right)X - \left(\frac{11'}{N^2}\right)X = \left(\frac{I}{N} - \frac{11'}{N^2}\right)X \tag{12.2.12}$$

If we substitute from the right side of Eq. (12.2.12) into the right side of Eq. (12.2.11), we have

$$X'\left(\frac{I}{N}\right)X - X'\left(\frac{11'}{N^2}\right)X = X'\left(\frac{I}{N} - \frac{11'}{N^2}\right)X \tag{12.2.13}$$

Substituting the right side of Eq. (12.2.13) into the right side of Eq. (12.2.11) gives us

$$\sigma^2 = X'\left(\frac{I}{N} - \frac{11'}{N^2}\right)X \tag{12.2.14}$$

Now from simple algebra we know that the term in brackets on the right of Eq. (12.2.14) can be written as

$$\left(\frac{I}{N} - \frac{11'}{N^2}\right) = \frac{1}{N}\left(I - \frac{11'}{N}\right) \tag{12.2.15}$$

The factor $1/N$ is a scalar that is commutative with all factors; therefore, substituting Eq. (12.2.15) into Eq. (12.2.14) gives us finally

$$\sigma^2 = \frac{1}{N} X'\left(I - \frac{11'}{N}\right)X \tag{12.2.16}$$

Equation (12.2.16) is simply Eq. (12.2.3) written in another form. For computational purposes one would use Eq. (12.2.3) rather than Eq. (12.2.16). We shall need Eq. (12.2.16), however, in the next section when we express the variance of a set of measures as a function of these values measured from their mean.

12.3   The Variance of a Set of Measures from Their Mean

The deviation score formula

If now we let x be a deviation measure, the standard equation for x in terms of the raw measure and the mean of these measures is given in scalar notation as

$$x = X - M_X \tag{12.3.1}$$

Then the vector of deviation measures, if we let x be the entire vector rather than simply a single element of it, is

$$x = X - 1M_X \tag{12.3.2}$$

where X now is a vector of raw measures rather than a single raw measure. We know that in terms of deviation measures the variance is given in scalar notation by

$$\sigma^2 = \frac{\Sigma x^2}{N} \tag{12.3.3}$$

In matrix notation the variance in terms of deviation measures is given by

$$\sigma^2 = \frac{x'x}{N} \tag{12.3.4}$$

Proof of equality with raw score formula

Let us prove now that the right sides of Eq. (12.3.4) and Eq. (12.2.16) are equal. First, we substitute from Eq. (12.1.7) into the second term on the right of Eq. (12.3.2) to get

$$x = X - 1\frac{1'X}{N} \tag{12.3.5}$$

Equation (12.3.5) can be written as

$$x = IX - \left(\frac{11'}{N}\right)X \tag{12.3.6}$$

Applying the distributive law to Eq. (12.3.6), we get

$$x = \left(I - \frac{11'}{N}\right)X \tag{12.3.7}$$

We are now ready to substitute the right side of Eq. (12.3.7) into the right side of Eq. (12.3.4), but to do this we need the transpose of the right side of Eq. (12.3.7). We have, therefore,

$$x' = \left[\left(I - \frac{11'}{N}\right)X\right]' \tag{12.3.8}$$

If on the right of Eq. (12.3.8), we take the product of the transposes in reverse order, we have

$$\left[\left(I - \frac{11'}{N}\right)X\right]' = X'\left(I - \frac{11'}{N}\right)' \tag{12.3.9}$$

But the transpose of a sum is equal to the sum of the transposes; therefore, we can write the second factor on the right of Eq. (12.3.9) as in

$$\left(I - \frac{11'}{N}\right)' = I' - \left(\frac{11'}{N}\right)' \tag{12.3.10}$$

But the transpose of the identity matrix is the identity matrix and the major product moment of any vector is symmetric. Therefore Eq. (12.3.10) may be written as

$$\left(I - \frac{11'}{N}\right)' = \left(I - \frac{11'}{N}\right) \tag{12.3.11}$$

We could have obtained Eq. (12.3.11) directly by observing that the sum of any two symmetric matrices is a symmetric matrix, and therefore the transpose is the original matrix.

We can now rewrite Eq. (12.3.9) as

$$\left[\left(I - \frac{11'}{N}\right)X\right]' = X'\left(I - \frac{11'}{N}\right) \tag{12.3.12}$$

Substituting Eq. (12.3.12) in Eq. (12.3.8) gives

$$x' = X'\left(I - \frac{11'}{N}\right) \tag{12.3.13}$$

Next we substitute Eq. (12.3.1) in Eq. (12.3.4) to get

$$\sigma^2 = \frac{X'\left(I - \frac{11'}{N}\right)\left(I - \frac{11'}{N}\right)X}{N} \tag{12.3.14}$$

Let us now multiply out the two middle terms in the numerator in Eq. (12.3.14) as

$$\left(I - \frac{11'}{N}\right)\left(I - \frac{11'}{N}\right) = II - \frac{I11'}{N} - \frac{11'I}{N} + \left(\frac{11'}{N}\right)\left(\frac{11'}{N}\right)$$

(12.3.15)

Equation (12.3.15) may be greatly simplified if we remember that the product of any matrix multiplied by the identity matrix is the matrix itself. Therefore Eq. (12.3.15) may be written as

$$\left(I - \frac{11'}{N}\right)\left(I - \frac{11'}{N}\right) = I - 2\frac{11'}{N} + \left(\frac{11'}{N}\right)\left(\frac{11'}{N}\right)$$

(12.3.16)

Now the last term on the right of Eq. (12.3.16) may be written as

$$\left(\frac{11'}{N}\right)\left(\frac{11'}{N}\right) = \frac{1(1'1)1'}{N^2}$$

(12.3.17)

But the middle product on the right of Eq. (12.3.17) is the minor product moment of the unit vector, which is a scalar equal to its order. In this case, the order is simply N. Therefore, we have

$$\left(\frac{11'}{N}\right)\left(\frac{11'}{N}\right) = \frac{N}{N^2}11' = \frac{1}{N}11'$$

(12.3.18)

If we substitute the right side of Eq. (12.3.18) in Eq. (12.3.16), we have

$$\left(I - \frac{11'}{N}\right)\left(I - \frac{11'}{N}\right) = \left(I - \frac{11'}{N}\right)$$

(12.3.19)

Equation (12.3.19) can be written as

$$\left(I - \frac{11'}{N}\right)^2 = \left(I - \frac{11'}{N}\right)$$

(12.3.20)

We have here a very interesting type of matrix that will often be useful.

If we premultiply each side of Eq. (12.3.20) by the matrix on the right, we have

$$\left(I - \frac{11'}{N}\right)\left(I - \frac{11'}{N}\right)^2 = \left(I - \frac{11'}{N}\right)^2$$

(12.3.21)

From Eqs. (12.3.20) and (12.3.21), we can also write

$$\left(I - \frac{11'}{N}\right)^3 = \left(I - \frac{11'}{N}\right)$$

(12.3.22)

In the same way, we can show that no matter to what positive integral power we raise the matrix, it is still the same as the first power. Therefore, we have the general rule given by Eq. (12.3.23), where k is a positive integral power.

$$\left(I - \frac{11'}{N}\right)^k = \left(I - \frac{11'}{N}\right)$$

(12.3.23)

A matrix is called *idempotent* if any positive integral power of it is the same as the matrix.

Let us now substitute Eq. (12.3.19) into Eq. (12.3.14), and we get

$$\sigma^2 = \frac{1}{N}X'\left(I - \frac{11'}{N}\right)X = \frac{x'x}{N} \qquad (12.3.24)$$

You see that Eq. (12.3.24) is the same as Eq. (12.2.16), which we derived by starting with the raw-score vector.

The sum of deviation scores

Let us note in passing that the sum of the deviation scores is 0. We know this is true by definition, but we shall prove it very simply by matrix algebra. We premultiply both sides of Eq. (12.3.5) by a row unit vector. This gives us

$$1'x = 1'\left(X - 1\frac{1'X}{N}\right) \qquad (12.3.25)$$

From Eq. (12.3.25) we get

$$1'x = 1'X - 1'\left(\frac{11'X}{N}\right) \qquad (12.3.26)$$

From Eq. (12.3.26) we get

$$1'x = 1'X - \left(\frac{1'1}{N}\right)1'X \qquad (12.3.27)$$

But we know that the factor in parentheses in the second term on the right of Eq. (12.3.27) may be given by

$$\frac{1'1}{N} = \frac{N}{N} = 1 \qquad (12.3.28)$$

Substituting Eq. (12.3.28) in Eq. (12.3.27), we have

$$1'x = 1'X - 1'X \qquad (12.3.29)$$

Thus

$$1'x = 0 \qquad (12.3.30)$$

We have therefore proved by matrix algebra that the sum of deviation scores is equal to 0.

12.4   The Covariance

Matrix notation for covariance

The covariance of two sets of measures, say x and y, is the mean of the sum of their products minus the product of their means. The scalar equation for the covariance is

$$C_{xy} = \frac{\Sigma XY}{N} - \frac{\Sigma X}{N}\frac{\Sigma Y}{N} \qquad (12.4.1)$$

If we let X be a vector of the X measure, and Y a vector of the Y measure, we have, in matrix notation, for the first term on the right of Eq. (12.4.1)

$$\frac{\Sigma XY}{N} = \frac{X'Y}{N} \tag{12.4.2}$$

The matrix equivalents of the factors in the second term on the right of Eq. (12.4.1) are given by Eqs. (12.4.3) and (12.4.4) respectively.

$$\frac{\Sigma X}{N} = \frac{X'1}{N} \tag{12.4.3}$$

$$\frac{\Sigma Y}{N} = \frac{1'Y}{N} \tag{12.4.4}$$

Substituting (12.4.2) to (12.4.4) inclusive in the right side of Eq. (12.4.1), we get

$$C_{xy} = \frac{X'Y}{N} - \frac{X'1}{N}\frac{1'Y}{N} \tag{12.4.5}$$

But the right side of Eq. (12.4.5) may be written as in

$$C_{xy} = \frac{1}{N}X'\left(I - \frac{11'}{N}\right)Y = \frac{1}{N}Y'\left(I - \frac{11'}{N}\right)X \tag{12.4.6}$$

Variance as a special case of covariance

It is easy to show that the variance is a special case of covariance. If we substitute X for Y in Eq. (12.4.6), we get

$$C_{xx} = \frac{1}{N}X'\left(I - \frac{11'}{N}\right)X \tag{12.4.7}$$

But if you compare Eq. (12.4.7) with Eq. (12.2.16), you see that the two are exactly the same.

12.5   The Pearson Product Moment Correlation Coefficient

Scalar notation for the formulas

There are a number of different so-called raw-score formulas for the Pearson product moment coefficient of correlations. One of these is Eq. (12.5.1).

$$r_{ij} = \frac{\dfrac{\Sigma X_i X_j}{N} - \dfrac{\Sigma X_i}{N}\dfrac{\Sigma X_j}{N}}{\sqrt{\dfrac{\Sigma X_i^2}{N} - \left[\dfrac{\Sigma X_i}{N}\right]^2}\sqrt{\dfrac{\Sigma X_j^2}{N} - \left[\dfrac{\Sigma X_j}{N}\right]^2}} \tag{12.5.1}$$

In this case, we have two sets of measures, X_i and X_j respectively. If we now multiply both the numerator and denominator by N, we get Eq. (12.5.2)

$$r_{ij} = \frac{\Sigma X_i X_j - \dfrac{\Sigma X_i \Sigma X_j}{N}}{\sqrt{\Sigma X_i^2 - \left[\dfrac{\Sigma X_i}{N}\right]^2}\sqrt{\Sigma X_j^2 - \left[\dfrac{\Sigma X_j}{N}\right]^2}} \tag{12.5.2}$$

If here again we multiply both numerator and denominator by N, we get Eq. (12.5.3).

$$r_{ij} = \frac{N\Sigma X_i X_j - \Sigma X_i \Sigma X_j}{\sqrt{N\Sigma X_i^2 - [\Sigma X_i]^2}\sqrt{\Sigma X_j^2 - [\Sigma X_j]^2}} \tag{12.5.3}$$

Equations (12.5.1), (12.5.2), and (12.5.3) all give the same result. The only difference is in the method of computation. Equation (12.5.3) is preferred by a good many computers because here the original measures are in whole numbers, and we therefore do not need to deal with decimal numbers until we begin to take the square roots of the denominator terms. On the other hand, if the number of cases is rather large, the first term in the numerator can become awkward to handle. It is not our purpose here to recommend any one of these three formulas above another.

Matrix notation for the formulas

Let us now write these three formulas in terms of matrix notation. We know that the sum of the cross products is the minor product of two vectors and that the sum of a set of measures is the minor product of its vector by the unit vector. Therefore, we can write Eq. (12.5.1) in matrix notation as Eq. (12.5.4).

$$r_{ij} = \frac{\dfrac{X'_{.i}X_{.j}}{N} - \dfrac{X'_{.i}1}{N}\dfrac{1'X_{.j}}{N}}{\sqrt{\dfrac{X'_{.i}X_{.i}}{N} - \dfrac{X'_{.i}1}{N}\dfrac{1'X_{.i}}{N}}\sqrt{\dfrac{X'_{.j}X_{.j}}{N} - \dfrac{X'_{.j}1}{N}\dfrac{1'X_{.j}}{N}}} \tag{12.5.4}$$

Equation (12.5.2) in matrix notation becomes

$$r_{ij} = \frac{X'_{.i}X_{.j} - \dfrac{X'_{.i}1}{N}1'X_{.j}}{\sqrt{X'_{.i}X_{.i} - \dfrac{X'_{.i}1}{N}1'X_{.i}}\sqrt{X'_{.j}X_{.j} - \dfrac{X'_{.j}1}{N}1'X_{.j}}} \tag{12.5.5}$$

For Eq. (12.5.3), we get

$$r_{ij} = \frac{NX'_{.i}X_{.j} - X'_{.i}1\,1'X_{.j}}{\sqrt{NX'_{.i}X_{.1} - X'_{.i}1\,1'X_{.i}}\sqrt{NX'_{.j}X_{.j} - X'_{.j}1\,1'X_{.j}}} \tag{12.5.6}$$

A comparison of Eqs. (12.5.1), (12.5.2), and (12.5.3) with the corresponding matrix equations (12.5.4), (12.5.5), and (12.5.6) leads us to conclude that there is no particular economy in expressing the scalar quantity r_{ij} in terms of matrix notation. We shall see, however, that there is great economy in the matrix notation when we are concerned with a matrix of correlation coefficients rather than a single correlation coefficient. This is the more common situation in behavioral science data. Matrix notation helps us to express a table of correlation coefficients in terms of the data matrix of raw measures.

12.6 The Correlation Matrix from Raw Scores

The raw covariance matrix

Suppose we let X be a raw-score matrix of order N by r where N is the number of entities and n is the number of variables. In general, N is much larger than n, so the natural order of the matrix X is vertical. The first step in the computation of the matrix of intercorrelations among the n variables is the computation of the raw-score covariance matrix. This is simply the minor product moment of the matrix X divided by N. If we let G be this product moment matrix of raw scores over N, we have

$$G = \frac{X'X}{N} \tag{12.6.1}$$

Expanding the term on the right into a matrix whose elements are the minor products of columns of X, we have Eq. (12.6.2).

$$G = \frac{1}{N} \begin{bmatrix} X'_{.1}X_{.1} & X'_{.1}X_{.2} & \cdots & X'_{.1}X_{.n} \\ X'_{.2}X_{.1} & X'_{.2}X_{.2} & \cdots & X'_{.2}X_{.n} \\ \cdots & \cdots & \cdots & \cdots \\ X'_{.n}X_{.1} & X'_{.n}X_{.2} & \cdots & X'_{.n}X_{.n} \end{bmatrix} \tag{12.6.2}$$

The correction term matrix

To get the correction terms for these raw-score product moments, we shall need products of corresponding means. We therefore compute a vector of these means as in

$$V'_M = \frac{1'X}{N} = \frac{1}{N}[1'X_{.1} \quad 1'X_{.2} \quad \cdots \quad 1'X_{.n}] \tag{12.6.3}$$

To get the matrix of correction terms for the raw-score product moment matrix, we need the major product moment of this vector of means. We indicate it as

$$M = V_M V'_M \tag{12.6.4}$$

Then the covariance matrix for deviation scores is

$$C = G - M \tag{12.6.5}$$

From Eqs. (12.6.3) and (12.6.4) we get Eq. (12.6.6).

$$M = \frac{1}{N^2} \begin{bmatrix} X'_{.1}1 \\ X'_{.2}1 \\ \cdots \\ X'_{.n}1 \end{bmatrix} [1'X_{.1} \quad 1'X_{.2} \quad \cdots \quad 1'X_{.n}] \tag{12.6.6}$$

The difference matrix

Expanding the right side of Eq. (12.6.6) and substituting ·in Eq. (12.6.5) together with Eq. (12.6.2), we get Eq. (12.6.7).

$$
C = \frac{\begin{bmatrix} X'_{.1}X_{.1} & X'_{.1}X_{.2} & \ldots & X'_{.1}X_{.n} \\ X'_{.2}X_{.1} & X'_{.2}X_{.2} & \ldots & X'_{.2}X_{.n} \\ \ldots & \ldots & & \ldots \\ X'_{.n}X_{.1} & X'_{.n}X_{.2} & \ldots & X'_{.n}X_{.n} \end{bmatrix}}{N} -
$$

$$
\frac{\begin{bmatrix} X'_{.1}1\ 1'X_{.1} & X'_{.1}1\ 1'X_{.2} & \ldots & X'_{.1}1\ 1'X_{.n} \\ X'_{.2}1\ 1'X_{.1} & X'_{.2}1\ 1'X_{.2} & \ldots & X'_{.2}1\ 1'X_{.n} \\ \ldots & \ldots & \ldots & \ldots \\ X'_{.n}1\ 1'X_{.1} & X'_{.n}1\ 1'X_{.2} & \ldots & X'_{.n}1\ 1'X_{.n} \end{bmatrix}}{N^2} \qquad (12.6.7)
$$

Writing the right-hand side of Eq. (12.6.7) as a single matrix, we have Eq. (12.6.8).

$$
C =
$$

$$
\begin{bmatrix} \dfrac{X'_{.1}X_{.1}}{N} - \dfrac{X'_{.1}1\ 1'X_{.1}}{N^2} & \dfrac{X'_{.1}X_{.2}}{N} - \dfrac{X'_{.1}1\ 1'X_{.2}}{N^2} & \ldots & \dfrac{X'_{.1}X_{.n}}{N} - \dfrac{X'_{.1}1\ 1'X_{.n}}{N^2} \\[2ex] \dfrac{X'_{.2}X_{.1}}{N} - \dfrac{X'_{.2}1\ 1'X_{.1}}{N^2} & \dfrac{X'_{.2}X_{.2}}{N} - \dfrac{X'_{.2}1\ 1'X_{.2}}{N^2} & \ldots & \dfrac{X'_{.2}X_{.n}}{N} - \dfrac{X'_{.2}1\ 1'X_{.n}}{N^2} \\[2ex] \ldots & \ldots & \ldots & \ldots \\[1ex] \dfrac{X'_{.n}X_{.1}}{N} - \dfrac{X'_{.n}1\ 1'X_{.1}}{N^2} & \dfrac{X'_{.n}X_{.1}}{N} - \dfrac{X'_{.n}1\ 1'X_{.2}}{N^2} & \ldots & \dfrac{X'_{.n}X_{.n}}{N} - \dfrac{X'_{.n}1\ 1'X_{.n}}{N^2} \end{bmatrix}
$$

$$
(12.6.8)
$$

You will notice now in Eq. (12.6.8) that the nondiagonal elements are of the same form as the numerator terms in Eq. (12.5.4), and the diagonal elements are of the same form as the terms under the radicals in the denominator terms of Eq. (12.5.4).

The diagonal matrix

The next step is to make up a matrix of the diagonal terms of Eq. (12.6.8), which we indicate by D_C as in Eq. (12.6.9).

$$
D_C =
$$

$$
\begin{bmatrix} \dfrac{X'_{.1}X_{.1}}{N} - \dfrac{X'_{.1}1\ 1'X_{.1}}{N^2} & 0 & \ldots & 0 \\[2ex] 0 & \dfrac{X'_{.2}X_{.2}}{N} - \dfrac{X'_{.2}1\ 1'X_{.2}}{N^2} & \ldots & 0 \\[2ex] \ldots & \ldots & \ldots & \ldots \\[1ex] 0 & 0 & \ldots & \dfrac{X'_{.n}X_{.n}}{N} - \dfrac{X'_{.n}1\ 1'X_{.n}}{N^2} \end{bmatrix}
$$

$$
(12.6.9)
$$

But Eq. (12.6.9) is simply a diagonal matrix whose elements are the variances of the variables in the data matrix X. We can therefore write it more simply as Eq. (12.6.10).

$$D_C = \begin{bmatrix} \sigma_1^2 & 0 & \cdots & 0 \\ 0 & \sigma_2^2 & \cdots & 0 \\ \cdots & \cdots & \cdots & \cdots \\ 0 & 0 & \cdots & \sigma_n^2 \end{bmatrix} \qquad (12.6.10)$$

If now we let the numerator terms in Eq. (12.5.4) be represented by C_{ij}, Eq. (12.5.4) can be rewritten as

$$r_{ij} = \frac{C_{ij}}{\sigma_i \sigma_j} \qquad (12.6.11)$$

Also we can write Eq. (12.6.8) more compactly as Eq. (12.6.12).

$$C = \begin{bmatrix} \sigma_1^2 & C_{12} & \cdots & C_{1n} \\ C_{12} & \sigma_2^2 & \cdots & C_{2n} \\ \cdots & \cdots & \cdots & \cdots \\ C_{1n} & C_{2n} & \cdots & \sigma_n^2 \end{bmatrix} \qquad (12.6.12)$$

Next let us take the square root of both sides of Eq. (12.6.10). This gives Eq. (12.6.13).

$$D_C^{1/2} = \begin{bmatrix} \sigma_1 & 0 & \cdots & 0 \\ 0 & \sigma_2 & \cdots & 0 \\ \cdots & \cdots & \cdots & \cdots \\ 0 & 0 & \cdots & \sigma_n \end{bmatrix} \qquad (12.6.13)$$

Then we take the inverse of both sides of Eq. (12.6.13) to get Eq. (12.6.14).

$$D_C^{-1/2} = \begin{bmatrix} \dfrac{1}{\sigma_1} & 0 & \cdots & 0 \\ 0 & \dfrac{1}{\sigma_2} & \cdots & 0 \\ \cdots & \cdots & \cdots & \cdots \\ 0 & 0 & \cdots & \dfrac{1}{\sigma_n} \end{bmatrix} \qquad (12.6.14)$$

Multiplication by a diagonal matrix

Next we premultiply Eq. (12.6.12) by the matrix in Eq. (12.6.14) to get Eq. (12.6.15).

$$D_C^{-1/2} = \begin{bmatrix} \dfrac{1}{\sigma_1} & 0 & \cdots & 0 \\ 0 & \dfrac{1}{\sigma_2} & \cdots & 0 \\ \cdots & \cdots & \cdots & \cdots \\ 0 & 0 & \cdots & \dfrac{1}{\sigma_n} \end{bmatrix} \begin{bmatrix} \sigma_1^2 & C_{12} & \cdots & C_{1n} \\ C_{12} & \sigma_2^2 & \cdots & C_{2n} \\ \cdots & \cdots & \cdots & \cdots \\ C_{1n} & C_{2n} & \cdots & \sigma_n^2 \end{bmatrix} \qquad (12.6.15)$$

Multiplying out the right side of Eq. (12.6.15), we get Eq. (12.6.16).

$$D_C^{-1/2}C = \begin{bmatrix} \sigma_1 & \dfrac{C_{12}}{\sigma_1} & \cdots & \dfrac{C_{1n}}{\sigma_1} \\[2mm] \dfrac{C_{12}}{\sigma_2} & \sigma_2 & \cdots & \dfrac{C_{2n}}{\sigma_2} \\[2mm] \cdots & \cdots & \cdots & \cdots \\[2mm] \dfrac{C_{1n}}{\sigma_n} & \dfrac{C_{2n}}{\sigma_n} & \cdots & \sigma_n \end{bmatrix} \tag{12.6.16}$$

Finally, we postmultiply both sides of Eq. (12.6.16) by the matrix in Eq. (12.6.14) as in Eq. (12.6.17).

$$[D_C^{-1/2}C]D_C^{-1/2} = \begin{bmatrix} \sigma_1 & \dfrac{C_{12}}{\sigma_1} & \cdots & \dfrac{C_{1n}}{\sigma_1} \\[2mm] \dfrac{C_{12}}{\sigma_2} & \sigma_2 & \cdots & \dfrac{C_{2n}}{\sigma_2} \\[2mm] \cdots & \cdots & \cdots & \cdots \\[2mm] \dfrac{C_{1n}}{\sigma_n} & \dfrac{C_{2n}}{\sigma_n} & \cdots & \sigma_n \end{bmatrix} \begin{bmatrix} \dfrac{1}{\sigma_1} & 0 & \cdots & 0 \\[2mm] 0 & \dfrac{1}{\sigma_2} & \cdots & 0 \\[2mm] \cdots & \cdots & \cdots & \cdots \\[2mm] 0 & 0 & \cdots & \dfrac{1}{\sigma_n} \end{bmatrix}$$

$$\tag{12.6.17}$$

Multiplying out the right side of Eq. (12.6.17) we get Eq. (12.6.18).

$$[D_C^{-1/2}C]D_C^{-1/2} = \begin{bmatrix} 1 & \dfrac{C_{12}}{\sigma_1\sigma_2} & \cdots & \dfrac{C_{1n}}{\sigma_1\sigma_n} \\[2mm] \dfrac{C_{12}}{\sigma_1\sigma_2} & 1 & \cdots & \dfrac{C_{2n}}{\sigma_2\sigma_n} \\[2mm] \cdots & \cdots & \cdots & \cdots \\[2mm] \dfrac{C_{1n}}{\sigma_1\sigma_n} & \dfrac{C_{12}}{\sigma_2\sigma_n} & \cdots & 1 \end{bmatrix} \tag{12.6.18}$$

But you see that the nondiagonal terms of Eq. (12.6.18) are the correlation coefficients as given by Eq. (12.6.11). Therefore the right side of Eq. (12.6.18) can be written as in Eq. (12.6.19).

$$[D_C^{-1/2}C]D_C^{-1/2} = \begin{bmatrix} 1 & r_{12} & \cdots & r_{1n} \\ r_{12} & 1 & \cdots & r_{2n} \\ \cdots & \cdots & \cdots & \cdots \\ r_{1n} & r_{2n} & \cdots & 1 \end{bmatrix} = r \tag{12.6.19}$$

Alternate matrix formulas

It should be noted that the entire development from Eq. (12.6.1) through Eq. (12.6.19) is based on the scalar formula, Eq. (12.6.1), for the correlation coefficient. You can easily see that if the procedure had been based on the formula of Eq. (12.6.2), the steps would have been as follows: first, the covariance matrix C would have been replaced by N times this covariance matrix as in Eq. (12.6.20). Then the diagonal term of the matrix NC would have been D_{NC}.

$$NC = NG - V_M 1'X \tag{12.6.20}$$

Corresponding to Eq. (12.6.19), we would have had

$$[D_{NC}^{-1/2} NC]D_{NC}^{-1/2} = D_C^{-1/2}CD_C^{-1/2} \tag{12.6.21}$$

To prove that Eq. (12.6.21) is an identity, we first observe that the diagonal matrix of NC is N times the diagonal matrix of C. This is indicated in

$$D_{NC} = ND_C \tag{12.6.22}$$

If we substitute the right side of Eq. (12.6.22) into the left side of Eq. (12.6.21), we have

$$[(ND_C)^{-1/2}NC]\,[ND_C]^{-1/2} = D_C^{-1/2}CD_C^{-1/2} \tag{12.6.23}$$

But notice that the terms in brackets to the $-1/2$ power in Eq. (12.6.23) may be written as

$$[ND_C]^{-1/2} = N^{-1/2}D_C^{-1/2} \tag{12.6.24}$$

Substituting the right side of Eq. (12.6.24) into the left side of Eq. (12.6.23), we have

$$[(N^{-1/2}D_C^{-1/2})NC]\,[N^{-1/2}D_C^{-1/2}] = D_C^{-1/2}CD_C^{-1/2} \tag{12.6.25}$$

Now since N is a scalar quantity, it is commutative with the matrices on the left of Eq. (12.6.25), and we can rewrite Eq. (12.6.25) as

$$N^{-1/2}N^{-1/2}ND^{-1/2}CD_C^{-1/2} = D_C^{-1/2}CD_C^{-1/2} \tag{12.6.26}$$

Obviously, the N's cancel on the left of Eq. (12.6.26), so the equation is an identity.

If we had used Eq. (12.5.6) as the basis for our computation, then the numerator term would have been N^2 times the covariance, and we would have had Eq. (12.6.27) instead of Eq. (12.6.19).

$$[D_{N^2C}]^{-1/2}\,[N^2C]\,[D_{N^2C}]^{-1/2} = r \tag{12.6.27}$$

It is easy to prove that Eq. (12.6.28) is an identity.

$$[D_{N^2C}]^{-1/2} = N^{-1}D_C^{-1/2} \tag{12.6.28}$$

Therefore, we can rewrite Eq. (12.6.27) as

$$N^{-1}D_C^{-1/2}N^2CN^{-1}D_C^{-1/2} = r \tag{12.6.29}$$

From this we get

$$D_C^{-1/2}CD_C^{-1/2} = r \tag{12.6.30}$$

12.7 The Covariance Matrix from Deviation Measures

Equation (12.3.7) shows a vector of deviation measures as a function of the vector of raw measures. Suppose now we wish to express an entire matrix of measures as deviation measures. We can write the same equation as Eq. (12.3.7), except that now X and x are of order N by n. We let this deviation score matrix be represented by

$$x = \left(I - \frac{11'}{N} \right) X \tag{12.7.1}$$

Now the deviation covariance matrix would then be the minor product moment of x divided by the number of cases.

$$\frac{x'x}{N} = C \tag{12.7.2}$$

But Eq. (12.7.2) can be written as

$$x'x = NC \tag{12.7.3}$$

If we substitute the right side of Eq. (12.7.1) into the left side of Eq. (12.7.3) we get

$$X'\left(I - \frac{11'}{N} \right)\left(I - \frac{11'}{N} \right)X = NC \tag{12.7.4}$$

But because of Eq. (12.3.19), we can rewrite Eq. (12.7.4) as

$$X'\left(I - \frac{11'}{N} \right)X = NC \tag{12.7.5}$$

But Eqs. (12.7.5) and (12.4.7) are the same except for the factor N. Therefore we have proved that the minor product moment of the deviation score matrix gives the same result as Eq. (12.6.5), where we take the raw-score minor product moment G and subtract from it the major product moment of the vector of means. In the next chapter we shall see how the correlation matrix may be computed from the raw-score matrix with convenient algebraic checks throughout, based on the associative and distributive laws of matrix multiplication.

SUMMARY

1. The mean of a vector of measures is expressed as

$$M_x = \frac{1'X}{N} = \frac{X'1}{N}$$

2. The variance of a vector of raw scores, X, may be expressed as

$$\sigma_x^2 = \frac{1}{N} X'\left(I - \frac{11'}{N} \right)X$$

3. The variance of a set of measures from their mean:
   a. A matrix of deviation measures x may be expressed in terms of the rawscore data matrix x by

$$x = \left(I - \frac{11'}{N}\right)X$$

   b. The matrix $I - \dfrac{11'}{N}$ is idempotent, that is,

$$\left(I - \frac{11'}{N}\right)^K = \left(I - \frac{11'}{N}\right)$$

   for positive integral values of k.
   c. Therefore,

$$\frac{x'x}{N} = \frac{1}{N}X'\left(I - \frac{11'}{N}\right)X = \sigma^2$$

4. The scalar covariance C:
   a. In matrix notation the scalar covariance C_{xy} of the vectors X and Y may be expressed as

$$C_{xy} = \frac{1}{N}X'\left(I - \frac{11'}{N}\right)Y = \frac{1}{N}Y'\left(I - \frac{11'}{N}\right)X$$

   b. The variance is a special case of the covariance when $X = Y$.

5. The Pearson product moment correlation coefficient may be expressed in matrix notation as a function of the raw-score vectors X_i, X_j as follows:

   a. $$r_{ij} = \frac{\dfrac{X_i'X_j}{N} - \dfrac{X_i'1}{N}\dfrac{1'X_j}{N}}{\sqrt{\dfrac{X_i'X_i}{N} - \dfrac{X_i'1}{N}\dfrac{1'X_i}{N}}\sqrt{\dfrac{X_j'X_j}{N} - \dfrac{X_j'1}{N}\dfrac{1'X_j}{N}}}$$

   b. $$r_{ij} = \frac{X_i'X_j - \dfrac{X_i'1\ 1'X_j}{N}}{\sqrt{X_i'X_i - \dfrac{X_i1\ 1'X_i}{N}}\sqrt{X_j'X_j - \dfrac{X_j1\ 1'X_j}{N}}}$$

   c. $$r_{ij} = \frac{NX_i'X_j - X_i1\ 1'X_j}{\sqrt{NX_i'X_i - X_i'1\ 1'X_i}\sqrt{NX_jX_j - X_{j1}'1'X_j}}$$

6. The correlation matrix from raw scores:
   a. The raw-score covariance matrix of the data matrix X is given by

$$G = \frac{X'X}{N}$$

   b. The vector of means is given by

$$V_M' = \frac{1'X}{N}$$

c. The covariance matrix is given by

$$C = G - V_M V'_M$$

d. The correlation matrix is given by

$$r = D_C^{-1/2} C D_C^{-1/2}$$

where D_C is the diagonal of C.

7. The covariance matrix may be expressed as a function of a matrix x of deviation measures by

$$C = \frac{x'x}{N}$$

EXERCISES

1. (a) If V is a vector of measures of order N, what is $\dfrac{1'V}{N}$?

   (b) Write the same quantity another way.

2. (a) What is $\dfrac{1}{N} V' \left(I - \dfrac{11'}{N} \right) V$?

   (b) Remove the brackets from the expression.

3. Which of the following is the best computational procedure?

   (a) $(V'1)(1'V)$ (b) $V'[(11')V]$ (c) $[V'(11')]V$

4. Which of the following is the best computational procedure?

   (a) $V' \left[\left(I - \dfrac{11'}{N} \right) V \right]$ (b) $V'V - V' \left(\dfrac{11'}{N} \right) V$

   (c) $V'V - \left(\dfrac{V'1}{N} \right) 1'V$ (d) $V'V - \dfrac{V'}{N} (11') V$

5. What are

   (a) $1' \left(I - \dfrac{11'}{1'1} \right)$ (b) $\left(I - \dfrac{11'}{1'1} \right) 1$

   if all unit vectors are equal?

6. (a) If V_1 and V_2 are vectors of order N, what is $\dfrac{1}{N} V'_1 \left(I - \dfrac{11'}{N} \right) V_2$?

   (b) If $V_1 = V_2$, what is it?

7. If $G = \dfrac{X'X}{N}$, what is G_{ij} in terms of vectors of X?

8. (a) If $NC = X' \left(I - \dfrac{11'}{N} \right) X$, what is NC_{ij} in terms of the vectors of X?

   (b) What is C_{ii} called?

9. If $C = \dfrac{1}{N} X'\left(I - \dfrac{11'}{N}\right)X$ and $r = D_C^{-1/2} C D_C^{-1/2}$

   (a) Write r_{ij} in terms of the vectors of X without the use of brackets.

   (b) What is the diagonal D_r of r?

ANSWERS

1. (a) The mean of the measures

   (b) $\dfrac{V'1}{N}$

2. (a) The variance of V

   (b) $\dfrac{1}{N} V'V - \dfrac{1}{N} V' \dfrac{11'}{N} V$

3. (a)

4. (c)

5. (a) A row null vector

   (b) A column null vector

7. $\dfrac{X'_{.i} X_{.j}}{N}$

8. (a) $X'_{.j}\left(I - \dfrac{11'}{N}\right)X_{.j}$

   (b) The variance of $X_{.i}$

9. (a) $r_{ij} = \dfrac{\dfrac{X'_{.i}X_{.j}}{N} - \dfrac{X'_{.i}1}{N}\dfrac{1'X_{.j}}{N}}{\sqrt{\dfrac{X'_{.i}X_{.i}}{N} - \dfrac{X'_{.i}1\,1'X_{.i}}{N^2}}\sqrt{\dfrac{X'_{.j}X_{ij}}{N} - \dfrac{X'_{.j}1\,1'X_{.j}}{N^2}}}$

   (b) I

Chapter 13

Computing the Correlation Matrix

13.1 The Role of the Correlation Matrix in the Analysis of Experimental Data

In the analysis of data involving large numbers of variables or attributes, the correlation matrix plays a fundamental role. As we have seen earlier, most of the problems in the behavioral sciences do involve large numbers of attributes, or to put it another way, data matrices with many columns. Since the correlation matrix is so important in all problems involving analysis and prediction in the social sciences, it is well for you to understand how it may be derived from a data matrix with least labor and greatest accuracy. From this it will be evident that matrix algebra is very well suited to describing the most efficient methods of computation for behavioral science data. If the computational procedures for the analysis of a set of data cannot be clearly indicated by matrix notation, they have probably not been efficiently and economically designed.

Correlation programs now available for high-speed computers make it possible to calculate large-order correlation matrices at great speed with relatively low cost. These facilities are not universally available, however, and in any case it is well to be familiar with efficient procedures suitable for desk calculators. This chapter will show how the notation of matrix algebra can be used to compute the correlation matrix from the data matrix in the most efficient manner. It will also show how, in terms of matrix notation, the associative and distributive laws of matrix algebra may be used to check the successive steps in the computation.

13.2 The Data Matrix

The row and column sums

The basic data for the computation of the correlation matrix are the elements of the data matrix, or the matrix of elements giving the value for each of a given number of entities on each of a given number of attributes. If we are given the data matrix X, we construct a supermatrix from it by adding one row and one column. The new row is simply a vector of column sums of X, or X premultiplied by the row unit vector. The new column is simply a vector of row sums of X, or the matrix X postmultiplied by a column unit vector. The element in the new row and column position is the sum of the elements in the new column, or the new column vector premultiplied by a row unit vector. The order in which each element of the supermatrix is obtained is shown in Fig. 13.2.1, steps 1 to 4.

$$\begin{bmatrix} 1 & 3 \\ X & X1 \\ \hline 1'X & 1'(X1) \\ 2 & 4 \end{bmatrix} \begin{matrix} (1'X)1 \\ \\ 5 \end{matrix}$$

FIG. 13.2.1

Checking the row and column sums

As a final check operation on row and column sums, we postmultiply the column sum vector by the unit vector and write it outside the matrix beside the element in the last row and last column, as step 5 in Fig. 13.2.1. As you see, this is the same value as the fourth entry in the matrix except for the order of multiplication. The two must be equal since matrix multiplication is associative. If the two are not equal, we may assume that a computational error has been made in one of the steps. The error may be in steps 2, 3, 4, or 5 or in several of these. In order to find out which step contains the error, it is well to recompute first the step requiring the smallest number of operations. Ordinarily, this would be step 5, which requires summing the elements in a vector whose order is equal to the number of attributes, whereas step 4 requires summing a vector whose order is equal to the number of entities, generally much larger than the number of attributes.

If recomputing step 5 gives the same results as before, step 4 is repeated. If again, the results are the same as before, we must assume that the row vector of column sums in step 2 or the column vector of row sums in step 3 is incorrect. If the computations are made on a calculating machine, the number of keyboard entries for both 2 and 3 will be the same, since each element of X is punched into the keyboard once for both steps 2 and 3. In step 2, however, the keyboard must be cleared only m times; in step 3, it must be cleared n times. Furthermore, in step 2, only m values are copied from the dials to the paper, while in step 3, n values are copied; therefore, the actual number of operations in step 2 is less than in step 3. We recommend, therefore, that step 2 be recomputed first. Finally, then, if necessary, step 3 would be recomputed.

13.3   The Minor Product Moment of the Data Matrix

The formula

The formula we shall consider first for calculating the elements of the correlation matrix is Eq. (12.5.6) in the previous chapter, that is,

$$r_{ij} = \frac{NX_i'X_j - X_i'1\ 1'X_j}{\sqrt{NX_i'X_i - X_i1\ 1'X_i}\ \sqrt{NX_j'X_j - X_j1\ 1'X_j}} \qquad (13.3.1)$$

The general matrix form for the scalar quantity given by Eq. (13.3.1) must be indicated in several steps. We let Eq. (13.3.2) be a matrix of numerator terms of the type given in Eq. (13.3.1).

$$H = NX'X - (X'1)(1'X) \qquad (13.3.2)$$

Then the matrix of correlation coefficients will be given by

$$r = D_H^{-1/2}HD_H^{-1/2} \qquad (13.3.3)$$

You remember, of course, that D with the subscript H means a diagonal matrix whose elements are the diagonal elements of the matrix H.

The first step is to calculate the minor product moment of X as given in the first term on the right side of Eq. (13.3.2). This product moment must be calculated with appropriate checks, accomplished as follows: first we set up the type III row supervector in which the first element is the data matrix X and the second element is a vector of its row sums. This is the first row of the supermatrix in Fig. 13.2.1. We then premultiply this supermatrix by its transpose, or to put it another way, we get the minor product moment of this supervector. This product is indicated in Eq. (13.3.4).

$$\left[\frac{X'}{1'X'}\right][X \mid X1] = \left[\begin{array}{c|c} \mathbf{1} & \mathbf{2} \\ X'X & X'(X1) \\ \hline (1'X')X & (1'X')(X1) \\ & \mathbf{3} \end{array}\right] \qquad (13.3.4)$$

The computational steps

The computations are made in the following steps indicated by the boldface numbers: Step **1** The minor product moment of X is computed to give the upper left-hand element in the supermatrix on the right. In the actual computation, it is not necessary to calculate the infradiagonal elements of $X'X$, since they are the same as the corresponding supradiagonal elements. Step **2** The sum vector is premultiplied by X' to give the column vector that is the second element in the first row of the supermatrix. Step **3** The third step is simply the minor product moment of the sum vector or the scalar quantity in the lower right-hand corner of the supermatrix. The simple matrix form of the supermatrix is symmetric. Therefore the first element in the second row need not be calculated, since it is the transpose of the second element in the first row.

The fourth step is to multiply the supermatrix on the right side of Eq. (13.3.4) by a type I supervector whose first element is a unit vector, and whose second element is a 0 scalar, as shown in Eq. (13.3.5).

$$
\begin{bmatrix}
\overset{1}{X'X} & \overset{2}{X'(X1)} \\
\hline
(1'X')X & \underset{3}{(1'X')(X1)}
\end{bmatrix}
\begin{bmatrix} 1 \\ - \\ 0 \end{bmatrix}
=
\begin{bmatrix}
\overset{4}{(X'X)1} \\
\underset{5}{[(1'X')X]1}
\end{bmatrix}
\begin{matrix} 1'[(X'X)1] \\ 6 \end{matrix}
\qquad (13.3.5)
$$

This, of course, amounts to summing the simple rows of the first column of the supermatrix product moment. Since, however, we have not entered the infradiagonal elements, our actual computation in Eq. (13.3.4) must consist of adding down a column and then across the corresponding row to get the element corresponding to the sum of elements in a given row.

Figure 13.3.3 shows diagrammatically how these additions are performed. The elements not entered are indicated by the crossed-out squares.

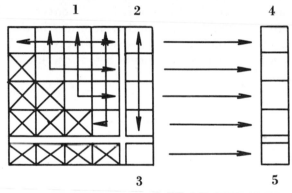

FIG. 13.3.3

Checking the computations

To check our computations, we compare the vectors obtained in steps **2** and **4**. If these two are equal, we can assume, except for possible compensating errors, that all operations have been correctly performed. If one or several corresponding elements in steps **2** and **4** differ, we cannot tell offhand whether the error is in **2**, in **4**, or in the rows of **1** from which **4** was obtained. If **3** and **5** are equal, then we may assume that **2** is correct. If they are not equal, however, we first recompute **5**. If it has been correctly computed, then we recalculate **3**. We also calculate **6**, which is the sum of the elements in the vector given by **4**. This obviously should also be equal to **3** and **5**. If it is equal to **3**, we can assume that **4** is correct, and if **4** is correct, we can also assume that **1** is correct. However, if we have verified that

3 is correct and that **6** has been correctly computed from **4**, then if **3** and **6** do not agree, we know that **4** is incorrect. If **5** does not agree with **3**, we know also that **2** is incorrect.

Our next step, then, would be to check each element in **4** by resuming the corresponding row in **1** that does not agree with the corresponding element in **2**. In this way, we may narrow down the errors by correcting errors that might have occurred in computing **4**.

Still assuming that **5** and **3** do not agree, we may next recalculate the elements in **2** that do not agree with the elements in **4**. In this way, we correct the elements in **2** that were incorrectly calculated. With these corrections made and **5** recomputed, **3** and **5** should now check. We may assume then that all remaining errors are in the product moment matrix or **1**.

If now there are only two remaining errors in **4**, we check the element in **1** corresponding to the row and column indicated by the two errors in **4**. For example, if the second and fifth elements of **4** are incorrect, then we might assume that the error would be in the second row and fifth column of **1**. This we would recalculate. However, if there is only one error in **4**, this is almost certain to be in the corresponding diagonal element since any nondiagonal error would involve two errors in **4**. Conversely, of course, we could have two errors in **4** simply by having errors in the corresponding diagonal elements. In the case of two errors, however, it is better to check the corresponding nondiagonal elements before checking the two corresponding diagonal elements.

In case of more than two errors in **4**, it is well to use the scheme illustrated in Fig. 13.3.4.

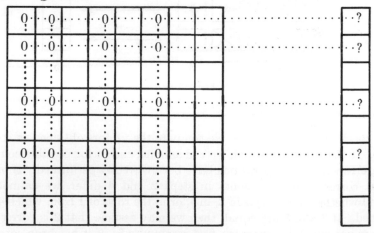

Fig. 13.3.4

Draw a grid on graph paper corresponding to the product-moment matrix. Indicate on a column to the right the position in which the errors are located.

Draw a dotted line through each row corresponding to the position of an error and then draw dotted lines through the corresponding columns. Each row corresponding to an error will then have in it as many intersections of dotted lines as there are errors, including the diagonal.

Now the first step, beginning with the first error, is to check the entry corresponding to the first nondiagonal intersection in the corresponding row. If an error is found in this position, we correct it and recompute the two corresponding elements for step 4. For example, if the second element in the first row of Fig. 13.3.4 has been found to be incorrect, and if the correct value substituted in the matrix 1 of Eq. (13.3.5) makes the first element of 4 agree with the first element of 2, then we may assume that only one error was made in the first row. But this was also an error in the second row, since it was in the second column of the first row. We therefore recompute the second element in operation 4, and see whether this now checks with the second element in 2. If so, we may now assume that all circled positions in the first and second rows and columns are correct and exclude from further consideration the first and second rows and columns of the grid in Fig. 13.3.4. We then proceed to the next uncorrected error in the same way.

Let us assume, however, that the correction of this first nondiagonal element does not result in agreement between the first elements in steps 2 and 4. We then go on to the next intersection, which in our example is the one in the fourth position, and recompute it. Then we recompute the first entry in step 4. In this way, we continue to check all possible errors in the first row and make the necessary corrections below the diagonal in each case of error. If all nondiagonal intersections in the first row turn out to be correct, then we check the diagonal element. The advantage of checking the nondiagonal elements first is that errors in these positions, if corrected, are likely also to correct other discrepancies in the step 4 vector. Once all errors in the first error row have been corrected, it is quite possible that one or more of the subsequent error rows will check. For all errors eliminated in the first row, corresponding rows and columns are deleted from the grid in Fig. 13.3.4, and the same process is repeated for subsequent rows that still fail to check.

The total number of discrepancies in steps 2 and 4 will be equal to or greater than the number of errors in the row of 1 that has the maximum number of errors. As skill in the operation of computing machines increases, the number of errors will decrease, and the operations suggested in connection with Fig. 13.3.4 for finding the errors will probably reveal but a single error in any one row or column. For a reasonably skillful operator, the diagonal elements are less likely to be incorrect than the nondiagonal, because the former is a sum of squares and the latter a sum of cross products. It is then to be expected that in general the correction of one error will correct two discrepancies in steps 2 and 4.

13.4   The Correction Matrix

The computations

There are several ways of obtaining the matrix H in Eq. (13.3.2), once we have computed the minor product moment of X. One method consists of obtaining the product in the second term on the right of Eq. (13.3.2) and the difference matrix H simultaneously, while the other method consists first in obtaining the second term on the right of 2 and then obtaining the difference matrix H. We shall first consider the latter procedure, which is best for those who have not developed facility in the use of computing machines. We assume that the vector of column sums is available from Fig. 13.2.1. Using the last row of the matrix in Fig. 13.2.1, we calculate the major product moment of this supervector as in Eq. (13.4.1). As in writing the product moment matrix of X, we do not write down entries below the diagonal, but use the same scheme as shown in Fig. 13.3.3.

$$\left[\begin{array}{c} X'1 \\ \hline 1'X'1 \end{array} \right] [1'X \mid 1'X1] = \left[\begin{array}{c|c} 1 & 2 \\ (X'1)(1'X) & (X'1)(1'X1) \\ (1'X'1)(1'X) & (1'X'1)(1'X1) \\ & 3 \end{array} \right] \quad (13.4.1)$$

The computational checks

To get our check column from Eq. (13.4.1), we sum rows in the first column of the right-hand side of Eq. (13.4.1), as in Eq. (13.4.2).

$$\left[\begin{array}{c|c} 1 & 2 \\ (X'1)(1'X) & (X'1)(1'X'1) \\ (1'X'1)(1'X) & (1'X'1)(1'X'1) \\ & 3 \end{array} \right] \left[\begin{array}{c} 1 \\ 0 \end{array} \right] =$$

$$\left[\begin{array}{c} 4 \\ [(X'1)(1'X)]1 \\ [(1'X'1)(1'X)]1 \\ 5 \end{array} \right] \begin{array}{c} 1'\{[(X'1)(1'X)]1\} \\ 6 \end{array} \quad (13.4.2)$$

In step 6 we sum the column vector obtained in step 4 and write this scalar below the scalar obtained in step 5. Now if 3 and 6 are equal, we can assume that 4 is correct, and if 4 is correct, we can assume that 1 is correct. If 3 and 6 are not equal, however, these two values should be rechecked. If rechecking shows they are still not equal, we may assume that 4 is incorrect.

Now if 5 and 3 are equal, we may assume that 2 is correct, and we may compare 2 and 4 to see which elements of 4 are incorrect. But if 3 and 5 are not equal, we know that 2 is also not correct. We then recalculate the elements in 2 that do not agree with those in 4. In this way, all errors in 2 should be corrected. We know when they have all been corrected because a recalculation of 5 should check with **3**. We next proceed to recalculate those elements of **4** that differ from the corresponding elements in **2**. If any of these have been incorrectly calculated from **1**, we correct them.

We are now safe in assuming that the remaining discrepancies between **2** and **4** are due to errors in **1**. The procedure for locating errors in **1** is the same as the procedure for locating errors in the minor product moment matrix of X. We identify the rows in which the errors are located and use a grid with dotted lines as in Fig. 13.3.4 to locate the errors in step **1** of Eq. (13.4.2).

13.5 The Difference Matrix

The computations

First we multiply the matrix given on the right of Eq. (13.3.4) by the scalar quantity N. This operation is shown in Eq. (13.5.1).

$$N\left[\begin{array}{c|c} X'X & X'(X1) \\ \hline (1'X)X & (1'X')(X1) \end{array}\right] = \left[\begin{array}{c|c} N(X'X) & N[X'(X1)] \\ \hline N[(1'X')X] & N[(1'X')(X1)] \end{array}\right] \quad (13.5.1)$$

To check these computations we postmultiply the right side of Eq. (13.5.1) by a supervector whose first element is the unit vector and whose last element is a 0 scalar, as in Eq. (13.5.2).

$$\left[\begin{array}{c|c} \overset{1}{N(X'X)} & \overset{2}{N[X'(X1)]} \\ \hline N[(1'X')X] & N[(1'X')(X1)] \\ & \overset{}{3} \end{array}\right] \left[\begin{array}{c} 1 \\ \hline 0 \end{array}\right] = \left[\begin{array}{c} \overset{4}{[N(X'X)]1} \\ \hline \{N[(1'X')X]\}1 \\ 5 \end{array}\right]$$

$$\frac{N\{1'[(X'X)1]\}}{6} \quad (13.5.2)$$

Next we multiply the vector obtained in step **4** by the unit vector to get the scalar in step **6**. If step **6** is equal to step **3**, we may assume that the computations are correct. (As in the computations of steps **3** and **4**, we do not have entries below the diagonal.) If **5** and **6** are the same, we may assume **2** and **4** are the same. The simplest way to check **1**, assuming that **2** and **4** do not agree, is as follows. First recalculate each element in **2** that disagrees with its corresponding element in **4**. Then recalculate each element in **4** that differs from its corresponding element in **2**. Next use the grid system represented by Fig. 13.3.4 to find errors in step **1**.

Next we subtract from the matrix on the right of Eq. (13.5.1) the matrix on the right of Eq. (13.4.1). This is illustrated by Eq. (13.5.3).

$$\begin{bmatrix} N(X'X) & N[X'(X1)] \\ N[(1'X')X] & N(1'X')(X1) \end{bmatrix} - \begin{bmatrix} (X'1)(1'X) & (X'1)(1'X'1) \\ (1'X'1)(1'X) & (1'X'1)(1'X'1) \end{bmatrix} =$$

$$\begin{bmatrix} N(X'X) - (X'1)(1'X) & N[X'(X1)] - (X'1)(1'X'1) \\ N[(1'X')X] - (1'X'1)(1'X) & N(1'X')(X1) - (1'X'1)(1'X'1) \end{bmatrix} \quad (13.5.3)$$

Let us now define new symbols by Eqs. (13.5.4) and (13.5.5).

$$G = NX'X \quad (13.5.4)$$

$$L = (X'1)(1'X) \quad (13.5.5)$$

Then Eq. (13.5.3) can be written as Eq. (13.5.6).

$$\begin{bmatrix} G & G1 \\ 1'G & 1'G1 \end{bmatrix} - \begin{bmatrix} L & L1 \\ 1'L & 1'L1 \end{bmatrix} = \begin{bmatrix} G - L & G1 - L1 \\ 1'G - 1'L & 1'G1 - 1'L1 \end{bmatrix} \quad (13.5.6)$$

Again, for the matrices in Eq. (13.5.6) we do not write in the entries below the diagonal.

The computational checks

Before checking the computations on the right side of Eq. (13.5.6), we first fill in the entries below the diagonal with the exception of the elements in the last row. That is to say, we do not copy the summation column on the left of the matrix on the right side of Eq. (13.5.6) into the row at the bottom of this supermatrix. It is implied, however, and we postmultiply the right side of Eq. (13.5.6) by our summing vector, with a 0 scalar in the last position, to get Eq. (13.5.7).

$$\begin{bmatrix} 1 & 2 \\ G - L & G1 - L1 \\ 1'G - 1'L & 1'G1 - 1'L1 \end{bmatrix}\begin{bmatrix} 1 \\ - \\ 0 \end{bmatrix} = \begin{bmatrix} \overset{4}{(G - L)1} \\ \hline (1'G - 1'L)1 \end{bmatrix} \overset{}{\underset{6}{1'[(G - L)1]}} \quad (13.5.7)$$

$$\qquad\qquad\qquad\quad 3 \qquad\qquad\qquad\qquad 5$$

Here the supervector to the right of the equation should be equal to the last column of the supermatrix on the left because of the distributive law of matrix multiplication. In step **6** we sum the column given in **4** and enter the result at the bottom of this column below step **5**. If **5** is equal to **6**, then both **2** and **4** must be correct, unless there are compensating errors. If **3** and **6** are equal then **4** must be correct. If **2** and **4** do not agree, we first check the entries in **2** that differ from the corresponding entries in **4**. When these entries have been corrected, we recalculate **5**, which must be the same as **3**.

Next, we recompute each element in **4** that differs from its corresponding element in **2**. If discrepancies between **2** and **4** still remain, we examine each row of **1** that has an error in the corresponding element of **4** to make sure that the infradiagonal elements have been properly copied from those above the diagonal. Having verified the accuracy of the copying, and made the necessary corrections in both **1** and **4**, we proceed to locate the remaining errors in **1** by testing those elements isolated by means of the grid technique in Fig. 13.3.4.

It should be noted that the checks used to verify the computations for the products of matrices are based on the associative law, whereas those used to verify the computations for the difference between matrices are based on the distributive law.

13.6 The Simultaneous Computations for the Difference Matrix

The computations

As stated in Section 13.4, it is not necessary to go through the three separate operations indicated by Eqs. (13.4.1), (13.5.1), and (13.5.3) to get the H matrix. We could immediately set up the product of two supermatrices, as in Eq. (13.6.1).

$$[NI \mid - X'1] \begin{bmatrix} X'X & (X'X)1 \\ \hline 1'X & 1'X1 \end{bmatrix} =$$

$$[N(X'X) - (X'1)(1'X) \quad N(X'X)1 - X'1(1'X1)] \quad (13.6.1)$$

Considering first the left-hand factor of the product on the left of Eq. (13.6.1), we see that the first term is simply a scalar matrix NI. The second term is a column vector of column sums obtained from step **2** in Fig. 13.2.1. Considering next the right-hand factor on the left of Eq. (13.6.1), the first row of this supermatrix is simply the first row of the right-hand side of Eq. (13.3.4), which we assumed to have been checked by Eq. (13.3.5) The second row of this matrix is simply the second row of the matrix in Fig. 13.2.1, which we presumed also to have been checked.

If now we sum rows of the left element in the type III supervector on the right of Eq. (13.6.1), we get Eq. (13.6.2).

$$[N(X'X) - (X'1)(1'X) \mid N(X'X)1 - (X'1)(1'X1)] \begin{bmatrix} 1 \\ 0 \end{bmatrix} =$$
$$\quad\quad\quad 1 \quad\quad\quad\quad\quad\quad 2 \quad\quad\quad\quad\quad\quad$$
$$3$$

$$[N(X'X) - (X'1)(1'X)]1 \quad (13.6.2)$$

But note that here the right side of the equation is the same as the second element of the first factor on the left, except for order of multiplication. Because of the distributive law, the third operation in **3**, if it agrees with the second operation, checks the first set of operations.

13.7 The Case When N Is Very Large

The formula

If the number of entities happens to be very large, then formula 1 (Eq. 13.3.1) would not be advisable as a basis for computing the correlation matrix. In this case, the preferred formula would be Eq. (13.7.1).

$$r_{ij} = \frac{X_i'X_j - (X_i1)\left(\dfrac{1'X_j}{N}\right)}{\sqrt{X_i'X_i - \dfrac{X_i'1\,1'X_i}{N}}\,\sqrt{X_j'X_j - \dfrac{X_j'1\,1'X_j}{N}}} \tag{13.7.1}$$

Then the matrix form for the numerator terms of Eq. (13.7.1) would be

$$K = X'X - (X'1)\left(\frac{1'X}{N}\right) \tag{13.7.2}$$

The matrix of correlation coefficients would be

$$r = D_K^{-1/2} K D_K^{-1/2} \tag{13.7.3}$$

Obviously, the relationship between Eqs. (13.3.2) and (13.7.2) is given by

$$H = NK \tag{13.7.4}$$

The vector of means

If Eqs. (13.7.1) and (13.7.2) are used as a basis for computing the correlation matrix, then we still use Eqs. (13.3.4) and (13.4.5) to get the minor product moment of X, but we cannot use Eq. (13.4.2), because the correlation matrix and its check require that we first compute the vector of means indicated by the second factor in the second term on the right side of Eq. (13.7.2). This vector we compute and check by first getting the reciprocal of N, then multiplying the last row of the matrix in Fig. 13.2.1 by this reciprocal.

$$\left[\frac{1}{N}\right][1'X \mid 1'X\,1] = \left[\frac{1'X}{N} \;\middle|\; \frac{1'X\,1}{N}\right] \tag{13.7.5}$$

The very simple check for this division is given by

$$\left[\frac{1'X}{N} \;\middle|\; \frac{1'X\,1}{N}\right]\begin{bmatrix}1\\0\end{bmatrix} = \left[\frac{1'X}{N}\right]1 \tag{13.7.6}$$

This simply means that the sum of the means must be equal to the total sum of all of the elements in X divided by N. In other words, the right-hand side of the equation must be equal to the second element in the first factor of the left-hand side.

The correction matrix

The correction matrix and its check is now given by Eq. (13.7.7).

$$
\begin{bmatrix} X'1 \\ \hline 1'X\ 1 \end{bmatrix}\begin{bmatrix} \dfrac{1'X}{N} & \bigg| & \dfrac{1'X\ 1}{N} \end{bmatrix} = \begin{bmatrix} (X'1)\left(\dfrac{1'X}{N}\right) & \bigg| & (X'1)\left(\dfrac{1'X\ 1}{N}\right) \\ \hline (1'X\ 1)\left(\dfrac{1'X}{N}\right) & \bigg| & (1'X\ 1)\left(\dfrac{1'X\ 1}{N}\right) \end{bmatrix}
$$

$$(13.7.7)$$

As in the preceding product moment matrices, only the entries above the diagonal are written into the right side of the equation. The check for this equation is Eq. (13.7.8), and the interpretations are the same as the checks for Eq. (13.4.2).

$$
\begin{bmatrix} \overset{1}{(X'1)\left(\dfrac{1'X}{N}\right)} & \bigg| & \overset{2}{(X'1)\left(\dfrac{1'X\ 1}{N}\right)} \\ \hline (1'X\ 1)\left(\dfrac{1'X}{N}\right) & \bigg| & (1'X\ 1)\left(\dfrac{1'X\ 1}{N}\right) \\ & 3 & \end{bmatrix}\begin{bmatrix} 1 \\ \hline 0 \end{bmatrix} = \begin{bmatrix} \overset{4}{\left[(X'1)\left(\dfrac{1'X}{N}\right)\right]1} \\ \hline \left[(1'X\ 1)\left(\dfrac{1'X}{N}\right)\right]1 \\ 5 \end{bmatrix}
$$

$$
1'\left\{\left[(X'1)\left(\dfrac{1'X}{N}\right)\right]1\right\}
$$
$$6 \qquad (13.7.8)$$

The difference matrix

To get the difference matrix, we require no operations comparable to those indicated in Eqs. (13.5.1) and (13.5.2), because the correction matrix may be subtracted directly from the minor product moment of X without multiplication by a scalar. We proceed, then, to find the difference equation directly, as in Eq. (13.7.9).

$$
\begin{bmatrix} X'X & \bigg| & X'(X\ 1) \\ \hline (1'X')X & \bigg| & (1'X')(X\ 1) \end{bmatrix} - \begin{bmatrix} (X'1)\left(\dfrac{1'X}{N}\right) & \bigg| & (X'1)\left(\dfrac{1'X\ 1}{N}\right) \\ \hline (1'X\ 1)\left(\dfrac{1'X}{N}\right) & \bigg| & (1'X\ 1)\left(\dfrac{1'X\ 1}{N}\right) \end{bmatrix} =
$$

$$
\begin{bmatrix} X'X - (X'1)\left(\dfrac{1'X}{N}\right) & \bigg| & X'(X\ 1) - (X'1)\left(\dfrac{1'X\ 1}{N}\right) \\ \hline (1'X')X - (1'X\ 1)\left(\dfrac{1'X}{N}\right) & \bigg| & (1'X')(X\ 1) - (1'X\ 1)\left(\dfrac{1'X\ 1}{N}\right) \end{bmatrix}
$$

$$(13.7.9)$$

The checks are then carried out as shown in Eq. (13.5.7) for Eqs. (13.5.3) and (13.5.6).

Direct computation of the K matrix

If we wish to calculate the K matrix given by Eq. (13.7.2) directly, without first computing the correction matrix, we set up an equation comparable to Eq. (13.6.1) as

$$[I \mid -X'1] \begin{bmatrix} X'X & (X'X)1 \\ \hline \dfrac{1'X}{N} & \dfrac{1'X\,1}{N} \end{bmatrix} = \tag{13.7.10}$$

$$\left[X'X - X'1\left(\frac{1'X}{N}\right) \;\middle|\; (X'X)1 - X'1\left(\frac{1'X\,1}{N}\right) \right] = [K \mid K\,1]$$

The check for the first matrix element, the matrix K, on the right side of this equation is the same as the check in Eq. (13.6.2). In other words, we simply add the elements in each row of the K matrix to see that the sum is equal to the corresponding element in the last column of this supervector. As in writing the H matrix, we copy the elements above the diagonal in Eq. (13.7.9) into their appropriate positions below the diagonal before checking the subtraction.

13.8 Calculating the Reduced Minor Product Moment Matrix Directly from the Data Matrix

If we wish to calculate the matrix K given by Eq. (13.7.2), rather than the matrix H given by Eq. (13.3.2), we can set up the product matrix given by Eq. (13.8.1).

$$\begin{bmatrix} X' & X'1 \\ \hline 1'X' & 1'X'1 \end{bmatrix} \begin{bmatrix} X \\ \hline \dfrac{-1'X}{N} \end{bmatrix} = \begin{bmatrix} X'X - (X'1)\left(\dfrac{1'X}{N}\right) \\ \hline 1'X'X - (1'X'1)\left(\dfrac{1'X}{N}\right) \end{bmatrix} \tag{13.8.1}$$

The first element in the type III vector on the right of Eq. (13.8.1) is the reduced minor product moment given by Eq. (13.7.2). The check is given by Eq. (13.8.2).

$$[1' \mid 0] \begin{bmatrix} X'X - X'1\left(\dfrac{1'X}{N}\right) \\ \hline (1'X')X - (1'X'1)\left(\dfrac{1'X}{N}\right) \end{bmatrix} = 1'(X'X) - 1'\left[(X'1)\left(\frac{1'X}{N}\right) \right]$$

$$\tag{13.8.2}$$

It should be noted that the first factor on the left of Eq. (13.8.1) is simply the transpose of the matrix in Fig. 13.2.1. The right-hand factor on the left of Eq. (13.8.1) is simply the data matrix to which has been added a row of negative column sums divided by N. Figure 13.8.1 is a numerical example of the computations indicated in Eqs. (13.8.1) and (13.8.2).

$$
\begin{bmatrix}
4 & 7 & 8 & 5 & | & 24 \\
6 & 3 & 4 & 3 & | & 16 \\
3 & 1 & 9 & 7 & | & 20 \\
\hline
13 & 11 & 21 & 15 & | & 60
\end{bmatrix}
\begin{bmatrix}
4 & 6 & 3 \\
7 & 3 & 1 \\
8 & 4 & 9 \\
5 & 3 & 7 \\
\hline
\frac{-24}{4} & \frac{-16}{4} & \frac{-20}{4}
\end{bmatrix}
=
\begin{bmatrix}
10 & -4 & 6 \\
-4 & 6 & -2 \\
6 & -2 & 40 \\
\hline
12 & 0 & 44
\end{bmatrix}
$$

$$
[1 \ \ 1 \ \ 1 \mid 0]
\begin{bmatrix}
10 & -4 & 6 \\
-4 & 6 & -2 \\
6 & -2 & 40 \\
\hline
12 & 0 & 44
\end{bmatrix}
= [12 \ \ 0 \ \ 44] \quad \leftarrow\text{check}
$$

FIG. 13.8.1

13.9 The Diagonal Multiplier

The computations

Whether we use Eqs. (13.3.2) and (13.3.3) or Eqs. (13.7.2) and (13.7.3) as a basis for computing the correlation matrix, the next set of operations is the same. We shall therefore use Eqs. (13.7.2) and (13.7.3) to indicate the method. The first step is to compute a vector of the square roots of the diagonal elements of the matrix K. Let us assume that we have the matrix K together with its row sums, as indicated in Eq. (13.7.9). We shall not be concerned with the last row of the right side of Eq. (13.7.9), so we merely indicate the first matrix row by

$$[K \mid K1]$$

FIG. 13.9.1

We now calculate a vector of the square roots of the diagonal elements of K and place it three columns to the left of the matrix K, as in Fig. 13.9.2.

$$
\begin{array}{cccc}
& 1 & 2 & 3 \\
[D_K^{1/2}2 \mid & & \mid & K \mid K1]
\end{array}
$$

FIG. 13.9.2

Now in column 3 to the left of K, we write a vector of the reciprocals of the elements in the first column of Fig. 13.9.2. This is a vector of the reciprocals of the square roots of the diagonal elements of K and is illustrated in Fig. 13.9.3.

$$\begin{array}{ccc} \mathbf{1} & \mathbf{2} & \mathbf{3} \\ [D_K^{1/2}\,| & |\;D_K^{-1/2}1\;|\;K\;|\;K1] \end{array}$$

FIG. 13.9.3

The computational checks

Next we multiply each element in column **3** of Fig. 13.9.3 by the corresponding diagonal elements from K and put the entry in column **2** of Fig. 13.9.3. This is illustrated by Fig. 13.9.4.

$$[D_K^{1/2}1\;|\;D_K D_K^{-1/2}1\;|\;D_K^{-1/2}1\;|\;K\;|\;K1]$$

FIG. 13.9.4

Now if the computations in columns **1**, **2**, and **3** of Fig. 13.9.4 have all been correct, then corresponding elements in columns **1** and **2** should be equal. This is true because

$$D_K D_K^{-1/2} = D_K^{(1-1/2)} = D_K^{1/2} \tag{13.9.1}$$

Therefore by means of columns **1**, **2**, and **3** we have obtained in column **3** the elements of the diagonal matrix by which K is pre- and postmultiplied in Eq. (13.7.3). These elements have also been checked, as we have noticed in column **2**.

13.10   Premultiplication by the Diagonal Matrix

The computations

First we premultiply the matrix K, augmented by a vector of row sums, by $D_K^{-1/2}$. Figure 13.9.4 indicates the column headings for a worksheet in convenient form for this multiplication. Each row of the K matrix, together with its sums, is preceded now by the value by which each of its elements is to be multiplied. We then calculate a new matrix, as in Eq. (13.10.1).

$$D_K^{-1/2}[K\;|\;K1] = [D_K^{-1/2}K\;|\;D_K^{-1/2}(K1)] \tag{13.10.1}$$

The computational checks

The check on the multiplication is given by

$$[D_K^{-1/2}K\;|\;D_K^{-1/2}(K1)] \begin{bmatrix} 1 \\ - \\ 0 \end{bmatrix} = [D_K^{-1/2}K]1 \tag{13.10.2}$$

$$\begin{array}{ccc} \mathbf{1} & \mathbf{2} & \mathbf{3} \end{array}$$

The right side of this equation is a column vector of row sums of the first element in the first factor of the left side of the equation or the row sums of K after having been premultiplied by $D_K^{-1/2}$. These row sums in step 3 should be equal to the elements in step 2. Figure 13.10.1 is a numerical example of the operations represented by columns 1, 2, and 3 of Fig. 13.9.4. Figure 13.10.2 is a numerical example of the operations represented by Eqs. (13.10.1) and (13.10.2).

$$
\left[\begin{array}{c|c|c}
\sqrt{10} & \sqrt{10} & \dfrac{1}{\sqrt{10}} \\[2ex]
\sqrt{6} & \sqrt{6} & \dfrac{1}{\sqrt{6}} \\[2ex]
\sqrt{40} & \sqrt{40} & \dfrac{1}{\sqrt{40}}
\end{array}\right]
\qquad
\left[\begin{array}{ccc|c}
10 & -4 & 6 & 12 \\
-4 & 6 & -2 & 0 \\
6 & -2 & 40 & 44
\end{array}\right]
$$

$$D_K^{1/2}1 \quad D_K D_K^{-1/2}1 \quad D_K^{-1/2}1 \qquad\qquad K \qquad\qquad K1$$

FIG. 13.10.1

$$
\left[\begin{array}{ccc}
\dfrac{1}{\sqrt{10}} & 0 & 0 \\[2ex]
0 & \dfrac{1}{\sqrt{6}} & 0 \\[2ex]
0 & 0 & \dfrac{1}{\sqrt{40}}
\end{array}\right]
\left[\begin{array}{ccc|c}
10 & -4 & 6 & 12 \\
-4 & 6 & -2 & 0 \\
6 & -2 & 40 & 44
\end{array}\right]
=
\left[\begin{array}{ccc|c}
\dfrac{10}{\sqrt{10}} & \dfrac{-4}{\sqrt{10}} & \dfrac{6}{\sqrt{10}} & \dfrac{12}{\sqrt{10}} \\[2ex]
\dfrac{-4}{\sqrt{6}} & \dfrac{6}{\sqrt{6}} & \dfrac{-2}{\sqrt{6}} & 0 \\[2ex]
\dfrac{6}{\sqrt{40}} & \dfrac{-2}{\sqrt{40}} & \dfrac{40}{\sqrt{40}} & \dfrac{44}{\sqrt{40}}
\end{array}\right]
$$

$$\qquad D_K^{-1/2} \qquad\qquad\qquad K \qquad\qquad K1 \qquad\qquad\qquad D_K^{-1/2}K \qquad\qquad D_K^{-1/2}K1$$

$$
\left[\begin{array}{ccc|c}
\dfrac{10}{\sqrt{10}} & \dfrac{-4}{\sqrt{10}} & \dfrac{6}{\sqrt{10}} & \dfrac{12}{\sqrt{10}} \\[2ex]
\dfrac{-4}{\sqrt{6}} & \dfrac{6}{\sqrt{6}} & \dfrac{-2}{\sqrt{6}} & 0 \\[2ex]
\dfrac{6}{\sqrt{40}} & \dfrac{-2}{\sqrt{40}} & \dfrac{40}{\sqrt{40}} & \dfrac{44}{\sqrt{40}}
\end{array}\right]
\left[\begin{array}{c}
1 \\ 1 \\ \hline 0
\end{array}\right]
=
\left[\begin{array}{c}
\dfrac{12}{\sqrt{10}} \\[2ex]
0 \\[2ex]
\dfrac{44}{\sqrt{40}}
\end{array}\right]
\longleftarrow \text{check}
$$

$$[D_K^{-1/2}K \qquad D_K^{-1/2}K1]\left[\dfrac{1}{0}\right] = [D_K^{-1/2}K]1$$

FIG. 13.10.2

13.11 Postmultiplication by the Diagonal Matrix

The computations

Let us indicate the new matrix given by step 1 in Eq. (13.10.2) as B. At the top of the B matrix we copy the elements from column 3 of Fig. 13.9.4 to get Fig. 13.11.1.

$$\left[\frac{1'D_K^{-1/2}}{B} \right]$$

FIG. 13.11.1

The next step is to postmultiply B by $D_K^{-1/2}$. This amounts to multiplying each element in a column of B by the corresponding element from the vector above it in Fig. 13.11.1. This final operation gives us the correlation matrix as

$$BD_K^{-1/2} = r \qquad (13.11.1)$$

The computational checks

There are two sets of checks for the computation in Eq. (13.11.1) These are

$$D_r = I \qquad (13.11.2)$$

and

$$r = r' \qquad (13.11.3)$$

Equation (13.11.2) says that the diagonal elements of r must all be unity. Equation (13.11.3) says that r must be a symmetric matrix, or that $r_{ij} = r_{ji}$.

The procedure for calculating the correlation matrix from the H given by Eq. (13.3.2) is precisely the same as for K, except that we use $D_H^{-1/2}$ instead of $D_K^{-1/2}$.

SUMMARY

1. The role of the correlation matrix:
   a. The correlation matrix is fundamental in the analysis of behavioral science data.
   b. Matrix notation is well suited to indicating efficient computational procedures.

2. The data matrix:
   a. Row and column sums of the data matrix are calculated for use in subsequent checking operations.
   b. The row and column sum vectors are checked by the associative law, thus,

$$1'[X1] = [1'X]1$$

3. The minor product moment of the data matrix:
   a. The computational formulas to consider first are

$$H = N(X'X) - (X'1)(1'X)$$

$$r = D_H^{-1/2}HD_H^{-1/2}$$

b. The minor product moment $X'X$ is calculated with the associative check

$$X' (X1) = (X'X)1$$

4. The correction matrix $(X'1)(1'X)$ is computed with the associative check

$$X'1(1'X'1) = [(X'1)(1'X)]1$$

5. The difference matrix:
   a. The product moment matrix is multiplied by N with the associative check

$$N[X'(X1)] = [N(X'X)]1$$

   b. With G and L defined as

$$G = N(X'X) \qquad L = (X'1)(1'X)$$

   the matrix $H = G - L$ is computed with the distributive check

$$G1 - L1 = (G - L)1$$

6. Simultaneous computations for the difference matrix:
   a. The computations for H may be computed directly by

$$H = [NI \mid -X'1]\left[\begin{array}{c|c} X'X & (X'X)1 \\ \hline -1'X & 1'X1 \end{array}\right]$$

   b. The check uses both the associative and distributive laws and is

$$N[(X'X)1] - (X'1)(1'X1) = [N(X'X) - (X'1)(1'X)]1$$

7. The case when N is large:
   a. The computational formulas are

$$K = X'X - (X'1)\left(\frac{1'X}{N}\right)$$

$$r = D_{\overline{K}}^{-1/2} K D_{\overline{K}}^{-1/2}$$

   b. The vector of means $\dfrac{1'X}{N}$ is calculated with the associative check

$$\frac{1'X1}{N} = \left(\frac{1'X}{N}\right)1$$

   c. The matrix $(X'1)\left(\dfrac{1'X}{N}\right)$ is calculated with the associative checks

$$(X'1)\left(\frac{1'X1}{N}\right) = \left[(X'1)\left(\frac{1'X}{N}\right)\right]1$$

   d. The difference matrix K is calculated with distributive checks as for H.

e. K may be calculated directly with associative and distributive checks by

$$[K \mid K1] = [I \mid -X'1] \begin{bmatrix} X'X & (X'X)1 \\ \hline \dfrac{1'X}{N} & \dfrac{1'X1}{N} \end{bmatrix}$$

8. The K matrix can be calculated directly from the data matrix with associative and distributive checks by

$$\begin{bmatrix} X' \mid X'1 \\ \hline 1'X' \mid 1'X'1 \end{bmatrix} \begin{bmatrix} X \\ \hline \dfrac{-1'X}{N} \end{bmatrix} = \begin{bmatrix} K \\ \hline 1'K \end{bmatrix}$$

9. The diagonal multiplier:
   a. $D_K^{1/2}1$ and $D_K^{-1/2}1$ are computed.
   b. The check for $D_K^{-1/2}1$ is

$$D_K[D_K^{-1/2}1] = D_K^{1/2}$$

10. The K matrix is premultiplied by $D_K^{-1/2}$ with the associative check

$$(D_K^{-1/2}K)1 = D_K^{-1/2}(K1)$$

11. Postmultiplication by the diagonal matrix:
   a. The matrix $D_K^{-1/2}K = B$ is postmultiplied by $D_K^{-1/2}$ to give

$$r = BD_K^{-1/2}$$

   b. The checks for r are

$$D_r = I \qquad r = r'$$

EXERCISES

In the following, assume Y is a data matrix and let

$$H = NY'Y - Y'1\,1'Y$$

$$K = Y'Y - \frac{Y'1\,1'Y}{N}$$

$$C = \frac{Y'Y}{N} - \frac{Y'1\,1'Y}{N^2}$$

1. Let H be expressed as the product of type III vectors. Indicate the postfactor for each of the following prefactors.

   (a) $[NI, Y'1]$ (b) $[NI, -Y']$ (c) $[I, Y'1\,1']$

2. Let K be expressed as the product of type III vectors. Indicate the prefactor for each of the following postfactors.

(a) $\begin{bmatrix} Y \\ 1\,1'Y \end{bmatrix}$ (b) $\begin{bmatrix} I \\ -1'Y \end{bmatrix}$ (c) $\begin{bmatrix} Y \\ 1'Y \\ N \end{bmatrix}$

3. Let r be expressed as the product of type III vectors. Indicate for each of the following factors the other factor.

(a) $\begin{bmatrix} \dfrac{1}{N}\,Y'Y & \dfrac{Y'1}{N} \end{bmatrix}$ (b) $\begin{bmatrix} Y \\ 1'Y \end{bmatrix}$ (c) $\begin{bmatrix} \dfrac{1}{N}I & \dfrac{Y'1}{N} \end{bmatrix}$

4. For each of Exs. 1, 2, and 3 above, which is the best computational procedure, assuming N large?

5. What is the relationship between the following?

(a) H and K (b) H and C (c) K and C

ANSWERS

1.
(a) $\begin{bmatrix} Y'Y \\ -1'Y \end{bmatrix}$ (b) $\begin{bmatrix} Y'Y \\ 1\,1'Y \end{bmatrix}$ (c) $\begin{bmatrix} NY'Y \\ -Y \end{bmatrix}$

2.
(a) $\begin{bmatrix} Y' \Big| \dfrac{-Y'}{N} \end{bmatrix}$ (b) $\begin{bmatrix} Y'Y & \dfrac{Y'1}{N} \end{bmatrix}$ (c) $[Y' \quad -Y'1]$

3.
(a) $\begin{bmatrix} I \\ \dfrac{-1'Y}{N} \end{bmatrix}$ (b) $\begin{bmatrix} \left(\dfrac{1}{N}\right)Y' & \dfrac{Y'1}{N^2} \end{bmatrix}$ (c) $\begin{bmatrix} Y'Y \\ \dfrac{-1'Y}{N} \end{bmatrix}$

4. 1(a), 2(b), 3(c).

5. (a) $H = NK$ (b) $H = N^2C$ (c) $K = NC$

Part III

The Structure of a Matrix

Chapter 14

Orthogonal Matrices

In the analysis of social science data an important class of matrices is in frequent use. We shall call it the *orthogonal* matrix, taking the term from the mathematician, but using it in a somewhat different sense. Under the heading of orthogonal matrices we shall distinguish a wider variety of matrices than is customary among mathematicians. Some matrices, closely related to the orthogonal matrix, are very useful in data analysis, and we shall call them orthogonal also.

14.1   Orthogonal Vectors

Two vectors are said to be orthogonal to one another if their minor product is 0. For example, if V_1 and V_2 are two vectors, then the fact that they are orthogonal will be expressed as in Fig. 14.1.1.

$$V_1'V_2 = 0$$

FIG. 14.1.1

This states that the minor product of the two vectors is 0. Examples of products of orthogonal vectors are given in Fig. 14.1.2.

$$[1 \quad 1 \quad 1 \quad 1]\begin{bmatrix} 1 \\ -1 \\ 1 \\ -1 \end{bmatrix} = 1 - 1 + 1 - 1 = 0$$

$$[3 \quad 1 \quad 2]\begin{bmatrix} 1 \\ 3 \\ -3 \end{bmatrix} = 3 + 3 - 6 = 0$$

FIG. 14.1.2

Before considering orthogonal matrices in general, you should recognize a simple type of vector that plays an important role in operations with orthogonal matrices. This is the *normal* vector.

309

14.2 Normal Vectors

A vector is called normal if the sum of the squares of its elements is equal to 1. If V_1 is a normal vector, then we can also indicate that the sum of the squares of its elements is equal to 1 by saying that its minor product moment is equal to 1. We can illustrate this relationship by considering a vector V_1 defined as normal by Fig. 14.2.1.

$$V_1'V_1 = 1$$

Fig. 14.2.1

Examples of normal vectors are given in Fig. 14.2.2.

$$\begin{bmatrix} \dfrac{1}{\sqrt{3}} & \dfrac{1}{\sqrt{3}} & \dfrac{1}{\sqrt{3}} \end{bmatrix} \begin{bmatrix} \dfrac{1}{\sqrt{3}} \\[2mm] \dfrac{1}{\sqrt{3}} \\[2mm] \dfrac{1}{\sqrt{3}} \end{bmatrix} = \dfrac{1}{3} + \dfrac{1}{3} + \dfrac{1}{3} = 1$$

$$\begin{bmatrix} \sqrt{\dfrac{2}{10}} & -\sqrt{\dfrac{3}{10}} & -\sqrt{\dfrac{1}{10}} & \sqrt{\dfrac{4}{10}} \end{bmatrix} \begin{bmatrix} \sqrt{\dfrac{2}{10}} \\[2mm] \sqrt{\dfrac{3}{10}} \\[2mm] \sqrt{\dfrac{1}{10}} \\[2mm] \sqrt{\dfrac{4}{10}} \end{bmatrix} = \dfrac{2}{10} + \dfrac{3}{10} + \dfrac{1}{10} + \dfrac{4}{10} = 1$$

Fig. 14.2.2

Any vector may be converted into a normal vector, or as we say, *normalized*, by dividing it by the square root of its minor product moment. This we can prove very easily. If we let V be any vector, and U be that vector divided by the square root of its minor product moment, we can express this relationship between U and V by

$$U = \frac{V}{\sqrt{V'V}} \tag{14.2.1}$$

We shall now prove that U in Eq. (14.2.1) is a normal vector, that is,

$$U'U = 1 \tag{14.2.2}$$

This we do readily by substituting the right-hand side of Eq. (14.2.1) into the left-hand side of Eq. (14.2.2) to get

$$\frac{V'}{\sqrt{V'V}} \frac{V}{\sqrt{V'V}} = \frac{V'V}{V'V} = 1 \tag{14.2.3}$$

14.3 The Definition of an Orthogonal Matrix

Our definition of an orthogonal matrix is somewhat different from that of the mathematicians. In our use of the term, an orthogonal matrix is one whose minor product moment is a diagonal matrix. If X, a vertical matrix, is also an orthogonal matrix, this fact can be stated as in Fig. 14.3.1.

$$X'X = D$$

or

$$
\begin{array}{llll}
X'_{.1}X_{.1} = D_1 & X'_{.1}X_{.2} = 0 & \ldots & X'_{.1}X_{.m} = 0 \\
& X'_{.2}X_{.2} = D_2 & \ldots & X'_{.2}X_{.m} = 0 \\
& \ldots & \ldots & \\
& & & X'_{.m}X_{.m} = D_m
\end{array}
$$

FIG. 14.3.1

In the first equation in Fig. 14.3.1, as is customary, we use D on the right to indicate a diagonal matrix. The second set of equations shows that each column vector of X is orthogonal to every other column vector of X, that is, the minor products of all pairs of distinct column vectors are 0.

A numerical example of an orthogonal matrix is Fig. 14.3.2.

$$
X = \begin{bmatrix} 1 & 1 \\ 1 & 2 \\ 1 & -3 \end{bmatrix}
$$

$$
X'X = \begin{bmatrix} 1 & 1 & 1 \\ 1 & 2 & -3 \end{bmatrix} \begin{bmatrix} 1 & 1 \\ 11 & 2 \\ 1 & -3 \end{bmatrix} = \begin{bmatrix} 3 & 0 \\ 0 & 14 \end{bmatrix}
$$

FIG. 14.3.2

14.4 Orthonormal Matrices

Definition

A special type of orthogonal matrix is the *orthonormal* matrix. We shall define an orthonormal matrix as one whose minor product moment is an identity matrix. This definition of an orthonormal matrix is illustrated in Fig. 14.4.1.

$$Q'Q = 1$$

or

$$
\begin{array}{llll}
Q'_{.1}Q_{.1} = 1 & Q'_{.1}Q_{.2} = 0 & \ldots & Q'_{.1}Q_{.m} = 0 \\
& Q'_{.2}Q_{.2} = 1 & \ldots & Q'_{.2}Q_{.m} = 0 \\
& \ldots & \ldots & \\
& & & Q'_{.m}Q_{.m} = 1
\end{array}
$$

FIG. 14.4.1

In Fig. 14.4.1, the first equation is the definition of an orthonormal matrix, the other equations show that each column vector of Q is normal and that each column vector of Q is orthogonal to every other one. We see therefore that an orthonormal matrix is a special case of an orthogonal matrix in which all of the column vectors are normal.

Constructing an orthonormal matrix from an orthogonal matrix

You will next see that an orthonormal matrix may be readily constructed from any orthogonal matrix. Suppose, for example, we have the orthogonal matrix X such that

$$X'X = D \tag{14.4.1}$$

Let us then define the matrix Q by

$$Q = XD^{-1/2} \tag{14.4.2}$$

We can easily prove now that Q is an orthonormal matrix. If we take the minor product moment of both sides of Eq. (14.4.2), we have

$$Q'Q = D^{-1/2}X'XD^{-1/2} \tag{14.4.3}$$

Next we substitute the right side of Eq. (14.4.1) into the right side of Eq. (14.4.3), and we get

$$Q'Q = D^{-1/2}DD^{-1/2} \tag{14.4.4}$$

Or if we apply the law of exponents to the right side of Eq. (14.4.4), we get

$$Q'Q = D^{[-1/2+1-1/2]} = D^0 = I \tag{14.4.5}$$

Since any diagonal matrix raised to the 0 power is equal to the identity matrix, Eq. (14.4.6) shows that Q is orthonormal.

It should be clear by now that in order to change an orthogonal matrix into an orthonormal matrix we simply normalize the column vectors of the orthogonal matrix by the method of Section 14.2.

14.5 The Major Product Moment of an Orthonormal Matrix

The major product moment of an orthonormal matrix has some interesting and useful properties. First, we may prove that it cannot be an identity matrix. If Q is an $n \times m$ matrix where n is greater than m, then by definition,

$$Q'Q = I \tag{14.5.1}$$

We shall prove that if Eq. (14.5.1) holds, the major product moment of Q cannot be the identity matrix, as given by the inequality

$$QQ' \neq I \tag{14.5.2}$$

Now, we know that the sum of the diagonal elements of the minor product moment of a matrix is equal to the sum of the diagonal elements of the major product moment. This we proved in Section 10.5. The relationship is

$$1'D_{Q'Q}1 = 1'D_{QQ'}1 \tag{14.5.3}$$

But by Eq. (14.5.1), we have

$$1'D_{Q'Q}1 = 1'I1 = 1'1 = m \tag{14.5.4}$$

But from Eq. (14.5.3) and Eq. (14.5.4), we must have

$$1'D_{QQ'}1 = m \tag{14.5.5}$$

Now according to Eq. (14.5.5), the sum of the diagonal elements of the major product moment of Q must be equal to m less than n, and since this major product moment has n elements in the diagonal, they cannot all be 1; some of them must be less than 1. Therefore, we have proved the inequality in Eq. (14.5.2). We shall see later on that the major product moment of Q must in general be distinct from any diagonal matrix. But this proof we shall reserve for the last section of this chapter.

The integral positive powers of the major product of an orthonormal matrix have another interesting property. We recall that an orthonormal matrix is defined by Eq. (14.5.1). We may prove now that any positive integral power of a major product moment of an orthonormal matrix is equal to that major product moment, that is, it is idempotent. We express this rule by

$$(QQ')^n = QQ' \tag{14.5.6}$$

Suppose first we let n equal 2. Then we have

$$(QQ')^2 = (QQ')(QQ') \tag{14.5.7}$$

Because of the associative law, we have

$$(QQ')(QQ') = Q(Q'Q)Q' \tag{14.5.8}$$

And because of Eq. (14.5.1), we have

$$Q(Q'Q)Q' = QQ' \tag{14.5.9}$$

and finally, we have

$$(QQ')^2 = QQ' \tag{14.5.10}$$

Next we let n equal 3. By the rule of exponents, we write

$$(QQ')^3 = (QQ')^2(QQ') \tag{14.5.11}$$

But from Eq. (14.5.10) we can rewrite this as

$$(QQ')^3 = (QQ')(QQ') \tag{14.5.12}$$

and substituting from Eq. (14.5.9) into the right side of Eq. (14.5.12), we have

$$(QQ')^3 = QQ' \qquad (14.5.13)$$

We may then continue this same procedure for any positive integral value of n so that we have proved Eq. (14.5.6) by induction.

14.6 Square Orthogonal Matrices

The product moment

So far we have been considering orthogonal and orthonormal matrices that are not square. The square orthogonal matrix has interesting and useful properties. We shall state without proof that only one product moment of a square orthogonal matrix is a diagonal matrix. This statement applies to the general orthogonal matrix and not to the square orthogonal matrix that is also orthonormal. The rule can be stated symbolically by the equation and two inequalities given in Eq. (14.6.1).

If $\qquad\qquad X'X = D$

and $\qquad\qquad D \neq I \qquad\qquad (14.6.1)$

then $\qquad\qquad XX' \neq D$

The numerical example in Fig. 14.6.1 illustrates the point.

$$X = \begin{bmatrix} \sqrt{2} & -\frac{1}{2} \\ \sqrt{2} & +\frac{1}{2} \end{bmatrix}$$

$$X'X = \begin{bmatrix} 4 & 0 \\ 0 & \frac{1}{2} \end{bmatrix} = D$$

$$XX' = \begin{bmatrix} \frac{9}{4} & \frac{7}{4} \\ \frac{7}{4} & \frac{9}{4} \end{bmatrix}$$

Fig. 14.6.1

Constructing a square orthonormal matrix

It should be clear that if we can construct a rectangular orthonormal matrix from an orthogonal matrix, we can also construct a square orthonormal matrix from a square orthogonal matrix. Let us consider the square orthogonal matrix X and its minor product moment D given in Fig. 14.6.1.

The orthonormal matrix is obtained from the orthogonal matrix as in Fig. 14.6.2.

$$XD^{-1/2} = Q$$

FIG. 14.6.2

But from Fig. 14.6.1, we have Fig. 14.6.3.

$$D^{-1/2} = \begin{bmatrix} \dfrac{1}{\sqrt{4}} & 0 \\ 0 & \dfrac{1}{\sqrt{\dfrac{1}{2}}} \end{bmatrix} = \begin{bmatrix} \dfrac{1}{2} & 0 \\ 0 & \sqrt{2} \end{bmatrix}$$

FIG. 14.6.3

Q is therefore given by Fig. 14.6.4.

$$Q = \begin{bmatrix} \sqrt{2} & -\dfrac{1}{2} \\ \sqrt{2} & \dfrac{1}{2} \end{bmatrix} \begin{bmatrix} \dfrac{1}{2} & 0 \\ 0 & \sqrt{2} \end{bmatrix} = \begin{bmatrix} \dfrac{1}{\sqrt{2}} & \dfrac{-1}{\sqrt{2}} \\ \dfrac{1}{\sqrt{2}} & \dfrac{1}{\sqrt{2}} \end{bmatrix}$$

FIG. 14.6.4

It is easy to see now that the minor product moment of Q given by Fig. 14.6.4 is the identity matrix, as shown in

$$Q'Q = \begin{bmatrix} \dfrac{1}{\sqrt{2}} & \dfrac{1}{\sqrt{2}} \\ \dfrac{-1}{\sqrt{2}} & \dfrac{1}{\sqrt{2}} \end{bmatrix} \begin{bmatrix} \dfrac{1}{\sqrt{2}} & \dfrac{-1}{\sqrt{2}} \\ \dfrac{1}{\sqrt{2}} & \dfrac{1}{\sqrt{2}} \end{bmatrix} = \begin{bmatrix} 1 & 0 \\ 0 & 1 \end{bmatrix}$$

FIG. 14.6.5

But let us see what happens if we take the major product moment of Q as in Fig. 14.6.6.

$$QQ' = \begin{bmatrix} \dfrac{1}{\sqrt{2}} & \dfrac{-1}{\sqrt{2}} \\ \dfrac{1}{\sqrt{2}} & \dfrac{1}{\sqrt{2}} \end{bmatrix} \begin{bmatrix} \dfrac{1}{\sqrt{2}} & \dfrac{1}{\sqrt{2}} \\ \dfrac{1}{\sqrt{2}} & \dfrac{1}{\sqrt{2}} \end{bmatrix} = \begin{bmatrix} 1 & 0 \\ 0 & 1 \end{bmatrix}$$

FIG. 14.6.6

The product moments of square orthonormals

We now state without proof a generalization from Figs. 14.6.5 and 14.6.6. If Q is a square orthonormal matrix, both of its product moments are equal to the identity matrix. This rule is expressed by

$$Q'Q = QQ' = I \tag{14.6.2}$$

For Q, a square orthonormal, Eq. (14.6.2) may be expressed as in Fig. 14.6.7.

$$\left. \begin{array}{llll} Q'_{.1}Q_{.1} = 1 & Q'_{.1}Q_{.2} = 0 & \ldots & Q'_{.1}Q_{.n} = 0 \\ & Q'_{.2}Q_{.2} = 1 & \ldots & Q'_{.2}Q_{.n} = 0 \\ & \ldots & \ldots & \ldots \\ & & & Q'_{.n}Q_{.n} = 1 \end{array} \right\}$$

$$\left. \begin{array}{llll} Q'_{1.}Q_{1.} = 1 & Q'_{1.}Q_{2.} = 0 & \ldots & Q'_{1.}Q_{n.} = 0 \\ & Q'_{2.}Q_{2.} = 1 & \ldots & Q'_{2.}Q_{n.} = 0 \\ & \ldots & \ldots & \ldots \\ & & & Q'_{n.}Q_{n.} = 1 \end{array} \right\}$$

Fig. 14.6.7

The first set of equations in Eq. (14.6.7) shows that all column vectors of Q are normal and mutually orthogonal. The second set shows that all row vectors are normal and mutually orthogonal.

Infinity of square orthonormals of given order

For any given order, there exists an infinite number of square orthonormal matrices. This fact is very important when we come to consider the products of matrices, and particularly in factor analysis. A general indication of the truth of the statement may be given, although this is not a rigorous proof. Because of Fig. 14.6.7, we know that the number of scalar equations necessary to define a square orthonormal matrix is $n(n + 1)/2$. The total number of elements in a square matrix is n^2. The number of elements we can choose at will for the square orthonormal is equal to the total number of elements minus the number of equations necessary to define a square orthonormal matrix. This is given by

$$n^2 - \frac{n(n + 1)}{2} = \frac{n(n - 1)}{2} \tag{14.6.3}$$

The left-hand side of Eq. (14.6.3) shows the number of elements we can choose at will in determining or constructing a square orthonormal of order n.

Let us consider specifically a second-order orthonormal matrix. If we substitute 2 for n in the right-hand side of Eq. (14.6.3), we get

$$\frac{2(2 - 1)}{2} = 1 \tag{14.6.4}$$

Therefore, even for a second-order square orthonormal, we can choose one element at will. Suppose we let a be a value lying between -1 and 1. We can then write a second-order orthonormal as in Eq. (14.6.5).

$$A = \begin{bmatrix} a & -\sqrt{1-a^2} \\ \sqrt{1-a^2} & a \end{bmatrix} \qquad (14.6.5)$$

It is easy now to verify that the right side of this equation is a square orthonormal, as shown in Eqs. (14.6.6) and (14.6.7).

$$A'A = \begin{bmatrix} a & \sqrt{1-a^2} \\ -\sqrt{1-a^2} & a \end{bmatrix}\begin{bmatrix} a & -\sqrt{1-a^2} \\ \sqrt{1-a^2} & a \end{bmatrix} = \begin{bmatrix} 1 & 0 \\ 0 & 1 \end{bmatrix}$$

$$(14.6.6)$$

$$AA' = \begin{bmatrix} a & -\sqrt{1-a^2} \\ \sqrt{1-a^2} & a \end{bmatrix}\begin{bmatrix} a & \sqrt{1-a^2} \\ -\sqrt{1-a^2} & a \end{bmatrix} = \begin{bmatrix} 1 & 0 \\ 0 & 1 \end{bmatrix}$$

$$(14.6.7)$$

Note that if we let a be $1/2$, we get the second-order square orthonormal given by Fig. 14.6.4. Note also that there is an infinite number of values of a lying between 1 and -1, and that any of these values will give a second-order orthonormal. Therefore, we obviously have an infinite number of square orthonormals of order 2.

14.7 Special Types of Square Orthonormals

The sign matrix

You are already familiar with several very simple types of square orthonormal matrices. One of these is the sign matrix, which when pre- or postmultiplied by its transpose gives the identity matrix. As you have already seen, either pre- or postmultiplication of the sign matrix by the transpose simply amounts to raising it to the second power. You already know that the sign matrix raised to any even power is the identity matrix.

The permutation matrix

You have also seen that the permutation matrix pre- or postmultiplied by its transpose yields the identity matrix. Therefore by definition a permutation matrix is a special case of a square orthonormal matrix.

A square orthonormal from a vector

Another type of square orthonormal matrix may be constructed by means of the identity matrix and any vector V, as shown in Eq. (14.7.1).

$$Q = I - \frac{2VV'}{V'V} \qquad (14.7.1)$$

It is easy to prove that if V is any vector of appropriate order, then Q is a square orthonormal. This means that pre- or postmultiplication by its transpose yields the identity matrix. We need show this only for the case of one of the product moments, since we can see that the right side of Eq. (14.7.1) is a symmetric matrix, and therefore its minor and major product moments are equal. Thus we can prove that the right side of Eq. (14.7.1) is orthonormal by squaring both sides of Eq. (14.7.1) as in Eq. (14.7.2).

$$Q^2 = \left(I - \frac{2VV'}{V'V}\right)\left(I - \frac{2VV'}{V'V}\right) \tag{14.7.2}$$

Expanding the right side of Eq. (14.7.2), we have

$$Q^2 = I - \frac{4VV'}{V'V} + \frac{4(VV')(VV')}{(V'V)^2} \tag{14.7.3}$$

But notice that the last term on the right of Eq. (14.7.3) can be written as

$$\frac{4(VV')(VV')}{(V'V)^2} = \frac{4(V'V)VV'}{(V'V)^2} = \frac{4VV'}{V'V} \tag{14.7.4}$$

Therefore, because of Eq. (14.7.4), the last two terms on the right side of Eq. (14.7.3) cancel each other, and we have

$$Q^2 = I \tag{14.7.5}$$

which proves that Q is a square orthonormal.

An interesting special case of the type of square orthonormal given in Eq. (14.7.1) is when V is the unit vector as shown in Fig. 14.7.1.

$$I - \frac{2\,1\,1'}{1'1} = I - \frac{2\,1\,1'}{n}$$

FIG. 14.7.1

An example in which the unit vector is of order 3 is given in Fig. 14.7.2.

$$\left[\begin{pmatrix} 1 & 0 & 0 \\ 0 & 1 & 0 \\ 0 & 0 & 1 \end{pmatrix} - 2\begin{pmatrix} 1 & 1 & 1 \\ 1 & 1 & 1 \\ \dfrac{1}{3} & 1 & 1 \end{pmatrix}\right] = \begin{bmatrix} \dfrac{1}{3} & -\dfrac{2}{3} & -\dfrac{2}{3} \\ -\dfrac{2}{3} & \dfrac{1}{3} & -\dfrac{2}{3} \\ -\dfrac{2}{3} & -\dfrac{2}{3} & \dfrac{1}{3} \end{bmatrix}$$

FIG. 14.7.2

14.8 Products of Orthonormal Matrices

The product of two square orthonormals

We shall now consider some interesting properties of products of orthonormal matrices. First, we shall prove that the product of two square orthonormals is a square orthonormal matrix. Suppose we have two square orthonormal matrices, X and Y. The fact that they are square orthonormals may be indicated by

$$XX' = X'X = I \qquad (14.8.1)$$

and

$$YY' = Y'Y = I \qquad (14.8.2)$$

Next let us consider a matrix Z that is the product of the two square orthonormal matrices X and Y; thus

$$Z = XY \qquad (14.8.3)$$

Let us now prove that Z is also a square orthonormal, or that

$$ZZ' = I \qquad (14.8.4)$$

and

$$Z'Z = I \qquad (14.8.5)$$

First we substitute the right side of Eq. (14.8.3) into the left side of Eq. (14.8.4). This gives us

$$XYY'X' - I \qquad (14.8.6)$$

We rewrite Eq. (14.8.6) as

$$X[YY']X' = I \qquad (14.8.7)$$

But from Eq. (14.8.2), we can rewrite Eq. (14.8.7) as

$$XX' = I \qquad (14.8.8)$$

But this is the same as Eq. (14.8.1), therefore we have proved Eq. (14.8.4). Similarly, we substitute Eq. (14.8.3) in Eq. (14.8.5) and get

$$Y'X'XY = I \qquad (14.8.9)$$

Or because of Eqs. (14.8.1) and (14.8.9), we get Eq. (14.8.10), which, because of Eq. (14.8.2), proves Eq. (14.8.5).

$$Y'Y = I \qquad (14.8.10)$$

The product of any number of square orthonormals

Next let us prove that the product of any number of square orthonormal matrices is a square orthonormal matrix. We let $X_1, X_2, \ldots X_n$ be square orthonormal matrices of the same order, and call the product Z; thus

$$Z = X_1X_2 \ldots X_n \qquad (14.8.11)$$

We let

$$X_1X_2 = Y_2 \qquad (14.8.12)$$

$$Y_2X_3 = Y_3 \qquad (14.8.13)$$

$$\ldots$$

$$Y_{n-1}X_n = Z \qquad (14.8.14)$$

Now since the product of the two square orthonormals is a square orthonormal, we know that Y_2 is an orthonormal. Therefore, we also know that Y_3 in Eq. (14.8.13) is an orthonormal. In this way, we show by induction finally that Z in Eq. (14.8.14) is an orthonormal, or that the product of any number of square orthonormal matrices is a square orthonormal matrix.

Powers of a major product of orthonormals

Next let us consider products of rectangular orthonormal matrices. First we shall consider the case of powers of a major product of orthonormals. We define two orthonormals X and Y by

$$X'X = I \tag{14.8.15}$$

$$XX' \neq I \tag{14.8.16}$$

$$Y'Y = I \tag{14.8.17}$$

$$YY' \neq I \tag{14.8.18}$$

We then define a major product of X and Y by

$$Z = XY' \tag{14.8.19}$$

First we can prove that any positive integral power of the minor product moment of Z is idempotent. This we indicate by

$$(Z'Z)^n = Z'Z \tag{14.8.20}$$

We substitute Eq. (14.8.19) in Eq. (14.8.20) and get

$$(YX'XY')^n = (YX'XY') \tag{14.8.21}$$

But because of Eq. (14.8.15) we can write Eq. (14.8.21) as

$$(YY')^n = YY' \tag{14.8.22}$$

We have already proved that any positive integral power of the major product moment of an orthonormal matrix is equal to that product moment; therefore we have proved Eq. (14.8.20).

Next we prove that the major product moment of Z is also idempotent or

$$(ZZ')^n = ZZ' \tag{14.8.23}$$

We substitute Eq. (14.8.19) in Eq. (14.8.23) to get

$$(XY'YX')^n = (XY'YX') \tag{14.8.24}$$

And from Eq. (14.8.24) we get

$$(XX')^n = (XX') \tag{14.8.25}$$

We have already proved Eq. (14.8.25) so Eq. (14.8.23) is proved.

You should notice that the minor product moment of Z is independent of X and the major product moment is independent of Y. In the first case the X matrices cancel out, and in the second the Y matrices cancel out.

The minor product of orthonormals

In general, the minor product of two orthonormal matrices does not yield a matrix with simple properties. Consider, for example, the minor product of the matrices Y and X,

$$U = Y'X \qquad (14.8.26)$$

In general, neither the major nor minor product moment of U has any simple properties. Take, for example, the square of the major product moment:

$$(UU')^2 = Y'XX'YY'XX'Y \qquad (14.8.27)$$

The right side of Eq. (14.8.27) does not, in general, simplify. Consider also the square of the minor product moment:

$$(U'U)^2 = X'YY'XX'YY'X \qquad (14.8.28)$$

Here also the right side does not simplify. It is well to remember when you are trying to simplify equations or computational procedures that only in major products of orthonormals do we get the simple relationships involving positive integral powers of product moments.

Postmultiplication of a rectangular by a square orthonormal

Next let us consider the product of a rectangular orthonormal and a square orthonormal. We shall see first that if we postmultiply a rectangular orthonormal by a square orthonormal, the product is a rectangular orthonormal. We let a rectangular orthonormal and a square orthonormal be defined respectively by Eqs. (14.8.29) and (14.8.30).

$$Q'Q = I \qquad QQ' \neq I \qquad (14.8.29)$$

$$H'H = I \qquad HH' = I \qquad (14.8.30)$$

We let

$$Z = QH \qquad (14.8.31)$$

We shall prove that Z is orthonormal, as indicated in

$$Z'Z = I \qquad (14.8.32)$$

We substitute the right side of Eq. (14.8.31) in the left of Eq. (14.8.32) and get

$$H'Q'QH = I \qquad (14.8.33)$$

Because of Eq. (14.8.29), we rewrite Eq. (14.8.33) as

$$H'H = I \qquad (14.8.34)$$

But Eq. (14.8.34) is the same as Eq. (14.8.30), therefore we have proved (14.8.32).

Premultiplication of a rectangular by a square orthonormal

Next we shall prove that premultiplication of a rectangular orthonormal by a square orthonormal gives a rectangular orthonormal. Again suppose the rectangular orthonormal Q and the square orthonormal K are given. Equation (14.8.29) indicates that Q is a rectangular orthonormal; Eq. (14.8.35) indicates that K is a square orthonormal.

$$KK' = K'K = I \qquad (14.8.35)$$

We get the product of the two matrices as

$$U = KQ \qquad (14.8.36)$$

We wish to prove now that U is an orthonormal, as indicated in

$$U'U = I \qquad (14.8.37)$$

Substituting Eq. (14.8.36) in Eq. (14.8.37), we get

$$Q'K'KQ = I \qquad (14.8.38)$$

Substituting from Eq. (14.8.35) in Eq. (14.8.38) gives Eq. (14.8.39), which is the same as Eq. (14.8.29), and therefore Eq. (14.8.37) is proved.

$$Q'Q = I \qquad (14.8.39)$$

14.9 Products of Matrices by Orthonormal Matrices

The factors of a matrix

We shall next consider the role that orthonormal matrices play in products with other matrices. Let us take, for example, the product of any two conformable matrices X and Y as given in

$$XY = Z \qquad (14.9.1)$$

Now we let a be a square orthonormal matrix. Then we define two new matrices by

$$Xa = u \qquad (14.9.2)$$

$$a'Y = v \qquad (14.9.3)$$

We shall prove that u postmultiplied by v is equal to Z, as indicated by

$$uv = Z \qquad (14.9.4)$$

First we substitute from Eqs. (14.9.2) and (14.9.3) in Eq. (14.9.4) to get

$$Xaa'Y = Z \tag{14.9.5}$$

But since a is by definition a square orthonormal, we have

$$aa' = I \tag{14.9.6}$$

Substituting Eq. (14.9.6) into the left side of Eq. (14.9.5), we have Eq. (14.9.7), which is the same as Eq. (14.9.1).

$$XY = Z \tag{14.9.7}$$

Therefore we have proved Eq. (14.9.4). But you will recall that an infinite number of square orthonormal matrices of any given order exist; therefore, the matrix a could have been any one of this infinite number, and we could have had an infinite number of pairs of different u and v matrices. Therefore, an infinite number of different pairs of matrices u and v exist whose product is Z.

The factors of a product moment

Next we let X be a nonhorizontal matrix and a and b two square orthonormals whose order is defined by

$$Y = Xa \tag{14.9.8}$$

$$Z = bX \tag{14.9.9}$$

From Eq. (14.9.8), we have, by postmultiplying each side by its transpose,

$$YY' = Xaa'X' \tag{14.9.10}$$

But because of Eq. (14.9.6), we have from this

$$YY' = XX' \tag{14.9.11}$$

From Eq. (14.9.11) we see that if a nonhorizontal matrix X is postmultiplied by a square orthonormal matrix, the major product moment of the product is the same as the major product moment of X.

In the same way, if we premultiply each side of Eq. (14.9.9) by its transpose, we get

$$Z'Z = X'b'bX \tag{14.9.12}$$

and since b is a square orthonormal, this may be written

$$Z'Z = X'X \tag{14.9.13}$$

From Eq. (14.9.11) we see that if a nonhorizontal matrix X is premultiplied by a square orthonormal, the minor product moment of the product is the same as X.

We can also state a general rule as follows: If a symmetric matrix S is either the major or minor product moment of a matrix, it is also the major or minor product moment of each of an infinite number of matrices. This fact has important implications for defining the primary variables of a science.*

Traces of product moments involving square orthonormals

Multiplication by a square orthonormal matrix does not affect the trace of the product moment of a matrix. You will recall that the trace of a matrix is the sum of its diagonal elements. You will also recall that because of this definition only square matrices have traces. We shall prove that the trace of the product moment of a matrix is invariant with respect to multiplication by a square orthonormal matrix. This means simply that if we either pre- or postmultiply a matrix by a square orthonormal matrix to get another matrix, the trace of either product moment of the second matrix will be the same as the trace of the product moment of the original matrix. Again assume that a is a square orthonormal matrix.

$$Xa = Y \qquad (14.9.14)$$

The rule we wish to prove can now be stated by

$$1'D_{xx'}1 = 1'D_{yy'}1 \qquad (14.9.15)$$

Because of Eq. (14.9.14), we have

$$XX' = YY' \qquad (14.9.16)$$

This you will recall we have just proved in the previous paragraph. But if we substitute Eq. (14.9.16) in Eq. (14.9.15), we have an identity, so the rule is proved for the major product of matrices.

In Section 10.5 we proved that the trace of the major product moment of a matrix is equal to the trace of the minor product moment; therefore because of Eq. (14.9.15), we can also write

$$1'D_{x'x}1 = 1'D_{xx'}1 \qquad (14.9.17)$$

For the same reason, we can also write

$$1'D_{y'y}1 = 1'D_{yy'}1 \qquad (14.9.18)$$

If we substitute the left-hand sides of Eqs. (14.9.17) and (14.9.18) into the left and right sides of Eq. (14.9.15) respectively, we have

$$1'D_{x'x}1 = 1'D_{y'y}1 \qquad (14.9.19)$$

This same rule holds if we premultiply a matrix by an orthonormal matrix. In this case, we consider the square orthonormal matrix b, which satisfies

$$b'b = bb' = I \qquad (14.9.20)$$

*Thurstone, L.L,, *Multiple Factor Analysis*, University of Chicago Press, 1947.

Then we define the new matrix Z by

$$bX = Z \qquad (14.9.21)$$

Using Eq. (14.9.20) and Eq. (14.9.21), we can prove

$$1'D_{x'x}1 = 1'D_{z'z}1 \qquad (14.9.22)$$

And remembering that the trace of a major and minor product moment are equal, we have finally

$$1'D_{xx'}1 = 1'D_{zz'}1 \qquad (14.9.23)$$

Column sums and square orthonormals

Next suppose that we postmultiply a matrix by a square orthonormal to get a new matrix, and then take the column sums of both matrices. We shall prove that the sum of the squares of the sums for the first matrix is equal to the sum of the squares of the sums for the second matrix. Again let us consider the product given by Eq. (14.9.14). We recall first that a matrix premultiplied by a row unit vector gives a row vector whose elements are the sums of the elements in the corresponding columns of the matrix. We recall also that the sums of squares of elements can be expressed as the minor product moment of a vector. Therefore, what we wish to prove can be stated

$$(1'X)(X'1) = (1'Y)(Y'1) \qquad (14.9.24)$$

Substituting the left-hand side of Eq. (14.9.14) in the right-hand side of Eq. (14.9.24), we get

$$(1'X)(X'1) = (1'Xa)(a'X'1) \qquad (14.9.25)$$

Or, because a is a square orthonormal matrix, Eq. (14.9.25) becomes the identity given by

$$(1'X)(X'1) = (1'X)(X'1) \qquad (14.9.26)$$

Row sums and square orthonormals

The sums of the squares of row sums are invariant under a left orthonormal transformation. If we represent this left square orthonormal transformation by Eq. (14.9.21), the rule may be stated as

$$(1'X')(X1) = (1'Z')(Z1) \qquad (14.9.27)$$

Substituting from Eq. (14.9.21) in Eq. (14.9.27), we have

$$(1'X')(X1) = (1'X'b')(bX1) = (1'X')(X1) \qquad (14.9.28)$$

This as you see, turns out to be an identity.

14.10   Matrices Orthogonal to One Another

Orthogonal matrices orthogonal to one another

We may have matrices that are orthogonal to one another in the sense that their minor products are zero matrices whether the matrices themselves are orthogonal or not. First let us consider orthogonal matrices that in themselves are orthogonal to one another. We may let an orthogonal matrix X be defined by

$$X'X = D \tag{14.10.1}$$

Now let us partition X by columns into submatrices, so that

$$X = [X_1 \quad X_2 \quad \ldots \quad X_s] \tag{14.10.2}$$

Substituting Eq. (14.10.2) in Eq. (14.10.1) and partitioning the diagonal matrix D on the right accordingly, we have Eq. (14.10.3).

$$\begin{bmatrix} X_1' \\ X_2' \\ \ldots \\ X_s' \end{bmatrix} [X_1 \quad X_2 \quad \ldots \quad X_s] = \begin{bmatrix} D_1 & 0 & \ldots & 0 \\ 0 & D_2 & \ldots & 0 \\ \ldots & \ldots & \ldots & \ldots \\ 0' & 0' & \ldots & D_s \end{bmatrix} \tag{14.10.3}$$

Expanding the left hand side of Eq. (14.10.3) we have Eq. (14.10.4).

$$\begin{bmatrix} X_1'X_1 & X_1'X_2 & \ldots & X_1'X_s \\ X_2'X_1 & X_2'X_2 & \ldots & X_2'X_s \\ \ldots & \ldots & \ldots & \ldots \\ X_s'X_1 & X_s'X_2 & \ldots & X_s'X_s \end{bmatrix} = \begin{bmatrix} D_1 & 0 & \ldots & 0 \\ 0' & D_2 & \ldots & 0 \\ \ldots & \ldots & \ldots & \ldots \\ 0' & 0' & \ldots & D_s \end{bmatrix} \tag{14.10.4}$$

From Eq. (14.10.4), we have therefore Eq. (14.10.5).

$$\left. \begin{aligned} X_1'X_2 = 0 \quad \ldots \quad X_1'X_s &= 0 \\ X_2'X_s &= 0 \\ \text{etc.} \end{aligned} \right\} \tag{14.10.5}$$

It should be clear then that by partitioning any orthogonal matrix by columns we get submatrices such that the minor product of any pair is a null matrix, that is, the submatrices are mutually orthogonal to one another.

Rectangular matrices orthogonal to one another

We may have two matrices neither of which are orthogonal and yet the two may be orthogonal to one another. Even though the matrices are not orthogonal, their minor product may be a null matrix. Again consider an orthogonal matrix X partitioned as indicated in

$$X = [X_1 \quad X_2] \tag{14.10.6}$$

By definition we have

$$X'X = D \tag{14.10.7}$$

From Eqs. (14.10.6) and (14.10.7), we get Eq. (14.10.8).

$$\begin{bmatrix} X_1'X_1 & X_1'X_2 \\ X_2'X_1 & X_2'X_2 \end{bmatrix} = \begin{bmatrix} D_1 & 0 \\ 0' & D_2 \end{bmatrix} \tag{14.10.8}$$

Next consider the square matrices a_1 and a_2, whose elements may take any values whatsoever. Then we define new matrices Y_1 and Y_2 by

$$X_1 a_1 = Y_1 \tag{14.10.9}$$

$$X_2 a_2 = Y_2 \tag{14.10.10}$$

First we will see that even though the X matrices are orthogonal, the Y matrices are not in general so. Consider the minor product moment of Y_1, which is given by

$$Y_1'Y_1 = a_1'X_1'X_1 a_1 \tag{14.10.11}$$

Because of Eq. (14.10.8) we can rewrite Eq. (14.10.11) as

$$Y_1'Y_1 = a_1'D_1 a_1 \tag{14.10.12}$$

Now since the square matrix a_1 was taken as any conformable matrix whatsoever, the right side of Eq. (14.10.12) is not in general a diagonal matrix, so that Y_1 is not in general an orthogonal matrix. In the same way we show by Eq. (14.10.13) that Y_2 is not in general an orthogonal matrix, thus,

$$Y_2'Y_2 = a_2'X_2'X_2 a_2 = a_2'D_2 a_2 \tag{14.10.13}$$

However, let us now examine the minor product of Y_1 and Y_2. From Eqs. (14.10.9) and (14.10.10) we get

$$Y_1'Y_2 = a_1'X_1'X_2 a_2 \tag{14.10.14}$$

But according to Eq. (14.10.8), X_1 and X_2 are mutually orthogonal. Therefore Eq. (14.10.14) can be written as

$$Y_1'Y_2 = a_1'0 a_2 = 0 \tag{14.10.15}$$

We see then that Y_1 and Y_2 are orthogonal to one another even though the matrices themselves are not orthogonal.

14.11 Dimensionality of Matrices Orthogonal to One Another

Restrictions on order

It is important to point out certain restrictions on the orders of matrices that are orthogonal to one another. The proof of these restrictions is beyond the scope of this book, but since they are fundamental to some of our later developments they will be stated without proof. A number of orthogonal matrices of equal height cannot all be mutually orthogonal if the sum of their widths is greater than their height. For example, if we have the orthogonal matrices X_{ba_1} X_{ba_2} ... X_{ba_s}, where the subscripts indicate order, then it is not possible that $X'_{ba_i}X_{ba_j} = 0$ for all distinct values of i and j if $\Sigma a_i > b$.

Major product moment of a vertical matrix

The implications of this rule for the major product moment of a vertical matrix are important. No major product moment of a vertical matrix can be diagonal. This rule can be proved as follows:

We let X' be any horizontal matrix partitioned as indicated by

$$X' = [X_{ba_1} \quad X_{ba_2} \quad ... \quad X_{ba_s}] \tag{14.11.1}$$

and assume the X_{ba_i} are all orthogonal matrices. The major product moment of Eq. (14.11.1) is given by Eq. (14.11.2).

$$XX' = \begin{bmatrix} X'_{ba_1}X_{ba_1} & X'_{ba_1}X_{ba_2} & ... & X'_{ba_1}X_{ba_s} \\ X'_{ba_2}X_{ba_1} & X'_{ba_2}X_{ba_2} & ... & X'_{ba_2}X_{ba_s} \\ ... & ... & ... & ... \\ X'_{ba_s}X_{ba_1} & X'_{ba_s}X_{ba_2} & ... & X'_{ba_s}X_{ba_s} \end{bmatrix} \tag{14.11.2}$$

Since X' is horizontal, $\Sigma a_i > b$, and therefore the off-diagonal elements of Eq. (14.11.2) cannot all be null matrices. Since XX' is a major product moment, we have proved the rule.

SUMMARY

1. **Orthogonal vectors.** Two vectors are said to be orthogonal if their minor product is 0.

2. **Normal vectors:**
   a. A vector is said to be normal if its minor product moment is unity.
   b. Any vector V may be transformed into a normal vector U by

$$U = \frac{V}{\sqrt{V'V}}$$

3. **Definition of orthogonal matrices:**
   a. A vertical matrix X is said to be orthogonal if its minor product moment is a diagonal matrix.
   b. In particular a square matrix is orthogonal if one of its product moments is diagonal.

4. **Orthonormal matrices:**
   a. An orthonormal matrix is one whose minor product moment is the identity matrix. Thus if X is orthonormal

$$X'X = I$$

b. An orthonormal matrix may be constructed from any orthogonal matrix. If

$$X'X = D, \quad Q = XD^{-1/2}$$

then Q is orthonormal.

5. The major product moment of a vertical orthonormal matrix Q
   a. Cannot be an identity matrix, that is, if

   $$Q'Q = I \quad QQ' \neq I$$

   b. It is an idempotent matrix, that is,

   $$(QQ')^m = QQ'$$

   for all positive integral values of m.

6. Square orthogonal matrices:
   a. If X is a square orthogonal but not orthonormal matrix such that $X'X = D \neq I$, then XX' cannot be diagonal.
   b. If Q is a square orthonormal matrix such that $Q'Q = I$, then also $QQ' = I$.
   c. An infinite number of square orthonormals of order m exist.

7. Special types of square orthonormals:
   a. The sign matrix i is a square orthonormal.
   b. The permutation matrix π is a square orthonormal.
   c. If $Q = I - \dfrac{2VV'}{V'V}$, then for any vector V, Q is a square orthonormal.

8. Products of orthonormal matrices:
   a. The product of any number of square orthonormal matrices is a square orthonormal matrix.
   b. Given the major product moment $Z = XY'$ of the vertical orthonormal matrices X and Y.

   (1) $(Z'Z)^m = Z'Z$

   (2) $(ZZ')^m = ZZ'$

   c. Neither of the product moments of the minor product of two vertical orthonormal matrices has simple properties.
   d. Pre- or postmultiplication of a rectangular orthonormal matrix by a square orthonormal matrix yields a rectangular orthonormal matrix.

9. Products of matrices:
   a. If $XY = Z$, an infinite number of pairs of factors $X_i Y_i$ exist whose product is Z.
   b. The factors of a product moment:

(1) If a nonhorizontal matrix X is postmultiplied by a square orthonormal matrix a, the *major* product moment of the product is the same as the major product moment of X.

(2) If a nonhorizontal matrix X is premultiplied by a square orthonormal matrix b, the *minor* product moment of the product is the same as the minor product moment of X.

(3) If a symmetric matrix S is either the major or minor product moment of a matrix, it is also the major or minor product moment of each of an infinite number of matrices.

c. If we pre- or postmultiply a matrix X by a square orthonormal matrix to get another matrix, the trace of either product moment of the second matrix will be the same as the trace of either product moment of X. Thus if a and b are orthonormal matrices, and $Xa = Y$, $bX = Z$, we have $tr\ X'X = tr\ XX' = tr\ Y'Y = tr\ YY' = tr\ Z'Z = tr\ ZZ'$.

d. If a matrix X is postmultiplied by a square orthonormal matrix a to get a matrix Y, the sum of the squares of the column sums of X and Y are equal; that is, if

$$Y = Xa \qquad \text{then} \qquad (1'X)(X'1) = (1'Y)(Y'1)$$

e. If a matrix X is premultiplied by a square orthonormal matrix b to get a matrix Z, the sum of the squares of row sums of X and Z are equal; that is, if $Z = bX$, then

$$(1'X')(X1) = (1'Z')(Z1)$$

10. Matrices orthogonal to one another:

a. If any orthogonal matrix X is partitioned into a type III row vector, the submatrices are mutually orthogonal. Thus if $X = [X_1 \ \ X_2 \ \ldots \ \ X_s]$ then $X_i'X_j = 0$ for all distinct values of i and j.

b. Two matrices Y_1 and Y_2 may be orthogonal ($Y_1'Y_2 = 0$), even though neither is an orthogonal matrix.

11. Dimensionality of matrices orthogonal to one another:

a. A number orthogonal of matrices of equal height cannot be mutually orthogonal to one another if the sum of their widths is greater than their height.

b. No major product moment of a vertical matrix can be a diagonal matrix.

EXERCISES

1. If U_1 and U_2 are orthogonal vectors, what is $U_1'U_2$?

2. (a) If U is a normal vector, what is $U'U$?

   (b) If $W = \dfrac{V}{\sqrt{V'V}}$, what is $W'W$?

3. If X is vertical and $X'X$ is diagonal, what kind of a matrix is X?

4. (a) If $X'X = I$, what kind of a matrix is X?
   (b) If $X'X$ is diagonal, D, how would you construct an orthonormal matrix Q from it?

5. Let $Q'Q = I$ and $QQ' \neq I$.
   (a) What do you know about the order of Q?
   (b) What is $[QQ']^5$?

6. Assume X square and $X'X = D$. What do you know about XX'?
   (a) If $D \neq I$
   (b) If $D = I$
   (c) What kind of a matrix is X if $D = I$?

7. (a) What is a zero-one square orthonormal matrix called? If $X = I - \dfrac{2UU'}{U'U}$, what do you know about

   (b) $X'X$ (c) XX' (d) X^2

8. (a) If $Q_1 = I - \dfrac{2V_1V_1'}{V_1V_1}$ and $Q_2 = I - \dfrac{2V_2V_2'}{V_2V_2}$ and $Q_1Q_2 = b$, what is bb'?

   (b) If C_1, C_2, and C_3 are square orthonormal, and $C = C_1C_2'C_3$, what kind of a matrix is C? Given $X'X = I, XX' \neq I, Y'Y = I, YY' \neq I$ and $Z = XY'$,
   (c) What is $(ZZ')^3$ (d) $(Z'Z)^2$

9. If f and g are square orthonormal matrices and $Xf = Y$, $gX = Z$, what is
   (a) YY' (b) $Z'Z$
   (c) What is the relationship among the traces of all the product moments of X, Y, and Z?

10. If $X = I - \dfrac{2VV'}{V'V} = [X_1 \quad X_2] = \begin{bmatrix} Y_1' \\ Y_2' \end{bmatrix}$,

    (a) What is $X_1'X_2$?   (b) $Y_1'Y_2$

11. Let subscripts indicate order and let

    $$X = [X_{9\,3} \quad X_{9\,2}] \qquad Y = [X_{10\,2} \quad X_{10\,5} \quad X_{10\,4}] \qquad Z = [X_{11\,4} \quad X_{11\,2} \quad X_{11\,3}]$$

    $$W = [W_{8\,2} \quad W_{8\,7}]$$

    Which of the following cannot be diagonal matrices?
    (a) $X'X$ (b) $Y'Y$ (c) $Z'Z$ (d) $W'W$

ANSWERS

1. $U_1'U_2 = 0$

2. (a) $U'U = 1$ (b) $W'W = 1$

3. Orthogonal

4. (a) Orthonormal   (b) $Q = XD^{-1/2}$

5. (a) Vertical   (b) $(QQ')^5 = QQ'$

6. (a) It is not diagonal   (b) $XX' = I$ (c) Square orthonormal

7. (a) Permutation matrix   (b) $X'X = I$ (c) $XX' = I$ (d) $X^2 = I$

8. (a) $bb' = I$ (b) Square orthonormal   (c) $(ZZ')^3 = ZZ'$
   (d) $(Z'Z)^2 = Z'Z$

9. (a) $YY' = XX'$ (b) $Z'Z = X'X$ (c) They are all equal

10. (a) $X_1'X_2 = 0$ (b) $Y_1'Y_2 = 0$

11. (b) $Y'Y$ (d) $W'W$

Chapter 15

The Rank of a Matrix

The concepts discussed in previous chapters lead us now to the idea of rank, one of the fundamental properties of a matrix. As you will see later on, the idea of rank is important not only as a concept in matrix algebra but also for its significance in scientific research. But before we define the rank of a matrix, we must first consider some very simple ideas closely related to it.

15.1 The Factors of a Matrix

Infinity of pairs of factors of a product

In scalar algebra or arithmetic it is always possible to express any number or quantity as the product of two other quantities. Figure 15.1.1 shows a few of the infinite number of ways in which you can express the number 12.

$$12 = 2 \times 6 \qquad 12 = 3 \times 4$$
$$12 = \frac{24}{3} \times \frac{3}{2} \qquad 12 = \frac{12}{2} \times 2$$

You will remember from Section 14.9 that if a matrix can be expressed as the product of a given pair of matrices, there is an infinite number of pairs of matrices that will yield that product. In fact, we can think of any matrix as the product of two other matrices. A matrix x_{nm}, with n rows and m columns, can always be expressed as the product of another n by m matrix postmultiplied by an m by m matrix. This statement may be expressed as

$$X_{nm} = Y_{nm}Z_{mm} \tag{15.1.1}$$

where the subscripts all indicate the order of the matrices.

Matrices as major products of factors

There are some matrices that can be expressed as the product of two matrices whose common order is less than the width of the matrix. For example, suppose we have the product indicated by Fig. 15.1.1.

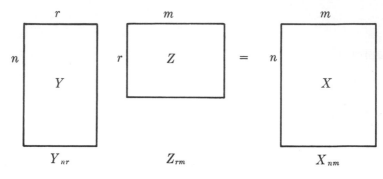

FIG. 15.1.1

Here the matrix X_{nm} can be expressed as a major product of two other matrices Y_{nr} and Z_{rm} with common order r less than m.

We know then that if a matrix X_{nm} can be expressed as the product of two matrices whose common order r is less than m, then an infinite number of pairs of matrices exist with common order r whose product is X_{nm}. You must keep in mind, however, that only *some* n by m matrices can be expressed as the major product of two matrices. A matrix X_{nm} cannot *in general* be expressed as the product of two matrices with common order less than m.

The concept of rank is based on this highly important fact: any matrix may be expressed as the product of two matrices whose common order is not greater than the smaller dimension of the matrix.

15.2 The Rank of a Matrix

First let us assume that the matrix X_{nm} can be expressed as the product of two matrices whose common order is r. But assume also that the matrix X_{nm} cannot be expressed as the product of any pair of matrices with a common order less than r. Then the *rank* of the matrix X_{nm} may be defined as r. From this definition, it is clear that the rank of a matrix is never greater than its smaller dimension.

An example of a matrix of rank 1 is Fig. 15.2.1.

$$\begin{bmatrix} 3 \\ 4 \\ 6 \end{bmatrix} \begin{bmatrix} 2 & 1 & 4 \end{bmatrix} = \begin{bmatrix} 6 & 3 & 12 \\ 8 & 4 & 16 \\ 12 & 6 & 24 \end{bmatrix}$$

FIG. 15.2.1

You see that the matrix on the right can be expressed as the product of two matrices (vectors) whose common order is 1. But since 1 is the smallest order any nonzero matrix can have, you know that the matrix cannot be expressed as the product of matrices with a common order less than 1. Therefore, the rank of the matrix is 1. The rule is then that the major product of any two vectors is a matrix of rank 1.

To summarize, the rank of a matrix is the smallest common order among all pairs of matrices whose product is the matrix.

15.3   The Rank of a Matrix with 0 Rows or Columns

We can prove rather easily now that the rank of a matrix cannot be greater than the number of columns or rows that do not consist of 0 elements. We shall prove this rule for 0 columns. If we prove it for columns, it can be proved for rows simply by taking the transpose, for the 0 rows of the matrix would then become columns in the transpose.

Suppose we have a matrix X_{nm} in which certain of the columns consist entirely of 0 elements. By interchanging columns, we can get another matrix with all 0 columns at the right and all non-0 columns at the left. This we can do by postmultiplying X by a permutation matrix. We can express this operation by

$$X_{nm}\pi = [Y_{nr} \quad 0_{nt}] \tag{15.3.1}$$

Here Y_{nr} is a submatrix made up of the r non-0 columns from X_{nm}, and 0_{nt} is a null matrix with t columns, where $r + t = m$. Now the right side of this equation can be expressed as the product of two matrices of common order r, as in

$$[Y_{nr} \quad 0_{nt}] = Y_{nr}[I_{rr} \quad 0_{rt}] \tag{15.3.2}$$

If we substitute the right side of Eq. (15.3.2) in the right of Eq. (15.3.1), we have

$$X_{nm}\pi = Y_{nr}[I_{rr} \quad 0_{rt}] \tag{15.3.3}$$

Next we postmultiply both sides of Eq. (15.3.3) by the transpose of π to get

$$X_{nm}\pi\pi' = Y_{nr}[I_{rr} \quad 0_{rt}]\pi' \tag{15.3.4}$$

But we know that the product moment of a permutation matrix is the identity matrix, or

$$\pi\pi' = I \tag{15.3.5}$$

We let

$$[I_{rr} \quad 0_{rt}]\pi' = Z_{rm} \tag{15.3.6}$$

where Z_{rm} is simply a matrix obtained by interposing 0 columns between certain columns of the identity matrix.

Substituting Eq. (15.3.5) and Eq. (15.3.6) in Eq. (15.3.4) gives

$$X_{nm} = Y_{nr}Z_{rm} \tag{15.3.7}$$

So you see that according to Eq. (15.3.7), the rank of X_{nm} cannot be greater than the common order r of the matrices on the right, which is the same as the number of non-0 columns in X_{nm}. A numerical example is given in Fig. 15.3.1.

$$X_{54} = \begin{bmatrix} 3 & 0 & 4 & 0 \\ 5 & 0 & 2 & 0 \\ 2 & 0 & 1 & 0 \\ 7 & 0 & 3 & 0 \\ 6 & 0 & 2 & 0 \end{bmatrix} \quad \text{then } \pi = \begin{bmatrix} 1 & 0 & 0 & 0 \\ 0 & 0 & 1 & 0 \\ 0 & 1 & 0 & 0 \\ 0 & 0 & 0 & 1 \end{bmatrix}$$

$$X_{54} = \underbrace{\begin{bmatrix} 3 & 0 & 4 & 0 \\ 5 & 0 & 2 & 0 \\ 2 & 0 & 1 & 0 \\ 7 & 0 & 3 & 0 \\ 6 & 0 & 2 & 0 \end{bmatrix}}_{X} \underbrace{\begin{bmatrix} 1 & 0 & 0 & 0 \\ 0 & 0 & 1 & 0 \\ 0 & 1 & 0 & 0 \\ 0 & 0 & 0 & 1 \end{bmatrix}}_{\pi} = \underbrace{\left[\begin{array}{cc|cc} 3 & 4 & 0 & 0 \\ 5 & 2 & 0 & 0 \\ 2 & 1 & 0 & 0 \\ 7 & 3 & 0 & 0 \\ 6 & 2 & 0 & 0 \end{array}\right]}_{[Y_{52} \quad 0_{52}]}$$

or

$$X_{54}\pi = \underbrace{\begin{bmatrix} 3 & 4 \\ 5 & 2 \\ 2 & 1 \\ 7 & 3 \\ 6 & 2 \end{bmatrix}}_{Y_{52}} \underbrace{\begin{bmatrix} 1 & 0 & 0 & 0 \\ 0 & 1 & 0 & 0 \end{bmatrix}}_{[I_{22} \quad 0_{22}]}$$

or

$$X_{54} = \underbrace{\begin{bmatrix} 3 & 4 \\ 5 & 2 \\ 2 & 1 \\ 7 & 3 \\ 6 & 2 \end{bmatrix}}_{Y_{52}} \underbrace{\left[\begin{array}{cc|cc} 1 & 0 & 0 & 0 \\ 0 & 1 & 0 & 0 \end{array}\right]}_{[I_{22} \quad 0_{22}]} \underbrace{\begin{bmatrix} 1 & 0 & 0 & 0 \\ 0 & 0 & 1 & 0 \\ 0 & 1 & 0 & 0 \\ 0 & 0 & 0 & 1 \end{bmatrix}}_{\pi'}$$

or

$$X_{54} = \underbrace{\begin{bmatrix} 3 & 4 \\ 5 & 2 \\ 2 & 1 \\ 7 & 3 \\ 6 & 2 \end{bmatrix}}_{Y_{52}} \underbrace{\begin{bmatrix} 1 & 0 & 0 & 0 \\ 0 & 0 & 1 & 0 \end{bmatrix}}_{Z_{24}} = \underbrace{\begin{bmatrix} 3 & 0 & 4 & 0 \\ 5 & 0 & 2 & 0 \\ 2 & 0 & 1 & 0 \\ 7 & 0 & 3 & 0 \\ 6 & 0 & 2 & 0 \end{bmatrix}}_{X_{54}}$$

Fig. 15.3.1

By taking the transpose of the matrix, you can show that the rank of a matrix cannot be greater than the number of its nonzero rows.

15.4   The Basic Matrix

There is a certain class of matrices to which mathematicians have given no distinguishing name. Since these matrices play a very important role in the analysis of social science data, we coin the term *basic matrix*. A basic matrix is one whose rank is equal to its width or smaller dimension.

According to this definition, any square matrix is basic if its rank is equal to its order. A square basic matrix is called a *nonsingular* matrix by the mathematicians.

Most data matrices are basic. As a matter of fact, it would be the sheerest coincidence if you found a data matrix of rank less than its width. One of the fundamental problems of science is to find data matrices that can be closely approximated by matrices that are not basic. Usually the problem is to find predictor attributes for a prediction data matrix such that very slight changes in the criterion attribute vector will yield a non-basic matrix. It is beyond the scope of this book to show the relationship between the rank of a matrix and techniques of scientific prediction. However, the reduction of basic data matrices to nonbasic matrices is fundamental in the kinds of prediction discussed in Chapter 1.

15.5   Types of Basic Matrices

Besides data matrices, which are basic except for rare exceptions, there are other important types of matrices that are always basic. In this section we shall merely list some of the most important types of basic matrices without proving that they are basic.

Orthogonal matrices

All orthogonal matrices are basic whether rectangular or square. Since orthonormal matrices are special cases of orthogonal matrices, all ortho-normal matrices are also basic, whether rectangular or square. Since the permutation matrix is a special case of a square orthonormal, all permutation matrices are basic.

Diagonal matrices

All diagonal matrices having all nonvanishing diagonal elements are basic. Since the scalar, the sign, and the identity matrix are all special cases of diagonal matrices with all nonzero diagonals, these are also basic.

Triangular matrices

All partial triangular matrices are basic. Since the triangular matrix is a special case of a partial triangular, all triangular matrices are also basic.

15.6   The Rank of a Supermatrix

The rank of a supermatrix is defined as the rank of the simple form of the matrix. We shall show that the rank of a supermatrix cannot be less than the rank of the submatrix of highest rank. Let us indicate with subscripts the orders of the elements in the supermatrix M as in Eq. (15.6.1).

$$M = \begin{bmatrix} M_{au} & M_{av} & \ldots & M_{az} \\ M_{bu} & M_{bv} & \ldots & M_{bz} \\ \ldots & \ldots & \ldots & \ldots \\ M_{fu} & M_{fv} & \ldots & M_{fz} \end{bmatrix} \qquad (15.6.1)$$

We can assume without loss of generality that M_{au} is the submatrix of largest rank, and that this rank is r. We can express the rank of M_{au} by

$$M_{au} = L_{ar}L_{ru} \qquad (15.6.2)$$

where the L's are basic matrices.

Now if Eq. (15.6.2) is true, it is possible for a supermatrix M to exist which is the major product of type III vectors whose simple common dimension is r. In this case, we could have Eq. (15.6.3).

$$\begin{bmatrix} L_{ar} \\ L_{br} \\ \ldots \\ L_{fr} \end{bmatrix} \begin{bmatrix} L_{ru} & L_{rv} & \ldots & L_{rz} \end{bmatrix} = \begin{bmatrix} M_{au} & M_{av} & \ldots & M_{az} \\ M_{bu} & M_{bv} & \ldots & M_{bz} \\ \ldots & \ldots & \ldots & \ldots \\ M_{fu} & M_{fv} & \ldots & M_{fz} \end{bmatrix} \qquad (15.6.3)$$

However, it is not possible to find type III vectors with common dimension *less* than r whose major product will yield M, for then M_{au} would have to be of rank less than r, also. Therefore, our rule is proved. The proof would be similar if we had assumed any other element to be of highest rank.

15.7   The Rank of Products of Matrices

The maximum rank of a product

The rank of the product of two or more matrices cannot be greater than the rank of the factor of lowest rank. If we show why this is true for three factors, you should be able to see why it is also true for two factors or more than three. Let us represent the product of three factors by

$$G = X_{ab}Y_{bc}Z_{ce} \qquad (15.7.1)$$

where the letter subscripts indicate the order of the matrices. We assume that neither X_{ab} nor Z_{ce} has a rank of less than that of Y_{bc}. We assume that the rank of Y_{bc} is r. This means, of course, that r is equal to or less than b or c, whichever is the smaller. But if the rank of Y_{bc} is r, then two basic matrices U_{br} and V_{rc} exist, such that

$$Y_{bc} = U_{br}V_{rc} \qquad (15.7.2)$$

Equation (15.7.2) is a way of indicating the rank of a matrix. We can say that Y_{bc} can be expressed as the product of two basic matrices whose common order is r. This tells us at once that the rank of Y_{bc} is r.

Now we substitute Eq. (15.7.2) in Eq. (15.7.1), and get

$$G = (X_{ab}U_{br})(V_{rc}Z_{ce}) \tag{15.7.3}$$

Next we define the two products in parentheses on the right of Eq. (15.7.3) by Eqs. (15.7.4) and (15.7.5) respectively.

$$X_{ab}U_{br} = \alpha_{ar} \tag{15.7.4}$$

$$V_{rc}Z_{ce} = \beta_{re} \tag{15.7.5}$$

Substituting Eqs. (15.7.4) and (15.7.5) in Eq. (15.7.3) gives

$$G = \alpha_{ar}\beta_{re} \tag{15.7.6}$$

Remember we assumed that r was not greater than the rank of either X_{ab} or Z_{ce}. Therefore, it cannot be greater than either a or e, and the rank of G cannot be greater than r. It could, of course, be less.

The minimum rank of a product

Let us next consider what is the minimum rank that the product of any two matrices can have. Without attempting to prove the rule now, we shall simply state that the rank of a product of two matrices cannot be less than the sum of their ranks less their common order. For example, suppose we have two matrices X_{nt} and Y_{tm} of rank r and s respectively. We indicate their product by

$$X_{nt}Y_{tm} = Z_{nm} \tag{15.7.7}$$

Then, according to our rule, the rank of Z_{nm} cannot be less than $r + s - t$.

By definition, of course, the rank of a product cannot be less than 0. Therefore, if $r + s - t$ is a quantity less than 0, the rank of the product can be 0. For example, if $r = 3$, $s = 2$ and $t = 6$, then $r + s - t = -1$ and the smallest possible rank for Z_{nm} is 0.

As you would guess, if the rank of the product of two matrices is 0, the product is a null matrix. Another way of saying the same thing is that the factors are orthogonal to one another.

15.8 Ranks of Special Kinds of Products

From the rules for the maximum and minimum ranks of products of matrices, let us work out some rules for the ranks of certain important types of products.

Maximum rank of product of two basic matrices

First let us consider the product of any two basic matrices. It is easy to prove that the rank of the product of two basic matrices cannot be greater than the smallest of the three dimensions. (This is true because the rank

of a basic matrix is equal to its smaller dimension, and the rank of a product
is not greater than the rank of the factor of smallest rank.) If we have the
product of

$$X_{ab}Y_{bc} = Z_{ac} \qquad (15.8.1)$$

then the rank of Z cannot be greater than the smallest of the three dimensions a, b, and c. If a is smallest, then the product is a horizontal matrix
and the rank cannot be greater than its height as shown in Fig. 15.8.1.

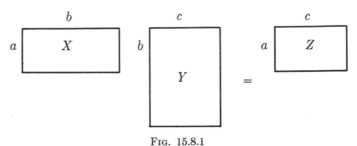

FIG. 15.8.1

If b is smallest, then the dimensions of the product do not indicate the
maximum possible rank, as shown in Fig. 15.8.2. This is a major product
moment of basic matrices, which automatically defines the rank.

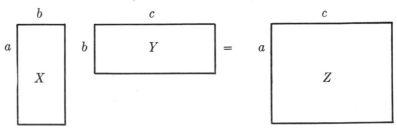

FIG. 15.8.2

If c is smallest, then the maximum rank is simply the width of the product,
as shown in Fig. 15.8.3.

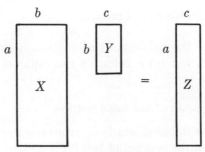

FIG. 15.8.3

Minimum rank of product of two basic matrices

Next let us determine the minimum rank of the product of two basic matrices. First remember that the rank of a basic matrix is its smaller dimension. Then recall that the rank of a product cannot be less than the sum of the ranks less their common dimension. You see then that the rank of the product of two basic matrices cannot be less than the sum of the smaller dimensions of each less their common dimension.

Minimum rank of major product of two basic matrices

Let us see how this rule works out for the major product of two basic matrices. This is the product of a vertical matrix postmultiplied by a horizontal matrix. Such a product is illustrated by Fig. 15.8.4.

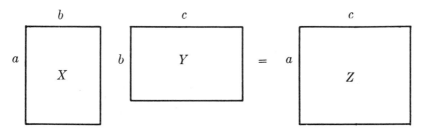

FIG. 15.8.4

Since b is the smaller dimension of both X and Y and since it is also the common dimension, then the rank of Z cannot be less than $b + b - b = b$. Or to put it otherwise, the rank of the major product of two basic matrices cannot be less than their common dimension. Another way of defining the rank of a matrix is by the common order of a major product of basic matrices that yield the matrix.

Minimum rank of the minor product of two basic matrices

The rule for the minor product of two basic matrices may be developed in much the same way. Figure 15.8.5 is an example of the minor product of two matrices.

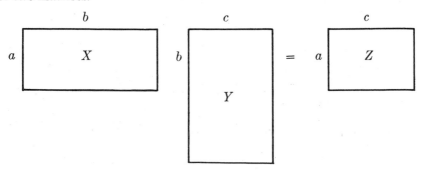

FIG. 15.8.5

You notice that the smaller dimensions of the two factors are their distinct dimensions, so according to the rule, the rank of the product could not be less than $a + c - b$. Therefore, the rank of the minor product of two basic matrices cannot be less than the sum of their distinct dimensions less their common dimension.

You will recall now that in a special case of the minor product of two matrices, the factors are two submatrices obtained by partitioning a square orthonormal matrix between two rows or two columns. For example, suppose we have the orthonormal matrix Q_{bb} partitioned as indicated in Fig. 15.8.6.

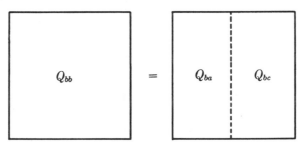

FIG. 15.8.6

Here we see that $a + c = b$. We can write the product of Q_{bb} and its transpose as in Eq. (15.8.2).

$$Q'_{bb}Q_{bb} = \begin{bmatrix} Q_{ab} \\ Q_{cb} \end{bmatrix} \cdot [Q_{ba} \quad Q_{bc}] = \begin{bmatrix} I_{aa} & 0_{ac} \\ 0_{ca} & I_{cc} \end{bmatrix} \tag{15.8.2}$$

where $Q_{ab} = Q'_{ba}$ and $Q_{cb} = Q'_{bc}$. From Eq. (15.8.2) we have

$$Q_{ab}Q_{bc} = 0 \tag{15.8.3}$$

According to the rule, the rank of a minor product cannot be less than $a + c - b$. But since $a + c = b$, in this case the rank cannot be less than $b - b$, or 0. We see then that the minor product of basic factors has a minimum rank when those factors are any two submatrices defining a square orthonormal.

Minimum rank for product of basic vertical matrices

Next let us consider the minimum rank of the product of two vertical basic matrices. Figure 15.8.7 illustrates such a product.

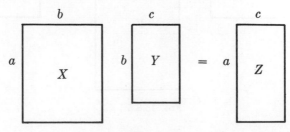

FIG. 15.8.7

In this example the smaller orders of the two factors are b and c respectively. The minimum rank is $b + c - b = c$. Therefore, the minimum rank of the product of two basic vertical matrices is the width of the product, or, what amounts to the same thing, the width of the postfactor. Therefore the product must be basic.

Minimum rank for product of basic horizontal matrices

In the same way you can also discover the rule for the minimum rank of the product of two basic horizontal matrices, as illustrated in Fig. 15.8.8.

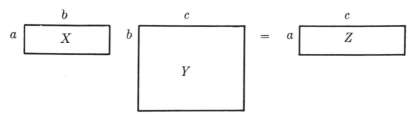

FIG. 15.8.8

Here you see that the smaller dimensions of the two matrices X and Y are a and b respectively. Therefore the rank of Z is $a + b - b = a$. Obviously, then the rank of the product of two basic horizontal matrices cannot be less than the height of the product, or, what amounts to the same thing, the height of the prefactor.

15.9 Ranks of Products Involving Square Basic Matrices

Basic matrix as a factor

Next let us see what happens to the maximum and minimum ranks if we multiply any basic matrix by a basic square matrix. First we shall consider the case of the maximum rank. But we may either pre- or postmultiply a rectangular matrix by a square matrix. If we premultiply, we have Fig. 15.9.1.

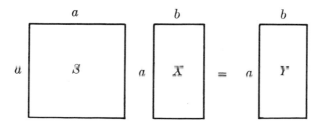

FIG. 15.9.1

If we postmultiply by a square matrix, we have Fig. 15.9.2.

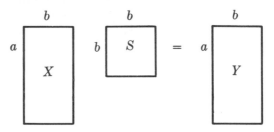

Fig. 15.9.2

But notice that in either case the rank of the product cannot exceed b, since this is the smaller dimension in both products. Therefore, the rank of the product of any basic matrix by any basic square matrix cannot be greater than the smaller order of the matrix.

Let us now refer to Fig. 15.9.1 to determine the minimum rank of such a product. The sum of the smaller dimensions less the common dimension is $a + b - a = b$. Therefore, if a vertical basic matrix is premultiplied by a square basic matrix, the rank of the product cannot be less than its width, which is the same as the width of the original matrix.

In the same way, we see by referring to Fig. 15.9.2 what the minimum rank of the product is if we postmultiply a basic vertical matrix by a basic square matrix. According to our rule this minimum rank is $b + b - b = b$. Therefore, if a basic vertical matrix is postmultiplied by a basic square, the rank of the product cannot be less than its width or the width of the original matrix.

We see, therefore, that if a vertical basic matrix is either pre- or post-multiplied by a basic square matrix, the rank of the product can be neither greater nor less than the width of the original matrix. By using the same logic we can show that if a horizontal basic matrix is pre- or postmultiplied by a basic square matrix, the rank of the product can be neither greater nor less than the height of the original matrix.

Since a square matrix is a special case of a rectangular matrix, we also see that the product of two square basic matrices is a basic matrix. Or in general, the product of any number of square basic matrices is a basic matrix.

Any matrix as a factor

Next let us see what happens to the rank if any matrix is multiplied by a basic square matrix. First we consider postmultiplication by a basic square. Suppose we have a matrix X_{nm} whose rank r, not less than m, is defined by

$$X_{nm} = X_{nr}X_{rm} \tag{15.9.1}$$

We let X_{mm} be a basic square matrix. Then from Eq. (15.9.1) we have

$$X_{nm}X_{mm} = X_{nr}X_{rm}X_{mm} \qquad (15.9.2)$$

Now we let

$$X_{rm}X_{mm} = Y_{rm} \qquad (15.9.3)$$

Since both X_{rm} and X_{mm} are basic, Y_{rm} is also basic, for we know that the product of any basic matrix by a square basic is basic. From Eqs. (15.9.2) and (15.9.3), we have

$$X_{nm}X_{mm} = X_{nr}Y_{rm} \qquad (15.9.4)$$

We know that r is less than m. Therefore the right side of Eq. (15.9.4) is the major product of two basic matrices, and its rank is the common dimension or r. Therefore, the rank of the left side of Eq. (15.9.4) is also r. We know then that the product of a vertical matrix of rank r postmultiplied by a square basic matrix is also of rank r.

Next let us premultiply X_{nm} by a basic square X_{nn}. We get from Eq. (15.9.1)

$$X_{nn}X_{nm} = X_{nn}X_{nr}X_{rm} \qquad (15.9.5)$$

We let

$$X_{nn}X_{nr} = Z_{nr} \qquad (15.9.6)$$

We know that Z_{nr} is basic because it is the product of a basic matrix by a square basic matrix. We substitute Eq. (15.9.6) in Eq. (15.9.5) to get

$$X_{nn}X_{nm} = Z_{nr}X_{rm} \qquad (15.9.7)$$

As before, we know that the rank of the product on the right is exactly r. Therefore premultiplication of any vertical matrix of rank r yields a matrix whose rank is also r.

We see then that pre- or postmultiplication of any vertical matrix of rank r by a basic square matrix yields a product whose rank is r. Simply by considering the transposes of the products and remembering that square matrices are special cases of rectangular matrices, we can state the general rule. Pre- or postmultiplication of any matrix of rank r by a basic square matrix yields a matrix whose rank is r.

15.10   Ranks of Product Moment Matrices

You may wonder by this time how the rank of the product moment of a matrix is related to the rank of the matrix. Without attempting to prove the rule now, we shall simply state it. The rank of the product moment, whether major or minor of any matrix is the same as the rank of the original matrix. From this rule it is clear that the rank of the minor product moment of a basic matrix is equal to its order.

Also the rank of the major product moment is simply the width of the original matrix. It is clear then that the major product moment of a non-square matrix cannot be basic.

15.11 Rank of a Sum of Matrices

Maximum rank for two matrices

We shall first consider the maximum rank possible for the sum of two matrices. We define a matrix U_1 by

$$X_1 Y_1' = U_1 \tag{15.11.1}$$

where X_1 and Y_1 are basic matrices and the common dimension on the left of Eq. (15.11.1) defines the rank of U_1. In the same way, we define the rank of U_2 by

$$X_2 Y_2' = U_2 \tag{15.11.2}$$

From Eqs. (15.11.1) and (15.11.2), we get

$$U_1 + U_2 = X_1 Y_1' + X_2 Y_2' \tag{15.11.3}$$

But the right side of Eq. (15.11.3) can be written as the minor product of type III supervectors as in Eq. (15.11.4).

$$U_1 + U_2 = [X_1 \quad X_2] \begin{bmatrix} Y_1' \\ Y_2' \end{bmatrix} \tag{15.11.4}$$

But the rank of Eq. (15.11.4) can be no greater than the common dimension of the factors on the right. And this common dimension is the sum of the widths of X_1 and X_2, which is by definition the sum of the ranks of U_1 and U_2. Figure 15.11.1 will help to illustrate this point. We assume that the rank of U_1 is b and that of U_2 is e.

FIG. 15.11.1 (Cont. on next page)

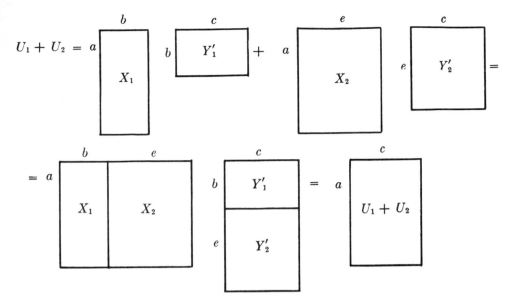

FIG. 15.11.1

Maximum rank for any number of matrices

Next we can show in very much the same manner the maximum rank of the sum of any number of matrices, knowing the rank of each. We define the ranks of n matrices $U_1, U_2, \ldots U_n$ all of the same order, by Eq. (15.11.5).

$$\left.\begin{array}{l} U_1 = X_1 Y_1' \\ U_2 = X_2 Y_2' \\ \cdots \qquad \cdots \\ U_n = X_n Y_n' \end{array}\right\} \tag{15.11.5}$$

The ranks $r_1, r_2, \ldots r_n$ are the widths of the matrices $X_1, X_2, \ldots X_n$ respectively. Then from Eq. (15.11.5) we have

$$U_1 + U_2 + \ldots + U_n = X_1 Y_1' + X_2 Y_2' + \ldots + X_n Y_n' \tag{15.11.6}$$

But the right side of Eq. (15.11.6) can be expressed as the minor product of type III supervectors, as in Eq. (15.11.7).

$$U_1 + U_2 + \ldots + U_n = [X_1 \quad X_2 \quad \ldots \quad X_n] \begin{bmatrix} Y_1' \\ Y_2' \\ \cdots \\ Y_n' \end{bmatrix} \tag{15.11.7}$$

The common dimension on the right here is the sum of the widths of the X matrices or the ranks of the U's. Therefore the rank of a sum of any number of matrices cannot exceed the sum of their ranks.

Minimum ranks

To find the minimum rank of the sum of two matrices, we begin by considering two matrices U_{ad} and V_{ad} whose ranks are defined by Eqs. (15.11.8) and (15.11.9) respectively.

$$U_{ad} = U_{ab}U_{bd} \qquad (15.11.8)$$

$$V_{ad} = V_{ac}V_{cd} \qquad (15.11.9)$$

Then we can write

$$U_{ad} + V_{ad} = U_{ab}U_{bd} + V_{ac}V_{cd} \qquad (15.11.10)$$

But this can be written on the left as the minor product of two type III supervectors as in Eq. (15.11.11).

$$U_{ad} + V_{ad} = [U_{ab} \quad V_{ac}]\left[\frac{U_{bd}}{V_{cd}}\right] \qquad (15.11.11)$$

For the purpose of our proof we assume that b is less than c. Next we let

$$X_{a[b+c]} = [U_{ab} \quad V_{ac}] \qquad (15.11.12)$$

and

$$Y_{[b+c]d} = \left[\frac{U_{bd}}{V_{cd}}\right] \qquad (15.11.13)$$

Then substituting Eqs. (15.11.12) and (15.11.13) in Eq. (15.11.11),

$$U_{ad} + V_{ad} = X_{a[b+c]}Y_{[b+c]d} \qquad (15.11.14)$$

Now we know that the rank of a supermatrix cannot be less than the rank of its submatrix of highest rank. Therefore, the ranks of X and Y cannot be less than c, since both V_{ac} and V_{cd} are basic. Matrices could exist, however, with ranks as low as c. Taking this minimum case, then, the rank of Eq. (15.11.14) cannot be less than $c + c - (b + c) = c - b$. Therefore, the rank of a sum of two matrices cannot be less than the absolute difference of their ranks.

SUMMARY

1. The factors of a matrix:
   a. Any matrix can be expressed as the product of two matrices whose common order is not greater than the smaller order of the matrix. This may be done in terms of an infinite number of pairs of such factors.
   b. Some matrices may be expressed as the major product of two matrices. This may be done in terms of an infinite number of pairs of such major products.

2. The rank of a matrix is the smallest common order among all pairs of matrices whose product is the matrix.

3. The rank of a matrix cannot be greater than the number of columns or rows which do not consist of 0 elements.

4. Basic matrices:
   a. *Definition:* A basic matrix is one whose rank is equal to its smaller dimension.
   b. Data matrices are usually basic.

5. Types of basic matrices: The following types of matrices are always basic:
   a. Orthogonal; rectangular, and square.
   b. Orthonormal; rectangular, square, and permutation.
   c. Diagonal; general, scalar, identity, and sign.
   d. Triangular; square and partial triangular.

6. The rank of a supermatrix cannot be less than the rank of its submatrix of highest rank.

7. The rank of products of matrices:
   a. The rank of the product of two or more matrices cannot be greater than the rank of the factor of lowest rank.
   b. The rank of the product of two matrices cannot be less than the sum of their ranks less their common order.

8. Ranks of special kinds of products:
   a. The rank of a product of two basic matrices cannot be greater than the smallest of the three dimensions. The maximum possible rank
      (1) For a horizontal minor product is the height of the prefactor.
      (2) For a major product is the common order.
      (3) For a product of vertical matrices is the width of the postfactor.
   b. The rank of a product of two basic matrices cannot be less than the sum of the smaller dimensions of each less their common order.
   c. The rank of a major product of basic matrices cannot be less than their common order.
   d. The rank of a minor product of two basic matrices cannot be less than the sum of their distinct dimensions less their common order.
   e. The product of two vertical basic matrices is basic.
   f. The product of two horizontal basic matrices is basic.

9. Ranks of products involving square basic matrices:
   a. If any basic matrix is pre- or postmultiplied by a square basic matrix, the product is basic.
   b. If a matrix of rank r is pre or postmultiplied by a square basic matrix, the product is of rank r.

10. Rank of a product moment:

   a. The rank of a major or minor product moment of a matrix is the rank of the matrix.

   b. The major product moment of a vertical matrix cannot be basic.

11. Rank of a sum of matrices:

   a. The rank of a sum of any number of matrices cannot be greater than the sum of their ranks.

   b. The rank of the sum of two matrices cannot be less than the absolute value of the difference between their ranks.

EXERCISES

1. Assume that X can be expressed as the major product of two factors Y and Z whose common order is F, but that neither of these two factors can be expressed as the product of matrices with common order less than F.

   (a) What kind of a matrix is Y?

   (b) What kind of a matrix is Z?

   (c) What is the rank of X?

2. Express in terms of basic factors

$$\text{(a)} \begin{bmatrix} 1 & 2 & 3 \\ 2 & 4 & 6 \\ 3 & 6 & 9 \end{bmatrix} \quad \text{(b)} \begin{bmatrix} 2 & 8 & 4 \\ 3 & 12 & 6 \\ 1 & 4 & 2 \\ 4 & 16 & 8 \end{bmatrix}$$

   (c) What is the rank of (a)? What is the rank of (b)?

3. Given

$$X = \begin{bmatrix} 3 & 0 & 6 \\ 2 & 0 & 4 \\ 4 & 0 & 8 \\ 1 & 0 & 1 \end{bmatrix}$$

   (a) Express X as the product of basic factors where one of the factors is a zero-one matrix.

   (b) What is the rank of X?

4. Assume that all matrices in the following products are basic and let the subscripts refer to their orders. Which products must be basic?

   (a) $X_{43}X_{32}$ (b) $X_{43}X_{35}$ (c) $X_{44}X_{42}$ (d) $X_{42}X_{22}$ (e) $X_{23}X_{35}$

   (f) $X_{34}X_{42}$ (g) $X_{34}X_{44}$ (h) $X_{52}X_{26}$ (i) $X_{39}X_{95}$ (j) $X_{22}X_{27}$

5. What are the ranks of the products in Ex. 4 which cannot be basic?

6. What are the maximum and minimum possible ranks for the products in Ex. 4 whose ranks are not known?

7. Let prescripts indicate rank in the following

   (a) $_1X + {}_3X + {}_2X$ (b) $_aX + {}_bX + {}_cX$ (c) $_4X + {}_bX$

What is the maximum possible rank for each of these sums?

8. Let prescripts indicate rank in the following

(a) $_3X + {}_2X$ (b) $_aX + {}_aY$ (c) $_3X - {}_2X$ (d) $_bY - {}_hX$

What is the minimum possible rank for each of the above?

9. Given the supermatrix

$$X = \begin{bmatrix} X_{au} & X_{av} & X_{aw} \\ X_{bu} & X_{bv} & X_{bw} \end{bmatrix}$$

What is the lowest possible rank of X if
(a) The rank of X_{aw} is 3 and the rank of the other elements is 2?
(b) The rank of X_{bw} is 3 and that of the other elements is 4?

10. Let X be a vertical matrix of rank r.
(a) What is the rank of $X'X$?
(b) Of XX'?
(c) If X is basic, what kind of matrix is $X'X$?
(d) If X is vertical and basic, what do you know about the rank of XX'?

11. Let

$$X = \begin{bmatrix} 3 & 0 & 0 \\ 2 & 4 & 0 \\ 1 & 4 & 3 \end{bmatrix} \qquad Y' = \begin{bmatrix} 4 & 3 & 2 & 3 & 2 \\ 0 & 1 & 2 & 1 & 6 \\ 0 & 0 & 3 & 4 & 1 \end{bmatrix} \qquad \pi_L = \begin{bmatrix} 0 & 1 & 0 & 0 \\ 0 & 0 & 0 & 1 \\ 1 & 0 & 0 & 0 \\ 0 & 0 & 1 & 0 \end{bmatrix}$$

$$\pi_R = \begin{bmatrix} 0 & 1 & 0 \\ 0 & 0 & 1 \\ 1 & 0 & 0 \end{bmatrix}$$

What are the ranks of
(a) YX (b) $\pi_L Y$ (c) $\pi_L YX$ (d) $Y'Y$ (e) $\pi_R X$ (f) $\pi_R XY'$
(g) YY' (h) $\pi_L YY'\pi'_L$

12. Which of the products in Ex. 11 are not basic?

ANSWERS

1. (a) basic   (b) basic   (c) F

2.
(a) $\begin{bmatrix} 1 \\ 2 \\ 3 \end{bmatrix} \begin{bmatrix} 1 & 2 & 3 \end{bmatrix}$ (b) $\begin{bmatrix} 2 \\ 3 \\ 1 \\ 4 \end{bmatrix} \begin{bmatrix} 1 & 4 & 2 \end{bmatrix}$ (c) one   (d) one

3. (a)
$$X = \begin{bmatrix} 3 & 6 \\ 2 & 4 \\ 4 & 8 \\ 1 & 1 \end{bmatrix} \begin{bmatrix} 1 & 0 & 0 \\ 0 & 0 & 1 \end{bmatrix}$$ (b) two

4, (a),   (c),   (d),   (e),   (g),   (j)

5. (b) 3,   (h) 2

6. (f) maximum 2, minimum 1,   (i) maximum 3, minimum 0

7. (a) 6   (b) $a + b + c$ (c) $4 + b$

8. (a) 1   (b) 0   (c) 1   (d) 0

9. (a) 3   (b) 4

10. (a) r (b) r (c) basic   (d) The rank is less than the order

11. Ranks are all 3

12. (g)   (h)

Chapter 16

Finding the Rank of a Matrix

16.1 The Triangular Factors of a Matrix

We have discussed in some detail in Chapter 15 how the ranks of sums and products of matrices are affected by the ranks of individual matrices. We shall now consider how the rank of any particular matrix of numbers may be determined. This can be done in a number of different ways. The one we shall use is computationally simple, and clearly accomplishes its purpose. We shall show that any matrix pre- and postmultiplied by suitable permutation matrices can be expressed as the major product of triangular-type matrices. This means that the prefactor is of the lower triangular form, and the postfactor the upper. We recall that the matrix multiplied by the permutation matrices has the same rank as the original matrix, because permutation matrices are square and basic. Now if the permuted matrix is expressed as the major product of triangular-type factors, the common order of these factors is the rank of the matrix. This is because triangular-type factors are basic, and the rank of a matrix is defined by the common order of the major product of factors that equal the matrix.

16.2 A Special Case of a Basic Matrix

The reduction procedure

Let us first consider a case in which we make special assumptions as indicated below. Suppose we have a matrix X. Then let us consider the product

$$_1L = \frac{X_{.1}X'_{1.}}{X_{11}} \tag{16.2.1}$$

where X_{11} is assumed different from 0.
Now the first row of $_1L$ will be

$$_1L'_{1.} = \frac{X_{11}X'_{1.}}{X_{11}} = X'_{1.} \tag{16.2.2}$$

The first column of $_1L$ will be

$$_1L_{.1} = \frac{X_{.1}X_{11}}{X_{11}} = X_{.1} \qquad (16.2.3)$$

Therefore the first row of $_1L$ will be the first row of X and the first column of $_1L$ will be the first column of X. We then subtract $_1L$ from X to get

$$Y = X - {_1L} \qquad (16.2.4)$$

But since the first row and column of $_1L$ are the same as the first row and column of X, the first row and column of Y must be null vectors. Next consider the product

$$_2L = \frac{Y_{.2}Y'_{2.}}{Y_{22}} \qquad (16.2.5)$$

where Y_{22} is assumed different from 0.
Now the first row of $_2L$ is

$$_2L'_{1.} = \frac{0Y'_{2.}}{Y_{22}} = 0 \qquad (16.2.6)$$

And the first column of $_2L$ is

$$_2L_{.1} = \frac{Y_{.2}0}{Y_{22}} = 0 \qquad (16.2.7)$$

Similarly the second row and columns of $_2L$ are given respectively by

$$_2L'_{2.} = \frac{Y_{22}Y'_{2.}}{Y_{22}} = Y'_{2.} \qquad (16.2.8)$$

and

$$_2L_{.2} = \frac{Y_{.2}Y_{22}}{Y_{22}} = Y_{.2} \qquad (16.2.9)$$

You see therefore that the first two rows and columns of $_2L$ are the same as the first two rows and columns of Y. We then subtract $_2L$ from Y to get

$$Y - {_2L} = Z \qquad (16.2.10)$$

Then since the first two rows and columns of Y and $_2L$ are the same, Eq. (16.2.10) shows that the first two rows and columns of Z are null vectors. This process is continued until the number of reductions is equal to the smaller dimension of X.

Constructing the triangular factors

Now suppose we use a slightly different notation so that the construction of the triangular factors becomes more obvious. We let $_1X$ be the original matrix, and as before $_1L$ is the major product of the first column and row vectors of $_1X$ divided by the first row and column element. Then

$$_1X - {_1L} = {_2X} \qquad (16.2.11)$$

where, as we have seen, the first row and column elements of $_2X$ are all 0.

Then $_2L$ is the major product of the second column and row vectors of $_2X$ divided by the second row and column element. Then

$$_2X - _2L = _3X \tag{16.2.12}$$

As we have seen in the case of Eq. (16.2.10), in $_3X$ in (16.2.12) the first two row and column elements are all 0. Then in general we have

$$_iX - _iL = _{i+1}X \tag{16.2.13}$$

Here $_iL$ is the major product of the ith column and row vectors of $_iX$ divided by the ith row and column element of $_iX$. The elements of the first i rows and columns of $_{i+1}X$ are all 0.

We next consider the definitions in Eqs. (16.2.14).

$$\left. \begin{aligned} _1L &= \frac{U_1V_1'}{a_1} \\ _2L &= \frac{U_2V_2'}{a_2} \\ \cdots \quad & \quad \cdots \\ _iL &= \frac{U_iV_i'}{a_i} \end{aligned} \right\} \tag{16.2.14}$$

Here U_i is the vector from the ith column of $_iX$, V_i' is the vector from the ith row of $_iX$, and a_i is the element from the ith row and column of $_iX$ (a_i is assumed to be different from 0).

Now since the first row and column of $_2X$ are null, the first element of both U_2 and V_2' must be 0. Similarly since the first $i-1$ elements of $_iX$ are 0, the first $i-1$ elements of U_i and V_i' must also be 0. If we consider Eqs. (16.2.11), (16.2.12), and (16.2.13), we have, dropping the prescript from X,

$$X_{nm} - _1L - _2L - \ldots - _iL = _{i+1}X \tag{16.2.15}$$

But if $_1X$ is nonhorizontal and of order $n \times m$, then i cannot be greater than m, for if $m = i$ then $_{i+1}X$ would be null, since its m rows and columns would all be 0. Therefore, we can write from Eq. (16.2.15)

$$X_{nm} = _1L + _2L + \ldots _mL \tag{16.2.16}$$

or substituting from Eq. (16.2.14) into Eq. (16.2.16),

$$X_{nm} = \frac{U_1V_1'}{a_1} + \frac{U_2V_2'}{a_2} + \ldots + \frac{U_mV_m'}{a_m} \tag{16.2.17}$$

But since the right side of this equation is the sum of major products of vectors, we can write it as Eq. (16.2.18).

$$X_{nm} = \begin{bmatrix} \dfrac{U_1}{a_1} & \dfrac{U_2}{a_2} & \cdots & \dfrac{U_m}{a_m} \end{bmatrix} \begin{bmatrix} V_1' \\ V_2' \\ \cdots \\ V_m' \end{bmatrix} \tag{16.2.18}$$

But the right side of this can be written as Eq. (16.2.19).

$$X_{nm} = [U_1 \quad U_2 \quad \ldots \quad U_m] \begin{bmatrix} \dfrac{1}{a_1} & 0 & \ldots & 0 \\ 0 & \dfrac{1}{a_2} & \ldots & 0 \\ \ldots & \ldots & \ldots & \ldots \\ 0 & 0 & \ldots & \dfrac{1}{a_m} \end{bmatrix} \begin{bmatrix} V_1' \\ V_2' \\ \ldots \\ V_m' \end{bmatrix} \quad (16.2.19)$$

Now remember that the first element of U_2 is 0, the first two of U_3 are 0, and so on. Therefore, the matrix made up of the U vectors is a partial triangular matrix. Also the matrix made up of the v row vectors on the extreme right of Eq. (16.2.19) is an upper triangular matrix. Equation (16.2.19) can be written more compactly as

$$X_{nm} = T_{nm} D_a^{-1} T_{mm}' \quad (16.2.20)$$

We have proved then for a special case that the matrix $_1X_{nm}$ can be expressed as the product of a lower-triangular-type matrix postmultiplied by an upper triangular matrix. Whether a triangular-type matrix is pre- or postmultiplied by a diagonal matrix, the product is still triangular.

16.3   A Special Case of a Nonbasic Matrix

Going back to Eq. (16.2.15), it is quite possible that we may get a matrix $_{i+1}X$ on the right which vanishes completely before $i = m$, that is, before the number of cycles through which we go is equal to the width of the matrix. In this case, the right and left matrices would be partial triangular. Such a case is illustrated in Fig. 16.3.1.

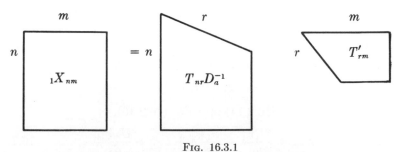

FIG. 16.3.1

But if a matrix can be expressed as the major product of two partial triangular matrices, the common order of the two is the rank of the matrix. We now have, therefore, a method of finding the rank of a matrix. The value of i for which $_{i+1}X$ becomes a null matrix is the rank of the matrix.

Let us now look carefully at the right side of Eq. (16.2.20). Notice that the diagonal matrix is raised to the -1 power, or in other words, it is an inverse. Suppose that one or more of the a's in D_a were 0. Then in D_a^{-1} the corresponding elements would be infinite. In this case, Eq. (16.2.20) would not make sense. We therefore consider the general case of the matrix pre- and postmultiplied by permutation matrices.

16.4   The Product of a General Matrix and Permutation Matrices

The reduction procedure

Suppose that in Eq. (16.2.1), X_{11} is 0. Then $_1L$ is infinite, and our calculations are stopped. We need not, however, start with the first row and first column. Let us look for the largest element in the matrix; we assume this is found in the ath row and bth column. Then we calculate the residual matrix as

$$Y = X - \frac{X_{.b}X'_{a.}}{X_{ab}} \tag{16.4.1}$$

This gives all 0's in the bth column and the ath row of Y. We next look for the element with the largest absolute value in Y. Let this be in the cth row and dth column. Then we calculate a second residual matrix as

$$Z = Y - \frac{Y_{.d}Y'_{c.}}{Y_{cd}} \tag{16.4.2}$$

In the residual matrix Z, we have all 0's in the ath and cth rows and in the bth and dth columns. This procedure we continue until we get a null residual matrix. If X is a basic matrix, then the number of residual matrices is equal to the width of the matrix. But if the matrix is of smaller rank, then the number of residual matrices, including the final null matrix, is equal only to the rank of X.

The permutation matrices

Now if we base our procedure on the largest element in each residual matrix, we are able to express X as the sum of major vector products just as in Eq. (16.2.17). However, when these vectors are arranged in matrix form as in Eq. (16.2.19), the matrices on the right and left of the right side of Eq. (16.2.19) are not in triangular form. It is possible, however, to get these matrices in triangular form as follows: For the original matrix and each residual matrix, we can designate the row and column number of its largest element. This is illustrated in Fig. 16.4.1.

	1	2	3	4	5
1					
2	4				
3				5	
4			2		
5					3
6		1			
7					
8					

Fig. 16.4.1

This shows that the largest element in the original matrix is row **6**, column **2**. The largest element in the first residual matrix is in row **4**, column **3**. The largest element in the second residual matrix is in row **5**, column **5**. Similarly, the largest elements for the third and fourth residual matrices are in row **2**, column **1**, and row **3**, column **4**, respectively. We may now premultiply X by a permutation matrix to interchange the rows so that the numbers are in sequence in the first five rows. This we illustrate in Fig. 16.4.2.

	1	2	3	4	5
1		1			
2			2		
3					3
4	4				
5				5	
6					
7					
8					

Fig. 16.4.2

The permutation matrix that will effect this interchange of rows let us call π_L. Next we postmultiply X by a permutation matrix to interchange the columns, so that the numbers in Fig. 16.4.2 are in sequence from left to right. This we illustrate in Fig. 16.4.3.

FIG. 16.4.3

The permutation that will effect this interchange of columns let us call π_R.
Suppose now that we had first set up the matrix as

$$\pi_L X \pi_R = W \tag{16.4.3}$$

simply by an interchange of rows and columns. Then the rule of selecting the largest element in each residual matrix would have resulted in factors of triangular type. We can say, therefore, that it is possible to find two permutation matrices π_L and π_R for any matrix X such that

$$\pi_L X_{nm} \pi_R = T_{nr} T'_{rm} \tag{16.4.4}$$

where the T's are of triangular type and the rank of the matrix X_{nm} is the common dimension, r, of T_{nr} and T_{rm}.

We are now ready to consider the actual computational procedure.

16.5 Computational Procedure

Given the matrix whose rank we wish to determine, we begin by calculating the row sum vector and the column sum vector. Both vectors are then summed to see that they yield the same total. These four sets of operations are shown in Fig. 16.5.1.

$$\begin{bmatrix} X & X1 \\ \hline 1'X & 1'(X1) \\ & (1'X)1 \end{bmatrix} = \begin{bmatrix} 2 & 3 & 5 \\ 1 & 3 & 4 \\ 2 & 1 & 3 \\ \hline 5 & 7 & 12 \end{bmatrix} \\ 12$$

FIG. 16.5.1

The reduction vectors

Next we should find the largest element in the matrix. For convenience in explanation, however, we shall take the element in the first row and column. We first calculate the row vector, $X'_{1.}/X_{11}$, although we could just as well calculate the column vector, $X_{.1}/X_{11}$. We must be sure to carry the sum at the right of the vector for our check. This step is shown in Fig. 16.5.2.

$$[X_{11}^{-1}X'_{1.} \mid X_{11}^{-1}(X'_{1.}1)] \qquad (X_{11}^{-1}X'_{1.})1$$

$$\left[\frac{(2 \quad 3)}{2} \; \middle| \; \frac{5}{2} \right] = [1 \quad 1.5 \mid 2.5]2.5$$

Fig. 16.5.2

Notice that each element of the vector is divided by 2 including the sum element. The number 2.5 at the right of the bracket is the sum of the elements, 1, 1.5, and shows that the division of each of the elements, 2 and 3, by 2 is correct.

The major vector product

The next step in the first cycle is to get the major product of the vectors as indicated in Fig. 16.5.3.

$$[X_{.1}(X_{11}^{-1}X'_{1.}) \mid X_{.1}[X_{11}^{-1}(X'_{1.}1)]] \qquad [X_{.1}(X_{11}^{-1}X'_{1.})]1$$

$$\begin{bmatrix} 2 \\ 1 \\ 2 \end{bmatrix} [1 \quad 1.5 \mid 2.5] = \begin{bmatrix} 2 & 3 & 5 \\ 1 & 1.5 & 2.5 \\ 2 & 3 & 5 \end{bmatrix} \quad \begin{bmatrix} 5 \\ 2.5 \\ 5 \end{bmatrix}$$

Fig. 16.5.3

Notice that the check column on the extreme right in Fig. 16.5.3 is the sum of row elements to the left of the partition.

The residual matrix

The final step in the first cycle is to subtract the major vector product from X as indicated in Fig. 16.5.4.

$$[X - X_{.1}(X_{11}^{-1}X'_{1.}) \quad X1 - X_{.1}[X_{11}^{-1}(X'_{1.}1)]] \qquad [X - X_{.1}(X_{11}^{-1}X'_{1.})]1$$

$$\begin{bmatrix} 2 & 3 & 5 \\ 1 & 3 & 4 \\ 2 & 1 & 3 \end{bmatrix} - \begin{bmatrix} 2 & 3 & 5 \\ 1 & 1.5 & 2.5 \\ 2 & 3 & 5 \end{bmatrix} = \begin{bmatrix} 0 & 0 & 0 \\ 0 & 1.5 & 1.5 \\ 0 & -2 & -2 \end{bmatrix} \quad \begin{bmatrix} 0 \\ 1.5 \\ -2 \end{bmatrix}$$

Fig. 16.5.4

16.6   The Second Residual Matrix

The reduction vectors

Since we have here only a 3×2 matrix, it is clear that the second residual matrix must be 0 and we can write directly

$$Y - \frac{Y_{.2}Y'_{2.}}{Y_{22}} = 0 \tag{16.6.1}$$

or

$$\begin{bmatrix} 0 & 0 \\ 0 & 1.5 \\ 0 & -2 \end{bmatrix} - \frac{\begin{bmatrix} 0 \\ 1.5 \\ -2 \end{bmatrix}[0 \quad 1.5]}{1.5} = \begin{bmatrix} 0 & 0 \\ 0 & 0 \\ 0 & 0 \end{bmatrix} \tag{16.6.2}$$

When X is of larger order, the first step in getting the second residual matrix is to get the row vector. This is the second row of Y, the first residual matrix, divided by the second row and column element of Y. The first residual matrix, you recall, is

$$Y = X - X_{.1}(X_{11}^{-1}X'_{1.}) \tag{16.6.3}$$

and

$$Y_{.1} = 0 \quad \text{and} \quad Y'_{1.} = 0' \tag{11.6.4}$$

First, then, we divide each element of $Y'_{2.}$ and also the scalar $Y'_{2.}1$ by Y_{22}^{-1} as shown in Fig. (16.6.1).

$$[Y_{22}^{-1}Y'_{2.} \mid Y_{22}^{-1}(Y'_{2.}1)] \qquad (Y_{22}^{-1}Y'_{2.})1$$

FIG. 16.6.1

Again remember that the scalar on the right in Fig. 16.6.1 is the sum of the elements to the left of the partition, and that this checks the division of all elements by Y_{22}.

The major vector product

The second step is to get the major vector product as shown in Fig. 16.6.2.

$$[Y_{.2}(Y_{22}^{-1}Y'_{2.}) \mid Y_{.2}[Y_{22}^{-1}(Y'_{2.}1)]] \qquad [Y_{.2}(Y_{22}^{-1}Y'_{2.})]1$$

FIG. 16.6.2

Here again the row sum check is given by the vector on the right. The last step is to subtract the major vector product from the first residual matrix to get the second residual matrix. This set of computations is indicated in Fig. 16.6.3.

$$[Y - Y_{.2}(Y_{22}^{-1}Y_{2.}')] \mid [Y1 - Y_{.2}[Y_{22}^{-1}(Y_{2.}'1)]] \qquad [Y - Y_{.2}(Y_{22}^{-1}Y_{2.}')]1$$

Fig. 16.6.3

The residual matrix

For the second residual matrix, then, we have

$$Z = Y - Y_{.2}(Y_{22}^{-1}Y_{2.}') \tag{16.6.5}$$

and

$$Z_{.1} = Z_{.2} = 0 \qquad Z_{1.}' = Z_{2.}' = 0 \tag{16.6.6}$$

so that now the first two rows and columns of the residual matrix consist of zero elements.

16.7 Successive Residual Matrices

The calculation of the third residual matrix is carried out just as in the case of the first two. For the third residual matrix, however, we get all zero elements in the first three rows and three columns. We continue to calculate residual matrices until we get a null residual. If the width of the X matrix is m and we get a null residual before the mth one, we know that the rank of X is less than m. For example, if X is a 7 by 5 matrix and the third residual matrix turns out to consist of zero elements, the rank of X is 3.

You may find, for certain matrices, that if you try to proceed systematically by reducing from the left to right and top to bottom as we have suggested, you have to divide by very small or nearly zero numbers. This causes trouble in that a great deal of decimal inaccuracy is apt to be introduced. You can, however, always base each reduction on the highest element in the preceding residual matrix and still follow the computational procedure outlined. The only difference is in the rows and columns where the 0's appear. Your checks work exactly the same. You must be careful to see that the 0's in your major vector products correspond to those in the residual matrix from which it is to be subtracted.

SUMMARY

1. The triangular factors of a matrix:
   a. Any matrix, pre- and postmultiplied by suitable permutation matrices, may be expressed as the major product of triangular-type matrices.
   b. The common order of the triangular-type factors is the rank of the matrix.

2. Special case of a basic matrix:
   a. Any matrix X may be reduced a row and column at a time by operations of the type

$$Y = X - \frac{X_{.1}X'_{11}}{X_{11}}$$

   provided

$$X_{ii} \neq 0$$

   b. This leads to the expression of X as a product of triangular-type factors thus

$$X_{nm} = T_{nm}D_a^{-1}T'_{mm}$$

   where D_a is diagonal.

3. In the special case of a nonbasic matrix both factors are partial triangular.

4. The product of a general matrix and permutation matrices:
   a. The reductions are of the type

$$Y = X - \frac{X_{.b}X'_{a.}}{X_{ab}}$$

   where X_{ab} is the largest element in the matrix or in a residual matrix.
   b. By appropriate selection of permutation matrices π_L and π_R, we can always find the triangular factors given by

$$\pi_L X_{nm} \pi_R = T_{nr}T'_{rm}$$

   where r is the rank of X_{nm}.

5. The computations for finding the rank of a matrix proceed through a series of steps with appropriate checks. First the residual matrix is

$$Y = X - \frac{X_{.b}X'_{a.}}{X_{ab}}$$

   calculated where X_{ab} is the element in X of largest absolute value.

6. Second, the residual matrix

$$Z = Y - \frac{Y_{.d}Y'_{c.}}{Y_{cd}}$$

   is calculated where Y_{cd} is the element of largest absolute value in Y.

7. Third, the process is continued until the residual matrix is null.

Chapter 17

The Basic Structure of a Matrix

17.1   The Basic Structure of a Matrix Defined

This chapter is concerned with a very useful concept, characteristic of all matrices, which we call the *basic structure*. Again, this term is not in general use by mathematicians, but it is fundamental to many of the analyses we conduct with behavioral science data and helps to make clear the relationships of matrices to one another. In order to define the basic structure of a matrix, let us first take any nonhorizontal matrix X having n rows and m columns as in Fig. 17.1.1.

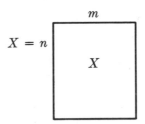

$$X = n \boxed{ X }$$

Fig. 17.1.1

Now let us consider three other matrices as shown in Fig. 17.1.2.

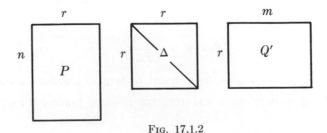

Fig. 17.1.2

In Fig. 17.1.2, the first matrix P is a nonhorizontal orthonormal with n rows and r columns. It has the same height as X in Fig. 17.1.1. For the time

364

being, all we know about r is that it is equal to or less than m, the width of matrix X. The second matrix in Fig. 17.1.2 is an $r \times r$ diagonal matrix Δ. The third matrix in Fig. 17.1.2 is Q', a nonvertical orthonormal matrix, with r rows and m columns. We are now ready to state the general rule that any nonhorizontal matrix X can always be expressed as the product of a non-horizontal orthonormal matrix P by a diagonal matrix Δ by the transpose of a nonvertical orthonormal matrix Q, as illustrated in Fig. 17.1.3.

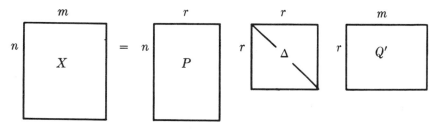

FIG. 17.1.3

The matrices on the right of Fig. 17.1.3, you will notice, have the same dimensions as those in Fig. 17.1.2.

We shall call P the left orthonormal of X, Δ the basic diagonal of X, and Q' the right orthonormal of X. Then the basic structure of a matrix consists of its right and left orthonormals and its basic diagonal. Because of the definition of orthonormal matrices, you remember that

$$P'P = I_{rr} \qquad (17.1.1)$$

and

$$Q'Q = I_{rr} \qquad (17.1.2)$$

However, we must remember that only if r is equal to m can the major product moment of Q be an identity matrix, and only if n, r and m are all equal can the major product moment of P be an identity matrix.

We shall not attempt to give a rigorous proof of the rule that a matrix can always be expressed as in Fig. 17.1.3. We shall merely appeal to a well-known rule in the theory of equations. If we have n unknowns to be determined, then precisely n independent restrictions on these unknowns must be specified in order for us to solve for them.

17.2   Proof of the Basic Structure of a Basic Matrix

If the matrix is a nonhorizontal basic matrix, then we can prove that it can be expressed as the product of a nonhorizontal orthonormal multiplied by a diagonal and by a square orthonormal as shown in Fig. 17.2.1.

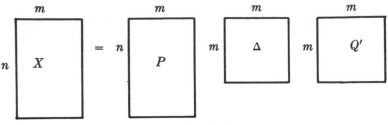

FIG. 17.2.1

In this Fig., you see that we let r of Fig. 17.1.2 be equal to m. Figure 17.2.1 can be expressed more compactly as

$$X = P\Delta Q' \qquad (17.2.1)$$

The number of unknowns

Let us see now how many unknowns we have in Eq. (17.2.1). We assume all of the values in X to be known and all of the elements in the factors on the right of Eq. (17.2.1) to be unknown. Therefore, the number of unknowns is simply the sum of the number of unknowns in each of the factors on the right. To indicate the total number of unknowns, we list them and take the total as in Fig. 17.2.2.

P has $\quad n \times m \quad$ unknowns

Q has $\quad m \times m \quad$ unknowns

Δ has $\qquad m \qquad$ unknowns

total $\quad m(n + m + 1)$ unknowns

FIG. 17.2.2

The number of restrictions

Now let us see how many restrictions we have in terms of scalar equations (17.1.1), (17.1.2) and (17.2.1) as given by Fig. 17.2.3.

equation 1 has $\quad \dfrac{m(m+1)}{2} \quad$ scalar equations

" 2 " $\quad m(m+1) \quad$ " "

" 3 " $\quad nm \quad$ " "

total $\quad m(n + m + 1)$ scalar equations

FIG. 17.2.3

Therefore, comparing the total number of unknowns in Fig. (17.2.2) with the total number of scalar equations in Fig. (17.2.3), we see that the two are exactly equal. Thus we should have just enough equations to solve for the unknowns. A more adequate proof would also have to prove that the solutions for the elements would all be real.*

17.3   The Basic Structure of Any Matrix

The basic structures of the basic factors

We can next prove that any matrix can be expressed as the product indicated in Fig. (17.2.3), even if the matrix is not basic. First let us indicate the rank of X as the product of two basic matrices with common dimension r as shown in

$$X_{nm} = Y_{nr}Z_{rm} \qquad (17.3.1)$$

Here Y and Z are basic matrices; therefore, we can indicate the basic structure of Y by Eq. (17.3.2), where now the order of Δ_y is r and the smaller dimensions of the H matrices are r.

$$Y_{nr} = H_1\Delta_y H_2' \qquad (17.3.2)$$

Furthermore, the H matrices are orthonormal by definition. In the same way, we indicate the basic structure of Z_{mr} by

$$Z_{mr} = F_1\Delta_z F_2' \qquad (17.3.3)$$

Or, if we take the transpose of Eq. (17.3.3), we have

$$Z_{rm} = F_2\Delta_z F_1' \qquad (17.3.4)$$

If we substitute the right sides of Eqs. (17.3.2) and (17.3.4) in the right side of Eq. (17.3.1), we have

$$X_{nm} = H_1\Delta_y H_2 F_2\Delta_z F_1' \qquad (17.3.5)$$

The basic structure of the basic square products

Now suppose we consider the product defined by

$$M = \Delta_y H_2' F_2\Delta_z \qquad (17.3.6)$$

You will notice that the right side of this equation is taken from the four middle factors of the right side of Eq. (17.3.5). Now M in Eq. (17.3.6) is a square basic matrix. This we know because each of the factors on the right of Eq. (17.3.6) is a square basic matrix, and we learned that the product of any number of square basic matrices is a square basic matrix. Therefore, we can write the basic structure of M in Eq. (17.3.6) as

$$M = g_1\Delta_1 g_1' \qquad (17.3.7)$$

*A more rigorous treatment is given by Carl Eckhart and Gale Young in "The Approximation of One Matrix by Another of Lower Rank," *Psychometrika*, Vol. 1, Sept. 1936.

where g_1 and g_2 are square orthonormals of order r.

Let us next substitute from Eq. (17.3.7) into Eq. (17.3.5) to get

$$X_{nm} = H_1 g_1 \Delta_r g_2' F_1' \qquad (17.3.8)$$

The basic orthonormals

Next let us define two matrices as

$$P = H_1 g_1 \qquad (17.3.9)$$

and

$$Q = F_1 g_2 \qquad (17.3.10)$$

It is easy to prove now that P and Q are both orthonormals. We prove this for P and the proof for Q is exactly the same. From Eq. (17.3.9), we have

$$P'P = g_1' H_1' H_1 g_1 \qquad (17.3.11)$$

But because H_1 is orthonormal, Eq. (17.3.11) becomes

$$P'P = g_1' g_1 \qquad (17.3.12)$$

And since g is a square orthonormal, Eq. (17.3.12) becomes

$$P'P = I \qquad (17.3.13)$$

The proof that Q is orthonormal is the same as for P.

Next we substitute equations (17.3.9) and (17.3.10) in (17.3.6) to get

$$X_{nm} = P \Delta_r Q' \qquad (17.3.14)$$

The right side of Eq. (17.3.12) is then in basic structure form, since P and Q are both orthonormals, and Δ is a diagonal matrix. Furthermore, the common orders of the matrices on the right are r. Therefore, we have proved that any nonhorizontal matrix may be expressed as the product from left to right of a nonhorizontal, a diagonal, and a nonvertical orthonormal, and the order of the diagonal is the rank of the matrix.

17.4   The Basic Diagonal Matrix

Signs of the elements

The basic diagonal plays a fundamental role in matrix algebra and in the analysis of social science data. We shall therefore prove some important rules about its properties. First we can prove that the diagonal elements of the basic diagonal can always be taken as positive. For the sake of argument, let us just suppose that some of these elements were negative. Let D be a basic diagonal that just happens to have several negative diagonal elements. We multiply D by a sign matrix in which the appropriate diagonal elements are negative so that we get a diagonal matrix all of whose diagonal elements are positive, so that

$$\Delta = iD \qquad (17.4.1)$$

In Eq. (17.4.1) i is the sign matrix, D is the basic diagonal with negative elements, and Δ is the basic diagonal with all positive elements. Now suppose that we had expressed X as the product

$$X_{nm} = kDq' \qquad (17.4.2)$$

where D is the diagonal matrix with some negative elements and k and q are orthonormal matrices. We could write Eq. (17.4.2) in two different ways as

$$X_{nm} = (ki)(iD)q' \qquad (17.4.3)$$

or

$$X_{nm} = k(Di)(iq') \qquad (17.4.4)$$

It is easy to see that Eqs. (17.4.3) and (17.4.4) are the same as Eq. (17.4.2) because the sign matrix multiplied by itself gives the identity matrix, so substituting the left side of Eq. (17.4.1) into Eqs. (17.4.3) and (17.4.4) gives respectively

$$X_{nm} = (ki)\Delta q' \qquad (17.4.5)$$

and

$$X_{nm} = k\Delta(iq') \qquad (17.4.6)$$

Now notice that in Eqs. (17.4.5) and (17.4.6) the diagonal matrix Δ has by definition all positive elements. Notice also in Eq. (17.4.5) that k post-multiplied by i is still an orthonormal matrix, because, as we have seen in Chapter 16, an orthonormal matrix multiplied by a square orthonormal gives a product that is orthonormal. You remember that the sign matrix is a special case of a square orthonormal. Similarly, in Eq. (17.4.6), q is pre-multiplied by i, but this product is also orthonormal. It should be clear, therefore, that if in Eq. (17.4.2) certain of the elements in D are negative, these may be changed to positive simply by changing the signs of the elements in the corresponding columns of k or in the corresponding rows of q'. We have proved, therefore, that the elements in the basic diagonal of a matrix may always be chosen positive.

Order of magnitude of the elements

Assuming now that all the elements in Δ are positive, let us next assume that they are all different in value; that is, we assume that no two of the diagonal elements are equal. We shall prove that the basic diagonal Δ may always be chosen so that its diagonal elements are in descending order of magnitude from upper left to lower right.* Let us assume that we have a

---

*By "descending order of magnitude" we mean that the Lth element of Δ is greater than or equal to the $(L+1)$th element.

basic diagonal Δ_a in which the elements are not in descending order of magnitude. We indicate the basic structure of X_{nm} by

$$X_{nm} = P_a \Delta_a Q_a' \tag{17.4.7}$$

where P_a and Q_a' are basic orthonormals corresponding to Δ_a. Equation (17.4.7) represents some particular arrangement or ordering of the diagonal elements of Δ that are not in general in descending order of magnitude. Now by pre- and postmultiplication of Δ_a by an appropriate permutation matrix and its transpose, we can get a diagonal matrix in which the elements are in descending order of magnitude. Suppose we indicate this reordering of the elements by

$$\Delta = \pi_a \Delta_a \pi_a' \tag{17.4.8}$$

As a simple illustration, let us assume that Δ_a and Δ are given by Fig. 17.4.1.

$$\Delta_a = \begin{bmatrix} 2 & 0 & 0 \\ 0 & 3 & 0 \\ 0 & 0 & 1 \end{bmatrix} \qquad \Delta = \begin{bmatrix} 3 & 0 & 0 \\ 0 & 2 & 0 \\ 0 & 0 & 1 \end{bmatrix}$$

Fig. 17.4.1

It is easy to show that the appropriate permutation matrix to change Δ_a to Δ is Fig. 17.4.2.

$$\pi_a = \begin{bmatrix} 0 & 1 & 0 \\ 1 & 0 & 0 \\ 0 & 0 & 1 \end{bmatrix} = \pi_a'$$

Fig. 17.4.2

In this particular example, π_a is a symmetric matrix. Figure 17.4.3 shows the steps by which we get the elements of Δ_a in descending order of magnitude.

$$\underbrace{\begin{bmatrix} 2 & 0 & 0 \\ 0 & 3 & 0 \\ 0 & 0 & 1 \end{bmatrix}}_{\Delta_a} \underbrace{\begin{bmatrix} 0 & 1 & 0 \\ 1 & 0 & 0 \\ 0 & 0 & 1 \end{bmatrix}}_{\pi_a'} = \underbrace{\begin{bmatrix} 0 & 2 & 0 \\ 3 & 0 & 0 \\ 0 & 0 & 1 \end{bmatrix}}_{\Delta_a \pi_a'}$$

$$\underbrace{\begin{bmatrix} 0 & 1 & 0 \\ 1 & 0 & 0 \\ 0 & 0 & 1 \end{bmatrix}}_{\pi_a} \underbrace{\begin{bmatrix} 0 & 2 & 0 \\ 3 & 0 & 0 \\ 0 & 0 & 1 \end{bmatrix}}_{\Delta_a \pi_a'} = \underbrace{\begin{bmatrix} 3 & 0 & 0 \\ 0 & 2 & 0 \\ 0 & 0 & 1 \end{bmatrix}}_{\pi_a \Delta_a \pi_a'}$$

Fig. 17.4.3

Now Eq. (17.4.7) can be written in the form

$$X_{nm} = P_a \pi_a' \pi_a \Delta_a \pi_a' \pi_a Q_a' \tag{17.4.9}$$

You will notice that the permutation matrices in Eq. (17.4.9) cancel, and therefore Eq. (17.4.9) is the same as Eq. (17.4.7). Because of Eq. (17.4.8) we can rewrite Eq. (17.4.9) as

$$X_{nm} = P_a \pi_a' \Delta \pi_a Q_a' \tag{17.4.10}$$

Next let us define the two matrices P and Q respectively as

$$P = P_a \pi_a' \tag{17.4.11}$$

$$Q = Q_a \pi_a' \tag{17.4.12}$$

Notice that P and Q are still orthonormal matrices because they are the product of orthonormal matrices by square orthonormal matrices. You recall that a permutation matrix is a special case of a square orthonormal matrix. Because of Eqs. (17.4.11) and (17.4.12), we can rewrite Eq. (17.4.10) as

$$X_{nm} = P \Delta Q'$$

Now Eq. (17.4.13) has all the elements of Δ in descending order of magnitude. Therefore we have proved that by an appropriate interchange of columns in P_a and rows in Q_a' the basic diagonal of a matrix can always be expressed with its diagonal elements in descending order of magnitude.

We have proved now that any nonhorizontal matrix may be expressed as the product from left to right of a nonhorizontal orthonormal, a diagonal, and a nonvertical orthonormal, where the diagonal has all positive elements in descending order of magnitude, and its order is the rank of the matrix.

You should remember that there is one and only one basic diagonal for any given matrix X. Furthermore, if all of the diagonal elements of the basic diagonal are different from one another, there is only one left orthonormal and only one right orthonormal in terms of which the basic structure can be expressed. The case is somewhat different, however, if the basic diagonal has some elements that are not distinct from one another.

One set of equal diagonal elements

Let us first consider the case in which all of the elements in Δ are distinct except a certain number which are all equal one to another. We could then write the basic structure of X in supermatrix notation as Eq. (17.4.14).

$$X = [P_0 \quad P_1] \begin{bmatrix} \Delta_0 & 0 \\ 0 & \Delta_1 \end{bmatrix} \begin{bmatrix} Q' \\ Q_1' \end{bmatrix} \tag{17.4.14}$$

This equation means that we have grouped the distinct diagonal elements of Δ in a submatrix Δ_0, and all of the equal elements in a submatrix Δ_1. Furthermore, the P and Q orthonormals have been partitioned to correspond with the submatrices Δ_0 and Δ_1 respectively. Now since all of the elements in Δ are equal, we can write

$$\Delta_1 = CI \tag{17.4.15}$$

where C is a scalar quantity equal to the elements in Δ_1. If now we multiply out the right side of Eq. (17.4.14), we have

$$X = P_0\Delta_0 Q_0' + P_1\Delta_1 Q_1' \tag{17.4.16}$$

Because of Eq. (17.4.15) we can write this as

$$X = P_0\Delta_0 Q_0' + CP_1 Q_1' \tag{17.4.17}$$

But notice that the second term on the right of this equation is the major product of two orthonormal matrices. Suppose we select any of an infinite number of square orthonormals h and write

$$P_1 = P_1 h \tag{17.4.18}$$

and

$$q_1 = Q_1 h \tag{17.4.19}$$

From Eqs. (17.4.18) and (17.4.19) we can write

$$P_1 = p_1 h' \tag{17.4.20}$$

and

$$Q_1 = q_1 h' \tag{17.4.21}$$

If now we substitute the right sides of Eqs. (17.4.20) and (17.4.21) into the right side of Eq. (17.4.17), we have

$$X = P_0\Delta_0 Q_0' + Cp_1' h' h q_1' \tag{17.4.22}$$

But since h was defined as a square orthonormal, Eq. (17.4.22) can be written as

$$X = P_0\Delta_0 Q_0' + Cp_1 q_1' \tag{17.4.23}$$

Or, to write (17.4.23) in the form of the product of super matrices, we get Eq. (17.4.24).

$$X = [P_0 \quad p_1] \begin{bmatrix} \Delta_0 & 0 \\ 0 & CI \end{bmatrix} \begin{bmatrix} Q_0' \\ q_1' \end{bmatrix} \tag{17.4.24}$$

But we remember that p_1 and q_1 in Eq. (17.4.24) can be obtained in an infinite number of ways simply by selecting a particular square orthonormal h. This means, therefore, that those vectors of the basic orthonormals that correspond to the repeated or equal elements in the basic diagonal are not unique, and that an infinite number of sets of such submatrices exist that will satisfy the original matrix X.

More than one set of equal diagonal elements

In general, if we have more than one set of equal elements in the basic diagonal, the general equation corresponding to Eq. (17.4.24) is given by Eq. (17.4.25).

$$X = [P_0 \quad p_1 \quad \ldots \quad p_n] \begin{bmatrix} \Delta_0 & 0 & \ldots & 0 \\ 0 & c_1 I & \ldots & 0 \\ \ldots & \ldots & \ldots & \ldots \\ 0 & 0 & \ldots & c_n I \end{bmatrix} \begin{bmatrix} Q'_0 \\ q'_1 \\ \ldots \\ q'_n \end{bmatrix}$$

$$(17.4.25)$$

None of the p and q submatrices on the right side of Eq. (17.4.25) are unique. There are as many pairs of submatrices that are not unique in the orthonormals as there are sets of repeated elements in the basic diagonal.

17.5 The Basic Structure of the Product Moment Matrix

Next we shall see what happens in the basic structure of a matrix when it is pre- or postmultiplied by its transpose. First let us express the matrix in its basic structure form, as indicated by

$$X = P_1 \Delta Q' \qquad (17.5.1)$$

The minor product moment

Let us now take the minor product moment of X. We premultiply each side of Eq. (17.5.1) by its transpose to get

$$X'X = Q\Delta P'P\Delta Q' \qquad (17.5.2)$$

But since P is an orthonormal, its minor product moment is an identity matrix, so Eq. (17.5.2) may be written as

$$X'X = Q\Delta^2 Q' \qquad (17.5.3)$$

Equation (17.5.3) shows first that the basic diagonal of the minor product moment of a matrix is equal to the square of the basic diagonal of the matrix Second, it shows that the left orthonormal of a minor product moment is the transpose of the right orthonormal of the matrix. And third, it shows that the right orthonormal of a matrix and its minor product moment are the same.

The major product moment

Next let us postmultiply each side of Eq. (17.5.1) by its transpose as in

$$XX' = P\Delta Q'Q\Delta P' \qquad (17.5.4)$$

But since Q is an orthonormal, its minor product moment is an identity and (17.5.4) may be written as

$$XX' = P\Delta^2 P' \qquad (17.5.5)$$

We see in Eq. (17.5.5) that the basic diagonal of the major product moment of a matrix is the square of the basic diagonal of the matrix. We also see from Eqs. (17.5.3) and (17.5.5) that the basic diagonals of the major and minor product moments of a matrix are equal. From Eq. (17.5.5) we also see that the left orthonormal of a matrix and its major product moment are the same. Finally we see that the right orthonormal of the major product moment is equal to the transpose of the left orthonormal of the matrix.

To summarize, we may say then that first, the basic diagonal of either product moment of a matrix is the square of the basic diagonal of the matrix; second, that the orthonormals of the major product moment are from the left orthonormal of the matrix; and third, the orthonormals of the minor product moment are from the right orthonormal of the matrix.

Powers of product moments

Next let us see what happens to the basic structure when we raise the product moment of a matrix to a power. First let us square both sides of Eq. (17.5.5). This gives

$$(X'X)^2 = (Q\Delta^2 Q_2')(Q_2\Delta^2 Q_2') \qquad (17.5.6)$$

We let

$$\partial = \Delta^2 \qquad (17.5.7)$$

Remembering that the minor product moment of Q is the identity matrix, and substituting from Eq. (17.5.7) in Eq. (17.5.6), we have

$$(X'X)^2 = Q\partial^2 Q' \qquad (17.5.8)$$

From Eq. (17.5.8) we see that the orthonormals of the square of the minor product moment are the same as the orthonormals of the minor product moment itself. The basic diagonal of the minor product moment is squared to get the basic diagonal of the square of the minor product moment. If now we multiply both sides of Eq. (17.5.8) by the corresponding sides of Eq. (17.5.3) and use Eq. (17.5.7), we get

$$(X'X)^3 = Q\partial^2 Q'Q\partial Q' \qquad (17.5.9)$$

Or from Eq. (17.5.9) we have

$$(X'X)^3 = Q\partial^3 Q' \qquad (17.5.10)$$

By continuing in the same way, we can write the general equation

$$(X'X)^n = Q\partial^n Q' \qquad (17.5.11)$$

Equation (17.5.11) shows that any positive integral power of the minor product moment of a matrix has the same orthonormals as the minor product moment itself, and the basic diagonal of the power of the product is the basic diagonal of the product moment raised to the same power.

The rule for the major product moment can be derived in exactly the same way. Since the entire demonstration depends on the fact that we get minor products of orthonormals that yield the identity matrix, we can directly write Eq. (17.5.12) corresponding to Eq. (17.5.11).

$$(XX')^n = P\partial^n P' \tag{17.5.12}$$

Equation (17.5.12) shows that the basic orthonormals of the major product moment raised to any integral power are the same as the basic orthonormals of the original major product moment. It also shows that the basic diagonal of the power of a product moment is obtained by raising the basic diagonal of the major product moment to the corresponding power. Equations (17.5.11) and (17.5.12) show that the basic diagonals for a given power of either the major or minor product moment are the same.

17.6 Matrices with the Same Orthonormals

The major product

Suppose we consider two matrices a and b, which have for their basic structure the same orthonormals but different basic diagonals. These we can illustrate by

$$a = P\Delta_a Q' \tag{17.6.1}$$

and

$$b = P\Delta_b Q' \tag{17.6.2}$$

It is obvious, of course, that if a and b as given here have the same orthonormals, they must be of the same order. Let us first then consider the major product of the matrices a and b given by

$$ab' = P\Delta_a Q' Q\Delta_b P'_1 \tag{17.6.3}$$

From this we have

$$ab' = P\Delta_a \Delta_b P' \tag{17.6.4}$$

Now let us take the transpose of this to get

$$ba' = P\Delta_b \Delta_a P' \tag{17.6.5}$$

But notice that since diagonal matrices are commutative, the right side of Eqs. (17.6.4) and (17.6.5) are equal. Therefore, we can write

$$ab' = ba' \tag{17.6.6}$$

Or to put Eq. (17.6.6) in words, we can say that if two matrices have the same basic orthonormals, their major product is a symmetric matrix.

The minor product

Also from Eqs. (17.6.1) and (17.6.2) we can write a minor product as in

$$a'b = Q\Delta_a P' P\Delta_b Q' \tag{17.6.7}$$

or from Eq. (17.6.7) we have

$$a'b = Q\Delta_a\Delta_bQ' \tag{17.6.8}$$

Taking the transpose of both sides of Eq. (17.6.8), we have

$$b'a = Q_2\Delta_b\Delta_aQ_2' \tag{17.6.9}$$

But the right sides of Eqs. (17.6.8) and (17.6.9) are equal, since diagonal matrices are commutative. Therefore, we may write

$$a'b = b'a \tag{17.6.10}$$

From this it follows that the minor product of two matrices with equal orthonormals is a symmetric matrix. Therefore, either the major or minor product of two matrices having the same orthonormals is a symmetric matrix.

Product moment matrices with same orthonormals

Next let us consider any two product moment matrices having the same orthonormals. The right orthonormal is the transpose of the left orthonormal. We can, therefore, write two such matrices as

$$S_a = Q\partial_aQ' \tag{17.6.11}$$

and

$$S_b = Q\partial_bQ' \tag{17.6.12}$$

Now from Eqs. (17.6.11) and (17.6.12) we can write

$$S_aS_b = Q\partial_aQ'Q\partial_bQ' \tag{17.6.13}$$

and

$$S_bS_a = Q\partial_bQ'Q\partial_aQ' \tag{17.6.14}$$

Or from Eqs. (17.6.13) and (17.6.14) we have respectively

$$S_aS_b = Q\partial_a\partial_bQ' \tag{17.6.15}$$

and

$$S_bS_a = Q\partial_b\partial_aQ' \tag{17.6.16}$$

But since diagonal matrices are commutative, the right sides of Eqs. (17.6.15) and (17.6.16) are equal; therefore, the left sides are equal, and we can write

$$S_aS_b = S_bS_a \tag{17.6.17}$$

Equation (17.6.17) shows, therefore, that any two product moment matrices that have the same orthonormals are commutative with each other.

Suppose we have another product moment matrix with the same orthonormal but with a different basic diagonal. We represent this matrix by

$$S_c = Q\partial_cQ' \tag{17.6.18}$$

From Eqs. (17.6.15) and (17.6.18) we can write

$$S_c S_a S_b = Q \partial_c \partial_a \partial_b Q' \tag{17.6.19}$$

and

$$S_a S_b S_c = Q \partial_a \partial_b \partial_c Q' \tag{17.6.20}$$

Also from Eqs. (17.6.16) and (17.6.18) we can write

$$S_c S_b S_a = Q \partial_c \partial_b \partial_a Q' \tag{17.6.21}$$

and

$$S_b S_a S_c = Q \partial_b \partial_a \partial_c Q' \tag{17.6.22}$$

The right sides of Eqs. (17.6.19) through (17.6.22) are all the same except for the order of the basic diagonals, and since these are all commutative, the four are equal. Therefore, all the products on the left-hand sides of these equations are also equal. In the same way, we could have interposed the matrix S_c between the other two matrices and shown that the product was not changed. Using the same procedure and remembering that the minor product moment of an othonormal matrix is the identity matrix, it is easy to show that all product moment matrices with the same orthonormals are commutative, one with another. We have also the rule that the product of any number of product moment matrices with the same basic orthonormal is equal to a matrix having the same basic orthonormal and a basic diagonal that is the product of the basic diagonals of the factors.

17.7 Special Powers of a Product Moment Matrix

Fractional powers

Let us consider now two product moment matrices whose basic structures are defined by

$$S_1 = Q \partial Q' \tag{17.7.1}$$

and

$$S_2 = Q \partial^{1/2} Q' \tag{17.7.2}$$

Squaring both sides of Eq. (17.7.2), we get

$$S_2^2 = Q \partial^{1/2} Q' Q \partial^{1/2} Q' \tag{17.7.3}$$

But this can be written as

$$S_2^2 = Q \partial^{1/2} \partial^{1/2} Q' \tag{17.7.4}$$

or

$$S_2^2 = Q \partial Q' \tag{17.7.5}$$

Notice that the right sides of Eqs. (17.7.1) and (17.7.5) are the same; therefore, we can write

$$S_2^2 = S_1 \tag{17.7.6}$$

Taking the square root of both sides of Eq. (17.7.6), we get

$$S_2 = S_1^{1/2} \tag{17.7.7}$$

In the same way, suppose we had let S_3 be defined by

$$S_3 = Q^{1/3}Q' \tag{17.7.8}$$

Then we could show that

$$S_3^3 = S_1 \tag{17.7.9}$$

is true. Or, from Eq. (17.7.9), we would have

$$S_3 = S_1^{1/3} \tag{17.7.10}$$

Suppose we have the general equation for the basic structure of a product moment matrix, as in

$$S = Q\partial Q' \tag{17.7.11}$$

Then the rule for the nth root of a product moment matrix can be stated as

$$S^{1/n} = Q\partial^{1/n}Q' \tag{17.7.12}$$

If a is any positive value, either fractional or integral, the general rule is expressed in

$$S^a = Q\partial^a Q' \tag{17.7.13}$$

The inverse

We shall next let the exponent a in Eq. (17.7.13) be any real number, either positive or negative. Therefore, if a takes the value -1 we have

$$S^{-1} = Q\partial^{-1}Q' \tag{17.7.14}$$

Suppose now we multiply each side of Eq. (17.7.14) by the corresponding side of Eq. (17.7.11) to get

$$SS^{-1} = Q\partial Q'Q\partial^{-1}Q' \tag{17.7.15}$$

From this we have

$$SS^{-1} = Q\partial\partial^{-1}Q' \tag{17.7.16}$$

or

$$SS^{-1} = QQ' \tag{17.7.17}$$

Now notice that if Q is square, that is, if S is a basic matrix, the right side of Eq. (17.7.17) is an identity, so that we have

$$SS^{-1} = I \tag{17.7.18}$$

In this particular case, S^{-1} is called the inverse of S. This subject will be considered more fully in Chapter 19.

17.8 The Basic Diagonal and Traces Involving Product Moments

The product moment

Another rule that we shall find convenient is that the sum of the elements in the basic diagonal of a product moment is the same as the trace of the major or minor product moment itself. This we can prove rather easily by considering the minor product moment of X, as given by

$$X'X = Q\partial Q' \tag{17.8.1}$$

Suppose now we define a matrix Z by

$$Z = Q\partial^{1/2} \tag{17.8.2}$$

From this we can rewrite (17.8.1) as

$$X'X = ZZ' \tag{17.8.3}$$

Then

$$tr\ X'X = tr\ ZZ' \tag{17.8.4}$$

but from Chapter 10 we know that

$$tr\ ZZ' = tr\ Z'Z \tag{17.8.5}$$

From (17.8.2) we have

$$tr\ Z'Z = tr\ \delta^{-1/2}Q'Q^{1/2} \tag{17.8.6}$$

or

$$tr\ Z'Z = tr\ \delta \tag{17.8.7}$$

Substituting Eq. (17.8.7) in Eq. (17.8.5), we get

$$tr\ ZZ' = tr\ \delta \tag{17.8.8}$$

and substituting Eq. (17.8.8) in Eq. (17.8.4) gives

$$tr\ X'X = tr\ \delta \tag{17.8.9}$$

But since the trace of a diagonal matrix is simply the sum of its elements, we have from Eq. (17.8.9)

$$tr\ X'X = 1'\delta 1 \tag{17.8.10}$$

And since the traces of a major and minor product moment are the same, we have

$$tr\ XX' = 1'\delta 1 \tag{17.8.11}$$

Therefore we have proved the rule.

Trace of powers of product moments

In the same way, we can write for any specified power of the major or minor product moment the relation between its trace and its basic diagonal by

$$tr(XX')^n = tr(X'X)^n = 1'\partial^n 1 \tag{17.8.12}$$

We have therefore the general rule that the trace of any power of a major or minor product moment is equal to the sum of the elements in the corresponding power of the basic diagonal.

17.9   The Product of a Matrix by a Square Orthonormal

If any matrix is either pre- or postmultiplied by a square orthonormal, the basic diagonal of the product is the same as for the original matrix. This can be shown very easily if we start with the basic structure form of the matrix X as given by

$$X = P\Delta Q' \tag{17.9.1}$$

Suppose now we premultiply (17.9.1) by a square orthonormal, say H_1, as in

$$H_1 X = H_1 P \Delta Q_2' \tag{17.9.2}$$

On the left of the basic diagonal in this equation, we have the orthonormal P premultiplied by a square orthonormal, which yields a product that is orthonormal. In a similar manner, we could postmultiply Eq. (17.9.1) by a square orthonormal H_2 as in

$$X H_2 = P \Delta Q' H_2 \tag{17.9.3}$$

On the right side of the basic diagonal Δ in Eq. (17.9.3), we again have an orthonormal multiplied by a square orthonormal, which yields a product that is orthonormal. Therefore, in both Eqs. (17.9.2) and (17.9.3), the right sides are in basic structure form and the basic diagonal Δ is still the same as it was in Eq. (17.9.1).

SUMMARY

1. The basic structure of a nonhorizontal matrix X is defined as the product of three matrices, thus

$$X = P\Delta Q'$$

where $P'P = Q'Q = I$, and Δ is diagonal.

2. In a basic matrix the number of unknown elements in the basic structure factors is precisely equal to the number of restrictions placed on these elements by the definition.

3. The basic structure of a nonbasic matrix:
   a. Every nonbasic matrix may be expressed as the major product of basic matrices each of which may be expressed as the product of its basic structure factors.

b. Therefore, any nonhorizontal matrix may be expressed as the product from left to right of a nonhorizontal orthonormal, a diagonal, and a nonvertical orthonormal matrix.

c.. The order of its basic diagonal is the rank of a matrix.

4. The basic diagonal:
   a. The signs of the basic diagonal elements may all be taken as positive.
   b. The basic diagonal elements may be taken in descending order of magnitude from upper left to lower right.
   c. The basic diagonal is unique for any given matrix.
   d. Any nonhorizontal matrix may be expressed as the product from left to right of a nonhorizontal orthonormal, a diagonal, and a nonvertical orthonormal, where the diagonal has all positive elements in descending order of magnitude, and its order is the rank of the matrix.
   e. If the basic diagonal includes sets of equal elements, only the submatrices from the basic orthonormals *not* corresponding to these sets are unique.

5. The basic structure of the product moment matrix. If the basic structure of X is $X = P\Delta Q'$, then the basic structure of
   a. The minor product moment is

   $$X'X = Q\Delta^2 Q'$$

   b. The major product moment is

   $$XX' = P\Delta^2 P'$$

   c. The power of a product moment is

   $$[X'X]^n = Q\Delta^{2n}Q'$$
   $$[XX']^n = P\Delta^{2n}P'$$

6. Matrices with the same orthonormals:
   a. The major and minor products of matrices with the same orthonormals are symmetric matrices.
   b. Product moment matrices with the same orthonormals are commutative with respect to multiplication.

7. Special exponents of product moment matrices.
   a. Any fractional power of a product moment matrix is given by a matrix with the same orthonormals and the corresponding fractional power of the basic diagonal. If $S = Q\delta Q'$

   $$S^{1/n} = Q\delta^{1/n}Q'$$

   b. The inverse of a basic product moment matrix $S = Q\delta Q'$ is given by

   $$S = Q\delta^{-1}Q'$$

8. The basic diagonal and traces involving product moments:

a. The sum of the elements in the basic diagonal of a product moment is the same as the trace of the product moment. If $X'X = Q\delta Q'$

$$tr(X'X) = tr(XX') = 1'\delta 1$$

b. The trace of any power of a product moment is given by

$$tr(X'X)^n = tr(XX')^n = 1'\delta^n 1$$

9. If any matrix is either pre- or postmultiplied by a square orthonormal matrix, the basic diagonal of the product is the same as for the original matrix.

Chapter 18

Finding the Basic Structure of a Matrix

18.1 The General Solution for the Basic Structure

Practical significance

You have seen in Section 17.1 that any matrix can be expressed as the product of two orthonormals and a diagonal matrix. This product we have called the basic structure of the matrix. A general area of interest centering around the basic structure of data matrices in psychology is actually the study we have referred to previously as factor analysis. In recent years other disciplines, such as sociology, anthropology, political science, economics, and even biology, have become interested in the application of factor analysis techniques to matrices of experimental data. One of the most important elements in factor analysis is determination of the basic structure of a matrix. It is therefore important for us to know how the basic orthonormals and the basic diagonal of a data matrix may be found. In this chapter we shall first consider the subject theoretically and then set up a computational routine.

The left orthonormal

You will recall that the general equation for the basic structure of a matrix is

$$X = P\Delta Q' \qquad (18.1.1)$$

where P and Q are orthonormals, and Δ is a diagonal matrix. Now if we knew the elements of the matrix Q, it would be quite easy to solve for P and Δ. We could postmultiply both sides of Eq. (18.1.1) by Q and get

$$XQ = P\Delta Q'Q \qquad (18.1.2)$$

But since Q is orthonormal, Eq. (18.1.2) would be written on the right as

$$XQ = P\Delta \qquad (18.1.3)$$

So you see that by simply multiplying X on the right by Q we would have on the right only P and Δ. We could readily solve for Δ by taking the minor product moment of both sides of Eq. (18.1.3) to get

$$Q'X'XQ = \Delta P'P\Delta \qquad (18.1.4)$$

But since P is also orthonormal, this could be written as

$$Q'X'XQ = \Delta^2 \qquad (18.1.5)$$

Equation (18.1.5) shows that the left side must be a diagonal matrix, since the right side is diagonal; therefore by taking the square root of the left side, we would solve for Δ by

$$(Q'X'XQ)^{1/2} = \Delta \qquad (18.1.6)$$

Then to solve for P, we would simply postmultiply both sides of Eq. (18.1.3) by Δ^{-1} to get

$$XQ\Delta^{-1} = P \qquad (18.1.7)$$

Or, if we substituted from the left of Eq. (18.1.6) into the left of Eq. (18.1.7), we would have

$$XQ_2(Q'X'XQ)^{-1/2} = P \qquad (18.1.8)$$

If only, then, we knew the elements of Q, Eq. (18.1.6) would enable us to find the basic diagonal Δ and Eq. (18.1.8) would enable us to find the left orthonormal P. Only rarely, however, do we have available the right orthonormal, Q, for any given data matrix, so our chief problem becomes one of first solving for the Q matrix.

18.2 Finding the Right Orthonormal of a Data Matrix

The method we shall outline for finding the right orthonormal consists of a step-by-step procedure in which we first get column 1 of Q together with the first element of the basic diagonal. Then we get column 2 of Q, and the corresponding second element of the basic diagonal. We proceed in this manner until all columns of Q and all elements of the basic diagonal have been solved for. The method assumes that all δ's are distinct, which will generally be the case with data matrices. We shall indicate the logic upon which the procedure is based and then show how the computations are carried out.

The minor product moment

First we get the minor product moment of both sides of Eq. (18.1.1) as in

$$X'X = Q\Delta P'P\Delta Q' \qquad (18.2.1)$$

But since P is orthonormal, the right side of Eq. (18.2.1) can be written as

$$X'X = Q\Delta^2 Q' \tag{18.2.2}$$

You are already familiar with this equation, which shows that the basic diagonal of the minor product moment of a matrix is the square of the basic diagonal of the matrix. Also it reminds you that the left orthonormal is simply the transpose of the right orthonormal. You see then that merely by taking the minor product moment of X, we have eliminated the left orthonormal.

For convenience we let

$$\partial = \Delta^2 \tag{18.2.3}$$

Equation (18.2.2) then becomes

$$X'X = Q\partial Q' \tag{18.2.4}$$

You will recall that without loss of generality we may assume that the elements in Δ are in descending order of magnitude from upper left to lower right.

Powers of the minor product moment

Now let us raise the minor product moment of X to some positive power n where n is an integer. This gives

$$(X'X)^n = Q\partial^n Q' \tag{18.2.5}$$

Equation (18.2.5) is taken from Section 17.0 and shows that in raising a product moment to any power, the basic orthonormals remain unchanged and the basic diagonal is raised to the same power as that of the product moment.

If n is large

Suppose now that we let n be a very large number and that we divide Eq. (18.2.5) on both sides by a number K, as in

$$\frac{1}{K}(X'X)^n = Q\frac{\partial^n}{K}Q' \tag{18.2.6}$$

Now let us look at this value K on the assumption that n is very large and that none of the elements in ∂ are equal. Suppose first we consider the basic diagonal of Eq. (18.2.6) in expanded form, divided by the constant K as in Eq. (18.2.7).

$$\frac{\partial^n}{K} = \begin{bmatrix} \dfrac{\partial_1^n}{K} & 0 & \cdots & 0 \\ 0 & \dfrac{\partial_2^n}{K} & \cdots & 0 \\ \cdots & \cdots & \cdots & \cdots \\ 0 & 0 & \cdots & \dfrac{\partial_m^n}{K} \end{bmatrix} \tag{18.2.7}$$

If we assume that K is equal to ∂_1^n, we can write Eq. (18.2.8)

$$\frac{\partial^n}{\partial_1^n} = \begin{bmatrix} 1 & 0 & \cdots & 0 \\ 0 & \left[\frac{\partial_2}{\partial_1}\right]^n & \cdots & 0 \\ \cdots & \cdots & \cdots & \cdots \\ 0 & 0 & \cdots & \left[\frac{\partial_m}{\partial_1}\right]^n \end{bmatrix} \tag{18.2.8}$$

Suppose now that the ratio of ∂_2 to ∂_1 is, let us say, .8. We could then select n large enough so that this ratio raised to the nth power would be 0 to any specified number of decimal places. For example, .8 to the 32nd power would be 0 in the first two decimal places. If the ∂'s were in descending order of magnitude, then for this particular example all elements in Eq. (18.2.8) would be 0 to at least two decimal places, except the first one, which would be unity. Suppose, however, that K was not exactly equal to ∂_1^n but was just roughly of that general order of magnitude. Even in this case, if we chose n large enough, we could have all elements but the first in Eq. (18.2.8) 0 to any specified number of decimal places. Provided then that we choose K properly, and specify the degree of decimal accuracy we desire, Eq. (18.2.7) can be written in Eq. (18.2.9) to any desired degree of decimal accuracy.

$$\frac{\partial^n}{K} = \frac{\partial_1^n}{K} \begin{bmatrix} 1 & 0 & \cdots & 0 \\ 0 & 0 & \cdots & 0 \\ \cdots & \cdots & \cdots & \cdots \\ 0 & 0 & \cdots & 0 \end{bmatrix} \tag{18.2.9}$$

But the matrix on the right of Eq. (18.2.9) can be written as the major product moment of an e_1 vector, as in

$$\frac{\partial^n}{K} = \frac{\partial_1^n}{K} e_1 e_1' \tag{18.2.10}$$

Substituting the right side of Eq. (18.2.10) into the right side of Eq. (18.2.6), we have

$$\frac{1}{K}(X'X)^n = Q\frac{\partial_1^n}{K}e_1 e_1' Q' \tag{18.2.11}$$

We can write Eq. (18.2.11) as

$$\frac{1}{K}(X'X)^n = \frac{\partial_1^n}{K}Q e_1 e_1' Q' \tag{18.2.12}$$

Next let us examine the product of the orthonormal Q by the e_1 vector on the right of Eq. (18.2.12). This is

$$Q e_1 = Q_{.1} \tag{18.2.13}$$

Equation (18.2.13) reminds us that when we postmultiply any matrix by an

e_i vector, the product is simply the ith column vector of that matrix. If we substitute the right side of Eq. (18.2.13) into the right side of Eq. (18.2.12), we get

$$\frac{1}{K}(X'X)^n = \frac{\partial_1^n}{K}Q_{.1}Q'_{.1} \tag{18.2.14}$$

Equation (18.2.14) shows that if we raise the minor product moment of a matrix to a sufficiently high power and divide it by an appropriate constant, the resulting matrix, to a specified number of decimal places, is proportional to the major product moment of the first column vector of its basic orthonormal.

The arbitrary vector

Next let us consider any arbitrary vector V that is not orthogonal to $Q_{.1}$. Suppose we postmultiply both sides of Eq. (18.2.14) by V to get

$$\frac{1}{K}(X'X)^n V = \frac{\partial_1^n}{K}Q_{.1}(Q'_{.1}V) \tag{18.2.15}$$

Because $Q'_{.1}V$ is a scalar, Eq. (18.2.15) can be written

$$\frac{1}{K}(X'X)^n V = \left[\frac{\partial_1^n}{K}Q'_{.1}V\right]Q_{.1} \tag{18.2.16}$$

Equation (18.2.16) shows, therefore, that if ∂_2 is less than δ_1, and if the minor product moment is raised to a sufficiently high power and divided by an appropriate constant, then multiplication by any vector that is not orthogonal to the first column of the basic orthonormal yields a vector proportional to the first column of the basic orthonormal. The power to which the matrix must be raised in order to give any specified degree of decimal accuracy depends on the ratio of the second to the first element in the basic diagonal. The smaller this ratio, the lower the power to which the matrix must be raised.

The solution for $Q_{.1}$

From Eq. (18.2.16), we can easily solve for $Q_{.1}$. First we rewrite Eq. (18.2.16) as

$$\frac{1}{K}(X'X)^n V = Y \tag{18.2.17}$$

Then $Q_{.1}$ will be given by

$$Q_{.1} = \frac{Y}{\sqrt{Y'Y}} \tag{18.2.18}$$

since $Q_{.1}$ is by definition normal.

The solution for δ

Once the value of $Q_{.1}$ is found, the value of δ_1 is readily obtained from Eq. (18.2.4). We premultiply both sides by $Q'_{.1}$ and postmultiply by $Q_{.1}$ to get

$$Q'_{.1}(X'X)Q_{.1} = Q'_{.1}Q\partial Q'Q_{.1} \qquad (18.2.19)$$

But since Q is an orthonormal, we have

$$Q'Q_{.1} = e_1 \qquad (18.2.20)$$

Equation (18.2.20) simply says that $Q_{.1}$ is orthogonal to each row of Q' except the first, which is $Q'_{.1}$, and therefore multiplication by $Q_{.1}$ yields unity. Substituting Eq. (18.2.20) in the right of Eq. (18.2.19), we have

$$Q'_{.1}(X'X)Q_{.1} = e'_1 \delta e_1 \qquad (18.2.21)$$

But we know that

$$e'_1 \delta e_1 = \delta_1 \qquad (18.2.22)$$

Substituting from Eq. (18.2.22) into the right of Eq. (18.2.21), we have

$$Q'_{.1}(X'X)Q_{.1} = \partial_1 \qquad (18.2.23)$$

Therefore, if we premultiply the product moment matrix by the transpose of the first column of its orthonormal, and postmultiply by the first column of the orthonormal, we get the first element of the basic diagonal.

18.3 The Iterative Solution for the First Vector of Q

Successive vector multiplications

Raising a matrix even to the second power can be pretty laborious. If it must be raised to a high power in order to give a large difference between the first and second elements of the basic diagonal, the work becomes prohibitive. But notice in Eq. (18.2.17) that we do not actually use the matrix itself raised to the nth power. We are concerned with this power of the matrix multiplied by some arbitrary vector. Let us therefore consider the following equations, where the V's are vectors.

$$(X'X)V_0 = V_1 \qquad (18.3.1)$$

$$(X'X)V_1 = V_2 \qquad (18.3.2)$$

$$(X'X)V_2 = V_3 \qquad (18.3.3)$$

Now notice that if we substitute the left side of Eq. (18.3.1) for V_1 in the left side of Eq. (18.3.2), we have

$$(X'X)^2 V_0 = V_2 \qquad (18.3.4)$$

Then if we substitute the left side of Eq. (18.3.4) for V_2, in the left side of Eq. (18.3.3), we have

$$(X'X)^3 V_0 = V_3 \tag{18.3.5}$$

In the same way, we could show

$$(X'X)^n V_0 = V_n \tag{18.3.6}$$

Equations (18.3.1) to (18.3.6), therefore, show that if we multiply the original product moment matrix by an arbitrary vector to get another vector and continue to multiply the product moment matrix each time by the last vector so obtained, this amounts to multiplying the original arbitrary vector by some corresponding power of the product moment matrix. Therefore, although it would be prohibitively laborious to raise a matrix to a fairly high power and then multiply by an arbitrary vector, the same results can be achieved in a very small fraction of the time by multiplying the original product moment matrix successively by the vectors obtained from the multiplications.

The scaling constant for iterated vectors

So far we have not considered how to obtain the value K, which we first used in Eq. (18.2.6). If we were simply to proceed as in Eqs. (18.3.1) through (18.3.4), the elements of the successive V vectors could become either very large or very small. Actually, therefore, we must adopt some procedure for keeping the size of the elements within reasonable limits. We therefore proceed as follows: we multiply the product moment matrix by an arbitrary vector V_0. In most cases the V_0 vector may be a unit vector. Our first multiplication is

$$S_0 V_0 = U_1 \tag{18.3.7}$$

where to simplify our notation we use S_0 for $X'X$. Instead of now multiplying S_0 again by U_1, we first divide the elements of U_1 by some appropriate constant. It has been convenient to use as this constant the largest absolute value in U_1.* If we indicate this largest absolute value by U_{1L}, then we define a new vector as in

$$\frac{U_1}{U_{1L}} = V_1 \tag{18.3.8}$$

The largest absolute value in V_1 will now be unity. Next we multiply S_0 by V_1 as in

$$S_0 V_1 = U_2 \tag{18.3.9}$$

*This method was first proposed by Harold Hotelling in "Analysis of a Complex of Statistical Variables," *Journal of Educational Psychology*, 1933, Vol. 24, pp. 417-441.

We then find the element with the largest absolute value in U_2 and calculate a new V_2 vector by

$$\frac{U_2}{U_{2L}} = V_2 \qquad (18.3.10)$$

The general equation for the multiplication of S_0 by a vector is

$$S_0 V_{i-1} = U_i \qquad (18.3.11)$$

The general equation for calculating the V vector from the U vector is

$$\frac{U_i}{U_{iL}} = V_i \qquad (18.3.12)$$

The $Q_{.1}$ vector

The process is continued until within the limit of decimal accuracy desired, the last V vector calculated is equal to the one immediately preceding, or until

$$V_{i-1} = V_i \qquad (18.3.13)$$

If the process has been continued until this equation is satisfied, we can write

$$\frac{1}{K} S_0^n V_0 = V_n = CQ_{.1} \qquad (18.3.14)$$

where

$$K = U_{1L} U_{2L} \ldots U_{nL} \qquad (18.3.15)$$

In other words, K is simply the product of all of the largest elements from the U vectors. The constant C in Eq. (18.3.14) is given by

$$C = \frac{\partial_1^n}{K} Q'_{.1} V_0 \qquad (18.3.16)$$

Actually, however, we rarely find it necessary to calculate either K or C given by Eqs. (18.3.15) and (18.3.16). The vector $Q_{.1}$ can be calculated directly from either the U_n or V_n vectors, since they are both proportional to one another and since we know that $Q_{.1}$ is a normal vector. Calculating $Q_{.1}$ from U_n therefore, we have

$$Q_{.1} = \frac{U_n}{\sqrt{U'_n U_n}} \qquad (18.3.17)$$

The δ_1 scalar

The value for ∂_1, the first element in the basic diagonal, is readily obtained as follows: assume that V_{n-1} is proportional to U_n within the limits of decimal error. We know then that V_{n-1} and U_n are both proportional to $Q_{.1}$. Let us begin then with

$$S_0 V_{n-1} = U_n \qquad (18.3.18)$$

We have said that both vectors V and U were proportional to $Q_{.1}$; therefore we could write

$$V_{n-1} = gQ_{.1} \qquad (18.3.19)$$

where g is a proportionality constant. If we substitute Eq. (18.3.19) in Eq. (18.3.18) and also use the basic structure form of S_0, we have

$$gQ\partial Q'Q_{.1} = U_n \qquad (18.3.20)$$

Or, because of Eq. (18.2.20), we can write Eq. (18.3.20) as

$$gQ\partial e_1 = U_n \qquad (18.3.21)$$

Because of Eqs. (18.2.23) and (18.2.24), we can write

$$g\partial_1 Qe_1 = U_n \qquad (18.3.22)$$

Because of Eq. (18.2.13), we can write Eq. (18.3.22) as

$$g\partial_1 Q_{.1} = U_n \qquad (18.3.23)$$

Because of Eq. (18.3.19), we can write Eq. (18.3.23) as

$$\partial_1 V_{n-1} = U_n \qquad (18.3.24)$$

But remember that according to our plan of computations, the largest element in each V vector is always unity. But if the largest element of V_{n-1} is unity, and U_n as given in Eq. (18.3.24) is simply the V vector multiplied by ∂_1, then the largest element U_{Ln} in U_n is ∂_1. Therefore, Eq. (18.3.24) shows that when a given U vector is proportional to the preceding V vector, the largest element in the U vector is the first element in the basic diagonal.

18.4 Successive Q Vectors and δ Elements

The first residual matrix

So far we have obtained the first column vector of the right orthonormal of X and the first element in the basic diagonal of the product moment, which, you will remember, is the square of the first element of the basic diagonal of X. Our next problem is to get the second vector of the orthonormal and the second element of the basic diagonal. We may write the basic structure of S_0 as a sum of major products of vectors, as in

$$S_0 = \partial_1 Q_{.1}Q'_{.1} + \partial_2 Q_{.2}Q'_{.2} + \ldots + \partial_n Q_{.n}Q'_{.n} \qquad (18.4.1)$$

Assuming now that we have already solved for $Q_{.1}$ and ∂_1, let us calculate a residual matrix given by

$$S_1 = S_0 - \partial_1 Q_{.1}Q'_{.1} \qquad (18.4.2)$$

From Eqs. (18.4.1) and (18.4.2), we have

$$S_1 = \partial_2 Q_{.2} Q'_{.2} + \ldots + \partial_n Q_{.n} Q'_{.n} \qquad (18.4.3)$$

We should note in passing that the rank of S_1 is one less than the rank of S_0. This is true because as you can see on the right of Eq. (18.4.2), the basic orthonormal of S_1 has one less column in it than the one in Eq. (18.4.1).

The second Q vector and δ element

We see, therefore, that to get S_1 we simply subtract from S_0 a matrix that is the major product moment of $Q_{.1}$ multiplied by ∂_1. Having calculated S_1, we then proceed to find $Q_{.2}$ and ∂_2 in exactly the same way as we found $Q_{.1}$ and ∂_1. We use Eqs. (18.3.7) through (18.3.17) to calculate $Q_{.2}$ except that now we use S_1 instead of S_0. To get ∂_2, we simply take the largest element in the final U vector.

Other residual matrices, Q vectors, and δ's

Having calculated $Q_{.2}$ and ∂_2, we then calculate a second residual matrix, as in

$$S_2 = S_1 - \partial_2 Q_{.2} Q'_{.2} \qquad (18.4.4)$$

To get $Q_{.3}$ and ∂_3 we proceed with S_2 exactly as we did with S_1 and S_0.

It should be noted, however, that for calculating $Q_{.2}$ and subsequent vectors, it is usually better not to start with V_0 as a unit vector. Ordinarily, in the residual matrices, about half the values will be plus and half minus. A good rule of thumb procedure is to find the column with the two largest absolute values in a nondiagonal position. Having found this column, note the sign of each element in the column. Then construct a sign vector giving the minus sign to the element corresponding to those elements which have minus signs in the selected column. This sign vector will then be the V_0 for any particular residual matrix S_i.

Ordinarily, for any matrix S_0 derived from a data matrix, you will have as many Q vectors and ∂ elements to solve for as there are columns in the data matrix. In practice, however, only the Q vectors corresponding to a relatively small number of the largest $\delta's$ will be required.

18.5   Finding the Left Orthonormal

You have already seen in Section 18.1 how the left orthonormal may be obtained by means of Eq. (18.1.8), assuming that the right orthonormal is available. In the procedures we have described, you will notice, however, that the diagonal elements of the basic diagonal are obtained simultaneously with the vectors of the right orthonormal. The procedure for computing the left orthonormal would therefore be as follows: first we get the square root of the basic diagonal ∂ of the product moment matrix that gives the basic diagonal Δ of the matrix X. This is given by

$$\Delta = \partial^{1/2} \qquad (18.5.1)$$

Next we get the reciprocal of this matrix and postmultiply Q by it. The product is premultiplied by X to get P. The complete set of operations for computing P is given by

$$X(Q\partial^{-1/2}) = P \qquad (18.5.2)$$

18.6 Computations for the Basic Structure

The product moment matrix

In getting the basic structure of a matrix X, we begin as we did for the correlation matrix. We calculate the minor product moment matrix of X exactly as we did in calculating the correlation matrix. In these computations, we have a sum of row and column elements as a result of our checking operations. We start, then, with the product moment matrix, at the bottom of which will be a row vector of column sums. Figure 18.6.1 shows the worksheet layout for getting the successive U and V vectors.

	$U_1 =$	$V_1 =$	$U_2 =$	$V_2 =$	$U_3 =$	$V_3 =$
S_0	$S_0 V_0$	$\dfrac{U_1}{U_{1L}}$	$S_0 V_1$	$\dfrac{U_2}{U_{2L}}$	$S_0 V_2$	$\dfrac{U_3}{U_{3L}}$
$1'S_0$	$(1'S_0)V_0$	$\dfrac{(1'S_0)V_0}{U_{1L}}$	$(1'S_0)V_1$	$\dfrac{(1'S_0)V_1}{U_{2L}}$	$(1'S_0)V_2$	$\dfrac{(1'S_0)V_2}{U_{3L}}$
	$1'U_1$	$1'V_1$	$1'U_2$	$1'V_2$	$1'U_3$	$1'V_3$

FIG. 18.6.1

The successive U and V vectors

In Fig. 18.6.1, the U_1 column is obtained by multiplying both the S_0 matrix and its sum vector by V_0. The last entry in the U_1 column is obtained merely by summing all of the U_1 elements, and should be equal to the entry immediately above it. The V_1 column is obtained by multiplying each element of U_1 by the reciprocal of the largest absolute value in the U_1 column. This multiplication also includes the element immediately below the V_1 vector. The final element in the column is obtained merely by summing the V_1 elements. This should equal the value immediately above it. The remaining U and V vectors are calculated in the same way. These operations continue until two successive V columns are equal.

Ordinarily, it is sufficient to carry the computations to three decimal places. Actually, for the first few approximations, a smaller number of decimals may be carried. Sometimes it is advisable to carry only one decimal until two successive V columns are equal, then two decimals until two successive V columns are equal, and finally three decimals until two successive V columns are equal. This procedure cuts down somewhat on the computational labor.

The $Q_{.1}$ vector

The next step is to calculate the Q vector from the final V vector. This we do by the operation indicated in Fig. 18.6.2.

	1	2	3
	V_n	$Q_1 =$ $\dfrac{V_n}{\sqrt{V'_n V_n}}$	$w =$ $Q_{.1}\sqrt{U_{nL}}$
a	$\dfrac{(1'S_0)V_{n-1}}{U_{nL}}$	$\dfrac{1'V_n}{\sqrt{V'_n V_n}}$	$(1'Q_{.1})\sqrt{U_{nL}}$
b	$1'V_n$	$1'Q_{.1}$	$1'w$
c	$V'_n V_n$	$Q'_{.1}Q_{.1} = 1$	$w'w = U_{nL}$
d	$\sqrt{V'_n V_n}$	$\sqrt{U_{nL}}$	
e	$\dfrac{1}{\sqrt{V'_n V_n}}$	$(\sqrt{U_{nL}})^2 = U_{nL}$	

Fig. 18.6.2

The first column in Fig. 18.6.2 is simply the final V vector. The rows at the bottom of vectors 1, 2, and 3 are labeled from a to e inclusively to facilitate explanation of what the various steps require.

In column 1 the entries in rows a and b are simply the check entries from Fig. 18.6.1. The entry in row c of column 1 is the sum of the squares of the elements in the V vector. The entry in row d is the square root of the entry in row c. The entry in row e is the reciprocal of the entry in row d.

Column 2 above and including row a is obtained by multiplying each element of column 1 above the row a by the entry in row e of column 1. The entry in row b of column 2 is obtained by summing the entries in column 2 above row a. The entry in row b of column 2 should equal the entry in row a of column 2. The entry in row c of column 2 is obtained by taking the sum of squares of elements in column 2 above row a. This entry should be equal to unity. The entry in row d of column 2 is the square root of the largest absolute value of U_n. The entry in row e of column 2 is the square of the entry in row d of column 2 and should be equal to the largest absolute value in U_n. This entry merely checks the square root operation.

The w vector

Column 3 above row **a** is obtained by multiplying every entry in column 2 above and including row **a** by the entry in row **d** of column 2. The entry in row **b** of column 3 is obtained by summing the elements in column 3 above row **a**. This entry should equal the entry in row **a** of column 3. The entry in row **c** of column 3 is the sum of the squares of the entries in column 3 above row **a**, and should equal the largest absolute value in the final U_n vector. This should give back the first element of the basic diagonal.

The residual matrix

The major product moment of the w vector in column 3 of Fig. 18.6.2 will be equal to the major product moment of the $Q_{.1}$ vector multiplied by the first element of the basic diagonal. This is the matrix we want to subtract from S_0 in order to get our first residual matrix, S_1. Figure 18.6.3 shows how we may perform these operations.

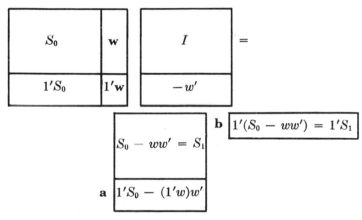

FIG. 18.6.3

We start with the product moment matrix S_0, which has been bordered at the bottom by a vector of column sums. On the right of this matrix we attach with a paper clip the w vector with its sum. This matrix is then postmultiplied by the identity matrix, bordered at the bottom by the w vector in row form with the sign of each of its elements reversed. This product is the first residual matrix S_1. The **a** row of the product is obtained merely by multiplying the sum vector at the bottom of the S_0 matrix by the bordered identity matrix on the right. Row **b** at the bottom of the product matrix is obtained by summing all of the columns above row **a**. Rows **a** and **b** should be equal. It is, of course, possible to calculate first the ww' matrix and then subtract this from the S matrix. The appropriate associative and distributive checks should be carried out.

Successive Q vectors

The operation shown in Fig. 18.6.3 gives the first residual matrix S_1 together with the summation row at the bottom. You now proceed with the S_1 matrix in exactly the same manner as you did with the S_0 matrix to get the second column vector $Q_{.2}$ of the basic orthonormal and the second element ∂_2 of the basic diagonal. The process is continued until you have the desired number of Q vectors and δ elements.

The left orthonormal

The first step in getting the left orthonormal is to calculate the reciprocals of the square roots of the basic diagonal elements. This means that you find the reciprocals of the values in row **d** of column **2** for each worksheet illustrated in Fig. 18.6.2. You will, of course, have a worksheet of this kind for each vector of Q. These reciprocals are entered in row **c** of the worksheet in Fig. 18.6.4.

	1	2		n
	$Y_{.1} =$	$Y_{.2} =$		$Y_{.n} =$
	$\dfrac{Q_{.1}}{\sqrt{\partial_1}}$	$\dfrac{Q_{.2}}{\sqrt{\partial_2}}$	etc.	$\dfrac{Q_{.n}}{\sqrt{\partial_n}}$
a	$\dfrac{1'Q_{.1}}{\sqrt{\partial_1}}$	$\dfrac{1'Q_{.2}}{\sqrt{\partial_2}}$	etc.	$\dfrac{1'Q_{.n}}{\sqrt{\partial_n}}$
b	$1'Y_{.1}$	$1'Y_{.2}$	etc.	$1'Y_{.n}$
c	$\dfrac{1}{\sqrt{\partial_1}}$	$\dfrac{1}{\sqrt{\partial_2}}$	etc.	$\dfrac{1}{\sqrt{\partial_n}}$
d	$\dfrac{\partial_1}{\sqrt{\partial_1}} = \sqrt{\partial_1}$	$\dfrac{\partial_2}{\sqrt{\partial_2}} = \sqrt{\partial_2}$	etc.	$\dfrac{\partial_n}{\sqrt{\partial_n}} = \sqrt{\partial_n}$

FIG. 18.6.4

The entries in row **c** are made before any other entries in this work sheet. You will notice now that we use the symbol ∂ instead of U_{nL}, since this is actually what these values are. It is very important that our calculations be correct; therefore we use row **d** as a check column for the reciprocals. To get each element in row **d**, we multiply the corresponding element in row **c** by its U_{nL} value, which you recall is the largest absolute value in the final U vector for a given cycle. Once the elements in row **c** of Fig. 18.6.4 have been checked by row **d**, we are ready to calculate the column entries above row **b**.

The entries in column 1 down through row **a** are obtained by multiplying the corresponding entries of column 2, Fig. 18.6.2, by the entry in Fig. 18.6.4, column 1, row **c**. To get the entry in row **b** of column 1 we simply add the elements in column 1 above row **a**. The entry in row **b** of column 1 should equal the entry in row **a** of column **1**. The remaining columns of Fig. 18.6.4 are obtained in the same way using the appropriate Fig. 18.6.2 worksheet. In this way we get a matrix of Y vectors that is equal to Q postmultiplied by the $-1/2$ power of the basic diagonal.

Figure 18.6.5 shows how from the data matrix X and the matrix Y obtained in Fig. 18.6.4 we finally get P, or the left orthonormal. Starting with the primary matrix X bordered at the bottom by a vector of column sums, we postmultiply this by the matrix Y. The **a** row in the product matrix is simply the row at the bottom of X postmultiplied by Y. The **b** row represents the column sums in the product matrix down to but not including the **a** row. The **b** row in the product should equal the **a** row. The product XY is the P matrix. As a final check, you should calculate the minor product moment of P to see that it does yield the identity matrix within limits of decimal accuracy.

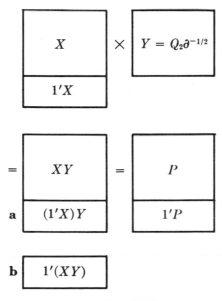

FIG. 18.6.5

18.7 The Basic Structure for Standard Measures

In most applications of basic structure analysis it is customary to find the Q and δ matrices of the correlation matrix rather than the minor product moment of the data matrix X. This means that instead of the basic structure of X we are concerned with the basic structure of a matrix Z of standard measures. This is a matrix obtained as follows:

$$x = X - 1\left(\frac{1'X}{N}\right) \tag{18.7.1}$$

$$Z = xD_{x'x}^{-1/2} \tag{18.7.2}$$

From Chapter 13, we know that

$$r = \frac{Z'Z}{N} \qquad (18.7.3)$$

We have then

$$r = Q\delta Q' \qquad (18.7.4)$$

From Eq. (18.7.4) we find Q and δ precisely as outlined for S_0.

$$\frac{Z}{\sqrt{N}} = P\delta^{1/2}Q' \qquad (18.7.5)$$

From (18.7.5)

$$\frac{Z}{\sqrt{N}}(Q\delta^{-1/2}) = P \qquad (18.7.6)$$

Actually, of course, we need not calculate Z as such. We get $D_{x'x}$ in the process of calculating the correlation matrix. Therefore from Eqs. (18.7.1), (18.7.2), and (18.7.6), we have

$$\left[X - 1\left(\frac{1'X}{N}\right) \right] \frac{D_{x'x}^{-1/2}}{\sqrt{N}} Q\delta^{1/2} = P \qquad (18.7.7)$$

We let

$$d = (ND_{x'x})^{-1/2} \qquad (18.7.8)$$

$$Y = Q\delta^{-1/2} \qquad (18.7.9)$$

Then we calculate

$$dY = M \qquad (18.7.10)$$

$$\left(\frac{1'X}{N}\right)M = v' \qquad (18.7.11)$$

From Eq. (18.7.7) through Eq. (18.7.11) we have

$$P = XM - 1v' \qquad (18.7.12)$$

It can be readily shown that Eqs. (18.7.10), (18.7.11), and (18.7.12) are much more economical than Eqs. (18.7.1), (18.7.2), and (18.7.6) for calculating P, particularly if the number of entities is much greater than the number of attributes.

SUMMARY

1. General solution for the basic structure:
   a. The basic structure of a data matrix is of great practical importance in the social and biological sciences.

b. If the right orthonormal Q' of X were known, the basic diagonal would be

$$\Delta = (Q'x'XQ)^{1/2}$$

and the left orthonormal would be

$$P = XQ\Delta^{-1}$$

2. Finding the right orthonormal of a matrix:
   a. The basic structure of the minor product moment is

$$X'X = Q\delta Q'$$

   b. The basic structure of a power of a product moment is

$$[X'X]^n = Q\delta^n Q'$$

   c. If n is sufficiently large and K is appropriately chosen, we can write

$$\frac{1}{K}(X'X)^n = \frac{\delta_1^n}{K} Q_{.1} Q'_{.1}$$

   d. If V is any vector not orthogonal to $Q_{.1}$ we have

$$\frac{1}{K}(X'X)^n V = \left(\frac{\delta_1^n}{K}Q'_{.1}\right)Q_{.1} = Y$$

   e. The solution for $Q_{.1}$ is

$$Q_{.1} = \frac{Y}{\sqrt{Y'Y}}$$

   f. The solutions for δ_1 is

$$\delta_1 = Q'_{.1}(X'X)Q_{.1}$$

3. Iterative solution for the first vector of Q:
   a. The product of any power n of $X'X = S_0$ by an arbitrary vector can be obtained by successive multiplication of the form

$$S_0 V_{i-1} = V_i = (X'X)^i V_0$$

   b. To keep the V_i vectors in hand, we have

$$S_0 V_{i-1} = U_i$$

$$V_i = \frac{U_i}{U_{iL}}$$

   where U_{iL} is the element of largest absolute value in U_i.
   c. The $Q_{.1}$ vector is calculated when $U_{n-1} = U_n$ to any desired degree of decimal accuracy by

$$Q_{.1} \doteq \frac{U_n}{\sqrt{U'_n U_n}}$$

d. The δ_1 element is simply U_{Ln}, the largest element in U_n.

4. Successive Q vectors and δ elements:
   a. The first residual matrix δ_1 is given by

$$S_1 = S_0 - \delta_1 Q_{.1} Q'_{.1}$$

   b. $Q_{.2}$ and δ_2 are obtained from S_1 as in 3c and 3d above.
   c. In general

$$S_i = S_{i-1} - \delta_i Q_{.i} Q'_{.i}$$

   and the solutions for $Q_{.i}$ and δ_i are as in 3c and d above:

5. The left orthonormal of X is given by

$$P = XQ\delta^{-1/2}$$

6. Computations for the basic structure proceed through the following steps:
   a. The product moment matrix $X'X$.
   b. The successive U and V vectors for $Q_{.1}$.
   c. The $Q_{.1}$ vector.
   d. The $w_{.1}$ vector.
   e. The residual matrix S_1.
   f. The successive $Q_{.i}$ and $w_{.i}$ vectors and S_i residual matrices.

7. Basic structure for standard measures:
   a. The correlation matrix r is computed.
   b. The Q and δ matrices are computed from

$$r = Q\delta Q'$$

   c. The P matrix is computed from

$$P = XM - 1v'$$

   where

$$M = (ND_{X'X})^{-1/2} Q\delta^{-1/2}$$

$$v' = \left(\frac{1'X}{N}\right)M$$

Part IV

Matrix Solutions

Chapter 19

The Inverse of a Matrix

19.1 The General Inverse and the Basic Structure

You have already seen in previous chapters how we define matrix addition, subtraction, and multiplication. You have seen that matrix addition and subtraction resemble very closely scalar addition and subtraction. You have also seen that matrix multiplication is a little more complicated than scalar multiplication, which is in fact a special case of matrix multiplication.

The inverse defined

We may now define an operation in matrix algebra that is analogous to division in scalar algebra. You remember that in scalar algebra, when we divide a dividend by a divisor this amounts to dividing 1 by the divisor, or taking the reciprocal of the divisor and then multiplying the dividend by the reciprocal. For example, if we want to divide 3 by 2, we can first divide 1 by 2 to get .5 and then multiply 3 by .5. In scalar algebra, if we multiply a number by its reciprocal, we get 1. Therefore, we consider for a given matrix whether we can find another such that if the two are multiplied together, we get the identity matrix, which as you know, takes the place of 1 or unity in matrix algebra. If, for example, we have a matrix a and can find another matrix b such that the product of the two yields the identity matrix, then we call b the inverse of a and a the inverse of b.

The general inverse and the basic structure

In order to clarify the definition of inverses, we shall start with the basic structure of the matrix. Although it is not common among mathematicians to treat the subject of inverses from this point of view, matrix inversion will be clearer to you in terms of the basic structure. Let us start with the basic structure of a nonhorizontal matrix X, as given by

$$X = P\Delta Q'$$ (19.1.1)

You recall that P and Q are the orthonormals of X, and Δ is its basic diagonal. Now consider the matrix Y as given by

$$Y = P\Delta^{-1}Q' \qquad (19.1.2)$$

You will notice that the right sides of Eqs. (19.1.1) and (19.1.2) are the same except that the basic diagonal of Eq. (19.1.2) is the inverse of the basic diagonal of Eq. (19.1.1). The orthonormals of X and Y are the same. Next consider the transpose of Eq. (19.1.2) given by

$$Y' = Q\Delta^{-1}P' \qquad (19.1.3)$$

You recall that so far we have placed no restrictions on X. It need not be square or basic. If it is not basic, then Q' will be a horizontal matrix.

We shall now give a most unusual but extremely useful definition. Matrix Y' as given in Eq. (19.1.3) will be called the general inverse of X as given in Eq. (19.1.1). We therefore define the general inverse of any matrix as the transpose of another matrix whose basic orthonormals are the same as the original, and whose basic diagonal is the inverse of that of the original. This is not the traditional mathematical definition of an inverse of a matrix. The traditional definition, as you will see, is in fact a special case of this definition.

The major product of a matrix and its general inverse

Let us see now what happens if we postmultiply the matrix X as given in Eq. (19.1.1) by its general inverse as given in Eq. (19.1.3), as in

$$XY' = P\Delta Q'Q\Delta^{-1}P' \qquad (19.1.4)$$

But the right side of this equation includes the minor product moment of an orthonormal matrix, so we get

$$XY' = P\Delta\Delta^{-1}P' \qquad (19.1.5)$$

Or, from this equation, since we have a diagonal matrix multiplied by its inverse, we have

$$XY' = PP' \qquad (19.1.6)$$

We see then by Eq. (19.1.6) that postmultiplication of a matrix by its general inverse yields the major product moment of an orthonormal matrix. Since the right side of the equation is a product moment matrix, it is symmetrical. Therefore a matrix postmultiplied by its general inverse yields a symmetric matrix, as indicated by

$$XY' = YX' \qquad (19.1.7)$$

We also know that if the major product moment of an orthonormal matrix is raised to any positive integral power, that power is equal to the original product moment. Therefore, if n is any positive integer, we have from Eq. (19.1.6)

$$(XY')^n = XY' \tag{19.1.8}$$

$$(YX')^n = YX' \tag{19.1.9}$$

The minor product of a matrix and its general inverse

Next let us see what happens if we premultiply the matrix X by its general inverse. From Eqs. (19.1.1) and (19.1.3) we have

$$Y'X = Q\Delta^{-1}(P'P)\Delta Q' \tag{19.1.10}$$

Or from Eq. (19.1.10) we get

$$Y'X = Q(\Delta^{-1}\Delta)Q' \tag{19.1.11}$$

And from Eq. (19.1.11) we get

$$Y'X = QQ' \tag{19.1.12}$$

We see then from Eq. (19.1.12) that premultiplication of a matrix by its general inverse yields the major product moment of its right orthonormal. Postmultiplication, as you saw by Eq. (19.1.6), yielded the major product moment of the left orthonormal. We see therefore, also, that if we raise the product of X premultiplied by the general inverse to any positive integral power, this power will be the same as the original product. This rule is given by

$$(Y'X)^n = Y'X \tag{19.1.13}$$

We should note in passing that the general inverse of the transpose of a matrix is equal to the transpose of the general inverse. This you can easily see from Eqs. (19.1.1), (19.1.2), and (19.1.3). If you take the transpose of Eq. (19.1.1), then the right side of Eq. (19.1.2) is the general inverse of this transpose. This, of course, is the transpose of Eq. (19.1.3).

The general inverse of a basic matrix

So far we have made no assumptions about the rank of the matrix X. Let us assume now that X is a basic matrix. If X is a basic vertical matrix, then its right orthonormal is square. But either product moment of a square orthonormal is the identity matrix, as you already know. Therefore if X is basic, then the right side of Eq. (19.1.12) yields the identity matrix. You see then that if a basic vertical matrix is premultiplied by its general inverse, the product is the identity matrix as indicated in

$$Y'X = I \tag{10.1.14}$$

It is most important to remember this rule since it is the foundation of scientific prediction studies.

19.2 The Regular Inverse of a Matrix

But notice now that if X is both square and basic, then both its right
and left orthonormals must be square, so the right side of Eq. (19.1.6)
must also yield the identity matrix. We therefore have the rule that if X
is both basic and square, either pre- or postmultiplication by its general
inverse yields the identity matrix. For this particular case we call Y' the
regular inverse of X. The rule is illustrated by

$$XY' = I \tag{19.2.1}$$

$$Y'X = I \tag{19.2.2}$$

But if X is both square and basic we normally indicate its regular inverse
by X^{-1} so that Eqs. (19.2.1) and (19.2.2) would be written as

$$XX^{-1} = I \tag{19.2.3}$$

$$X^{-1}X = I \tag{19.2.4}$$

Therefore, for X both basic and square, we write the inverse from Eq.
(19.1.3) as in

$$X^{-1} = Q\Delta^{-1}P' \tag{19.2.5}$$

From Eqs. (19.2.3) and (19.2.4), a square basic matrix is commutative
with its inverse, as shown in

$$XX^{-1} = X^{-1}X = I \tag{19.2.6}$$

It is important to remember then that only square basic matrices have
regular inverses. Frequently we shall drop the term regular and refer to
the general inverse of a square basic matrix simply as its inverse. As a
matter of fact, mathematicians traditionally do not distinguish between
different kinds of inverses. They refer to the regular inverse of a matrix
simply as its inverse. They call a square basic matrix a nonsingular matrix
and all other matrices they call singular matrices. In general, they say
that only nonsingular matrices have inverses. However, this limited use of
the term inverse greatly restricts its usefulness in the human sciences.

19.3 The Inverse of the Product of Square Basic Matrices

We may now prove an important rule about the inverse of the product of
square basic matrices. We begin by considering the product of four basic
matrices a, b, c, and d as in

$$X = abcd \tag{19.3.1}$$

Let us now premultiply both sides of Eq. (19.3.1) by the inverse of a, as in

$$a^{-1}X = (a^{-1}a)bcd \tag{19.3.2}$$

But by definition the term in parentheses on the right side here is the identity matrix, so we write

$$a^{-1}X = bcd \qquad (19.3.3)$$

Next we premultiply this by the inverse of b. This gives

$$b^{-1}a^{-1}X = (b^{-1}b)cd \qquad (19.3.4)$$

But from this we have

$$b^{-1}a^{-1}X = cd \qquad (19.3.5)$$

Then we premultiply both sides of this equation by c^{-1} to get

$$c^{-1}b^{-1}a^{-1}X = d \qquad (19.3.6)$$

Finally we premultiply this by the inverse of d to get

$$d^{-1}c^{-1}b^{-1}a^{-1}X = I \qquad (19.3.7)$$

Now we postmultiply both sides of this by the inverse of X to get

$$d^{-1}c^{-1}b^{-1}a^{-1}XX^{-1} = X^{-1} \qquad (19.3.8)$$

Or, from this, we have

$$d^{-1}c^{-1}b^{-1}a^{-1} = X^{-1} \qquad (19.3.9)$$

But from Eq. (19.3.1) we get

$$X^{-1} = (abcd)^{-1} \qquad (19.3.10)$$

Finally from Eqs. (19.3.10) and (19.3.9), we have

$$(abcd)^{-1} = d^{-1}c^{-1}b^{-1}a^{-1} \qquad (19.3.11)$$

Now we could have used this same procedure no matter how many factors we had in the product X. Therefore, Eq. (19.3.11) shows that the inverse of a product of square basic matrices is equal to the product of the inverses in reverse order.

According to Eq. (19.3.11), it is also clear that the product of a number of matrices cannot have a regular inverse unless each of the factors has a regular inverse. This must be true because, as is shown on the left of Eq. (19.3.11), the inverse of the products involved the regular inverse of each of the individual factors.

19.4   Inverses of Special Types of Matrices

As you will see in the next chapter, the calculation of the inverse of a matrix is considerably more complicated than the operation of subtraction, addition, and even multiplication. There are, however, certain special types of matrices whose inverses are very easy to find.

Diagonal matrices

The inverse of a diagonal matrix is one you encountered very early in this book. As you recall, the inverse of a diagonal matrix is simply another diagonal matrix whose elements are the reciprocals of the diagonal elements of the original matrix.

Orthonormal matrices

Since the regular inverse of a matrix is simply a matrix such that when it is used as a pre- or postmultiplier, the product is the identity matrix, you know at once that the inverse of a square orthonormal is the same as the transpose of a square orthonormal. If Q is a square orthonormal, this rule is

$$Q^{-1} = Q' \tag{19.4.1}$$

Equation (19.4.1), of course, refers to the regular inverse of a square orthonormal. However, we can be perfectly consistent with our definition of the general inverse of an orthonormal matrix if we regard both the basic diagonal and the left orthonormal of a general orthonormal matrix as the identity matrix. For example, if we go back to Eqs. (19.1.1) and (19.1.3) and take the special case where Δ and P are identity matrices, then we have

$$X = PII \tag{19.4.2}$$

$$Y' = IIP' \tag{19.4.3}$$

Or simply

$$X = P \tag{19.4.4}$$

$$Y' = P' \tag{19.4.5}$$

We may then say that the general inverse of any orthonormal matrix is its transpose, and if the orthonormal is square, then the general inverse becomes a regular inverse.

You should also remember that a permutation matrix is a special case of a square orthonormal matrix, and therefore its inverse is equal to its transpose, as shown in

$$\pi' = \pi^{-1} \tag{19.4.6}$$

Next let us consider several kinds of matrices whose inverses are equal to the matrix itself. Quite obviously, the inverse of the identity matrix is the identity matrix. Also the inverse of a sign matrix is the sign matrix itself. But we have a more interesting case of the general inverse of a symmetric nonbasic matrix. If the matrix is the major product moment of a nonsquare orthonormal, then its general inverse may be defined as the matrix itself. We know that the right orthonormal of a symmetric matrix is the transpose of the left orthonormal. Suppose we return to Eq. (19.1.1), where X is symmetric. Then assuming also that the basic diagonal is an identity matrix, the basic structure of X is

$$X = QIQ' \qquad (19.4.7)$$

Now notice that Eq. (19.1.3) becomes

$$Y' = (QI^{-1}Q')' \qquad (19.4.8)$$

But this is simply

$$Y' = QIQ' \qquad (19.4.9)$$

which is the same as (19.4.7). Therefore the general inverse of the major product moment of an orthonormal is that major product moment itself.

Orthogonal matrices

Now let us consider the general inverse of a general orthogonal matrix. We recall that an orthogonal matrix P is defined as

$$P'P = D \qquad (19.4.10)$$

where the right side is a diagonal matrix. But we have seen previously that an orthogonal matrix may be expressed as the product of an orthonormal and a diagonal matrix. This is indicated by

$$P = QD^{1/2} \qquad (19.4.11)$$

In Eqs. (19.4.11) and (19.4.10), D is the same diagonal matrix. But we may regard the right side of Eq. (19.4.11) as the basic structure of P where the right orthonormal is the identity matrix. Therefore, we may write the general inverse of P as in

$$Y' = D^{-1/2}Q' \qquad (19.4.12)$$

From Eq. (19.4.11) we get

$$PD^{-1/2} = Q \qquad (19.4.13)$$

Substituting the left side of Eq. (19.4.13) into the right side of Eq. (19.4.12), we get

$$Y' = D^{-1}P' \qquad (19.4.14)$$

Therefore, according to Eq. (19.4.14), we see that the general inverse of an orthogonal matrix is obtained by premultiplying the transpose of that matrix by the inverse of its minor product moment. If P is square, then Eq. (19.4.14) becomes the regular inverse. It should be remembered in passing that for a square orthogonal matrix, the natural order is defined so that the minor product moment yields the diagonal matrix. This means that the column vectors are mutually orthogonal. You also recall that for any matrix, if the minor product moment is a diagonal matrix other than the identity, then the major product moment cannot be a diagonal, even though P is square. It is clear that if we premultiply P by the right side of Eq. (19.4.14), we get

$$D^{-1}P'P = D^{-1}D = I \qquad (19.4.15)$$

If we postmultiply P by the right side of Eq. (19.4.14), we get

$$PD^{-1}P' = QD^{1/2}D^{-1}D^{1/2}Q' \qquad (19.4.16)$$

The right side of Eq. (19.4.16) comes from Eq. (19.4.11). The diagonal factors on the right of Eq. (19.4.16) cancel each other, so we have

$$PD^{-1}P' = QQ' \qquad (19.4.17)$$

If P is not square, then postmultiplication of P by its general inverse gives the major product moment of an orthonormal, as in Eq. (19.4.17). If it is square, then the orthonormal must be square, and the right side of Eq. (19.4.17) becomes an identity matrix.

19.5 How to Solve for the Inverse of a Triangular Matrix

Since all triangular matrices are square and basic, provided no diagonal element is 0, they all have regular inverses. It is clear from Eqs. (19.1.1) and (19.1.3) that we can always get the general inverse of any matrix simply by solving for its basic structure. In particular, we could get the regular inverse of a square basic matrix by solving for its basic structure and then getting the inverse of its basic diagonal. This procedure, however, as you have seen in Chapter 18, can become quite laborious if the dimensions of X are large. We shall see, however, that if we know how to solve for the inverse of a triangular matrix, we shall be able to solve for the general inverse of any matrix, and the regular inverse of square basic matrices in particular, with considerably less labor than is required for finding the basic structure.

The matrix equations

First we shall consider a general procedure for finding the inverse of a triangular matrix. We let T' be an upper triangular matrix and D_T a matrix of its diagonal elements. We define the matrix T' therefore as

$$T' = D_T + t' \qquad (19.5.1)$$

where t' on the right is the matrix of the elements of T' exclusive of the diagonal terms. Therefore the diagonal elements of t' are all 0. Next we define a matrix B as the inverse of T', as in

$$B = T'^{-1} \qquad (19.5.2)$$

Postmultiplying both sides of Eq. (19.5.1) by the corresponding sides of Eq. (19.5.2), we get

$$(D_T + t')B = I \tag{19.5.3}$$

From this we have

$$D_T B + t'B = I \tag{19.5.4}$$

Premultiplying both sides of this by D_T^{-1} we have

$$B + D_T^{-1}t'B = D_T^{-1} \tag{19.5.5}$$

From this we get

$$-D_T^{-1}t'B + D_T^{-1} = B \tag{19.5.6}$$

But Eq. (19.5.6) may be expressed as the product of supervectors, as in

$$[-D_T^{-1}t' \mid D_T^{-1}] \begin{bmatrix} B \\ - \\ I \end{bmatrix} = B \tag{19.5.7}$$

Suppose now we define the elements in the left factor of the left side of Eq. (19.5.7) by

$$-D_T^{-1}t' = v \tag{19.5.8}$$

and

$$D_T^{-1} = D_u \tag{19.5.9}$$

Substituting Eqs. (19.5.8) and (19.5.9) in Eq. (19.5.7) gives

$$[v \mid D_u] \begin{bmatrix} B \\ - \\ I \end{bmatrix} = B \tag{19.5.10}$$

The expanded equation

In order to see how Eq. (19.5.10) enables us to solve for B or the inverse of T', let us write it out in expanded notation for the case of four variables as in Eq. (19.5.11).

$$\tag{19.5.11}$$

$$\begin{bmatrix} 0 & v_{12} & v_{13} & v_{14} & u_1 & 0 & 0 & 0 \\ 0 & 0 & v_{23} & v_{24} & 0 & u_2 & 0 & 0 \\ 0 & 0 & 0 & v_{34} & 0 & 0 & u_3 & 0 \\ 0 & 0 & 0 & 0 & 0 & 0 & 0 & u_4 \end{bmatrix} \begin{bmatrix} B_{11} & B_{12} & B_{13} & B_{14} \\ B_{21} & B_{22} & B_{23} & B_{24} \\ B_{31} & B_{32} & B_{33} & B_{34} \\ B_{41} & B_{42} & B_{43} & B_{44} \\ \hline 1 & 0 & 0 & 0 \\ 0 & 1 & 0 & 0 \\ 0 & 0 & 1 & 0 \\ 0 & 0 & 0 & 1 \end{bmatrix} = \begin{bmatrix} B_{11} & B_{12} & B_{13} & B_{14} \\ 0 & B_{22} & B_{23} & B_{24} \\ 0 & 0 & B_{33} & B_{34} \\ 0 & 0 & 0 & B_{44} \end{bmatrix}$$

Now remember that only the v's and u's in Eq. (19.5.11) are known. It is the B's we wish to solve for. We shall see, however, that even though the B's are involved on the left side of Eq. (19.5.11), we need not worry about the fact that they are unknown. We carry out the multiplication on the left of Eq. (19.5.11) somewhat differently than we ordinarily do. First we premultiply the first column of the second factor by the last row of the first factor. We see that this product is 0 and enter the value 0 in the last row, first column, on the right of Eq. (19.5.11). Next we premultiply the first column of the second factor by the next to last or third row of the first factor. Here again, the product is 0, which we enter in the third row, first column, of both of the B matrices. In the same way we find that premultiplication of the first column of the second factor by the second row of the first factor also gives 0, which we enter in the second row and first column of both of the B matrices. However, multiplication of the first column of the second factor by the first row of the first factor does not give 0, but rather u_1, which is the element in the first row and column of the product matrix, namely the element B_{11}.

We now go on to the second column of the postfactor and premultiply it by the last row of the prefactor, which yields 0. Similarly, premultiplication of the second column of the postfactor by the third row of the prefactor gives 0, which we enter in the B matrix. However, premultiplication of the second column of the postfactor by the second row of the prefactor gives u_2 which we write as B_{22} in the product matrix. Finally B_{12} is the product of the first row of the prefactor by the second column of the postfactor which is simply v_{12} times B_{22}. But we have already solved for B_{22}, so we can now solve for B_{12}.

In the same way, we solve for the elements in the third column of the B matrix. You should see then that the method of solution outlined in Eq. (19.5.11) depends on the fact that solving for the first unknown element requires no knowledge of any of the unknowns. Solving for subsequent elements of the B matrix requires a knowledge of only those B elements that have been previously solved for.

By the method of solution indicated in Eq. (19.5.11), we have proved that the inverse of an upper triangular matrix is also an upper triangular matrix. Since the inverse of a transpose is the transpose of an inverse, we see also that the inverse of a lower triangular matrix is a lower triangular matrix.

Numerical example

Let us now take a simple example to show how we solve for the inverse of a triangular matrix. We let T' be the matrix in Fig. 19.5.1.

$$T' \doteq \begin{bmatrix} 2 & 1 & 1 \\ 0 & 3 & 1 \\ 0 & 0 & 2 \end{bmatrix}$$

FIG. 19.5.1

Then because of Eqs. (19.5.8) and (19.5.9), we define the matrix v as in

$$
v = \begin{bmatrix} -\dfrac{1}{2} & 0 & 0 \\[6pt] 0 & -\dfrac{1}{3} & 0 \\[6pt] 0 & 0 & -\dfrac{1}{2} \end{bmatrix} \begin{bmatrix} 0 & 1 & 1 \\ 0 & 0 & 1 \\ 0 & 0 & 0 \end{bmatrix} = \begin{bmatrix} 0 & -\dfrac{1}{2} & -\dfrac{1}{2} \\[6pt] 0 & 0 & -\dfrac{1}{3} \\[6pt] 0 & 0 & 0 \end{bmatrix}
$$

FIG. 19.5.2

Corresponding to Eq. (19.5.11), we have

$$
\begin{bmatrix} 0 & -\dfrac{1}{2} & -\dfrac{1}{2} & \Big| & \dfrac{1}{2} & 0 & 0 \\[6pt] 0 & 0 & -\dfrac{1}{3} & \Big| & 0 & \dfrac{1}{3} & 0 \\[6pt] 0 & 0 & 0 & \Big| & 0 & 0 & \dfrac{1}{2} \end{bmatrix} \begin{bmatrix} \dfrac{1}{2} & -\dfrac{1}{6} & -\dfrac{1}{6} \\[6pt] 0 & \dfrac{1}{3} & -\dfrac{1}{6} \\[6pt] 0 & 0 & \dfrac{1}{2} \\ \hline 1 & 0 & 0 \\ 0 & 1 & 0 \\ 0 & 0 & 1 \end{bmatrix} = \begin{bmatrix} \dfrac{1}{2} & -\dfrac{1}{6} & -\dfrac{1}{6} \\[6pt] 0 & \dfrac{1}{3} & -\dfrac{1}{6} \\[6pt] 0 & 0 & \dfrac{1}{2} \end{bmatrix}
$$

FIG. 19.5.3

Figure 19.5.4 shows that the right side of Fig. 19.5.3 is actually the inverse of Fig. 19.5.1.

$$
\begin{bmatrix} 2 & 1 & 1 \\ 0 & 3 & 1 \\ 0 & 0 & 2 \end{bmatrix} \begin{bmatrix} \dfrac{1}{2} & -\dfrac{1}{6} & -\dfrac{1}{6} \\[6pt] 0 & \dfrac{1}{3} & -\dfrac{1}{6} \\[6pt] 0 & 0 & \dfrac{1}{2} \end{bmatrix} = \begin{bmatrix} 1 & 0 & 0 \\ 0 & 1 & 0 \\ 0 & 0 & 1 \end{bmatrix}
$$

FIG. 19.5.4

19.6 How to Solve for the Inverse of a Basic Square Matrix

The matrix equations

We have seen how to solve for the inverse of a triangular matrix. Now let us see how to use this procedure to solve for the inverse of a basic square matrix. We recall from Chapter 15 that any basic square matrix with rows and columns suitably ordered can be expressed as the product of a lower triangular matrix postmultiplied by an upper triangular matrix. This rule is indicated by

$$X = T_L T_u \tag{19.6.1}$$

We learned in Chapter 16 how to find these triangular factors if a matrix is square and basic, and how to find the partial triangular factors if the matrix is not square or basic. Assume now we have found the triangular factors in Eq. (19.6.1). Then the inverse on both sides is

$$X^{-1} = (T_L T_u)^{-1} \tag{19.6.2}$$

But you learned in Section 19.3 that the inverse of a product is equal to the product of the inverses in reverse order, so the right side of Eq. (19.6.2) may be written as

$$X^{-1} = T_u^{-1} T_L^{-1} \tag{19.6.3}$$

We then find the inverse of X simply by finding its triangular factors, then finding the inverse of these factors, and finally, postmultiplying the inverse of the upper triangular by the inverse of the lower triangular matrix. We shall see, however, that the process can be simplified somewhat, as follows: First let us consider a supermatrix in which the first matrix element is the matrix X and the second the identity matrix I as indicated in Eq. (19.6.4).

$$a = \begin{bmatrix} X \\ - \\ I \end{bmatrix} \tag{19.6.4}$$

Now suppose from Eq. (19.6.1.) and (19.6.4) we write Eq. (19.6.5).

$$a = \begin{bmatrix} T_L T_u \\ - \\ T_u^{-1} T_u \end{bmatrix} \tag{19.6.5}$$

But this can be written as Eq. (19.6.6).

$$a = \begin{bmatrix} T_L \\ - \\ T_u^{-1} \end{bmatrix} T_u \tag{19.6.6}$$

Now if we had started with the matrix a and factored it so that the left side was a lower partial triangular and the right side an upper triangular, the result would be exactly as in Eq. (19.6.6). Therefore, in using the triangular factoring method on the supermatrix in Eq. (19.6.4), we end with a supermatrix as the left partial triangular factor, such that the upper matrix element is the lower triangular factor of X, and the lower element is the inverse of the upper triangular element. We then find the inverse of the upper element T_L, which we already know how to do. Finally using Eq. (19.6.3) we get the product of the two inverses, which is the inverse of the matrix X.

Numerical example

To illustrate the procedure, let us start with the simple matrix a as indicated in Fig. 19.6.1.

$$a = \begin{bmatrix} X \\ \hline I \end{bmatrix} = \left[\begin{array}{ccc} 2 & 4 & 2 \\ 1 & 5 & 10 \\ 1 & 3 & 6 \\ \hline 1 & 0 & 0 \\ 0 & 1 & 0 \\ 0 & 0 & 1 \end{array}\right]$$

FIG. 19.6.1

Using the method described in Section 16.5, we get the first residual matrix as in Fig. 19.6.2.

$$a_1 = \left[\begin{array}{ccc} 2 & 4 & 2 \\ 1 & 5 & 10 \\ 1 & 3 & 6 \\ \hline 1 & 0 & 0 \\ 0 & 1 & 0 \\ 0 & 0 & 1 \end{array}\right] - \left[\begin{array}{c} 2 \\ 1 \\ 1 \\ \hline 1 \\ 0 \\ 0 \end{array}\right] [1 \quad 2 \quad 1] = \left[\begin{array}{ccc} 0 & 0 & 0 \\ 0 & 3 & 9 \\ 0 & 1 & 5 \\ \hline 0 & -2 & -1 \\ 0 & 1 & 0 \\ 0 & 0 & 1 \end{array}\right]$$

FIG. 19.6.2

The second residual matrix is given by Fig. 19.6.3.

$$a_2 = \left[\begin{array}{ccc} 0 & 0 & 0 \\ 0 & 3 & 9 \\ 0 & 1 & 5 \\ \hline 0 & -2 & -1 \\ 0 & 1 & 0 \\ 0 & 0 & 1 \end{array}\right] - \left[\begin{array}{c} 0 \\ 3 \\ 1 \\ \hline -2 \\ 1 \\ 0 \end{array}\right] [0 \quad 1 \quad 3] = \left[\begin{array}{ccc} 0 & 0 & 0 \\ 0 & 0 & 0 \\ 0 & 0 & 2 \\ \hline 0 & 0 & 5 \\ 0 & 0 & -3 \\ 0 & 0 & 1 \end{array}\right]$$

FIG. 19.6.3

Although it is not necessary to calculate the last residual matrix, we do so in Fig. 19.6.4 for the sake of completeness.

$$a_3 = \frac{\begin{bmatrix} 0 & 0 & 0 \\ 0 & 0 & 0 \\ 0 & 0 & 2 \\ \hline 0 & 0 & 5 \\ 0 & 0 & -3 \\ 0 & 0 & 1 \end{bmatrix}}{} - \frac{\begin{bmatrix} 0 \\ 0 \\ 2 \\ \hline 5 \\ -3 \\ 1 \end{bmatrix}}{} \begin{bmatrix} 0 & 0 & 1 \end{bmatrix} = \frac{\begin{bmatrix} 0 & 0 & 0 \\ 0 & 0 & 0 \\ 0 & 0 & 0 \\ \hline 0 & 0 & 0 \\ 0 & 0 & 0 \\ 0 & 0 & 0 \end{bmatrix}}{}$$

FIG. 19.6.4

From the left side of Figs. 19.6.2, 19.6.3, and 19.6.4, we now collect the column vectors for the left lower partial triangular matrix, and the row vectors for the right upper triangular matrix. This we show in Fig. 19.6.5.

$$\begin{bmatrix} T_L \\ \hline T_u^{-1} \end{bmatrix} = \begin{bmatrix} 2 & 0 & 0 \\ 1 & 3 & 0 \\ 1 & 1 & 2 \\ \hline 1 & -2 & 5 \\ 0 & 1 & -3 \\ 0 & 0 & 1 \end{bmatrix} \quad T_u = \begin{bmatrix} 1 & 2 & 1 \\ 0 & 1 & 3 \\ 0 & 0 & 1 \end{bmatrix}$$

FIG. 19.6.5

You will notice that we were foresighted enough to select the matrix X so that its left lower triangular factor is the transpose of the upper triangular matrix in Fig. 19.5.1. We already solved for the inverse of this transpose in Fig. 19.5.3, so we may take the transpose of this solution to get the inverse of T_L simply by taking the transpose as in Fig. 19.6.6.

$$T_L^{-1} = \begin{bmatrix} \frac{1}{2} & 0 & 0 \\ -\frac{1}{6} & \frac{1}{3} & 0 \\ -\frac{1}{6} & -\frac{1}{6} & \frac{1}{2} \end{bmatrix}$$

FIG. 19.6.6

Remember that the inverse of the upper triangular matrix is given in the lower submatrix in Fig. 19.6.5. Going back to Eq. (19.6.2) and using the matrices from Figs. 19.6.5 and 19.6.6, we have the inverse of the matrix X given in Fig. 19.6.7.

$$X^{-1} = T_u^{-1} T_L^{-1} = \begin{bmatrix} 1 & -2 & 5 \\ 0 & 1 & -3 \\ 0 & 0 & 1 \end{bmatrix} \begin{bmatrix} \frac{1}{2} & 0 & 0 \\ -\frac{1}{6} & \frac{1}{3} & 0 \\ \frac{1}{6} & -\frac{1}{6} & \frac{1}{2} \end{bmatrix} = \begin{bmatrix} 0 & -\frac{3}{2} & \frac{5}{2} \\ \frac{1}{3} & \frac{5}{6} & -\frac{3}{2} \\ -\frac{1}{6} & -\frac{1}{6} & \frac{1}{2} \end{bmatrix}$$

$$\qquad\qquad\qquad T_u^{-1} \qquad\qquad\quad T_L^{-1} \qquad\quad = \qquad X^{-1}$$

FIG. 19.6.7

To show that the right side of Fig. 19.6.7 is actually the inverse of X we multiply X as given in the upper submatrix of Fig. 19.6.1 by this inverse, as in Fig. 19.6.8.

$$XX^{-1} = \begin{bmatrix} 2 & 4 & 2 \\ 1 & 5 & 10 \\ 1 & 3 & 6 \end{bmatrix} \begin{bmatrix} 0 & -\frac{3}{2} & \frac{5}{2} \\ \frac{1}{3} & \frac{5}{6} & -\frac{3}{2} \\ -\frac{1}{6} & -\frac{1}{6} & \frac{1}{2} \end{bmatrix} = \begin{bmatrix} 1 & 0 & 0 \\ 0 & 1 & 0 \\ 0 & 0 & 1 \end{bmatrix}$$

$$\qquad\qquad X \qquad\qquad\qquad X^{-1} \qquad\qquad = \qquad I$$

FIG. 19.6.8

19.7 How to Solve for the Inverse of a Basic Symmetric Matrix

The matrix equations

If X happens to be a basic symmetric matrix, then the solution for the inverse is somewhat more simple than it is for a general square basic matrix. If X can be considered as a minor product moment of a basic matrix with real elements, then a triangular factoring of this matrix can result in upper and lower triangular factors that are transposes of one another. For example, if X is a product moment matrix, then Eq. (19.6.1) can be written as

$$X = TT' \qquad\qquad (19.7.1)$$

Then the equation corresponding to Eq. (19.6.5) would be given by Eq. (19.7.2).

$$a = \begin{bmatrix} X \\ \hline I \end{bmatrix} = \begin{bmatrix} TT' \\ \hline T'^{-1}T' \end{bmatrix} \qquad\qquad (19.7.2)$$

But the right side of this can be written as Eq. (19.7.3).

$$a = \begin{bmatrix} T \\ \hline T'^{-1} \end{bmatrix} T' \qquad\qquad (19.7.3)$$

Therefore, if we carry out a triangular-factoring procedure on the super-matrix that has the symmetric product moment as the upper element and the identity matrix as the lower element, Eq. (19.7.3) shows that the left partial triangular factor consists of an upper element that is the lower triangular factor, and a lower element that is the transpose of the inverse of this triangular factor. For the inverse of the matrix X in Eq. (19.7.1), we get

$$X^{-1} = T'^{-1}T^{-1} \tag{19.7.4}$$

Numerical example

We shall present a numerical example of this procedure. First we define the supermatrix a by Fig. 19.7.1.

$$a = \begin{bmatrix} X \\ - \\ I \end{bmatrix} = \begin{bmatrix} 1 & 2 & 1 \\ 2 & 5 & 5 \\ 1 & 5 & 11 \\ \hline 1 & 0 & 0 \\ 0 & 1 & 0 \\ 0 & 0 & 1 \end{bmatrix}$$

FIG. 19.7.1

Now if we carry out a triangular factoring of this supermatrix the result is given in Fig. 19.7.2.

$$a = \begin{bmatrix} 1 & 0 & 0 \\ 2 & 1 & 0 \\ 1 & 3 & 1 \\ \hline 1 & -2 & 5 \\ 0 & 1 & -3 \\ 0 & 0 & 1 \end{bmatrix} \begin{bmatrix} 1 & 2 & 1 \\ 0 & 1 & 3 \\ 0 & 0 & 1 \end{bmatrix} = \begin{bmatrix} X \\ - \\ I \end{bmatrix}$$

$$\begin{bmatrix} T_L \\ - \\ T_u^{-1} \end{bmatrix} \qquad [T_u]$$

FIG. 19.7.2

The lower submatrix in the prefactor of Fig. 19.7.2 is the inverse of the upper triangular factor, as indicated in Fig. 19.7.3.

$$T'^{-1} = \begin{bmatrix} 1 & -2 & 5 \\ 0 & 1 & -3 \\ 0 & 0 & 1 \end{bmatrix}$$

FIG. 19.7.3

Therefore, using the matrix of Fig. 19.7.3 in Eq. (19.7.4), we get Fig. 19.7.4 as the inverse of X.

$$X^{-1} = \begin{bmatrix} 1 & -2 & 5 \\ 0 & 1 & -3 \\ 0 & 0 & 1 \end{bmatrix} \begin{bmatrix} 1 & 0 & 0 \\ -2 & 1 & 0 \\ 5 & -3 & 1 \end{bmatrix} = \begin{bmatrix} 30 & -17 & 5 \\ -17 & 10 & -3 \\ 5 & -3 & 1 \end{bmatrix}$$

FIG. 19.7.4

Then if we multiply the matrix X as given in Fig. 19.7.1 by the inverse as given in Fig. 19.7.4, we see that the product is actually the identity matrix. This we show in

$$\begin{bmatrix} 1 & 2 & 1 \\ 2 & 5 & 5 \\ 1 & 5 & 11 \end{bmatrix} \begin{bmatrix} 30 & -17 & 5 \\ -17 & 10 & -3 \\ 5 & -3 & 1 \end{bmatrix} = \begin{bmatrix} 1 & 0 & 0 \\ 0 & 1 & 0 \\ 0 & 0 & 1 \end{bmatrix}$$

FIG. 19.7.5

19.8 How to Solve for the Product of a Matrix by the Inverse of a Basic Square Matrix

The matrix equations

In many cases, when we wish to solve a problem involving many variables and many cases, we are not interested in the inverse of a matrix as such, but rather in the product of some matrix by the inverse of another matrix. When this is true, we need not first get the inverse of the matrix before multiplying the other matrix by it. For example, suppose we have the matrix equation

$$Xa = Y \tag{19.8.1}$$

Here we assume that X and Y are known matrices and that a is a matrix to be determined. Now presumably we could premultiply both sides of Eq. (19.8.1) by the inverse of X, assuming that X has a regular inverse. This would give

$$a = X^{-1}Y \tag{19.8.2}$$

According to Eq. (19.8.2), we would find the inverse of X and then premultiply Y by it to find the unknown matrix a. However, we shall proceed somewhat differently. First we write Eq. (19.8.1) in the form of the product of supermatrices, as in Eq. (19.8.3).

$$[X \mid Y] \begin{bmatrix} a \\ \hline -I \end{bmatrix} = 0 \tag{19.8.3}$$

Next we write the identities

$$X = T_L T_u \tag{19.8.4}$$

and

$$Y = T_L Z \tag{19.8.5}$$

You notice that in Eq. (19.8.4) the matrix X is expressed as the product of triangular factors. We substitute Eqs. (19.8.4) and (19.8.5) in Eq. (19.8.3) to get Eq. (19.8.6)

$$[T_L T_u \mid T_L Z] \begin{bmatrix} a \\ \hline -I \end{bmatrix} = 0 \qquad (19.8.6)$$

You notice that a triangular factoring of the prefactor on the left of (19.8.6) gives Eq. (19.8.7).

$$T_L[T_u \mid Z] \begin{bmatrix} a \\ \hline -I \end{bmatrix} = 0 \qquad (19.8.7)$$

The triangular factoring yields an upper partial triangular matrix, the left submatrix T_u of which is the upper triangular factor of X and the right submatrix Z is given by

$$Z = T_L^{-1} Y \qquad (19.8.8)$$

as can be shown directly from Eq. (19.8.5). Therefore, the triangular factoring of the supermatrix in the prefactor on the left of Eq. (19.8.3) or Eq. (19.8.6) automatically yields a submatrix which gives the matrix Y premultiplied by the inverse of the lower triangular factor of X. We can therefore premultiply Eq. (19.8.7) by the inverse of this lower triangular factor to get Eq. (19.8.9).

$$[T_u \mid Z] \begin{bmatrix} a \\ \hline -I \end{bmatrix} = 0 \qquad (19.8.9)$$

We now define the upper triangular matrix T_u as in

$$T_u = D_T + t' \qquad (19.8.10)$$

The first term on the right of Eq. (19.8.10) is simply the diagonal matrix made up of the diagonal elements of T_u and the right term on the right is the same as T_u except that all of its diagonal elements are 0. If we substitute Eq. (19.8.10) in Eq. (19.8.9), we have Eq. (19.8.11).

$$[D_T + t' \mid Z] \begin{bmatrix} a \\ \hline -I \end{bmatrix} = 0 \qquad (19.8.11)$$

Then if we premultiply this equation by D_T^{-1}, we get Eq. (19.8.12).

$$[I + D_T^{-1} t' \mid D_T^{-1} Z] \begin{bmatrix} a \\ \hline -I \end{bmatrix} = 0 \qquad (19.8.12)$$

Next we define

$$v = -D_T^{-1} t' \qquad (19.8.13)$$

$$w = D_T^{-1} Z \qquad (19.8.14)$$

If we substitute Eqs. (19.8.13) and (19.8.14) in Eq. (19.8.12), we have

$$[I - v \mid w] \begin{bmatrix} a \\ \overline{} \\ -I \end{bmatrix} = 0 \qquad (19.8.15)$$

But multiplying out the left side of this equation we get

$$a - va - w = 0 \qquad (19.8.16)$$

Or, from this we get

$$va + w = a \qquad (19.8.17)$$

which may be written as Eq. (19.8.18).

$$[v \mid w] \begin{bmatrix} a \\ - \\ I \end{bmatrix} = a \qquad (19.8.18)$$

The expanded notation

Now let us write Eq. (19.8.18) in expanded notation. This we show in Eq. (19.8.19), for the case of X, a 4 by 4 matrix, and a, a 4 by 3 matrix.

$$\begin{bmatrix} 0 & v_{12} & v_{13} & v_{14} & w_{11} & w_{12} & w_{13} \\ 0 & 0 & v_{23} & v_{24} & w_{21} & w_{22} & w_{23} \\ 0 & 0 & 0 & v_{34} & w_{31} & w_{32} & w_{33} \\ 0 & 0 & 0 & 0 & w_{41} & w_{42} & w_{43} \end{bmatrix} \begin{bmatrix} a_{11} & a_{12} & a_{13} \\ a_{21} & a_{22} & a_{23} \\ a_{31} & a_{32} & a_{33} \\ a_{41} & a_{42} & a_{43} \\ \hline 1 & 0 & 0 \\ 0 & 1 & 0 \\ 0 & 0 & 1 \end{bmatrix} = \begin{bmatrix} a_{11} & a_{12} & a_{13} \\ a_{21} & a_{22} & a_{23} \\ a_{31} & a_{32} & a_{33} \\ a_{41} & a_{42} & a_{43} \end{bmatrix}$$

$$(19.8.19)$$

In Eq. (19.8.19) as in Eq. (19.5.11) the solutions for the a's are made possible because the v submatrix has all zeros in the diagonal and below. The essential difference between Eq. (19.5.11) and Eq. (19.8.19) is that in Eq. (19.5.11) the right submatrix of the prefactor is diagonal whereas in Eq. (19.8.19) the right submatrix of the prefactor is a general rectangular matrix. The known and unknown submatrices in both equations come in the same positions. For example, in both Eq. (19.5.11) and Eq. (19.8.19), we have on the right side of the equation the unknown matrix, which is the same as the upper submatrix of the postfactor on the left. In both Eq. (19.5.11) and Eq. (19.8.19) it does not matter which column of the unknown matrix we solve for first. In the case of Eq. (19.5.11), we started arbitrarily with the first column vector of the B matrix and then proceeded on to the remaining column vectors. This order of solution, is not important. But it is important that whatever column we start with, the solution proceed from the bottom upward. We must always start by premultiplying a column vector of the postfactor by the last row of the prefactor. We then go on to the second from the last row of the prefactor, and so on up to the first row of the prefactor. This is, of course, so that we can always solve for an unknown element in terms of an element or elements that have previously been solved for.

Essentially, then, the procedure for getting the product of one matrix by the inverse of another consists first of setting up a supermatrix in which the upper matrix element is the one whose inverse is one of the factors and the lower matrix element is the other factor. This supermatrix is then subjected to a triangular factoring and yields a partial triangular matrix as one of the factors. By means of this partial triangular matrix, we then solve for the product of the matrix Y by the inverse of X, according to the procedure indicated by Eq. (19.8.19).

We shall not here give a numerical example of the procedure. In most cases when we wish to multiply a matrix by an inverse, the matrix whose inverse is implied is a product-moment matrix that is symmetrical. The computational procedure for multiplying any matrix by the inverse of a product-moment matrix will be outlined in Chapter 20.

19.9 How to Solve for the General Inverse of a Nonsquare Basic Matrix

The solution

We have seen how to solve for the regular inverse of a square basic matrix. Now let us consider the general inverse of any basic matrix. As has been indicated previously, according to the definition of a general inverse, it could be found very easily once we had found the basic structure of the matrix. However, we remember that finding the basic structure can in many cases be quite laborious. We shall therefore consider a somewhat simpler method of solving for the general inverse of any basic matrix. Again let us start by repeating Eq. (19.1.1) for the basic structure of X.

$$X = P\Delta Q' \qquad (19.9.1)$$

We shall prove now that if X' is premultiplied by the inverse of its minor product moment, the resulting matrix is the general inverse of X. This rule is

$$Y' = (X'X)^{-1}X' \qquad (19.9.2)$$

The proof

First let us recall that the basic structure of a minor product moment has the same right orthonormal as the matrix itself, its left orthonormal is the transpose of the right orthonormal, and its basic diagonal is the square of the basic diagonal of the original matrix. Thus we repeat

$$X'X = Q\Delta^2 Q' \tag{19.9.3}$$

Now let us take the inverse of both sides of Eq. (19.9.3), This gives

$$(X'X)^{-1} = Q\Delta^{-2}Q' \tag{19.9.4}$$

since we have already learned that the inverse of any basic product moment raised to any power — positive, negative, fractional, or integral — has the same orthonormals as the original product moment, and the basic diagonal is that of the original raised to the power under consideration. Now if we substitute Eqs. (19.9.1) and (19.9.4) in Eq. (19.9.2), we have

$$Y' = Q\Delta^{-2}(Q'Q)\Delta P' \tag{19.9.5}$$

But from Eq. (19.9.5), we have

$$Y' = Q(\Delta^{-2}\Delta)P' \tag{19.9.6}$$

And from (19.9.6) we have

$$Y' = Q\Delta^{-1}P' \tag{19.9.7}$$

But notice that Eq. (19.9.7) is exactly the same as Eq. (19.1.3) which has been defined as the general inverse of the matrix X. We have therefore proved the rule given by Eq. (19.9.2).

It is of interest to note that in Eq. (19.9.2) if X is both square and basic the term in brackets can be written as

$$(X'X)^{-1} = X^{-1}X'^{-1} \tag{19.9.8}$$

Then if we substitute the right of Eq. (19.9.8) in the right of Eq. (19.9.2), we have

$$Y' = X^{-1}(X'^{-1}X') \tag{19.9.9}$$

or simply

$$Y' = X^{-1} \tag{19.9.10}$$

19.10 The Major Product Moment of the Left Orthonormal of a Basic Matrix

The major product moment as a function of the matrix

Some problems in the human sciences require use of the major product moment of the left orthonormal of a basic matrix. This product moment can be expressed as a function of the matrix X by

$$PP' = X(X'X)^{-1}X' \tag{19.10.1}$$

To prove this equation, we first rewrite it as

$$PP' = X[(X'X)^{-1}X']$$ (19.10.2)

On the right side of Eq. (19.10.2), we substitute for X its value given in Eq. (19.1.1). For the brackets we substitute the right side of Eq. (19.9.6). This gives

$$PP' = P\Delta(Q'Q)\Delta^{-1}P'$$ (19.10.3)

But this can be rewritten as

$$PP' = P(\Delta\Delta^{-1})P'$$ (19.10.4)

and this can be written as the identity

$$PP' = PP'$$ (19.10.5)

We have therefore proved Eq. (19.10.1).

The computational equation

Let us now solve for this major product moment of the left orthonormal without actually solving for either the inverse of the minor product moment of X or for Q directly.

First we define a supermatrix by Eq. (19.10.6).

$$G = \begin{bmatrix} X'X \\ \hline X \end{bmatrix}$$ (19.10.6)

The upper submatrix in Eq. (19.10.6) is the minor product moment of X, whereas the lower submatrix is simply X itself. Now the minor product moment can be expressed as the major product moment of a triangular matrix, as in

$$X'X = TT'$$ (19.10.7)

Let us also define a new matrix Z as in

$$Z = XT'^{-1}$$ (19.10.8)

Then from this we have

$$ZT' = X$$ (19.10.9)

Next let us substitute Eqs. (19.10.7) and (19.10.9) in Eq. (19.10.6). This gives Eq. (19.10.10).

$$G = \begin{bmatrix} T'_u T_u \\ \hline Z T_u \end{bmatrix}$$ (19.10.10)

But by factoring out T' on the right, this can be written as Eq. (19.10.11).

$$G = \begin{bmatrix} T \\ - \\ Z \end{bmatrix} T' \qquad (19.10.11)$$

Now Eq. (19.10.11) shows that by applying the triangular factoring procedure to Eq. (19.10.6), we get a left partial triangular factor consisting of two submatrices as follows: The upper submatrix is the lower triangular factor and the lower submatrix Z is simply the matrix X postmultiplied by the transpose of the inverse of this triangular factor. Having solved for Z by this triangular factoring, we can now prove that the major product moment of Z is equal to the major product moment of the left orthonormal of X. From Eq. (19.10.8), we get

$$ZZ' = (XT'^{-1})(T^{-1}X') \qquad (19.10.12)$$

But this may be written as

$$ZZ' = X(TT')^{-1}X' \qquad (19.10.13)$$

And this, because of Eq. (19.10.7), may be written

$$ZZ' = X(X'X)^{-1}X' \qquad (19.10.14)$$

But the right side of Eq. (19.10.14) is the same as the right side of Eq. (19.10.1), which we have already proved to be the major product moment of the left orthonormal of X. You see, therefore, that by getting the left partial triangular matrix of Eq. (19.10.5) and then getting the major product moment of the lower submatrix of this factor, you automatically get the major product moment of the left orthonormal of X.

The orthonormal matrix Z

It is also of interest to note that the minor product moment is an identity matrix, as of course it must be if the major product moment is a major product moment of an orthonormal. We can show this, however, very simply by taking the minor product moment of both sides of Eq. (19.10.8) as in

$$Z'Z = T^{-1}(X'X)T'^{-1} \qquad (19.10.15)$$

But if we substitute from Eq. (19.10.7) in Eq. (19.10.15), we get

$$Z'Z = (T^{-1}T)(T'T'^{-1}) \qquad (19.10.16)$$

or simply

$$Z'Z = 1 \qquad (19.10.17)$$

19.11   How to Solve for the General Inverse if X Is not Basic

Let us next find a simple way of solving for the general inverse of a non-basic matrix without first solving for its basic structure. We already know that any power of a matrix multiplied by its general inverse is equal to the product of the matrix and its inverse. This rule is given by the two equations

$$(Y'X)^n = Y'X \qquad (19.11.1)$$

and

$$(XY')^n = XY' \qquad (19.11.2)$$

If, therefore, we can find a matrix Y which satisfies Eqs. (19.11.1) and (19.11.2), we have solved for the general inverse of X.

We let X be defined by the basic factors given in

$$X = uv' \qquad (19.11.3)$$

The u and v matrices on the right of Eq. (19.11.3) may be obtained by the triangular factoring method of Chapter 16. We shall now prove that the general inverse of Eq. (19.11.3) is given by Eq. (19.11.4), no matter how the basic factors in Eq. (19.11.3) are arrived at.

$$Y' = v(v'v)^{-1}(u'u)^{-1}u' \qquad (19.11.4)$$

From Eqs. (19.11.3) and (19.11.4), we get

$$Y'X = v(v'v)^{-1}(u'u)^{-1}(u'u)v' \qquad (19.11.5)$$

But this can be written as

$$Y'X = v(v'v)^{-1}v' \qquad (19.11.6)$$

Now from Eqs. (19.11.3) and (19.11.4), we also get

$$XY' = uv'v(v'v)^{-1}(u'u)^{-1}u' \qquad (19.11.7)$$

But Eq. (19.11.7) can be written as

$$XY' = u(u'u)^{-1}u' \qquad (19.11.8)$$

Notice, however, that Eqs. (19.11.6) and (19.11.8) on the right are of the same type as Eq. (19.10.1); therefore, they are the major products of orthonormals and they do not change, no matter to what power they are raised. Thus the right sides of Eqs. (19.11.6) and (19.11.8) satisfy Eqs. (19.11.1) and (19.11.3) respectively, and Y', as given in Eq. (19.11.4), is the general inverse of X.

Ordinarily the solution for Y', as given in Eq. (19.11.4), would proceed somewhat as follows: First we would find the u and v factors of X, presumably by the simple triangular method of factoring. Defining two matrices

$$f = (v'v)^{-1}v' \qquad (19.11.9)$$

and

$$g = (u'u)^{-1}u \qquad (19.11.10)$$

we would solve for f and g by the method indicated in Section 19.8 rather than actually finding the inverse of the minor product moments of v and u respectively.

19.12 The Regular Inverse of Certain Special Sums of Matrices

There are certain special types of sums of matrices that you will meet very frequently in the analysis of data. The inverse of the matrices is frequently required and may be obtained quite simply in many cases.

The sum of a basic square matrix and a major product

First let us consider the type of sum indicated in

$$X = (A + uv') \tag{19.12.1}$$

We assume X to be square and basic, and we also assume that A is square and basic, so that it has a regular inverse. However, we assume the matrices u and v' to have a common dimension less than the order of A. We also assume that the inverse of A is already known. The question is then, can we find the inverse of X without actually going through the addition indicated in Eq. (19.12.1) and taking the inverse of the new sum? If the common dimension of u and v' is much less than the order of A, we can find a simpler procedure for getting the inverse of X.

To derive this procedure we first postmultiply both sides of Eq. (19.12.1) by X^{-1} giving

$$(A + uv')X^{-1} = I \tag{19.12.2}$$

From this we get

$$AX^{-1} = I - uv'X^{-1} \tag{19.12.3}$$

We premultiply both sides of Eq. (19.12.3) by A^{-1} to get

$$X^{-1} = A^{-1} - A^{-1}uv'X^{-1} \tag{19.12.4}$$

Next we premultiply this by v' to get

$$v'X^{-1} = v'A^{-1} - v'A^{-1}uv'X^{-1} \tag{19.12.5}$$

From this we get

$$v'X^{-1} + v'A^{-1}uv'X^{-1} = v'A^{-1} \tag{19.12.6}$$

Factoring on the left of Eq. (19.12.6), we get

$$(I + v'A^{-1}u)v'X^{-1} = v'A^{-1} \tag{19.12.7}$$

Premultiplying this by the inverse of the parenthesis on the left, we get

$$v'X^{-1} = (I + v'A^{-1}u)^{-1}v'A^{-1} \qquad (19.12.8)$$

Substituting the right side of Eq. (19.12.8) into the right side of Eq. (19.12.4), we have

$$X^{-1} = A^{-1} - A^{-1}u(I + v'A^{-1}u)^{-1}v'A^{-1} \qquad (19.12.9)$$

Now notice that the right side of Eq. (19.12.9) involves only the inverse of the matrix A and the inverse of the parentheses, whose order is the common order of u and v'. If this common order is much less than the order of A, then the inverse of the parentheses on the right of Eq. (19.12.9) would be much easier to get than to solve for the inverse of X in Eq. (19.12.1), assuming that the inverse of A is readily available.

The special case of a major product of vectors

As a special example, we might take the case frequently encountered in the analysis of social science data in which u and v are vectors. In this case, the parentheses in Eq. (19.12.9) reduce to a scalar quantity and we have

$$X^{-1} = A^{-1} - \frac{A^{-1}uv'A^{-1}}{1 + v'A^{-1}u} \qquad (19.12.10)$$

The last term on the right of this equation in the case of a vector is relatively easy to compute. Therefore, Eq. (19.12.10) is a very simple way of finding the inverse of the matrix X, provided that we already have the inverse of A and that u and v are vectors. In particular, if A is a diagonal matrix and u and v are vectors, then Eq. (19.12.10) is a very simple way to get the inverse of such a matrix. Suppose, for example, we have the matrix in Fig. 19.12.1.

$$X = \underset{A}{\begin{bmatrix} 1 & 0 & 0 \\ 0 & 3 & 0 \\ 0 & 0 & 2 \end{bmatrix}} + \underset{u}{\begin{bmatrix} 1 \\ 3 \\ 2 \end{bmatrix}} \underset{v'}{\begin{bmatrix} 3 & 1 & 2 \end{bmatrix}}$$

$$X = \underset{A}{\begin{bmatrix} 1 & 0 & 0 \\ 0 & 3 & 0 \\ 0 & 0 & 2 \end{bmatrix}} + \underset{uv'}{\begin{bmatrix} 3 & 1 & 2 \\ 9 & 3 & 6 \\ 6 & 2 & 4 \end{bmatrix}}$$

$$X = \begin{bmatrix} 4 & 1 & 2 \\ 9 & 6 & 6 \\ 6 & 2 & 6 \end{bmatrix}$$

Fig. 19.12.1

From Eq. (19.12.10), then, we have

$$
X^{-1} = \underbrace{\begin{bmatrix} 1 & 0 & 0 \\ 0 & \dfrac{1}{3} & 0 \\ 0 & 0 & \dfrac{1}{2} \end{bmatrix}}_{A^{-1}} - \underbrace{\dfrac{\begin{bmatrix} 1 \\ 1 \\ 1 \end{bmatrix}\begin{bmatrix} 3 & \dfrac{1}{3} & 1 \end{bmatrix}}{1 + 6}}_{\dfrac{(A^{-1}u)(v'A^{-1})}{1 + v'A^{-1}u}}
$$

$$
X^{-1} = \begin{bmatrix} 1 & 0 & 0 \\ 0 & \dfrac{1}{3} & 0 \\ 0 & 0 & \dfrac{1}{2} \end{bmatrix} - \begin{bmatrix} \dfrac{3}{7} & \dfrac{1}{21} & \dfrac{1}{7} \\ \dfrac{3}{7} & \dfrac{1}{21} & \dfrac{1}{7} \\ \dfrac{1}{7} & \dfrac{1}{21} & \dfrac{1}{7} \end{bmatrix} = \begin{bmatrix} \dfrac{4}{7} & -\dfrac{1}{21} & -\dfrac{1}{7} \\ -\dfrac{3}{7} & \dfrac{6}{21} & -\dfrac{1}{7} \\ -\dfrac{3}{7} & -\dfrac{1}{21} & \dfrac{5}{14} \end{bmatrix}
$$

FIG. 19.12.2

A sum of two matrices

Another type of sum involving inverses is also frequently encountered in the analysis of data. Consider the sum given by

$$X - a^{-1} - (a + b)^{-1}ba^{-1} \tag{19.12.11}$$

and also

$$Y = a^{-1} - ba^{-1}(a + b)^{-1} \tag{19.12.12}$$

It is easy to show that the right sides of Eqs. (19.12.11) and (19.12.12) can both be written as

$$X = (a + b)^{-1} \tag{19.12.13}$$

$$Y = (a + b)^{-1} \tag{19.12.14}$$

To prove Eq. (19.12.13), we simply substitute the right side of Eq. (19.12.13) on the left of Eq. (19.12.11) as in

$$(a + b)^{-1} = a^{-1} - (a + b)^{-1}ba^{-1} \tag{19.12.15}$$

We premultiply Eq. (19.12.15) by $a + b$ to get

$$I = (a + b)a^{-1} - ba^{-1} \tag{19.12.16}$$

But from this we have

$$I = I + ba^{-1} - ba^{-1} = I \tag{19.12.17}$$

In the same way, we can show by substituting Eq. (19.12.14) in Eq. (19.12.12) that an identity results. This substitution we make in

$$(a + b)^{-1} = a^{-1} - ba^{-1}(a + b)^{-1} \tag{19.12.18}$$

Equations (19.12.15) and (19.12.18) can also be modified by a reversal of signs to give respectively

$$(a - b)^{-1} = a^{-1} + (a - b)^{-1}ba^{-1} \qquad (19.12.19)$$

and

$$(a - b)^{-1} = a^{-1} + ba^{-1}(a - b)^{-1} \qquad (19.12.20)$$

The sum of a matrix and the identity matrix

When equations of the type given by Eqs. (19.12.15), (19.12.18), (19.12.19), and (19.12.20) are found in actual practice, the matrix a usually is the identity matrix. For this particular case, the four equations become respectively

$$(I + b)^{-1} = I - (I + b)^{-1}b \qquad (19.12.21)$$

$$(I + b)^{-1} = I - b(I + b)^{-1} \qquad (19.12.22)$$

$$(I - b)^{-1} = I + (I - b)^{-1}b \qquad (19.12.23)$$

$$(I - b)^{-1} = I + b(I - b)^{-1} \qquad (19.12.24)$$

It is convenient to be able to recognize the identities in Eqs. (19.12.21) to (19.12.24) immediately. Ordinarily, the forms encountered are those on the right of these four equations.

19.13   The Inverse of a Second-Order Matrix

The inverse of a second-order matrix is much easier to compute than that of higher order matrices. Therefore, we shall show how the inverse of a second-order matrix may be written at once. If we let

$$a = \begin{bmatrix} a_{11} & a_{12} \\ a_{21} & a_{22} \end{bmatrix} \qquad (19.13.1)$$

then it is easy to prove that the inverse of a is given by Eq. (19.13.2)

$$a^{-1} = \begin{bmatrix} a_{22} & -a_{12} \\ -a_{21} & a_{11} \end{bmatrix} \frac{1}{a_{11}a_{22} - a_{12}a_{21}} \qquad (19.13.2)$$

We simply multiply corresponding sides of Eqs. (19.13.1) and (19.13.2), as in Eq. (19.13.3).

$$aa^{-1} = \begin{bmatrix} a_{11} & a_{12} \\ a_{21} & a_{22} \end{bmatrix} \begin{bmatrix} a_{22} & -a_{12} \\ -a_{21} & a_{11} \end{bmatrix} \frac{1}{a_{11}a_{22} - a_{12}a_{21}} \qquad (19.13.3)$$

Multiplying out the right side of Eq. (19.13.3), we have Eq. (19.13.4)

$$aa^{-1} = \frac{\begin{bmatrix} a_{11}a_{22} - a_{12}a_{21} & -a_{11}a_{12} + a_{12}a_{11} \\ a_{21}a_{22} - a_{22}a_{21} & -a_{21}a_{12} + a_{22}a_{11} \end{bmatrix}}{a_{11}a_{22} - a_{12}a_{21}} \qquad (19.13.4)$$

or

$$aa^{-1} = \begin{bmatrix} 1 & 0 \\ 0 & 1 \end{bmatrix} \qquad (19.13.5)$$

Therefore the rule for writing down the inverse of a second-order matrix can be stated as follows:

(1) Write down a new matrix whose diagonal elements are the diagonal elements of a in reverse order and whose nondiagonal elements are opposite in sign to the nondiagonal elements of a.

(2) Subtract the product of the nondiagonal elements of a from the product of its diagonal elements.

(3) Divide the elements of the matrix in step (1) by the number calculated in step (2). This is the inverse of a.

A numerical example follows. If we have

$$a = \begin{bmatrix} 3 & 2 \\ -1 & 4 \end{bmatrix}$$

then steps (1), (2), and (3) are

$$(1) \quad \begin{bmatrix} 4 & -2 \\ 1 & 3 \end{bmatrix}$$

$$(2) \quad 3 \times 4 - 1 \times (-2) = 14$$

$$(3) \quad a^{-1} = \begin{bmatrix} \dfrac{4}{14} & -\dfrac{2}{14} \\ \dfrac{1}{14} & \dfrac{3}{14} \end{bmatrix}$$

SUMMARY

1. The general inverse and the basic structure:
   a. The inverse b of a is defined as a matrix such that $ab = ba = I$.
   b. The general inverse of $X = P\Delta Q'$ is defined as $Y' = Q\Delta^{-1}P'$.
   c. The major product of a matrix X and its general inverse Y' is $XY' = PP'$.
   d. The minor product of a matrix and its general inverse is $Y'X = QQ'$.
   e. If X is basic the minor product of X with its general inverse is the identity matrix, that is, $Y'X = I$.

2. If a matrix X is square and basic it is said to have a regular inverse or merely an inverse X^{-1}. In this case $Y'X = XY'$ and we use the notation $Y' = X^{-1}$.

3. The inverse of the product of any number of square basic matrices is the product of their inverses in reverse order, for example, $(a\,b\,c\,d)^{-1} = d^{-1}c^{-1}b^{-1}a^{-1}$.

4. Inverses of special types of matrices:
   a. *Diagonal matrices:*

$$\text{If } D = \begin{bmatrix} d_1 & \cdots & 0 \\ \cdots & \cdots & \cdots \\ 0 & \cdots & d_n \end{bmatrix} \text{ then } D^{-1} = \begin{bmatrix} \dfrac{1}{d_1} & \cdots & 0 \\ \cdots & \cdots & \cdots \\ 0 & \cdots & \dfrac{1}{d_n} \end{bmatrix}$$

b. $I^{-1} = I$.

c. For i a sign matrix, $i^{-1} = i$.

d. If Q is a square orthonormal, $Q^{-1} = Q'$.

e. If Q is a vertical orthonormal, its general inverse is $Y' = Q'$.

f. If P is an orthogonal matrix such that $P'P = D$ is diagonal, its general inverse is $Y' = D^{-1}P'$.

5. If T' is an upper triangular matrix and

$$T' = D_T + t' \qquad -D_T^{-1}t' = v \qquad D_T^{-1} = D_u$$

the inverse B of T' may be solved from the equation

$$[v \mid D_u]\begin{bmatrix} B \\ - \\ I \end{bmatrix} = B$$

6. If X is a square basic matrix with triangular factors $X = T_L T'_u$ and

$$a = \begin{bmatrix} X \\ - \\ I \end{bmatrix} \qquad \text{then}$$

the triangular factoring of a is

$$a = \begin{bmatrix} T_L \\ \overline{} \\ T_u^{-1} \end{bmatrix} T_u$$

The inverse of X is then $X^{-1} = T_u^{-1}T_L^{-1}$.

7. If X is a basic product moment matrix it may be expressed as the major product moment of a triangular matrix; thus $X = TT'$. If

$$a = \begin{bmatrix} X \\ - \\ I \end{bmatrix}$$

the triangular factors are

$$a = \begin{bmatrix} T \\ \overline{} \\ T'^{-1} \end{bmatrix} T'$$

and $X^{-1} = T'^{-1}T^{-1}$.

8. The product of a matrix by the inverse of a basic square matrix may be computed as follows: Let $a = X^{-1}Y$ and the triangular factors of X be $X = T_L T'_u$, let $Z = T_L^{-1}Y$, $T_u = D_T + t'$, $v = -D_T^{-1}t'$, $w = D_T^{-1}Z$. The solution for a is obtained from

$$[v \mid w]\begin{bmatrix} a \\ - \\ I \end{bmatrix} = a$$

9. The general inverse of a vertical basic matrix X is given by

$$Y' = (X'X)^{-1}X'$$

10. The major product moment of the left orthonormal of a vertical basic matrix

a. If $X = P\Delta Q'$ then $PP' = X[X'X]^{-1}X'$.

b. If $G = \begin{bmatrix} X'X \\ \overline{} \\ X \end{bmatrix}$ and $X'X = TT'$, then the triangular factoring of G is

$$G = \begin{bmatrix} T \\ - \\ Z \end{bmatrix} T' \text{ and } PP' = ZZ'.$$

c. If Z is defined as in (b) then $Z'Z = I$.

11. The general inverse Y' if X is not basic may be computed as follows: Let the partial triangular factoring of X be $X = uv'$. Then

$$Y' = v(v'v)^{-1}(u'u)^{-1}u'.$$

12. The regular inverse of special sums of matrices:
a. If A is square and basic and $X = (A + uv')$ then

$$X^{-1} = A^{-1} - A^{-1}u(I + v'A^{-1}u)^{-1}v'A^{-1}.$$

b. If u and v in (a) are vectors then

$$X^{-1} = A^{-1} - \frac{A^{-1}uv'A^{-1}}{1 + v'A^{-1}u}$$

c. If $X = a^{-1} - (a + b)^{-1}ba^{-1}$ and $Y = a^{-1} - ba^{-1}(a + b)^{-1}$ then $X = Y = (a + b)^{-1}$.

13. If $a = \begin{bmatrix} a_{11} & a_{12} \\ a_{21} & a_{22} \end{bmatrix}$ then $a^{-1} = \dfrac{\begin{bmatrix} a_{22} & -a_{12} \\ -a_{21} & a_{11} \end{bmatrix}}{a_{11}a_{22} - a_{12}a_{21}}$

Chapter 20

Computations Involving Inverses

20.1 Introduction

You have seen in the preceding chapter what is meant by the inverse of a matrix and how, by means of matrix algebra, we can solve for the inverse of various kinds of matrices. You have also seen how we can solve for the product of a matrix by the inverse of another matrix without actually first solving for its inverse.

In this chapter we shall describe how the computational worksheets may be drawn up and how the computations may be organized so as to make the solutions as routine and mechanical as possible. As indicated in Chapter 19, computations using the inverse of a matrix are in general somewhat more complex than those using the other three operations of addition, subtraction, and multiplication.

Many computational schemes are available for these problems. Some are more compact than the methods to be described here. Others require considerably more worksheet space and more writing down of figures than those we shall outline. The major considerations in the methods that follow are (1) an adequate system of algebraic checks shall be available throughout, (2) the number of operations involving scalar division shall be a minimum, (3) the computations shall be so organized that the maximum number of operations shall consist merely of getting the scalar product of vectors or the sum of products of numbers, (4) the mechanical layout of the worksheets shall be such that in the cumulative product operation, the two elements in a product will appear either side by side or one directly above the other, (5) the worksheet layout shall be such that the probability of writing down a calculated number in the wrong place shall be as small as possible.

These principles of computational procedure are all directed toward making the operations as simple and routine as possible so as to require a minimum of attention, concentration, and other higher mental processes. It has been our experience that the computational procedures can be readily mastered by high school graduates of normal intelligence. The chief skill

434

required is that of finger dexterity, such as is needed in typing or playing the piano. As a matter of fact, this type of skill is much more important than abstract intelligence or the intellectual skills and knowledge acquired in college.

It is not necessary to have a knowledge of either matrix or scalar algebra in order to master the computational procedures to be described, even though such a knowledge would in general provide greater interest and motivation.

20.2 A New Kind of Vector Multiplication

The most important principle to be mastered in these computations is very simple though perhaps unfamiliar. This is the principle of row-by-row or column-by-column multiplication. In matrix algebra you have become accustomed to the concept of row-by-column multiplication. The concept of row-by-row or column-by-column multiplication is one developed purely for computational convenience. Up to now you have been finding the scalar product of two vectors, where a row vector is postmultiplied by a column vector, as in Fig. 20.2.1.

$$[3 \quad 2 \quad 4 \quad 1 \quad 5] \begin{bmatrix} 2 \\ 4 \\ 3 \\ 1 \\ 6 \end{bmatrix} = 57$$

FIG. 20.2.1

Here, of course, the product is the sum of products of corresponding elements of the two vectors. However, in actually carrying out the multiplication of these elements, you must be careful to pick out corresponding elements from each vector. For example, you must be sure that you do not multiply the third element in the row vector by the fourth element in the column vector. It is not difficult to avoid this kind of mistake if you have only fifth-order vectors or less to work with, although if you have many scalar products to compute you can readily make such errors. The greater the order of the vector, however, the easier it is to multiply the wrong pairs of elements together.

To avoid making errors of this kind, we can first write the postfactor as a row directly above the prefactor. Using the example in Fig. 20.2.1, we then have

$$\begin{array}{ccccc} 2 & 4 & 3 & 1 & 6 \\ 3 & 2 & 4 & 1 & 5 \end{array}$$

FIG. 20.2.2

We now get the cumulative product of corresponding numbers in the two rows and it is almost impossible to get the wrong pair of numbers in a product since one is immediately above the other.

Another way we can avoid multiplying the wrong numbers together is to write the prefactor in Fig. 20.2.1 as a column immediately before the postfactor, as in Fig. 20.2.3.

$$
\begin{array}{cc}
3 & 2 \\
2 & 4 \\
4 & 3 \\
1 & 1 \\
5 & 6
\end{array}
$$

FIG. 20.2.3

Again, it would be very difficult to get the wrong numbers in a product because the two numbers to be multiplied together are always side by side. Purely for computational convenience, therefore, we may write down two vectors whose scalar product is required either as one row above another, or one column beside another. Then when we speak of multiplying one row by another we mean simply that we sum the products of corresponding elements of the two rows. In the same way when we speak of multiplying two columns together we mean that we sum the products of corresponding elements in the two columns.

Now one of the guiding principles in designing the worksheets for the computations in this chapter is that all elements be written down in such a way that (1) all scalar products of vectors may be handled as either row by row multiplications or column by column multiplications, (2) most of the operations will be column-by-column multiplication rather than row-by-row multiplications, and (3) vectors that are to be multiplied together can be readily placed one immediately above the other when they are row vectors, or one immediately beside the other when they are column vectors. In our discussion here of computational principles we have omitted one that is frequently mentioned. This principle is that the computational procedure should strive to reduce to a minimum the amount of copying from computing machine dials to worksheets. This objective is sometimes achieved at the expense of introducing complex machine operations. For relatively unskilled computors, such operations greatly increase the possibility of error. The act of copying numbers calls for a much lower order of skill than that of manipulating the various operating keys of a typical calculating machine.

We shall consider four major types of computational problems using inverses. All of these are concerned with basic square matrices, which, as you know, have regular inverses. These are (1) the product of a matrix b by the inverse of a basic square matrix a; (2) the inverse of a basic square matrix a; (3) the product of matrix b by the inverse of a symmetric basic matrix a; and (4) the inverse of a symmetric basic matrix a.

In illustrating the worksheets for each of these four problems we shall assume for the sake of concreteness that the square matrix a is of order four. We shall also assume in cases (1) and (3) that the matrix b has four rows and three columns. If you know how to make the calculations for matrices of this order you will know how to make them for matrices of any order. The methods outlined may be used to advantage for any problem in which the order of a is three or more and in which b has any number of columns, including the case in which b is simply a column vector.

If a is only a second-order matrix, it is better to calculate its inverse directly as indicated in Section 19.3. The matrix can then be multiplied by this inverse.

20.3   The Product of a Matrix by the Inverse of a Square Matrix

In a large proportion of problems involving the inverse of a matrix, the inverse as such is not required, but only the product of some other matrix or vector by the inverse. In such cases, it is not necessary, as you saw in Chapter 19, to compute separately the inverse of the matrix and then to multiply another matrix by it. To indicate the computational procedure, we shall first show the required worksheets in symbolic form. There will be three of these worksheets, which we shall call I, II, and III respectively. These are given in Fig. 20.3.1.

Worksheet I

a	b	C_1	
S_a'	S_b'	C_T	
v	Z	C_2	S_2

Worksheet II

I		
	1	
w'	C_2	S_2

Worksheet III

R	Y	X	C_1	S_1
	F'	$-I$	0	G

Fig. 20.3.1

So that the computations will be easier, it is important that each of the three worksheets be on a separate sheet of paper. This is true even though the number of variables involved is small, no greater than three.

The meanings of the symbols in each of the worksheets is as follows:
Worksheet I

a is the matrix whose inverse is involved in the computations.

b is the matrix to be premultiplied by the inverse of a.

C_1 is a column of row sums of both a and b.

S_a' is a row of column sums of a.

S_b' is a row of column sums of b.

C_T is the total of C_1 and also of S_a' and S_b' combined.

v is an upper triangular matrix whose diagonal elements are all -1; $-v$ is the upper triangular factor of a.

Z is b premultiplied by the inverse of the lower triangular factor of a, or may be defined as the negative of the rectangular submatrix of the partial triangular factor of the supermatrix $[a \mid b]$.

C_2 is a check column obtained by methods to be described later.

S_2 is a column of row sums of both v and Z and should be the same as C_2.
Worksheet II

I is simply the identity matrix of the order of a.

1 is simply the scalar number 1.

w' is the transpose of the lower triangular factor of a.

C_2 is a check column to be explained later.

S_2 is a column of row sums of w' and should be the same as C_2.
Worksheet III

R is a column of the reciprocals of the corresponding diagonal elements of Y.

Y is a matrix closely related to v in worksheet I which we shall explain in more detail shortly.

X is a matrix closely related to Z in worksheet I, also to be explained more fully.

C_1 is a check column to be explained later.

S_1 is a column of row sums of Y and X combined and should be the same as C_1.

F' is the transpose of $a^{-1}b$; that is, the transpose of b multiplied by the inverse of a. This is the end result of our computations.

$-I$ is the negative of the identity matrix whose order is the width of b.

0 is simply a column of zeros.

G is a final check column to be explained presently. Its elements should all be zero.

The worksheets I, II, and III in expanded notation and with the appropriate rulings are shown in Fig. 20.3.2. It is important that the rulings be properly drawn in so as to make alignments of rows or columns as easy as possible in row-by-row or column-by-column multiplications.

Worksheet I

i	1A	2A	3A	4A	1B	2B	3B	C	S
1A	a_{11}	a_{12}	a_{13}	a_{14}	b_{11}	b_{12}	b_{13}	C_1	
2A	a_{21}	a_{22}	a_{23}	a_{24}	b_{21}	b_{22}	b_{23}	C_2	
3A	a_{31}	a_{32}	a_{33}	a_{34}	b_{31}	b_{32}	b_{33}	C_3	
4A	a_{41}	a_{42}	a_{43}	a_{44}	b_{41}	b_{42}	b_{43}	C_4	
C	S_{a1}	S_{a2}	S_{a3}	S_{a4}	S_{b1}	S_{b2}	S_{b3}	C_T	S_T
1B	-1	v_{12}	v_{13}	v_{14}	Z_{11}	Z_{12}	Z_{13}	C_{12}	S_{12}
2B		-1	v_{23}	v_{24}	Z_{21}	Z_{22}	Z_{23}	C_{22}	S_{22}
3B			-1	v_{34}	Z_{31}	Z_{32}	Z_{33}	C_{32}	S_{32}
4B				-1	Z_{41}	Z_{42}	Z_{43}	C_{42}	S_{42}

Worksheet II

i	1	2	3	4	C	S
1A	1	0	0	0		
2A	0	1	0	0		
3A	0	0	1	0		
4A	0	0	0	1		
C					1	
1B	w_{11}	w_{21}	w_{31}	w_{41}	C_{12}	S_{12}
2B		w_{22}	w_{32}	w_{42}	C_{22}	S_{22}
3B			w_{33}	w_{43}	C_{32}	S_{32}
4B				w_{44}	C_{42}	S_{42}

Worksheet III

R	i	1A	2A	3A	4A	1B	2B	3B	C	S
R_1	1A	Y_{11}	Y_{12}	Y_{13}	Y_{14}	X_{11}	X_{12}	X_{13}	C_{11}	S_{11}
R_2	2A		Y_{22}	Y_{23}	Y_{24}	X_{21}	X_{22}	X_{23}	C_{21}	S_{21}
R_3	3A			Y_{33}	Y_{34}	X_{31}	X_{32}	X_{33}	C_{31}	S_{31}
R_4	4A				Y_{44}	X_{41}	X_{42}	X_{43}	C_{41}	S_{41}
	1B	F_{11}	F_{21}	F_{31}	F_{41}	-1	0	0	0	G_1
	2B	F_{12}	F_{22}	F_{32}	F_{42}	0	-1	0	0	G_2
	3B	F_{13}	F_{23}	F_{33}	F_{43}	0	0	-1	0	G_3

FIG. 20.3.2

Notice carefully now that the vertical rulings of worksheet I should have exactly the same spacing as those of III. Furthermore, for purposes of alignment, the sheets of paper on which I and III are drawn should be of the same width, and the left margins of the two should be the same. Hence, if the vertical rulings of the two are the same, the right margins will also be the same.

Notice also that the horizontal rulings of worksheet I should be the same as the horizontal rulings of worksheet II. For purposes of alignment, the sheets of paper on which I and II are drawn should be of the same length and the top margin of the two should be the same. Therefore, if the horizontal rulings of the two are the same, the lower margins will also be the same.

The basic plan for the rulings of the worksheets is determined by worksheet I. This can be described in terms of its vertical and horizontal rulings. The vertical rulings are spaced as follows from left to right:

(1) a dummy column to correspond to a functional column in worksheet III

(2) a column of row designations

(3) a succession of columns equal in number to the order of the matrix a or the height of the matrix b that is to be premultiplied by the inverse of a

(4) a succession of columns equal in number to the width of the matrix b. If b is a column vector as it often may be, this would be but a single column.

(5) a check column C

(6) a column of row sums S to compare with the C column

The horizontal rulings are spaced as follows from top to bottom:

(1) a row of column designations

(2) a succession of rows equal in number to the height of the matrix b and the order of the matrix a

(3) a row of column sums for all entries above it

(4) a succession of rows equal in number to the height of b and the order of a. This is the same as (2) above.

For worksheet II the vertical rulings are spaced from left to right as follows:

(1) a column of row designations

(2) a succession of columns equal in number to the order of a and the height of b

(3) a check column C

(4) a column of row sums S

The horizontal rulings of II, you remember, are exactly the same as the horizontal rulings of I.

For worksheet III, you recall, the vertical rulings are exactly the same as the vertical rulings of I. However, the horizontal rulings are from top to bottom as follows:

(1) a row of column designations

(2) a succession of rows equal in number to the height of b and order of a

(3) a succession of rows equal in number to the width of b

Perhaps the most convenient paper to use for most computational work is that which has quarter-inch horizontal rulings and half-inch vertical rulings. The size of sheet required will depend, of course, upon the order of the matrices involved. Usually, however, such paper is available in both $8\frac{1}{2}$ by 11 inches and 11 by 17 inches. One of these two sizes can usually be adapted to handle most computational projects.

Once the three worksheets have been ruled up according to the specifications of the problem, it is well to cross out with diagonal rulings the spaces that are not needed. This takes only a few minutes and will help to prevent you from making entries in the wrong positions. Figure 20.3.2 shows where these diagonal rulings are drawn.

The next step is to write in the row and column designations in the row and column labeled **i** for each of the three worksheets. In each of the three worksheets certain entries can be made that do not depend on the elements of the matrices a and b. These are all unit elements as follows:

(1) In the lower left hand section of worksheet I enter -1 in each diagonal position.

(2) In the upper section of worksheet II enter 1 in each diagonal position. Also enter 1 in the cell where the **C** row and **C** column intersect.

(3) In the lower right hand section of worksheet III enter -1 in each diagonal position. Also in this worksheet enter zeros in the lower section of the C column.

You are now ready to write in the elements of the a and b matrices in the upper left and right sections respectively of worksheet I. Once these are written in, you are ready to proceed with the computations $a^{-1}b = F$. The steps consist of two main groups of operations, the *forward* solution and the *back* solution.

The forward solution

Cycle 1

(1) In I sum all entries in row **1A** of I and enter as C_1 in column **C**. This summation includes both the a and the b entries. In the same way, calculate and make the entries for C_2 through C_4.

(2) In I sum all entries in column **1A** above row **C** and enter as S_{a1} in row **C**. In the same way calculate and make entries for S_{a2} through S_{a4} and also for S_{b1} through S_{b3}.

(3) In I calculate the sum of the entries in column **C** of I and enter as C_T in the intersection of row **C** and column **C**.

(4) In I, calculate the sum of the entries in row **C** up to but not including column **C** and enter as S_T in the intersection of row **C** and column **S**. This entry S_T should be the same as C_T beside it.

(5) From I copy *all* of row **1A** into the corresponding positions of row **1A** on worksheet III. Therefore, the first row of Y's is the same as the first row of a's and the first row of X's is the same as the first row of b's.

(6) From I copy column **1A** *through* row **C** into row **1B** of worksheet II. The first column of a's then becomes the first row of w's and S_{a1} is the same as C_{12}.

(7) Next with paper clips fasten worksheet III to I so that the vertical rulings are in alignment and so that the upper edge of worksheet III is between rows **1B** and **2B** of I.

(8) Calculate the reciprocal of Y_{11} from row **1A**, column **1A** of III and enter it in column R, row **1A** of III as R_1.

(9) Lock the entry obtained in step (8) into the keyboard of the machine. Multiply Y_{11} by R_1 to see that it gives 1. Then multiply succeeding values in row **1A** of III by R_1, and without regard to sign, enter the products in the corresponding column positions of row **1B** in I. Do this for each entry in the row up to but not including the entry in the S column.

(10) Give each entry in row **1B** of I the sign opposite to that of the corresponding entry in row **1A** of III if Y_{11} is positive, or the same sign if Y_{11} is negative.

(11) Calculate the sum of all entries in row **1B** of I except C_{12} and enter in column S as S_{12}. This entry should be the same as C_{12}.

Cycle 2

(12) Next fold and crease worksheet II vertically between columns **2** and **3** and fold the right-hand section under so that only columns **1** and **2** are showing, with column **2** at the extreme right.

(13) Lay column **2** of worksheet II to the left and directly beside column **2A** of worksheet I. Be sure that the horizontal rulings of I and II are in alignment.

(14) Multiply column **2A** of worksheet I by column **2** of worksheet II and enter the result in row **2A** of column **2A**, worksheet III, which is still clipped to I. This entry will be Y_{22}. Note that the only entries currently in column 2 of worksheet II are 1 and w_{21}. Therefore, we have for the Y_{22} entry $Y_{22} = a_{22} + w_{21}v_{12}$.

(15) Slide column **2** of worksheet II one column to the right on worksheet I so that it lies immediately to the left of column **3A** of I.

(16) Multiply column **3A** of I by column 2 of II and enter the result in row **2A**, column **3A**, of III, as entry Y_{23}.

(17) Continue sliding column **2** of II to the right one column at a time and get the successive entries on row **2A** of III in the same way you got Y_{22} and Y_{23}. Do this for all of the **A** and **B** columns and also for the **C** column.

(18) Sum *all* of the entries in row **2A** of III to, but not including, the entry C_{21} in the **C** column, and enter in the S column as S_{21}. This should be the same as C_{21} immediately to its left.

(19) Copy Y_{22} from worksheet III into row **2B**, column 2 of worksheet II.

(20) Remove worksheet III from I, and turning I face down, fold and crease between columns **2A** and **3A**. Only columns **1A** and **2A** will be showing on I. These will be at the right, with **2A** at the extreme right.

(21) With worksheet II turned face up, lay column **2A** of worksheet I immediately to the left of column **3** of II. Be sure the horizontal rulings cf the two sheets are in alignment.

(22) Multiply column **3** of II by column **2A** of I and enter the result in row **2B** column **3** of II. This will be w_{32}. Currently there are only two entries in *all* of column **3** of II. These are 1 and w_{31}. Therefore $w_{32} = a_{32} + v_{12}w_{31}$.

(23) If n is greater than 3, slide column **2A** of I to the right on II and get the entries w_{42} and c_{22} in row **2B** of II. This entry c_{22} is obtained in exactly the same way as the other entries in row **2B** of II. It is column **C** of II multiplied by column **2A** of I. Inspection of the entries in these two columns shows that $c_{22} = S_{a2} + v_{12}c_{12}$.

(24) Sum the entries in row **2B** of II up to but not including c_{22} and enter in the **S** column as S_{22}. This should be the same as c_{22}.

(25) Calculate the reciprocal of Y_{22} in row **2A** of III and enter in the **R** column of III as R_2.

(26) Clip worksheet III to I so that the top edge of III lies between rows **2B** and **3B** of I.

(27) Lock R_2 in the keyboard and multiply it by Y_{22} to see that it gives 1.

(28) Multiply each of the remaining entries to but not including the **S** column in row **2A** of III, by R_2 and enter in row **2B** of I without regard to sign.

(29) Give each entry in row **2B** of I the sign opposite to that of the corresponding entry in row **2A** of III if Y_{22} is positive, or the same sign if Y_{22} is negative.

(30) Calculate the sum of all entries in row **2B** of I except C_{22} and enter in column **S** as S_{22}. This entry should be the same as C_{22}.

Cycle 3

(31) Next repeat steps (12) through (18) to get row **3A** of worksheet III except that you begin with columns **3A** of I and **3** of II. Note that now you have one more entry at the bottom of the columns of I and II than you had previously because of the addition of row **2B** on both I and III.

(32) Copy Y_{33} from worksheet III into row **3B**, column **3** of worksheet II.

(33) Repeat steps (20) through (24) to get row **3B** of worksheet II except that now you begin with column **3A** of I and **4A** of II.

(34) Repeat steps (25) through (30) to get row **3B** of I except that now you begin with row **3A** of III and clip worksheet III to I so that its top edge lies immediately below row **3B** of I.

Cycle 4 and Subsequent Cycles

(35) Note that each cycle provides in order, first, a new row in the upper section of III, second, a new row in the lower section of I, and third, a new row in the lower section of II.

For each new cycle, the folds in I and II are advanced one column to the right. To get a new **A**-row in III, always begin with the columns in I and II corresponding to this row. To get the first entry in a **B**-row of II, merely copy the first entry of the corresponding **A**-row in III. To get subsequent entries, multiply the corresponding columns of II by the column of I corresponding to the row.

Note that for each new cycle the columns of I and II corresponding in number to that cycle, and all columns to the right, each have one more element at the bottom because of the **B** row that has been added in the preceding cycle.

The total number of cycles will be equal to the order of the a matrix. Notice, however, that in the last cycle no calculations are required for the last row of II. There is only one entry w_{44} and this is Y_{44} from III. No check is required since it has already been checked in III.

The computations carried out so far constitute what is commonly known as the forward solution. We now proceed with the back solution. For this we require only the lower sections of worksheets I and III. Remember that up to now we have made no entries in the lower left section of worksheet III. These are the F' entries, which, as you recall, are those we started out to solve for.

The back solution

(1) From row **4B** of worksheet I, copy Z_{41} with sign reversed as F_{41} into row **1B** column **4A** of III. Also copy Z_{42} with sign reversed from I into column **4A** as F_{42}. Continue until the last row of Z's in the lower right section of I has been copied with signs reversed as the last column of F's in the lower left section of III.

Cycle 1

(2) Fold and crease worksheet III between rows **4A** and **1B** so that only the lower section of III is showing, and place III so that row **1B** lies immediately below row **3B** of worksheet I. Be sure the vertical rulings of I and III are in alignment.

(3) Multiply row **3B** of I by row **1B** of III and enter the result as F_{31} in row **1B** of III. As you can see, $F_{31} = F_{41}v_{34} - Z_{31}$.

(4) Slide worksheet III up one row so that row **1B** of III lies immediately below row **2B** of I.

(5) Multiply row **2B** of I by row **1B** of III and enter the result in row **1B** of III as F_{21}. As you can see, $F_{21} = F_{31}v_{23} + F_{41}v_{24} - Z_{21}$.

(6) Slide worksheet III up one row so that row **1B** of III lies immediately below row **1B** of I.

(7) Multiply row **1B** of I by row **1B** of III and enter the result as F_{11} in row **1B** of III. The first row of the f' matrix is now completed.

(8) Slide worksheet III up one row so that row **1B** of III lies immediately below row **C** of I.

(9) Multiply row **C** of I by row **1B** of III and enter the result as G_1 at the extreme right end of row **1B** of III. This entry should be 0 or the same as

the one immediately to its left. This constitutes a check on the first row of the F' matrix. If the A matrix had been of order greater than four, the same procedures would have continued until all elements in the first row of F' had been calculated.

Cycle 2

(10) Fold and crease worksheet III between rows **1B** and **2B** so that only rows including **2B** and lower are showing. Place III so that row **2B** lies immediately below **3B** of worksheet I.

(11) Multiply row **3B** of I by row **2B** of III and enter the result as F_{32} in row **2B** of III.

(12) Continue as in steps (4) through (9) to fill in and check the remaining elements of row **2B** of III.

Cycle 3

(13) Fold and crease worksheet III between rows **2B** and **3B** so that only row **3B** is showing. Proceed as in steps (3) through (9).

Subsequent Cycles

(14) If there had been more than three rows to F' or more than three columns to b, subsequent cycles would have followed the same routines. For each cycle, worksheet III is folded one row lower than for the preceding cycle. Otherwise the operations within one cycle are the same as for any other.

In many problems, the matrix b consists of a single column vector. In such cases, of course, F' has only one row so that there is only one cycle to the back solution.

A numerical example of the method outlined in this section is given in Fig. 20.3.3 for the case in which the a matrix is of order four and the b matrix has four rows and three columns.

Worksheet I

i	1A	2A	3A	4A	1B	2B	3B	C	S
1A	2.120	.780	1.160	.920	1.880	.960	2.040	9.860	
2A	3.460	2.200	.660	1.580	1.000	.980	1.120	11.000	
3A	1.740	1.820	4.040	2.220	3.060	1.760	1.520	16.160	
4A	.860	2.180	4.000	1.160	1.020	1.040	1.060	11.320	
C	8.180	6.980	9.860	5.880	6.960	4.740	5.740	48.340	48.340
1B	−1.000	− .368	− .547	− .434	− .887	− .453	− .962	−4.651	−4.651
2B		−1.000	1.330	− .084	2.232	.633	2.383	5.493	5.494
0D			−1.000	− .290	.001	.000	.571	3.124	3.124
4B				−1.000	− .828	− .342	1.100	−1.071	−1.070

Fig. 20.3.3 (Cont. on next page)

Worksheet II

i	1	2	3	4	C	S
1A	1	0	0	0		
2A	0	1	0	0		
3A	0	0	1	0		
4A	0	0	0	1		
C					1	
1B	2.120	3.460	1.740	.860	8.180	8.180
2B		.927	1.180	1.864	3.970	3.971
3B			4.658	6.009	10.666	10.667
4B				−1.130	−1.129	−1.130

Worksheet III

R	i	1A	2A	3A	4A	1B	2B	3B	C	S
.47170	1A	2.120	.780	1.160	.920	1.880	.960	2.040	9.860	9.860
1.07875	2A		.927	−1.233	.078	−2.069	− .587	−2.209	−5.092	−5.093
.21468	3A			4.658	1.366	4.150	1.719	2.658	14.549	14.551
−.88496	4A				−1.130	− .936	− .387	1.243	−1.213	−1.210
	1B	.703	−1.440	.648	.828	−1	0	0	0	− .003
	2B	.269	− .304	.269	.342	0	−1	0	0	.002
	3B	1.357	−1.103	.893	−1.100	0	0	−1	0	− .002

FIG. 20.3.3. (Cont.) Product of a Matrix by Inverse of a Square Matrix

20.4 The Inverse of a General Square Matrix

To get the inverse of a general square matrix, the computational procedure is exactly the same as in the previous section, where we multiplied a matrix b by the inverse of a matrix a. Getting the inverse of a is simply a special case of getting the product of a matrix b by the inverse of a. In this special case, the matrix b is the identity matrix. Obviously the product of the identity matrix by the inverse of a is the inverse of a.

Worksheet II for getting the inverse of a is exactly the same as in the preceding section. Since the b matrix is the identity matrix, worksheets I and III have several characteristics they do not have in the preceding section. These are illustrated in Fig. 20.4.1 for a fourth-order matrix.

Worksheet I

i	1A	2A	3A	4A	1B	2B	3B	4B	C	S
1A	a_{11}	a_{12}	a_{13}	a_{14}	1	0	0	0	C_1	
2A	a_{21}	a_{22}	a_{23}	a_{24}	0	1	0	0	C_2	
3A	a_{31}	a_{32}	a_{33}	a_{34}	0	0	1	0	C_3	
4A	a_{41}	a_{42}	a_{43}	a_{44}	0	0	0	1	C_4	
C	S_{a1}	S_{a2}	S_{a3}	S_{a4}	1	1	1	1	C_T	S_T
1B	-1	v_{12}	v_{13}	v_{14}	Z_{11}	0	0	0	C_{12}	S_{12}
2B		-1	v_{23}	v_{24}	Z_{21}	Z_{22}	0	0	C_{22}	S_{22}
3B			-1	v_{34}	Z_{31}	Z_{32}	Z_{33}	0	C_{32}	S_{32}
4B				-1	Z_{41}	Z_{42}	Z_{43}	Z_{44}	C_{42}	S_{42}

Worksheet III

R	i	1A	2A	3A	4A	1B	2B	3B	4B	C	S
R_1	1A	Y_{11}	Y_{12}	Y_{13}	Y_{14}	1	0	0	0	C_{11}	S_{11}
R_2	2A		Y_{22}	Y_{23}	Y_{24}	X_{21}	1	0	0	C_{21}	S_{21}
R_3	3A			Y_{33}	Y_{34}	X_{31}	X_{32}	1	0	C_{31}	S_{31}
R_4	4A				Y_{44}	X_{41}	X_{42}	X_{43}	1	C_{41}	S_{41}
	1B	F_{11}	F_{21}	F_{31}	F_{41}	-1	0	0	0	0	G_1
	2B	F_{12}	F_{22}	F_{32}	F_{42}	0	-1	0	0	0	G_2
	3B	F_{13}	F_{23}	F_{33}	F_{43}	0	0	-1	0	0	G_3
	4B	F_{14}	F_{24}	F_{34}	F_{44}	0	0	0	-1	0	G_4

FIG. 20.4.1

You will notice that in both worksheets I and III the number of **B** columns in the right sections is equal to the number of **A** columns in the left sections. This must be true since the matrix b is now the identity matrix which is of course square. For the same reason, the number of **B** rows in the lower section of III is the same as the number of **A** rows in the upper section. Therefore, all of the matrices in each of the four quadrants of I and II are square and of the same order. Furthermore, the matrix of Z's in the lower right quadrant of I and the matrix of X's in the upper right quadrant of III are each lower triangular, the latter having diagonal elements of unity.

Once the worksheets have been ruled up for a matrix of given order, it is possible to write in the entries of unity at once, irrespective of what the elements of the a matrix might be, as follows:

In worksheet I, you can write in the 1's in the diagonal positions in the upper right section and the 1's immediately below in the C row. You can also write in the diagonal of -1's in the lower left quadrant, just as you did in the preceding section.

Worksheet 1

i	1A	2A	3A	4A	1B	2B	3B	4B	C	S
1A	2.120	.780	1.160	.920	1	0	0	0	5.980	
2A	3.460	2.200	.660	1.580	0	1	0	0	8.900	
3A	1.740	1.820	4.040	2.220	0	0	1	0	10.820	
4A	.860	2.180	4.000	1.160	0	0	0	1	9.200	
C	8.180	6.980	9.860	5.880	1	1	1	1	34.900	34.900
1B	-1	$-.368$	$-.547$	$-.434$	$-.472$	0	0	0	-2.821	-2.821
2B		-1	1.330	$-.084$	1.761	-1.079	0	0	.929	.928
3B			-1	$-.293$	$-.270$	.273	$-.215$	0	-1.504	-1.505
4B				-1	1.110	$-.328$	-1.143	.885	$-.477$	$-.476$

Worksheet II

i	1	2	3	4	C	S
1A	1	0	0	0		
2A	0	1	0	0		
3A	0	0	1	0		
4A	0	0	0	1		
C					1	
1B	2.120	3.460	1.740	.860	8.180	8.180
2B		.927	1.180	1.864	3.970	3.971
3B			4.658	6.009	10.666	10.667
4B				-1.130	-1.129	-1.130

Worksheet III

R	i	1A	2A	3A	4A	1B	2B	3B	4B	C	S
.47170	1A	2.120	.780	1.160	.920	1	0	0	0	5.980	5.980
1.07875	2A		.927	-1.233	.078	-1.633	1	0	0	$-.861$	$-.861$
.21468	3A			4.658	1.366	1.257	-1.273	1	0	7.006	7.008
$-.88496$	4A				-1.130	1.254	$-.371$	-1.292	1	$-.538$	$-.539$
	1B	.951	$-.876$	.595	-1.110	-1	0	0	0	0	.005
	2B	$-.147$	.561	$-.369$	.328	0	-1	0	0	0	$-.004$
	3B	$-.336$	$-.256$	$-.120$	1.143	0	0	-1	0	0	.002
	4B	.088	.419	.259	$-.885$	0	0	0	-1	0	$-.006$

FIG. 20.4.2. Inverse of a General Square Matrix

Worksheet II, you recall, is identical to II of the preceding section.

In worksheet III you can write in the 1's in the diagonal of the upper right quadrant. As in the preceding section, you can also write in the -1's in the diagonal of the lower right quadrant.

In getting the forward solution for the F' matrix, follow exactly the instructions given from (1) to (35) in the preceding section. However, because b is an identity matrix, the computations are somewhat simpler.

The back solution for the entries in F' is also exactly the same as for the back solution instructions in steps (1) through (14) of the previous section. Here again the computations are somewhat simplified because the matrix of Z's in the lower right quadrant of I consists of all zeros above the diagonal.

Precisely the same system of checks is used in both this and the previous section. A numerical example of the solution of the inverse of a fourth-order matrix is given in Fig. 20.4.2.

20.5 The Product of a Matrix by the Inverse of a Triangular Matrix

Only rarely do we require the product of a matrix by the inverse of a triangular matrix except in the process of getting the product of a matrix by the inverse of a square matrix. The procedure is much simpler for a triangular than for a square matrix. Two worksheets are required. Worksheet I is identical in vertical and horizontal ruling with that of worksheet III in Section 20.2. Worksheet II is the same in vertical and horizontal ruling to the lower section of worksheet I in Section 20.3. These worksheets are illustrated in Fig. 20.5.1 for a, a fourth-order matrix and b, a 4×3 matrix.

Worksheet I

R	i	1A	2A	3A	4A	1B	2B	3B	C	S
R_1	1A	a_{11}	a_{12}	a_{13}	a_{14}	b_{11}	b_{12}	b_{13}	C_1	
R_2	2A		a_{22}	a_{23}	a_{24}	b_{21}	b_{22}	b_{23}	C_2	
R_3	3A			a_{33}	a_{34}	b_{31}	b_{32}	b_{33}	C_3	
R_4	4A				a_{44}	b_{41}	b_{42}	b_{43}	C_4	
	1B	F_{11}	F_{21}	F_{31}	F_{41}	-1	0	0	0	G_1
	2B	F_{12}	F_{22}	F_{32}	F_{42}	0	-1	0	0	G_2
	3B	F_{13}	F_{23}	F_{33}	F_{43}	0	0	-1	0	G_3

Fig. 20.5.1 (Cont. on next page)

Worksheet II

i	1A	2A	3A	4A	1B	2B	3B	C	S
C	S_{a1}	S_{a2}	S_{a3}	S_{a4}	S_{b1}	S_{b2}	S_{b3}	C_T	S_T
1	-1	v_{12}	v_{13}	v_{14}	Z_{11}	Z_{12}	Z_{13}	C_1	S_1
2		-1	v_{23}	v_{24}	Z_{21}	Z_{22}	Z_{23}	C_2	S_2
3			-1	v_{34}	Z_{31}	Z_{32}	Z_{33}	C_3	S_3
4				-1	Z_{41}	Z_{42}	Z_{43}	C_4	S_4

Fig. 20.5.1 (Cont.)

After the worksheets are ruled up as in Fig. 20.5.1, the unused spaces are crossed out as indicated and the row and column designations are written in. Then a diagonal of -1's is entered in the lower right-hand section of worksheet I and the left-hand section of II. Next the triangular matrix a is copied into the upper left quadrant of I and the matrix b is copied into the upper right quadrant of I. Then the computations may be made for $a^{-1}b = F$. The elements of F will finally be entered in transposed form in the lower left quadrant of I. The steps in the computations are as follows:

(1) Sum all elements of row **1A** of I and enter as C_1 in column **C** of 1. In the same way sum the remaining rows of I and enter as C_2 to C_4.

(2) Sum each column of I and enter as the corresponding element of row **C** of worksheet II. This will provide all of the S_a and S_b entries as well as the C_T entry.

(3) Sum all the entries in row **C** of II except C_T and enter as S_T. This should be the same as C_T.

(4) Calculate the reciprocal of each left-hand element in the a matrix in I and enter in the corresponding position in the **R** column at the extreme left of I.

(5) Lock the reciprocal R_1 into the keyboard as a constant multiplier. Multiply each element in row **1A** of I including the **C** column by R_1 and enter the product without regard to sign in the corresponding position in row **1** of II. Be sure that $R_1 a_{11} = 1$.

(6) To each element in row **1** of II, give the sign opposite to that of the corresponding element in row **1A** of I.

(7) Sum the entries in row **1** of II and enter as S_1. This should be the same as C_1 beside it.

(8) Repeat steps (5) through (7) to get the remaining rows of II.

(9) To get the F entries in the lower left quadrant of I, follow the steps for the back solution, in Section 20.3. However, remember that what is referred to as worksheet III in Section 20.3 now becomes worksheet I, and

what is referred to as worksheet I in Section 20.3 becomes worksheet II. Furthermore,. the references to rows **1B**, **2B**, etc. of worksheet I are now merely rows **1**, **2**, etc. of worksheet II.

A numerical example of the computations is given in Fig. 20.5.3.

Worksheet I

R	i	1A	2A	3A	4A	1B	2B	3B	C	S
.47170	1A	2.120	.780	1.160	.920	1.880	.960	2.040	9.860	
.45455	2A		2.200	.660	1.580	1.000	.980	1.120	7.540	
.24752	3A			4.040	2.220	3.060	1.760	1.520	12.600	
.86207	4A				1.160	1.020	1.040	1.060	4.280	
	1B	.451	− .258	.274	.879	−1	0	0	0	.001
	2B	.161	− .182	− .056	.897	0	−1	0	0	.005
	3B	.674	− .109	− .126	.914	0	0	−1	0	.000

Worksheet II

i	1A	2A	3A	4A	1B	2B	3B	C	S
C	2.120	2.980	5.860	5.880	6.960	4.740	5.740	34.280	34.280
1	−1	− .368	− .547	− .434	− .887	− .453	− .962	−4.651	−4.651
2		−1	− .300	− .718	− .455	− .445	− .509	−3.427	−3.427
3			−1	− .549	− .757	− .436	− .376	−3.119	−3.118
4				−1	− .879	− .897	− .914	−3.690	−3.690

Fig. 20.5.3. Product of a Matrix by the Inverse of a Triangular Matrix

20.6 The Inverse of a Triangular Matrix

Although the inverse of a triangular matrix as such is rarely needed in the analysis of data from the social sciences, we nevertheless present a convenient worksheet layout for this purpose. The procedure is the same as in Section 20.5 except that we take a special case of the b matrix, namely the identity matrix. This results in certain simplifications. Two worksheets are required, as in Section 20.5. These are illustrated by Fig. 20.6.1.

Worksheet I

i	1	2	3	4	C
1A	a_{11}	a_{12}	a_{13}	a_{14}	C_1
2A		a_{22}	a_{23}	a_{24}	C_2
3A			a_{33}	a_{34}	C_3
4A				a_{44}	C_4
1B	R_1	0	0	0	1
2B	F_{12}	R_2	0	0	1
3B	F_{13}	F_{23}	R_3	0	1
4B	F_{14}	F_{24}	F_{34}	R_4	1

Worksheet II

i	1	2	3	4	C	S
C	S_{a1}	S_{a2}	S_{a3}	S_{a4}	C_T	S_T
1	-1	v_{12}	v_{13}	v_{14}	C_1	S_1
2		-1	v_{23}	v_{24}	C_2	S_2
3			-1	v_{34}	C_3	S_3
4				-1	C_4	S_4

Fig. 20.6.1

The elements of the triangular matrix a are written into the upper section of worksheet I and the computations proceed according to the following steps:

(1) Sum the rows of a in I and enter as C_1 to C_4 respectively in column **C** of I.

(2) Sum the columns of a including the **C** column and enter as row C of II.

(3) Sum the entries in row **C** of II up to but not including C_T and enter S_T. This should be the same as C_T.

(4) Calculate the reciprocals of the diagonal entries in the upper section of I and enter as R_1 to R_4 in the diagonal positions in the lower section of I.

(5) Multiply each element in row **1A** of I by R_1 and enter in row **1** of II without regard to sign. Be sure that $a_{11}R_1 = 1$.

(6) Give each element in row **1** of II the sign opposite to its corresponding element in row **1A** of I.

(7) Sum the entries in row **1** of II up to but not including C_1 and enter as S_1. This should be the same as C_1.

(8) As in step (7), calculate the remaining rows of II by using the corresponding R from I.

You are now ready to compute the F entries below the diagonal, which will complete the computation of $a^{1^{-1}}$. Note that the inverse will be given in transpose form.

Row **1B**

Row **1B** is already given since it consists only of R_1 and zeros.

Row **2B**

To compute F_{12} take the product of row **2B** of I by row **1** of II. But this is merely $F_{12} = R_2 v_{12}$.

Row **3B**

First calculate F_{23} by taking the product of row **3B** of I by row **2** of II. Then calculate F_{13} by taking the product of row **3B** of I by row **1** of II, remembering that now F_{23} is a part of row **3B** of I.

Row **4B**

First calculate F_{34} by taking the product of row **4B** of I by row **3** of II. Using row **4B** of I as a constant multiplier, move up from row **3** of II a row at a time to get F_{24}, then F_{14}.

Succeeding rows of F

If a is of larger order, the procedure for getting successive elements of each row from right to left is the same. You use the row of I as a constant multiplier and move up a row at a time on the rows of II to complete the row of I. For a large matrix, it is well to fold and crease between rows of the lower section of I, so that you can lay the row of I directly above (or below) the row of II that is being multiplied.

The check on the rows of the F' matrix is obtained by multiplying each row by row **C** of II. For each row, the product should be 1, as indicated in the lower section of column C of I. You should compute this check for each row of F' before starting to compute the next row.

A numerical illustration of the procedure is given in Fig. 20.6.2.

Worksheet I

i	1	2	3	4	C	
1A	2.120	.780	1.160	.920	4.980	
2A		2.200	.660	1.580	4.440	
3A			4.040	2.220	6.260	
4A				1.160	1.160	
1B	.47170	0	0	0	1	
2B	$-$ .167	.45455	0	0	1	
3B	$-$ .108	$-$ .074	.24752	0	1	
4B	.061	$-$ .477	$-$ .474	.86207	1	

Fig. 20.6.2. (Cont. on next page)

Worksheet II

i	1	2	3	4	C	S
C	2.12	2.98	5.86	5.88	16.84	16.84
1	−1	− .368	− .547	− .434	−2.349	−2.349
2		−1	− .300	− .718	−2.018	−2.018
3			−1	− .549	−1.549	−1.549
4				−1	−1.000	−1.000

Fig. 20.6.2. (Cont.) The Inverse of a Triangular Matrix

20.7 Multiplication of a Matrix by the Inverse of a Symmetric Matrix

The computational procedures for multiplying a matrix by the inverse of a symmetric matrix may be simplified somewhat from those involving the inverse of a general square matrix. In the social sciences, when a prediction study requires multiplication of a matrix by the inverse of a matrix, the inverse is ordinarily that of a symmetric matrix. In fact, it is usually a vector derived from the criterion attribute or variable in the primary matrix. If you know how to carry out the computations for a matrix of any order, however, you can certainly do so for a vector.

You have three worksheets as in Section 20.3, but they are quite different in design. The three worksheets are given in symbolic form in Fig. 20.7.1.

Worksheet I

	a	b	C_1	
	S_a'	S_b'	C_2	S_2
R	Y	X	C_2	S_2

Worksheet II

I			
v	Z	C_2	S_2

Worksheet III

F'		I	0	G_1

Fig. 20.7.1

The meanings of the symbols in each of the worksheets are as follows:

Worksheet I

a is the symmetric matrix whose inverse is involved in the computations.

b is the matrix to be premultiplied by the inverse of a.

C_1 is a column of row sums of both a and b.

S_a' is a row of column sums of a.

S_b' is a row of column sums of b.

C_T is the total of C_1 and also of S_a' and S_b' combined.

R is a column of the reciprocals of the corresponding diagonal elements of Y.

Y is an upper triangular matrix closely related to a lower triangular factor of a.

X is a matrix to be computationally defined.

C_2 is a check column to be computationally defined.

S_2 is a column of row sums of both X and Y and should be the same as C_2.

Worksheet II

I is simply the identity matrix whose order is the same as a.

v is an upper triangular matrix closely related to Y in worksheet I.

Z is a matrix closely related to X in worksheet I.

C_2 is a check column.

S_2 is a column of row sums of w' and should be the same as C_2.

Worksheet III

F' is the transpose of $a^{-1}b$ which is the end result of our computations.

I is the identity matrix whose order is the width of b.

0 is a column of zeros.

G is a final check column whose elements are also zeros.

Next let us assume that the a matrix whose inverse is involved is of order four and the matrix b that is to be premultiplied by the inverse of a has four rows and three columns. The three required worksheets are shown in Fig. 20.7.2.

Worksheet I

R	**i**	**1A**	**2A**	**3A**	**4A**	**1B**	**2B**	**3B**	**C**	**S**
	1A	a_{11}	a_{12}	a_{13}	a_{14}	b_{11}	b_{12}	b_{13}	C_1	
	2A		a_{22}	a_{23}	a_{24}	b_{21}	b_{22}	b_{23}	C_2	
	3A			a_{33}	a_{34}	b_{31}	b_{32}	b_{33}	C_3	
	4A				a_{44}	b_{41}	b_{42}	b_{43}	C_4	
	C	S_{a1}	S_{a2}	S_{a3}	S_{a4}	S_{b1}	S_{b2}	S_{b3}	C_T	S_T
R_1	1B	Y_{11}	Y_{12}	Y_{13}	Y_{14}	X_{11}	X_{12}	X_{13}	C_{12}	S_{12}
R_2	2B		Y_{22}	Y_{23}	Y_{24}	X_{21}	X_{22}	X_{23}	C_{22}	S_{22}
R_3	3B			Y_{33}	Y_{34}	X_{31}	X_{32}	X_{33}	C_{32}	S_{32}
R_4	4B				Y_{44}	X_{41}	X_{42}	X_{43}	C_{42}	S_{42}

FIG. 20.7.2 (Cont. on next page)

Worksheet II

	i	1A	2A	3A	4A	1B	2B	3B	C	S
	1A	1	0	0	0					
	2A	0	1	0	0					
	3A	0	0	1	0					
	4A	0	0	0	1					
	1B	-1	v_{12}	v_{13}	v_{14}	Z_{11}	Z_{12}	Z_{13}	C_{12}	S_{12}
	2B		-1	v_{23}	v_{24}	Z_{21}	Z_{22}	Z_{23}	C_{22}	S_{22}
	3B			-1	v_{34}	Z_{31}	Z_{32}	Z_{33}	C_{32}	S_{32}
	4B				-1	Z_{41}	Z_{42}	Z_{43}	C_{42}	S_{42}

Worksheet III

	i	1A	2A	3A	4A	1B	2B	3B	Check	
	1	F_{11}	F_{21}	F_{31}	F_{41}	-1	0	0	0	G_1
	2	F_{12}	F_{22}	F_{32}	F_{42}	0	-1	0	0	G_2
	3	F_{13}	F_{23}	F_{33}	F_{43}	0	0	-1	0	G_3

FIG. 20.7.2 (Cont.)

In ruling up the worksheets for this problem, make sure the rulings are in alignment, just as you did for the problems of the previous sections. The rulings for worksheet I determine the rulings for II and III. Both the vertical and horizontal rulings of worksheet I are exactly the same as for worksheet I in Section 20.3. Therefore, in drawing up worksheet I for this section, follow exactly the specifications as given in that section.

For worksheet II, both the vertical and horizontal rulings are spaced exactly as for worksheet I.

The vertical rulings of worksheet III are exactly the same as for I and II. The horizontal rulings provide simply for a row of column designations and a succession of rows equal in number to the width of b, which in the example is three.

After the three worksheets have been ruled up, you should cross out with diagonal rulings those spaces on each of the three worksheets in which no entries are to be made, as shown in Fig. 20.7.2. Then you will write in the row and column designations in the row and column labeled i for each of the three worksheets.

Next write in the entries in worksheets II and III that do not depend on the elements of the matrices a and b. In II these will be a diagonal of -1's in the lower left-hand quadrant. In III they will be a diagonal of -1's in the right-hand section and a column of 0's in the next-to-last column.

You now write in the elements of a and b matrices in the upper left and right sections respectively of worksheet I. Since a is symmetric, write in only the elements in and above the diagonal. Once the a and b elements have been entered, you proceed with the computations for $a^{-1}b = F$. As in Section 20.3, the steps consist of two main groups of operations, the forward solution and the back solution.

The forward solution

(1) Sum all entries in row **1A** of worksheet I and enter as C_1 in column **C**. This summation includes both the a and b entries. In the same way, calculate and make the entries for C_2 through C_4. But remember that the entries to the left of the diagonal of the matrix a are implied. The sums in column **C** are to include these entries. Since a is symmetrical, the entries to the left of the diagonal entry are the same as the corresponding entries above the diagonal. Therefore, to get the sum of the entries in a row of a, including the unrecorded entries, get the sum total of the recorded elements in the row and the entries above the diagonal.

(2) In I sum all entries in column **1A** above row **C** and enter as S_{a1} in row **C**. In the same way, calculate and make entries for S_{a2} through S_{a4} and also for S_{b1} through S_{b3}. In calculating the S_a entries, remember that the entries below the diagonal of a are implied and must be included in the column sums. Therefore, to get the sum of the entries in a column of a including the unrecorded entries, get the sum total of the recorded elements in the column and the entries to the right of the diagonal.

(3) Calculate the sum of the entries in column **C** of I and enter as C_T in the intersection of row **C** and column **C**.

(4) In I calculate the sum of entries in row C up to but not including column **C** and enter as S_T in the intersection of row **C** and column **S**. This entry S_T should be the same as C_T immediately to its left.

Cycle 1

(5) In I copy all of row **1A** into the corresponding positions of row **1B** of I. Therefore, the first row of Y's is the same as the first row of a's. the first row of X's is the same as the first row of b's, and C_{12} is the same as C_1.

(6) Calculate the reciprocal of Y_{11} in row **1B** column **1A**, of I, and enter as R_1 in column R, row **1B** at the extreme right of I.

(7) Fold and crease worksheet I between lines **1B** and **2B** so that everything below line **1B** is turned under and line **1B** is the last line showing at the bottom of worksheet I.

(8) With paper clips, fasten worksheet I over II so that the vertical rulings are in alignment and so that line **1B** of I lies immediately above **1B** of II.

(9) Lock the reciprocal R_1 calculated in step (6) into the keyboard of the machine. Multiply Y_{11} by R_1, which is in the keyboard, to see that the product is 1. Then multiply succeeding values in row **1B** of I by R_1, and without regard to sign, enter the products in the corresponding column

positions of row **1B** in II. Do this for each entry in the row up to but not including the entry in the S column.

(10) Give each entry in row **1B** of II the sign opposite to that of the corresponding entry in row **1B** of I.

(11) Calculate the sum of all entries in row **1B** of II except C_{12} and enter in column S as S_{12}. This entry should be the same as C_{12}.

Cycle 2

(12) Unfasten worksheet I from II and unfold I. Then fold and crease II between column **2A** and **3A** and turn face down so that all columns to the right of **2A** are turned under and **2A** is showing at the extreme right.

(13) Place column **2A** of II immediately to the left of column **2A** of I and be sure the horizontal rulings of both worksheets are in alignment. Multiply these two columns together and enter the result as Y_{22} in row **2B**, column **2A** of I. Currently there are only two nonzero entries in column **2A** of II. Therefore, $Y_{22} = 1 \times a_{22} + v_{12}Y_{12}$.

(14) Move column **2A** of II one column to the right on I so that it lies immediately to the left of column **3A** of I. Multiply these two columns together and enter the result as Y_{23} in row **2B** of I.

(15) By continuing to slide column **2A** of II one column to the right on I calculate all but S_{22} of the remaining entries in row **2B** of worksheet I.

(16) Sum all entries in row **2B** of I except C_{22} and enter as S_{22} in column **S**. This entry should be the same as C_{22}.

(17) Calculate the reciprocal of Y_{22} in row **2B**, column **2A**, of I and enter as R_2 in column R, row **2B**, at the extreme right of I.

(18) Fold and crease worksheet I between lines **2B** and **3B** so that everything below line **2B** is turned under and line **2B** is the last line showing at the bottom of worksheet I.

(19) As in step (8), fasten I over II, except that now line **2B** of I lies immediately above line **2B** of II.

(20) Lock the reciprocal of Y_{22} calculated in step (17) into the machine. Multiply Y_{22} by R_1, which is in the keyboard, to see that the product is 1. Then multiply succeeding values in row **2B** of I by R_2, and without regard to sign, enter the products in the corresponding column positions of row **2B** in II. Do this for each entry in the row up to, but not including, the entry in the **S** column.

(21) Give each entry in row **2B** of II the sign opposite that of the corresponding entry in row **2B** of I.

(22) Calculate the sum of all entries in row **2B** of II except C_{22} and enter in column **S** as S_{22}. This entry should be the same as C_{22}.

Cycle 3

(23) Next repeat steps (12) through (16) to get line **3B** of worksheet I, except that now you begin with column **3A** of both I and II.

(24) Repeat steps (17) through (22) to get row **3B** of II, except now you fold I between lines **3B** and **4B** so that you can fasten line **3B** of I immediately above the line **3B** of II.

Cycle 4 and subsequent cycles

(25) Note that each new cycle provides a new row in the lower section of I and a new row in the lower section of II. For each new cycle the fold is advanced one column to the right on II. To get a new **B** row for I, always begin with the columns of I and II corresponding to this row. Each new **B** row of II is derived from the latest **B** row of I.

In the example given, you have only four cycles, since a is a fourth-order matrix. The computations for successive cycles would continue as outlined, however, no matter how large the order of a.

The back solution

After all the rows in the lower sections of I and II have been calculated, you will have finished with the forward solution and will be ready to start the back solution. The back solution is carried out as in Section 20.3, but since the design of the worksheets is somewhat different for a symmetric matrix, the instructions are repeated with the appropriate modifications.

(1) From row **4B** of worksheet II, copy Z_{41} with sign reversed as F_{41} into row 1, column **4A** of worksheet III. Also copy Z_{42} with sign reversed from II into column **4A** of III as F_{42}. Continue until the last row of Z's in the lower right quadrant of II has been copied with signs reversed as the last column of F's in the left section of II.

Cycle 1

(2) Fold and crease worksheet III between row 1 and the column designation row so that the latter is folded under, and place worksheet III so that row 1 lies immediately below row **3B** of worksheet II. Be sure the vertical rulings of II and III are in alignment.

(3) Multiply row **3B** of II by row 1 of III, and enter the result as F_{31} in row 1 of III. As you can see, $F_{31} = F_{41}v_{34} - Z_{31}$.

(4) Slide worksheet III up one row so that row 1 of III lies immediately below row **2B** of II.

(5) Multiply row **2B** of II by row 1 of III and enter the result as F_{21} in row 1 of III. As you can see, $F_{21} = F_{31}v_{23} + F_{41}v_{24} - Z_{21}$.

(6) Slide worksheet III up one row so that row 1 of III lies immediately below row **1B** of II.

(7) Multiply row **1B** of II by row 1 of III and enter the result as F_{11} in row 1 of III. The first row of the F' matrix is now completed.

(8) Next place row 1 of worksheet III immediately below the **C** row of worksheet I.

(9) Multiply row **C** of I by row 1 of III and enter the result as G_1 at the extreme right entry of row 1 of III. This entry should be 0 within limits of decimal error, or the same as the entry immediately to its left. This constitutes a check on the first row of the F' matrix.

If the a matrix is of order greater than four, the same procedures continue until all elements in the first row of F' have been calculated.

Cycle 2

(10) Fold and crease worksheet III between rows **1** and **2** so that row **2** is the first row showing at the top of the sheet. Place III so that row **2** lies immediately below row **3B** of II.

(11) Multiply row **3B** of II by row **2** of III and enter the result as F_{32} in row **2** of III.

(12) Continue as in steps (4) through (9) to fill in and check the remaining elements of row **2** of III.

Cycle 3

(13) Fold and crease worksheet III between rows **2** and **3** so that only row **3** is showing. Proceed as in steps (3) through (9).

Subsequent cycles

(14) If there are more than three rows of F', or more than three columns in b, subsequent cycles follow the same routines. For each cycle, worksheet III is folded one row lower than the preceding cycle. Otherwise the operations within one cycle are the same as for any other.

A numerical example of the method outlined in this section is given in Fig. 20.7.3.

Worksheet I

R	i	1A	2A	3A	4A	1B	2B	3B	C	S
	1A	2.120	.780	1.160	.920	1.880	.960	2.040	9.860	
	2A	.780	2.200	.660	1.580	1.000	.980	1.120	8.320	
	3A	1.160	.660	4.040	2.220	3.060	1.760	1.520	14.420	
	4A	.920	1.580	2.220	1.160	1.020	1.040	1.060	9.000	
	C	4.980	5.220	8.080	5.880	6.960	4.740	5.740	41.600	41.600
.47170	1B	2.120	.780	1.160	.920	1.880	.960	2.040	9.860	9.860
.52274	2B		1.913	.233	1.241	.308	.627	.369	4.692	4.691
.29612	3B			3.377	1.565	1.994	1.158	.359	8.454	8.453
−1.30039	4B				− .769	− .919	− .320	− .231	−2.239	−2.239

Worksheet II

i	1A	2A	3A	4A	1B	2B	3B	C	S
1A	1	0	0	0					
2A	0	1	0	0					
3A	0	0	1	0					
4A	0	0	0	1					
1B	−1	− .368	− .547	− .434	− .887	− .453	− .962	−4.651	−4.65
2B		−1	− .122	− .649	− .161	− .328	− .193	−2.452	−2.45
3B			−1	− .463	− .590	− .343	− .106	−2.503	−2.50
4B				−1	−1.195	− .416	− .300	−2.912	−2.91

Fig. 20.7.3 (Cont. on next page)

Worksheet III

i	1A	2A	3A	4A	1B	2B	3B		Check
1	.576	− .619	.037	1.195	−1	0	0	0	.003
2	.176	.040	.150	.416	0	−1	0	0	.003
3	.849	.002	− .033	.300	0	0	−1	0	− .004

FIG. 20.7.3. (Cont.) Product of a Matrix by the Inverse of a Symmetric Matrix

In most problems involving the multiplication of a matrix by the inverse of another, a is a symmetric matrix and b is a vector. If b is a vector, worksheet III has only one row in addition to the row of column designations. Rather than draw up a separate worksheet, append this row to the bottom of worksheet I. In this case, the back solution consists simply of the first cycle in the back solution we have just outlined.

20.8 The Inverse of a Symmetric Matrix

The computations for the inverse of a symmetric matrix may be simplified somewhat from those of Section 20.4. We need only two worksheets as shown in Fig. 20.8.1.

Worksheet I

R	i	1A	2A	3A	4A	1B	2B	3B	4B	C	S
	1A	a_{11}	a_{12}	a_{13}	a_{14}	1	0	0	0	C_1	
	2A		a_{22}	a_{23}	a_{24}	0	1	0	0	C_2	
	3A			a_{33}	a_{34}	0	0	1	0	C_3	
	4A				a_{44}	0	0	0	1	C_4	
R_1	1B	Y_{11}	Y_{12}	Y_{13}	Y_{14}	1	0	0	0	C_{12}	S_{12}
R_2	2B		Y_{22}	Y_{23}	Y_{24}	X_{21}	1	0	0	C_{22}	S_{22}
R_3	3B			Y_{33}	Y_{34}	X_{31}	X_{32}	1	0	C_{32}	S_{32}
R_4	4B				Y_{44}	X_{41}	X_{42}	X_{43}	1	C_{42}	S_{42}

Worksheet II

i	1A	2A	3A	4A	1B	2B	3B	4B	C	S
1A	1	0	0	0	F_{11}	F_{12}	F_{13}	F_{14}	C_{11}	S_{11}
2A	0	1	0	0	F_{12}	F_{22}	F_{23}	F_{24}	C_{12}	S_{12}
3A	0	0	1	0	F_{13}	F_{23}	F_{33}	F_{34}	C_{13}	S_{13}
4A	0	0	0	1	F_{14}	F_{24}	F_{34}	F_{44}	C_{14}	S_{14}
1B	−1	v_{12}	v_{13}	v_{14}	Z_{11}				C_{21}	S_{21}
2B		−1	v_{23}	v_{24}	Z_{21}	Z_{22}			C_{22}	S_{22}
3B			−1	v_{34}	Z_{31}	Z_{32}	Z_{33}		C_{23}	S_{23}
4B				−1	Z_{41}	Z_{42}	Z_{43}	Z_{44}	C_{24}	S_{24}

FIG. 20.8.1 (Cont.)

The vertical rulings of worksheet I are as follows from left to right:

(1) The R column

(2) the column of row designations

(3) a succession of columns equal in number to the order of the matrix a

(4) a second series of columns equal in number to the order of a

(5) a check column C

(6) a sum column S

The horizontal rulings of worksheet I are as follows from top to bottom:

(1) a row of column designations

(2) a series of rows equal in number to the order of a

(3) a second series of rows equal in number to the order of a

Both the vertical and horizontal rulings of worksheet II are exactly the same as for worksheet I.

First write in the row and column designations and then cross out the spaces in which no entries are to be made, as indicated in Fig. 20.8.1. Then write in the elements that are independent of the elements of a. These will all be 1's or -1's as follows:

(1) a diagonal of 1's in the upper right quadrant of worksheet I

(2) a diagonal of 1's in the lower right section of I

(3) a diagonal of 1's in the upper left quadrant of worksheet II

(4) a diagonal of -1's in the lower left quadrant of II

You may now write in the elements of the a matrix in the upper left quadrant of worksheet I. Since a is symmetric, write in only the elements in and above the diagonal. Once the elements of a are entered, you proceed with the computation of the inverse of a. The first part of the solution is very much the same as the forward solution for Section 20.6. This consists of getting the entries for the lower sections of both worksheets I and II. The steps are as follows:

(1) Sum all entries in row **1A** of worksheet I and enter as C_1 in column **C**. This summation includes in addition to the a entries, the 1 in column **1B**. In the same way, calculate and make the entries for C_2 through C_4. As for step (1) of the forward solution in Section 20.6, remember that entries to the left of the diagonal of the matrix a are implied. The sums in the C column are to include these entries. Therefore, to get the sum of the entries in a row of a, you get the sum total of the recorded elements in the row and the entries above its diagonal.

(2) Now proceed exactly as in steps (5) through (25) for the forward solution in Section 20.6. The elements in the upper right quadrant of I are a special case of a b matrix, namely an identity matrix. You will see as you proceed through the various cycles that because b is an identity matrix, you get in the lower right quadrant of I all 0's above the diagonal and all 1's in the diagonal. The 1's you have already entered in advance. After you have calculated and checked each of the rows in the lower sections of I and II, you are ready for the final group of steps, which yield the inverse of a. These consist in computing the elements of the F matrix in the upper right quadrant of II.

(3) Fold and crease worksheet I between columns **1B** and **2B** so that column **1B** will be showing at the extreme right, and place column **1B** of I immediately to the left of **1B** of II.

(4) Multiply column **1B** of II by column **1B** of I and enter the result *with sign reversed* as F_{11} in row **1A**, column **1B** of II. Note that this entry is the negative of the product of the first column of the Z matrix by the first column of the X matrix.

(5) Slide column **1B** of I one space to the right, so that it lies immediately to the left of column **2B** of II. Multiply column **2B** of II by column **1B** of I and enter the result with sign reversed as F_{12} in row **1A**, column **2B**, of II.

(6) Moving column **1B** of I one column to the right each time, calculate the remaining entries in row **1A** of II. Do this through the **C** column, but not for the **S** column of II. Remember each time to reverse the sign.

(7) Sum the elements in row **1A** of II, including the 1 in column **1A**, but not including C_{11}. Enter this sum as S_{11} in the **S** column of II. This entry should be the same as C_{11}.

(8) Copy all of the F elements to the right of F_{11} in row **1A** of II into the corresponding positions in column **1B** of II immediately below F_{11}. You do this because F is symmetric and you need not compute the elements below the diagonal.

(9) Next fold worksheet I between columns **2B** and **3B** so that **3B** is showing at the extreme left. Beginning with column **2B** of II, compute the entries in row **2A** of II from F_{22} through C_{12}, just as in steps (4) through (7).

(10) Copy all of the F elements to the right of F_{22} in row **2A** of II into the corresponding positions in column **2B** of II immediately below F_{22}.

(11) Advance the fold in worksheet I a column to the right each time, and calculate the remaining rows of the F matrix.

A numerical example of the computations is given in Fig. 20.8.2.

Worksheet I

R	i	1A	2A	3A	4A	1B	2B	3B	4B	C	S
	1A	2.120	.780	1.160	.920	1	0	0	0	5.980	
	2A	.780	2.200	.660	1.580	0	1	0	0	6.220	
	3A	1.160	.660	4.040	2.220	0	0	1	0	9.080	
	4A	.920	1.580	2.220	1.160	0	0	0	1	6.880	
.47170	1B	2.120	.780	1.160	.920	1	0	0	0	5.980	5.980
.52274	2B		1.913	.233	1.241	− .368	1	0	0	4.019	4.019
.29612	3B			3.377	1.565	− .502	− .122	1	0	5.319	5.318
−1.30039	4B				− .769	.037	− .593	− .463	1	− .786	− .788

FIG. 20.8.2 (Cont. on next·page)

Worksheet II

i	1A	2A	3A	4A	1B	2B	3B	4B	C	S
1A	1	0	0	0	.615	− .146	− .126	− .048	1.295	1.295
2A	0	1	0	0	− .146	.070	− .393	.771	1.301	1.302
3A	0	0	1	0	− .126	− .393	.017	.602	1.100	1.100
4A	0	0	0	1	− .048	.771	.602	−1.300	1.025	1.025
1B	−1	− .368	− .547	− .434	− .472	0	0	0	−2.821	−2.821
2B		−1	− .122	− .649	.192	− .523	0	0	−2.101	−2.102
3B			−1	− .463	.149	.036	− .296	0	−1.575	−1.574
4B				−1	.048	− .771	− .602	1.300	−1.025	−1.025

FIG. 20.8.2. (Cont.) The Inverse of a Symmetric Matrix

Chapter 21

The Inverse of a Supermatrix

21.1 Conditions Required for a Regular Inverse

The simple form of the supermatrix

You have seen that just as we can take the transposes and add, subtract, and multiply simple matrices, so we can also perform these operations with supermatrices. You will doubtless wonder, then, whether we can also take the inverse of a supermatrix. We can, of course, do anything with a supermatrix that we can with a simple matrix if we first reduce the matrix to simple form. However, as you have seen, we can perform the operations of transposition, addition, subtraction, and multiplication with the matric elements of supermatrices, provided we observe certain rules. It is also true that we can express the inverse of a supermatrix in terms of its matric elements, provided certain conditions are satisfied. We shall consider only the regular inverse of a supermatrix.

In referring to the inverse of a supermatrix we mean the regular inverse. It should be clear, then, that a supermatrix cannot have an inverse unless its simple form has an inverse. But we know that only basic square matrices have regular inverses; therefore, in discussing the inverses of supermatrices, we shall restrict ourselves to supermatrices whose simple forms are square and basic.

Restrictions on the supermatrix

But we shall also place further restrictions on the supermatrices whose inverses we shall express as functions of their matric elements. In the first place the supermatrix must also be square, which means that the number of matric rows and columns must be equal. And if the supermatrix is square, then it has, of course, a principal diagonal of matric elements. In the second place, we shall require that the simple form of each of the diagonal elements also be a square matrix. Third, we specify that each diagonal matric element be basic.

Because of the second condition, you can readily see that the supermatrix whose inverse we shall express must be expressible as a symmetrically partitioned square simple matrix. If the simple matrix were not symmetrically partitioned, then of course the diagonal submatrices would not be square.

We shall consider in detail only the inverse of second-order supermatrices, for if we can express the inverse of a second-order supermatrix in terms of its matric elements, it is possible, though quite cumbersome, to express the inverse of a higher-order supermatrix in terms of its matric elements.

21.2   The Inverse of a Second-Order Supermatrix

We shall begin with the supermatrix given by Eq. (21.2.1).

$$A = \begin{bmatrix} A_{11} & A_{12} \\ A_{21} & A_{22} \end{bmatrix} \tag{21.2.1}$$

Remember that both A_{11} and A_{22} are square and basic submatrices, but they are not necessarily of the same order. We let the inverse of A be represented by

$$A^{-1} = \begin{bmatrix} \alpha_{11} & \alpha_{12} \\ \alpha_{21} & \alpha_{22} \end{bmatrix} \tag{21.2.2}$$

where α_{11} and α_{22} are the same order as A_{11} and A_{22} respectively.

Derivation of the inverse

Our problem now is to express the matric elements in the right side of Eq. (21.2.2) in terms of the matric elements in the right side of Eq. (21.2.1). Because of Eqs. (21.2.1) and (21.2.2), we have by definition Eq. (21.2.3).

$$\begin{bmatrix} A_{11} & A_{12} \\ A_{21} & A_{22} \end{bmatrix} \begin{bmatrix} \alpha_{11} & \alpha_{12} \\ \alpha_{21} & \alpha_{22} \end{bmatrix} = \begin{bmatrix} I_1 & 0 \\ 0 & I_2 \end{bmatrix} \tag{21.2.3}$$

where I_1 and I_2 are identity matrices of the order of A_{11} and A_{22} respectively. Expanding the left side of Eq. (21.2.3), we get the following equations:

$$A_{11}\alpha_{11} + A_{12}\alpha_{21} = I_1 \tag{21.2.4}$$

$$A_{11}\alpha_{12} + A_{12}\alpha_{22} = 0 \tag{21.2.5}$$

$$A_{21}\alpha_{11} + A_{22}\alpha_{21} = 0 \tag{21.2.6}$$

$$A_{21}\alpha_{12} + A_{22}\alpha_{22} = I_2 \tag{21.2.7}$$

By means of Eqs. (21.2.4) through (21.2.7) we shall now solve for the α's in terms of the A's. First we premultiply Eq. (21.2.4) by A_{11}^{-1}, and after transposing we get

$$\alpha_{11} = A_{11}^{-1}(I - A_{12}\alpha_{21}) \tag{21.2.8}$$

Next we premultiply Eq. (21.2.5) by A_{11}^{-1} and transpose to get

$$\alpha_{12} = -A_{11}^{-1}A_{12}\alpha_{22} \tag{21.2.9}$$

Then we premultiply Eq. (21.2.6) by A_{22}^{-1} and transposing we get

$$\alpha_{21} = -A_{22}^{-1}A_{21}\alpha_{11} \tag{21.2.10}$$

We also premultiply Eq. (21.2.7) by A_{22}^{-1} and transpose to get

$$\alpha_{22} = A_{22}^{-1}(I - A_{21}\alpha_{12}) \tag{21.2.11}$$

Now we substitute the right side of Eq. (21.2.8) in the right side of Eq. (21.2.10), which gives

$$\alpha_{21} = -A_{22}^{-1}A_{21}A_{11}^{-1}(I - A_{12}\alpha_{21}) \tag{21.2.12}$$

We also substitute the right side of Eq. (21.2.9) in the right side of Eq. (21.2.11) and get

$$\alpha_{22} = A_{22}^{-1}(I + A_{21}A_{11}^{-1}A_{12}\alpha_{22}) \tag{21.2.13}$$

From Eq. (21.2.12), we get, after a few simple manipulations,

$$(A_{22} - A_{21}A_{11}^{-1}A_{12})\alpha_{21} = -A_{21}A_{11}^{-1} \tag{21.2.14}$$

From Eq. (21.2.13) we can also get with a few simple operations

$$(A_{22} - A_{21}A_{11}^{-1}A_{12})\alpha_{22} = I \tag{21.2.15}$$

Premultiplying Eq. (21.2.14) by the inverse of the expression in brackets, we have

$$\alpha_{21} = -(A_{22} - A_{21}A_{11}^{-1}A_{12})^{-1}A_{21}A_{11}^{-1} \tag{21.2.16}$$

We also premultiply Eq. (21.2.15) by the inverse of the term in brackets, and get

$$\alpha_{22} = (A_{22} - A_{21}A_{11}^{-1}A_{12})^{-1} \tag{21.2.17}$$

If now we substitute the right side of Eq. (21.2.16) in Eq. (21.2.8), we get

$$\alpha_{11} = A_{11}^{-1} + A_{11}^{-1}A_{12}(A_{22} - A_{21}A_{11}^{-1}A_{12})^{-1}A_{21}A_{11}^{-1} \tag{21.2.18}$$

Substituting Eq. (21.2.17) in Eq. (21.2.9) gives

$$\alpha_{12} = -A_{11}^{-1}A_{12}(A_{22} - A_{21}A_{11}^{-1}A_{12})^{-1} \tag{21.2.19}$$

We now let

$$M_2 = A_{22} - A_{21}A_{11}^{-1}A_{12} \tag{21.2.20}$$

Then if we substitute Eqs. (21.2.16) through (21.2.20) in Eq. (21.2.2), we have

$$A^{-1} = \begin{bmatrix} A_{11}^{-1} + A_{11}^{-1}A_{12}M_2^{-1}A_{21}A_{11}^{-1} & -A_{11}^{-1}A_{12}M_2^{-1} \\ -M_2^{-1}A_{21}A_{11}^{-1} & M_2^{-1} \end{bmatrix} \tag{21.2.21}$$

Equation (21.2.21) then gives us the matric elements of the inverse of A in terms of the matric elements of A. Notice that Eqs. (21.2.16) through (21.2.19) assume that M_2 given by Eq. (21.2.20) has a regular inverse.

Alternate formulas for the inverse

Next let us consider another solution for the inverse of A, which we shall show to be identical with Eq. (21.2.21), but which is expressed somewhat differently. We begin by considering a permutation of the matric rows and columns of A, as in Eq. (21.2.22).

$$A_\pi = \begin{bmatrix} 0 & I_2 \\ I_1 & 0 \end{bmatrix} \begin{bmatrix} A_{11} & A_{11} \\ A_{21} & A_{22} \end{bmatrix} \begin{bmatrix} 0 & I_1 \\ I_2 & 0 \end{bmatrix} \qquad (21.2.22)$$

Multiplying out the right side of Eq. (21.2.22), we have Eq. (21.2.23).

$$A_\pi = \begin{bmatrix} A_{22} & A_{21} \\ A_{12} & A_{11} \end{bmatrix} \qquad (21.2.23)$$

We let

$$\pi = \begin{bmatrix} 0 & I_2 \\ I_1 & 0 \end{bmatrix} \qquad (21.2.24)$$

Then Eq. (21.2.22) can be written more compactly as

$$A_\pi = \pi A \pi' \qquad (21.2.25)$$

To get the inverse of A_π in Eq. (21.2.23), we might have used exactly the same method as to get the inverse of A in Eq. (21.2.1). Notice that if we substitute the subscript 1 for 2 and vice versa in Eq. (21.2.1), we have precisely the right side of Eq. (21.2.23). Therefore, if we use the same method for getting the inverse of Eq. (21.2.23) as for Eq. (21.2.1), we end with Eqs. (21.2.20) and (21.2.21), except that the subscripts are reversed. We can therefore write instead of Eq. (21.2.20)

$$M_1 = A_{11} - A_{12}A_{22}^{-1}A_{21} \qquad (21.2.26)$$

Similarly, instead of Eq. (21.2.21), we have

$$A_\pi^{-1} = \begin{bmatrix} A_{22}^{-1} + A_{22}^{-1}A_{21}M_1^{-1}A_{12}A_{22}^{-1} & -A_{22}^{-1}A_{21}M_1^{-1} \\ -M_1^{-1}A_{12}A_{22}^{-1} & M_1^{-1} \end{bmatrix} \qquad (21.2.27)$$

But from Eq. (21.2.25) we have

$$(A_\pi)^{-1} = (\pi A \pi')^{-1} \qquad (21.2.28)$$

Remember that the inverse of a product of matrices is the product of their inverses in reverse order, and that the inverse of a permutation matrix is its transpose. Therefore the right side of Eq. (21.2.28) can be written as

$$A_\pi^{-1} = \pi A^{-1}\pi' \qquad (21.2.29)$$

But if we premultiply both sides of Eq. (21.2.29) by π' and postmultiply by π, we have

$$A^{-1} = \pi' A_\pi^{-1}\pi \qquad (21.2.30)$$

Notice now that the right side of Eq. (21.2.30) means simply that we interchange matric rows and columns of A_π^{-1}. We have therefore, from Eqs. (21.2.27) and (21.2.30), Eq. (21.2.31).

$$A^{-1} = \begin{bmatrix} M_1^{-1} & -M_1^{-1}A_{12}A_{22}^{-1} \\ -A_{22}^{-1}A_{21}M_1^{-1} & A_{22}^{-1} + A_{22}^{-1}A_{21}M_1^{-1}A_{12}A_{22}^{-1} \end{bmatrix} \quad (21.2.31)$$

This equation is therefore an alternate form of Eq. (21.2.21) for expressing the inverse of a matrix in terms of its matric elements.

Proof of the identity of both inverses

We should be able to prove that corresponding elements of Eqs. (21.2.21) and (21.2.31) are identical. The identities we shall prove are given by the following equations:

$$M_1^{-1} = A_{11}^{-1} + A_{11}^{-1}A_{12}M_2^{-1}A_{21}A_{11}^{-1} \quad (21.2.32)$$

$$M_1^{-1}A_{12}A_{22}^{-1} = A_{11}^{-1}A_{12}M_2^{-1} \quad (21.2.33)$$

$$A_{22}^{-1}A_{21}M_1^{-1} = M_2^{-1}A_{21}A_{11}^{-1} \quad (21.2.34)$$

$$A_{22}^{-1} + A_{22}^{-1}A_{21}M_1^{-1}A_{12}A_{22}^{-1} = M_2^{-1} \quad (21.2.35)$$

First we shall prove Eq. (21.2.33). We premultiply both sides of Eq. (21.2.33) by M_1 and postmultiply both sides by M_2. This gives

$$A_{12}A_{22}^{-1}M_2 = M_1A_{11}^{-1}A_{12} \quad (21.2.36)$$

We substitute Eqs. (21.2.20) and (21.2.26) in Eq. (21.2.36) and get

$$A_{12}A_{22}^{-1}(A_{22} - A_{21}A_{11}^{-1}A_{12}) = (A_{11} - A_{12}A_{22}^{-1}A_{21})A_{11}^{-1}A_{12} \quad (21.2.37)$$

Multiplying out both sides of Eq. (21.2.37) gives

$$A_{12} - A_{12}A_{22}^{-1}A_{21}A_{11}^{-1}A_{12} = A_{12} - A_{12}A_{22}^{-1}A_{21}A_{11}^{-1}A_{12} \quad (21.2.38)$$

This is, of course, an identity.

Next we shall prove Eq. (21.2.34). We premultiply by M_2 and postmultiply by M_1 to get

$$M_2A_{22}^{-1}A_{21} = A_{21}A_{11}^{-1}M_1 \quad (21.2.39)$$

We substitute Eqs. (21.2.20) and (21.2.26) in Eq. (21.2.39) and get

$$(A_{22} - A_{21}A_{11}^{-1}A_{12})A_{22}^{-1}A_{21} = A_{21}A_{11}^{-1}(A_{11} - A_{12}A_{22}^{-1}A_{21}) \quad (21.2.40)$$

From this we have

$$A_{21} - A_{21}A_{11}^{-1}A_{12}A_{22}^{-1}A_{21} = A_{21} - A_{21}A_{11}^{-1}A_{12}A_{22}^{-1}A_{21} \quad (21.2.41)$$

We see that this is an identity.

To prove Eq. (21.2.32), we first substitute the left side of Eq. (21.2.33) into the second term of the right side of Eq. (21.2.32), and get

$$M_1^{-1} = A_{11}^{-1} + M_1^{-1}A_{12}A_{22}^{-1}A_{21}A_{11}^{-1} \quad (21.2.42)$$

Next we pre- and postmultiply this by M_1 and A_{11} respectively to get

$$A_{11} = M_1 + A_{12}A_{22}^{-1}A_{21} \qquad (21.2.43)$$

Substituting Eq. (21.2.26) in Eq. (21.2.43) gives

$$A_{11} = (A_{11} - A_{12}A_{22}^{-1}A_{21}) + A_{12}A_{22}^{-1}A_{21} \qquad (21.2.44)$$

or

$$A_{11} = A_{11}$$

which proves the identity of Eq. (21.2.32).

Finally we prove Eq. (21.2.35) by first substituting the right side of Eq. (21.2.34) into the second term on the left of Eq. (21.2.35). This gives

$$A_{22}^{-1} + M_2^{-1}A_{21}A_{11}^{-1}A_{12}A_{22}^{-1} = M_2^{-1} \qquad (21.2.45)$$

We pre- and postmultiply Eq. (21.2.45) by M_2 and A_{22} respectively to get

$$M_2 + A_{21}A_{11}^{-1}A_{12} = A_{22} \qquad (21.2.46)$$

Then we substitute Eq. (21.2.20) in (21.2.46), which gives us

$$(A_{22} - A_{21}A_{11}^{-1}A_{12}) + A_{21}A_{11}^{-1}A_{12} = A_{22} \qquad (21.2.47)$$

or

$$A_{22} = A_{22}$$

which proves Eq. (21.2.35).

Applications of the method

For a square basic matrix, it is possible to get the inverse by symmetrically partitioning the matrix into a second-order supermatrix and performing the computations indicated by Eq. (21.2.21) or Eq. (21.2.31), according to methods outlined in Chapter 20. For the general square basic matrix, however, it is probably just as economical of computational time to apply the methods of Chapter 20 directly to the simple form of the matrix. It is only in certain special cases that the methods indicated in this section are particularly useful for computational purposes. They may also be used to develop certain useful rules as indicated in Section 21.4. You will notice, for example, that in Eq. (21.2.21) two inverses are required, namely A_{11}^{-1} and M_2^{-1}. The sum of the orders of these two matrices is clearly equal to the simple order of A. Since much matrix multiplication is necessary in getting the matric elements of Eq. (21.2.21) in addition to the two inverses, you may well doubt whether there is any advantage in using the supermatrix method, even though the required inverses are of smaller order. In certain special problems, however, it may happen that A_{11} and A_{22} are matrices whose inverses are easy to compute. For example, it may be that one or both are diagonal matrices. It may also be that A_{12} and A_{21} are special types of matrices, such as the major product of two vectors. In such cases, Eq. (21.2.21) could simplify the operation considerably, and might be much simpler computationally than working directly with the simple form of the matrix. Such examples can be found in certain experimental designs in the analysis of variance.

21.3 Solution When One Diagonal Element Is a Scalar Quantity

The general solution

A special case of the inverse of a second-order supermatrix arises when one of the diagonal elements is a scalar quantity. This means that A_{12} and A_{21} are vectors. Let us suppose, for example, that A_{22} is a scalar. Then A_{12} would have to be a column vector and A_{21} a row vector. This we illustrate in Fig. 21.3.1.

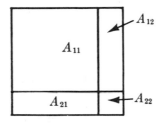

Fig. 21.3.1

Let us now simplify the notation of Eq. (21.2.21) as follows

$$A_{12} = v \qquad (21.3.1)$$

$$A_{21} = u' \qquad (21.3.2)$$

$$A_{22} = d \qquad (21.3.3)$$

$$A_{11}^{-1}A_{12} = V \qquad (21.3.4)$$

$$A_{21}A_{11}^{-1} = U' \qquad (21.3.5)$$

Substituting Eqs. (21.3.2) through (21.3.4) in Eq. (21.2.20), and dropping the subscript 2 on M_2, we have

$$M = \overset{.}{d} - u'V \qquad (21.3.6)$$

Now M is a scalar quantity, as you can see from the right side of Eq. (21.3.6). Therefore, by means of Eqs. (21.3.1) through (21.3.6), we can write Eq. (21.2.21) as in Eq. (21.3.7).

$$A^{-1} = \begin{bmatrix} A_{11}^{-1} & \bigm| & \dfrac{VU'}{M} & \dfrac{-V}{M} \\ & \dfrac{-U'}{M} & \dfrac{1}{M} \end{bmatrix} \qquad (21.2.7)$$

Suppose that we already had the inverse of the matrix A_{11}. Assume then we increased the order of A_{11} by adding a row and a column, and we wished to find the inverse of this enlarged matrix. Equation (21.3.7) shows us how this can be done readily by using the inverse of A_{11}.

Application for the inversion of simple matrices

As a matter of fact, therefore, Eq. (21.3.7) suggests a method for getting the inverse of any basic square matrix. Suppose we define matrices in terms of scalar elements, as in Eq. (21.3.8).

$$A_1 = [a_{11}]$$

$$A_2 = \begin{bmatrix} a_{11} & a_{12} \\ a_{21} & a_{22} \end{bmatrix} \qquad (21.3.8)$$

$$A_3 = \begin{bmatrix} a_{11} & a_{12} & a_{13} \\ a_{21} & a_{22} & a_{23} \\ a_{31} & a_{32} & a_{33} \end{bmatrix}$$

$$\text{etc.}$$

We also define column and row vectors as in Eqs. (21.3.9) and (21.3.10).

$$v_2 = a_{12} \qquad v_3 = \begin{bmatrix} a_{13} \\ a_{23} \end{bmatrix} \qquad v_4 = \begin{bmatrix} a_{14} \\ a_{24} \\ a_{34} \end{bmatrix} \qquad \text{etc.} \qquad (21.3.9)$$

$$\left. \begin{aligned} u_2' &= a_{21} \\ u_3' &= [a_{31} \quad a_{32}] \\ u_4' &= [a_{41} \quad a_{42} \quad a_{43}] \end{aligned} \right\} \qquad (21.3.10)$$

$$\text{etc.}$$

Now let us see how we could find the inverse of A_n, represented by the nth equation in Eq. (21.3.8). From Eqs. (21.3.8), (21.3.9), and (21.3.10), we have Eq. (21.3.11).

$$\left. \begin{aligned} A_2 &= \begin{bmatrix} A_1 & v_2 \\ u_2' & a_{22} \end{bmatrix} \\ A_3 &= \begin{bmatrix} A_2 & v_3 \\ u_3' & a_{33} \end{bmatrix} \end{aligned} \right\} \qquad (21.3.11)$$

or we have, in general, Eq. (21.3.12).

$$A_i = \begin{bmatrix} A_{i-1} & v_i \\ u_i' & a_{ii} \end{bmatrix} \qquad (21.3.12)$$

Suppose now we let

$$A_{i-1}^{-1} v_i = V_i \qquad (21.3.13)$$

and

$$u_i' A_{i-1}^{-1} = U_i' \qquad (21.3.14)$$

Then we would have

$$M_i = a_{ii} - U_i'V_i \qquad (21.3.15)$$

Taking the inverse of both sides of Eq. (21.3.12) and using Eqs. (21.3.7), (21.3.13), (21.3.14), and (21.3.15), we have Eq. (21.3.16).

$$A_i^{-1} = \begin{bmatrix} A_{i-1}^{-1} + \dfrac{V_i U_i'}{M_i} & \dfrac{-V_i}{M_i} \\[2ex] \dfrac{-U_i'}{M_i} & \dfrac{1}{M_i} \end{bmatrix} \qquad (21.3.16)$$

Equation (21.3.16) shows that we can let i take in turn the values 2 to n, and so solve for the inverse of any order matrix. We start with $A_1^{-1} = 1/a_{11}$ and solve for each higher order A_i in sequence according to Eq. (21.2.27).

This procedure is extremely useful in the behavioral sciences when we wish to study the effect of including additional variables with a set of variables already selected.

21.4 The Inverse of a Lower Order Matrix from One of Higher Order

Deletion of the last row and column

Suppose now that the inverse of a matrix A_n of order n is given. We wish to find what the inverse would be of a matrix obtained from A_n by deleting certain rows and an equal number of columns. First we consider the case in which the last row and last column of A_n has been deleted. We rewrite Eq. (21.3.16), using n rather than i as a subscript, as Eq. (21.4.1).

$$A_n^{-1} = \begin{bmatrix} A_{n-1}^{-1} + \dfrac{V_n U_n'}{M_n} & \dfrac{-V_n}{M_n} \\[2ex] \dfrac{-U_n'}{M_n} & \dfrac{1}{M_n} \end{bmatrix} \qquad (21.4.1)$$

We let

$$B = A_n^{-1} \qquad (21.4.2)$$

Now from the right side of Eq. (21.4.1) we have Eq. (21.4.3).

$$\left. \begin{aligned} B_{.n} &= \frac{1}{M_n} \begin{bmatrix} -V_n \\ \hline 1 \end{bmatrix} \\[2ex] B_{n.}' &= \frac{1}{M_n} [\ U_n' \mid 1] \\[2ex] B_{nn} &= \frac{1}{M_n} \end{aligned} \right\} \qquad (21.4.3)$$

From this we have

$$\frac{B_{.n}}{\sqrt{B_{nn}}} \frac{B'_{n.}}{\sqrt{B_{nn}}} = \frac{\begin{bmatrix} -V_n \\ 1 \end{bmatrix}}{\sqrt{M_n}} \frac{[-U'_n \mid 1]}{\sqrt{M_n}} \tag{21.4.4}$$

Or expanding the right side of Eq. (21.4.4), we get Eq. (21.4.5).

$$\frac{B_{.n}B'_{n.}}{B_{nn}} = \begin{bmatrix} \dfrac{V_n U'_n}{M_n} & \dfrac{-V_n}{M_n} \\ \dfrac{-U'_n}{M_n} & \dfrac{1}{M_n} \end{bmatrix} \tag{21.4.5}$$

Subtracting Eq. (21.4.5) from Eq. (21.4.1), and using Eq. (21.4.2), we get

$$B - \frac{B_{.n}B'_{n.}}{B_{nn}} = \begin{bmatrix} A_{n-1}^{-1} & 0 \\ 0 & 0 \end{bmatrix} \tag{21.4.6}$$

If then we have the inverse of a matrix A_n, and we wish to find the inverse of a matrix obtained by deleting the last row and column of A_n, Eq. (21.4.6) shows how this may be done rather simply. We divide each element in the last row and column of B or A_n^{-1} by the square root of the last diagonal element. We postmultiply the resulting column vector by the resulting row vector and subtract this matrix from B. This gives zeros in the last row and column. The remaining nonzero part of the matrix is the inverse of the matrix obtained by deleting the last row and column of A_n.

Deletion of any row and column

But suppose that we are not interested in finding the inverse of the matrix with the last row and column of A_n deleted. Rather we wish to consider the general case in which the kth row and ith column have been deleted.

We let π_k be a permutation matrix that puts the kth row of A last and moves each row below it one row up, Similarly π_i is a permutation matrix that puts the ith column last and moves each column to its right one column to the left. The resulting matrix is given by

$$C_n = \pi_k A_n \pi_i \tag{21.4.7}$$

Now what was the kth row of A is the last row of C, and what was the ith column is the last column. From Eq. (21.4.7) we have

$$C_n^{-1} = \pi'_i A_n^{-1} \pi'_k \tag{21.4.8}$$

We let

$$C_n^{-1} = E \tag{21.4.9}$$

Now we have, analogous to Eq. (21.4.6), Eq. 21.4.10.

$$E - \frac{E_{.n}}{\sqrt{E_{nn}}} \frac{E'_{n.}}{\sqrt{E_{nn}}} = \begin{bmatrix} C_{n-1}^{-1} & 0 \\ 0 & 0 \end{bmatrix} \tag{21.4.10}$$

But notice that C_{n-1} is the matrix C with its last row and column deleted. Therefore, it is also the matrix A_n with its kth row and ith column deleted. Then C_{n-1}^{-1} on the right of Eq. (21.4.10) is the inverse of the matrix obtained by deleting the kth row and ith column of A_n. Substituting Eqs. (21.4.2) and (21.4.9) in Eq. (21.4.8), we have

$$E = \pi_i' B \pi_k' \tag{21.4.11}$$

We see from this that E is obtained from B by putting the ith row and kth column of B last so that they become the nth row and column. Therefore the nth row of E is the same as the ith row of B and the nth column of E is the same as the kth column of B. Let us then write

$$B - \frac{B_{.k}B_{i.}'}{B_{ik}} = A_{[ki]}^{-1} \tag{21.4.12}$$

Then the right side of Eq. (21.4.12) is the same as the right side of Eq. (21.4.10) except for the position of the zeros. The column of zeros in Eq. (21.4.12) is in the kth column and the row of zeros is in the ith row. Therefore, if we compress the right side of Eq. (21.4.12), we have the inverse of a matrix obtained by deleting the kth row and ith column of A_n.

Deletion of any number of rows and columns

Suppose now we have the inverse of a matrix A_m, and we wish to find what the inverse would be if we had deleted not only the kth row and ith column but also the pth row and qth column. We could calculate first

$$B - \frac{B_{.k}B_{i.}'}{B_{ik}} = G = A_{[ki]}^{-1} \tag{21.4.13}$$

and then

$$G - \frac{G_{.p}G_{q.}'}{G_{qp}} = A_{[kp][iq]}^{-1} \tag{21.4.14}$$

In a similar manner, we can calculate from the inverse of a matrix another matrix that is the inverse of the original matrix with any number of rows and columns deleted.

This procedure is useful in the behavioral sciences when we wish to study the effect of eliminating variables from a set of variables previously selected.

SUMMARY

1. Conditions required for the regular inverse of a supermatrix:
   a. The simple form of the matrix must be square and basic.
   b. The supermatrix must be a symmetrically partitioned simple matrix.
   c. The diagonal submatrices must be basic.
2. The inverse of a second order supermatrix: Let

$$A = \begin{bmatrix} A_{11} & A_{12} \\ A_{21} & A_{22} \end{bmatrix}$$

where A_{11} and A_{22} are square and basic. Then the inverse of A is given by the two alternate expressions.

a.
$$A = \begin{bmatrix} A_{11}^{-1} + A_{11}^{-1}A_{12}M_2^{-1}A_{21}A_{11}^{-1} & -A_{11}^{-1}A_{12}M_2^{-1} \\ -M_2^{-1}A_{21}A_{11}^{-1} & M_2^{-1} \end{bmatrix}$$

where $M_2 = A_{22} - A_{21}A_{11}^{-1}A_{12}$, and

b.
$$A = \begin{bmatrix} M_1^{-1} & -M_1^{-1}A_{12}A_{22}^{-1} \\ -A_{22}^{-1}A_{21}M_1^{-1} & A_{22}^{-1} + A_{22}^{-1}M_1^{-1}A_{12}A_{22}^{-1} \end{bmatrix}$$

where $M_1 = A_{11} - A_{12}A_{22}^{-1}A_{21}$

c. The formulas in a and b are most useful when either or both A_{11} and A_{22} are diagonal, or when A_{12} is a vector.

3. The case when one diagonal element is a scalar quantity:
   a. *The general solution.* Assume A_{22} a scalar. Let $A_{12} = v$, $A_{21} = u'$, $A_{22} = d$, $A_{11}^{-1}A_{12} = V$, $A_{21}A_{11}^{-1} = U'$, $M = d - u'V$. Then

$$A^{-1} = \begin{bmatrix} A_{11}^{-1} + \dfrac{VU'}{M} & \dfrac{-V}{M} \\ \dfrac{-U'}{M} & \dfrac{1}{M} \end{bmatrix}$$

   b. *Application for the inversion of simple matrices.* Let

$$A_i = \begin{bmatrix} a_{11} & \cdots & a_{1i} \\ \cdots & \cdots & \cdots \\ a_{i1} & \cdots & a_{ii} \end{bmatrix}, \quad v_i = \begin{bmatrix} a_{1i} \\ \cdots \\ a_{[i-1]i} \end{bmatrix}, \quad u'_i = [a_{i1} \quad \cdots \quad a_{i[i-1]}],$$

$$V_i = A_{i-1}^{-1}v_i, \qquad U'_i = u_iA_{i-1}^{-1}, \qquad M_i = a_{ii} - u'_iV_i$$

   Then the formula

$$A_i^{-1} = \begin{bmatrix} A_{i-1}^{-1} + \dfrac{V_iU'_i}{M_i} & \dfrac{-V_i}{M_i} \\ \dfrac{-U'_i}{M_i} & \dfrac{1}{M_i} \end{bmatrix}$$

   may be used iteratively for $i = 2$ to n to calculate the inverse of a simple matrix.

4. The inverse of a lower order matrix from one of higher order:
   a. *Deletion of last row and column.* If $B = A_n^{-1}$ and A_{n-1} is obtained from A_n by deletion of its last row and column, then

$$B - \frac{B_{.n}B'_{n.}}{B_{nn}} = \begin{bmatrix} A_{n-1}^{-1} & 0 \\ 0 & 0 \end{bmatrix}$$

   b. *Deletion of any row and column.* The kth column and ith row of

$$A_{[ki]}^{-1} = B - \frac{B_{.k}B'_{i.}}{B_{ik}}$$

   are 0. If we compress $A_{[ki]}^{-1}$ to eliminate this column and row we have the inverse of the matrix obtained by deleting the kth row and ith column from A_n.

Chapter 22

The General Matrix Reduction Theorems

22.1 The Reduction Equation

We shall now consider a theorem that is extremely useful because of its many practical applications. It is the basis for most of the methods used in factor analysis, a technique applicable in all the sciences whose fundamental variables are not generally agreed upon. It serves primarily to aid in discovering the fundamental variables of a science.

In the following development, subscripts are used to indicate the order of the matrices, the first subscript indicating the number of rows and the second the number of columns. We begin with a general matrix whose rank r is defined by

$$X_{Nn} = X_{Nr}X_{rn} \tag{22.1.1}$$

The arbitrary multipliers

Next we consider two other matrices X_{SN} and X_{nS}. We call X_{SN} the *premultiplier of the general matrix* X_{Nn}, and X_{nS} its *postmultiplier*. Aside from the restrictions of order that we place on these multipliers of the general matrix, the only restriction we impose is that the triple product $X_{SN} X_{Nn} X_{nS}$ be basic. Then, of course, it must have a regular inverse. This means also that S cannot be greater than r. Since these are the only restrictions we place on the multipliers of X_{Nn}, we may also call them the *arbitrary pre- and postfactors*.

The reduction equation

We shall next define a new matrix Y_{Nn} by

$$Y_{Nn} = X_{Nn} - (X_{Nn}X_{nS})(X_{SN}X_{Nn}X_{nS})^{-1}(X_{SN}X_{Nn}) \tag{22.1.2}$$

Study the right side of Eq. (22.1.2) carefully. It looks rather complicated,

477

but actually the rules for setting up the equation are simple. You see, of course, that the first term on the right is simply the general matrix with which we started. The second term on the right, which is to be subtracted from our general matrix, consists of a product of three different products. The product in the first parentheses is the general matrix by its postfactor. The product in the middle parentheses is the general matrix by its pre- and postfactor. Notice, however, that we take the inverse of this triple product. The last parentheses include the general matrix by its prefactor.

22.2   The Orthogonality Theorem and Its Proof

The first theorem we shall prove is that Y_{Nn} on the left of Eq. (22.1.2) is orthogonal to both the arbitrary multipliers of X_{Nn}, or that

$$X_{SN}Y_{Nn} = 0 \tag{22.2.1}$$

$$X_{Nn}X_{nS} = 0 \tag{22.2.2}$$

The prefactor

First we shall prove Eq. (22.2.1). We premultiply both sides of Eq. (22.1.2) by X_{SN} to get

$$X_{SN}Y_{NS} = X_{SN}X_{Nn} - [X_{SN}(X_{Nn}X_{nS})](X_{SN}X_{Nn}X_{nS})^{-1}X_{SN}X_{Nn} \tag{22.2.3}$$

But notice that the parentheses with the inverse on the right of Eq. (22.2.3) is cancelled by the product in brackets on its left. Therefore, we have from Eq. (22.2.3)

$$X_{SN}Y_{NS} = X_{SN}X_{Nn} - X_{SN}X_{Nn} \tag{22.2.4}$$

or simply

$$X_{SN}Y_{NS} = 0 \tag{22.2.5}$$

Notice that Eq. (22.2.5) is the same as Eq. (22.2.1), which we set out to prove.

The postfactor

Next we prove Eq. (22.2.2) in a similar manner. This time we post-multiply Eq. (22.1.2) by X_{nS} and get

$$Y_{Nn}X_{nS} = X_{Nn}X_{nS} - (X_{Nn}X_{nS})(X_{SN}X_{Nn}X_{nS})^{-1}[(X_{SN}X_{Nn})X_{nS}] \tag{22.2.6}$$

Now the middle parentheses on the right of Eq. (22.2.6) cancels the product following it, and we have

$$Y_{Nn}X_{nS} = X_{Nn}X_{nS} - X_{Nn}X_{nS} \tag{22.2.7}$$

or simply

$$Y_{Nn}X_{nS} = 0 \tag{22.2.8}$$

which is the same as Eq. (22.2.2).

22.3 The Minimum Rank of a Product of Two Matrices

You may recall that in Section 15.7 we stated that the rank of a product of two matrices cannot be less than the sum of their ranks less their common dimension. At the time we did not prove this theorem. Since it plays such an important part in the results of this chapter, we now prove it.

The minimum rank of the minor product of two basic matrices

Let us express the rank t of the minor product of two basic matrices by

$$X_{ae}X_{ec} = -X_{at}X_{tc} \qquad (22.3.1)$$

where, by definition, the ranks of X_{ae} and X_{ec} are a and c respectively.

We wish now to find what is the smallest possible value t can have in terms of the dimensions of X_{ae} and X_{ec}. We get from Eq. (22.3.1)

$$X_{ae}X_{ec} + X_{at}X_{tc} = 0 \qquad (22.3.2)$$

But Eq. (22.3.2) may be written

$$[X_{ae} \quad X_{at}]\begin{bmatrix}X_{ec}\\X_{tc}\end{bmatrix} = 0 \qquad (22.3.3)$$

We define

$$[X_{ae} \quad X_{at}] = X_{ab} \qquad (22.3.4)$$

$$\begin{bmatrix}X_{ec}\\X_{tc}\end{bmatrix} = X_{bc} \qquad (22.3.5)$$

where

$$b = e + t \qquad (22.3.6)$$

Now X_{ab} and X_{bc} are basic, since the rank of a supermatrix cannot be less than the rank of its submatrix of highest rank (Section 15.6).

Substituting the right sides of Eqs. (22.3.4) and (22.3.5) in Eq. (22.3.3) gives

$$X_{ab}X_{bc} = 0 \qquad (22.3.7)$$

Let us see now what restrictions, if any, must be placed on the dimensions of X_{ab} and X_{bc} in order for Eq. (22.3.7) to be possible.

First let us define the basic structures of these two factors by

$$X_{ab} = Q_{aa}\Delta_a Q_{ab} \qquad (22.3.8)$$

$$X_{bc} = Q_{bc}\Delta_c Q_{cc} \qquad (22.3.9)$$

Notice that since the matrices are basic, the left orthonormal of X_{ab} and the right orthonormal of X_{bc} are both square.

Next we substitute the right sides of Eqs. (22.3.8) and (22.3.9) into the left of Eq. (22.3.7) and get

$$Q_{aa}\Delta_a Q_{ab} Q_{bc} \Delta_c Q_{cc} = 0 \tag{22.3.10}$$

We shall first prove that if the minor product of two basic matrices is zero then the product of their adjacent orthonormals must also be zero. We pre- and postmultiply Eq. (22.3.10) by Q'_{aa} and Q'_{cc} respectively and get

$$\Delta_a Q_{ab} Q_{bc} \Delta_b = Q'_{aa} 0 Q'_{cc} = 0 \tag{22.3.11}$$

Again we pre- and postmultiply Eq. (22.3.11) by Δ_a^{-1} and Δ_b^{-1} respectively and get

$$Q_{ab} Q_{bc} = \Delta_a^{-1} 0 \Delta_b^{-1} = 0 \tag{22.3.12}$$

which proves that the product of the adjacent orthonormals is 0.

Next we consider under what restrictions on the dimensions of the two orthonormals, Eq. (22.3.12), can be true. We define a matrix

$$Q_{(a+c)b} = \begin{bmatrix} Q_{ab} \\ Q_{cb} \end{bmatrix} \tag{22.3.13}$$

where

$$Q_{cb} = Q'_{bc} \tag{22.3.14}$$

Next consider the product

$$G = Q_{(a+c)b} Q_{b(a+c)} \tag{22.3.15}$$

where

$$Q_{b(a+c)} = Q'_{(a+c)b} \tag{22.3.16}$$

From Eq. (22.3.13) and (22.3.15), we get Eq. (22.3.17).

$$Q_{(a+c)b} Q_{b(a+c)} = \begin{bmatrix} Q_{ab} \\ Q_{cb} \end{bmatrix} \begin{bmatrix} Q_{ba} & Q_{bc} \end{bmatrix} \tag{22.3.17}$$

where

$$Q_{ab} = Q'_{ba} \tag{22.3.18}$$

Multiply out the right side of Eq. (22.3.16)

$$Q_{(a+c)b} Q_{b(a+c)} = \begin{bmatrix} Q_{ab} Q_{ba} & Q_{ab} Q_{bc} \\ Q_{cb} Q_{ba} & Q_{cb} Q_{bc} \end{bmatrix} \tag{22.3.19}$$

But since the Q's are orthonormals, and because of Eq. (22.3.12), we can write Eq. (22.3.19) simply.

$$Q_{(a+c)b} Q_{b(a+c)} = \begin{bmatrix} I_{aa} & 0_{ac} \\ 0_{ca} & I_{cc} \end{bmatrix} \tag{22.3.20}$$

Now the right side of Eq. (22.3.20) is a diagonal matrix, and the left side is a product moment matrix. But we know from Section 15.8 that the major product moment of a vertical matrix cannot be diagonal. This means that the distinct dimension on the left of Eq. (22.3.20) cannot be greater than the common dimension, or

$$b \geq a + c \qquad (22.3.21)$$

Therefore Eq. (22.3.12) cannot hold without Eq. (22.3.21) or the minor product of two orthonormals cannot vanish unless their common order is greater than or equal to the sum of their distinct orders.

But the major product of two basic matrices can be 0 only if the product of their basic orthonormals is 0. Therefore, the minor product of two basic matrices cannot be 0 unless their common order is greater than or equal to the sum of their distinct orders.

From Eqs. (22.3.6) and (22.3.21)

$$e + t \geq a + c \qquad (22.3.22)$$

or

$$e + t - e \geq a + c - e \qquad (22.3.23)$$

or

$$t \geq a + c - e \qquad (22.3.24)$$

But remember that t is the rank of the product given by Eq. (22.3.1). Therefore, Eq. (22.3.24) shows that the rank of the minor product of two basic matrices cannot be less than the sum of their distinct dimensions less their common dimension.

The general case for the product of any two matrices

We next show that the rank of the product of any two matrices cannot be less than the sum of their ranks less their common order. Let us define the ranks of X_{ab} and X_{bc} by

$$X_{ab} = X_{as}X_{sb} \qquad (22.3.25)$$

and

$$X_{bc} = X_{bt}X_{tc} \qquad (22.3.26)$$

where now X_{ab} may be either vertical or horizontal as may also X_{bc}. From Eqs. (22.3.25) and (22.3.26) we have

$$X_{ab}X_{bc} = X_{as}X_{sb}X_{bt}X_{tc} \qquad (22.3.27)$$

Let

$$X_{sb}X_{bt} = Y_{st} \qquad (22.3.28)$$

Then according to our newly proved theorem, the minimum rank of Y_{st} is given by

$$r_m \geq s + t - b \tag{22.3.29}$$

Assuming Y_{st} of minimum rank, we can then define the rank of Y_{st} by

$$Y_{st} = Y_{sr} Y_{rt} \tag{22.3.30}$$

Because of Eqs. (22.3.27), (22.3.28), and (22.3.30),

$$X_{ab} X_{bc} = X_{as} Y_{sr} Y_{rt} X_{tc} \tag{22.3.31}$$

Now let

$$Z_{ar} = X_{as} Y_{sr} \tag{22.3.32}$$

$$Z_{rc} = Y_{rt} X_{tc} \tag{22.3.33}$$

We show that Z_{ar} in Eq. (22.3.32) is basic by considering a row permutation π of X_{as} such that

$$\pi X_{as} = L_{as} = \begin{bmatrix} L_{ss} \\ L_{(a-s)s} \end{bmatrix} \tag{22.3.34}$$

and such that L_{ss} is basic.

From Eqs. (22.3.32) and (22.3.34) we have

$$\pi X_{as} Y_{sr} = \begin{bmatrix} L_{ss} & Y_{sr} \\ L_{(a-s)s} & Y_{sr} \end{bmatrix} = \pi Z_{ar} \tag{22.3.35}$$

But $L_{ss} Y_{sr}$ in the brackets in Eq. (22.3.35) is a limiting case of the minor product of two basic matrices. Hence its rank is exactly r and it is therefore basic. Since it is a submatrix of πZ_{ar} the rank of the latter cannot be less than r, hence it is also basic. But the interchange of rows or columns cannot alter the rank of a matrix, therefore Z_{ar} is also basic. Similarly we show that Z_{rc} in Eq. (22.3.34) is basic. Then from Eqs. (22.3.32) and (22.3.33) we rewrite Eq. (22.3.31) as

$$X_{ab} X_{bc} = Z_{ar} Z_{rc} \tag{22.3.36}$$

where the right of Eq. (22.3.34) defines the minimum rank of the product on the left as r. And according to Eq. (22.3.29), r cannot be less than the sum of the ranks s and t of X_{ab} and X_{ac} less their common order b. This is what we set out to prove.

22.4   The Rank Theorem and Its Proof

Now we may prove the main theorem of this chapter. This theorem is simply that the rank of Y_{Nn}, defined by Eq. (22.1.2) is exactly equal to the rank r of X_{Nn} less the smaller order, S, of the arbitrary multiplier. More simply, the rank of Y_{Nn} is exactly $r - S$.*

First, let us substitute the right side of Eq. (22.1.1) for X_{Nn} in Eq. (22.1.2). This gives

$$Y_{Nn} = X_{Nr}X_{rn} - X_{Nr}X_{rn}X_{nS}(X_{SN}X_{Nr}X_{rn}X_{nS})^{-1}X_{SN}X_{Nr}X_{rn} \quad (22.4.1)$$

On the right side of Eq. (22.4.1) you can now factor X_{Nr} on the left and X_{rn} on the right. This gives

$$Y_{Nn} = X_{Nr}[I_{rr} - X_{rn}X_{nS}(X_{SN}X_{Nr}X_{rn}X_{nS})^{-1}X_{SN}X_{Nr}]X_{rn} \quad (22.4.2)$$

To simplify our notation, we let

$$X_{rn}X_{nS} = U_{rS} \quad (22.4.3)$$

$$X_{SN}X_{Nr} = V_{Sr} \quad (22.4.4)$$

Substituting Eqs. (22.4.3) and (22.4.4) in Eq. (22.4.2) gives

$$Y_{Nn} = X_{Nr}[I_{rr} - U_{rS}(V_{Sr}U_{rS})^{-1}V_{Sr}]X_{rn} \quad (22.4.5)$$

Next we let

$$G_{rr} = U_{rS}(V_{Sr}U_{rS})^{-1}V_{Sr} \quad (22.4.6)$$

and

$$W_{rr} = I_{rr} - G_{rr} \quad (22.4.7)$$

From (22.4.6) and (22.4.7) we can rewrite (22.4.5) as in

$$Y_{Nn} = X_{Nr}W_{rr}X_{rn} \quad (22.4.8)$$

Let us see now whether we can determine the rank of W_{rr} exactly. First we find the rank of G_{rr} in Eq. (22.4.6). Since the parentheses on the right of Eq. (22.4.6) has, by definition, a regular inverse, it is basic and its inverse is also basic and of rank S. Also U_{rs} and V_{Sr} must be basic, or the parentheses could not have a regular inverse. Now the rank of $U_{rS}(V_{Sr}U_{rS})^{-1}$ cannot be less than $S + S - S$, or S. Therefore, it is S. It is easy to show, therefore, from the theorems of minimum and maximum ranks of products, that G_{rr} is of rank S.

Next let us find the rank t of W_{rr} given by Eq. (22.4.7). We know from Section 15.9 that the rank of a sum or difference of two matrices cannot be less than the absolute difference between their ranks. Since the rank of I_{rr} is r and the rank of G_{rr} is S, we have, therefore,

$$t \geq r - S \quad (22.4.9)$$

The inequality, Eq. (22.4.9), gives us the minimum rank that W_{rr} can have.

*A proof of this theorm relying heavily on the use of determinants is given in Guttman, Louis, "General theory and methods of matrix factoring," *Psychometrika*, Vol. 9 (1944), pp. 1–16.

To find the maximum rank, we first premultiply Eq. (22.4.7) by V_{Sr} to get

$$V_{Sr}W_{rr} = V_{Sr} - V_{Sr}G_{rr} \tag{22.4.10}$$

But from (22.4.6)

$$V_{Sr}G_{rr} = V_{Sr}U_{rS}(V_{Sr}U_{rS})^{-1}V_{Sr} \tag{22.4.11}$$

or

$$V_{Sr}G_{rr} = V_{Sr} \tag{22.4.12}$$

Substituting the left side of Eq. (22.4.12) for the second term on the right of Eq. (22.4.10), we have

$$V_{Sr}W_{rr} = 0 \tag{22.4.13}$$

Now we know that the rank of a product of two matrices must be equal to or greater than the sum of their ranks less their common order. But the rank of the product in Eq. (22.4.13) is obviously 0, the rank of V_{Sr} is S, and the rank of W_{rr} is t. Since the common order of the two factors is r, we have, therefore,

$$0 \geq S + t - r \tag{22.4.14}$$

or

$$S + t - r \leq 0 \tag{22.4.15}$$

Subtracting $S - r$ from both sides of Eq. (22.4.15)

$$t \leq r - S \tag{22.4.16}$$

But the only condition that will satisfy both Eq. (22.4.9) and Eq. (22.4.16) is

$$t = r - S \tag{22.4.17}$$

Therefore, the rank of W_{rr} is exactly $r - S$. We indicate the rank of W_{rr} by

$$W_{rr} = W_{rt}W_{tr} \tag{22.4.18}$$

and substitute in Eq. (22.4.8) to get

$$Y_{Nn} = X_{Nr}W_{rt}W_{tr}X_{rn} \tag{22.4.19}$$

We let

$$Z_{Nt} = X_{Nr}W_{rt} \tag{22.4.20}$$

$$Z_{tn} = W_{tr}X_{rn} \tag{22.4.21}$$

We readily prove that Z_{Nt} and Z_{tn} are basic by showing that their maximum and minimum ranks are t. Substituting Eqs. (22.4.20) and (22.4.21) in Eq. (22.4.19)

$$Y_{Nn} = Z_{Nt}Z_{tn} \tag{22.4.22}$$

But Eq. (22.4.22) defines the rank of Y_{Nn} since the right side is the major product of two basic matrices. Therefore, we have proved that the rank of Y_{Nn} is exactly $t = r - S$ which we set out to prove.

22.5 Some Special Cases of the Matrix Reduction Theorem

Let us now consider some interesting special cases of the general matrix reduction theorem. You are familiar with one of these, although it was not identified as such (Chapter 16).

e_i vectors for the arbitrary matrices

Let us consider, first, a very simple and special case of the arbitrary factors. We shall let

$$X_{SN} = e_i' \tag{22.5.1}$$

$$X_{nS} = e_j \tag{22.5.2}$$

Substituting Eqs. (22.5.1) and (22.5.2) in Eq. (22.1.2), we have

$$Y_{Nn} = X_{Nn} - X_{Nn}e_j(e_i'X_{Nn}e_j)^{-1}e_i'X_{Nn} \tag{22.5.3}$$

Dropping subscripts on X_{Nn}, we have

$$Xe_j = X_{.j} \tag{22.5.4}$$

$$e_i'X = X_{i.}' \tag{22.5.5}$$

$$e_i'Xe_j = X_{ij} \tag{22.5.6}$$

Substituting Eqs. (22.5.4), (22.5.5), and (22.5.6) in Eq. (22.5.3)

$$Y = X - \frac{X_{.j}X_{i.}'}{X_{ij}} \tag{22.5.7}$$

You will at once recognize Eq. (22.5.7) as the method outlined in Chapter 16 for factoring a matrix in order to determine its rank. You recall that Y in Eq. (22.5.7) has zeros in its ith row and jth column.

Basic structure vectors for the arbitrary matrices

Another interesting case of the theorem you have already met is closely related to the procedure for finding the basic structure of a matrix. We let

$$X_{Nn} = X'X \tag{22.5.8}$$

That is, X_{Nn} is the minor product moment of a matrix X. But we may write $X'X$ in terms of its basic structure, thus,

$$X'X = Q\delta Q' \tag{22.5.9}$$

Now we let

$$X_{SN} = Q_{.1}' \tag{22.5.10}$$

$$X_{nS} = Q_{.1} \tag{22.5.11}$$

Substituting Eqs. (22.5.9), (22.5.10), and (22.5.11) in Eq. (22.1.2), we have

$$Y_{Nn} = Q\delta Q' - Q\delta Q'Q_{.1}(Q'_{.1}Q\delta Q'Q_{.1})^{-1}Q'_{.1}Q\delta Q' \qquad (22.5.12)$$

But we can readily show that

$$Q\delta Q'Q_{.1} = \delta_1 Q_{.1} \qquad (22.5.13)$$

$$Q'_{.1}Q\delta Q' = \delta_1 Q'_{.1} \qquad (22.5.14)$$

$$Q'_{.1}Q\delta Q'Q_{.1} = \delta_1 \qquad (22.5.15)$$

Substituting Eqs. (22.5.13), (22.5.14) and (22.5.15) in Eq. (22.5.12)

$$Y = Q\delta Q' - \delta_1 Q_{.1}[\delta_1]^{-1}\delta_1 Q'_{.1} \qquad (22.5.16)$$

or

$$Y = Q\delta Q' - \delta_1 Q_{.1} Q'_{.1} \qquad (22.5.17)$$

or, substituting the left side of Eq. (22.5.9) in the right of Eq. (22.5.17)

$$Y = X'X - \delta_1 Q_{.1} Q'_{.1} \qquad (22.5.18)$$

You recall that Eq. (22.5.18) gives the first residual matrix in the solution for the basic structure of X as outlined in Chapter 18.

Unit vectors for the arbitrary matrices

A third simple application of the reduction theorem is closely related to one of the simpler methods of factor analysis known as the *centroid* method. As indicated in previous chapters, factor analysis is of great value to all sciences whose basic variables are not clearly defined. This includes most of the social and biological sciences, as well as some of the physical sciences.

In most factor-analysis procedures, a matrix of correlation coefficients is usually calculated from a data matrix and the factor analysis is conducted on this matrix. The analysis consists in obtaining from the correlation matrix a factor matrix. The factor matrix is calculated in such a way that its major product moment is approximately equal to the correlation matrix. One method of factor analysis is called the *centroid method*. In the centroid method of factor analysis, the first vector of the factor matrix is called the first centroid of factor loadings. To get this first centroid vector by means of the reduction theorem, we let X_{Nn} be the matrix of correlation coefficients indicated by r, thus,

$$X_{Nn} = r \qquad (22.5.19)$$

The arbitrary factors are simply

$$X_{SN} = 1' \qquad (22.5.20)$$

$$X_{nS} = 1 \qquad (22.5.21)$$

Substituting Eqs. (22.5.19), (22.5.20), and (22.5.21) in Eq. (22.1.2) gives

$$Y = r - r1(1'r1)^{-1}1'r \qquad (22.5.22)$$

or

$$Y = r - \left(\frac{r1}{\sqrt{1'r1}}\right)\left(\frac{1'r}{\sqrt{1'r1}}\right) \qquad (22.5.23)$$

The first centroid vector is, then, $r1/\sqrt{1'r1}$ or simply a vector made up of row (or column) sums of r and divided by the square root of the sum of all the elements in r. Equation (22.5.23) gives the first residual matrix, which is obtained as you see by subtracting the major product moment of the first centroid vector from the correlation matrix. This residual matrix is of rank one less than the rank of the correlation matrix, since the smaller dimension of the arbitrary factors is in this case simply 1.

The reduction theorem is the foundation of most factor analysis techniques, but these are beyond the scope of this book.

SUMMARY

1. The reduction equation: Given the matrix X_{Nn} with rank defined by $X_{Nn} = X_{Nr}X_{rn}$. Given also the arbitrary basic matrices X_{nS} and X_{SN} such that $S \lessgtr r$ and $(X_{SN}X_{Nn}X_{nS})^{-1}$ exists. A matrix Y_{Nn} is defined by

$$Y_{Nn} = X_{Nn} - X_{Nn}X_{nS}(X_{SN}X_{Nn}X_{nS})^{-1}X_{SN}X_{Nn}$$

2. The matrix Y_{Nn} is orthogonal to the arbitrary matrices. That is,

$$X_{SN}Y_{Nn} = 0$$

$$Y_{Nn}X_{nS} = 0$$

3. The minimum rank of the product of two matrices:
   a. The product of two basic matrices cannot be less than the sum of their distinct orders less their common order.
   b. The rank of the product of any two matrices cannot be less than the sum of their ranks less their common order.

4. The rank theorem. It is proved that the rank of Y_{Nn} in paragraph 1 is precisely $S - r$.

5. Special cases of the rank reduction theorem:
   a. If we let $X_{SN} = e'_i$ and $X_{NS} = e_j$, then

$$Y = X - \frac{X'_{\cdot j}X'_{i\cdot}}{X_{ij}}$$

   b. If we let $X_{Nn} = X'X = Q\delta Q'$, $X_{SN} = Q'_{\cdot 1}$, and $X_{nS} = Q_{\cdot 1}$, then

$$Y = X'X - \delta_1 Q_{\cdot 1}Q'_{\cdot 1}$$

   c. If we let $X_{Nn} = r$, $X_{SN} = 1'$, and $X_{nS} = 1$, then

$$Y = r - \left(\frac{r1}{\sqrt{1'r1}}\right)\left(\frac{1r}{\sqrt{1'r1}}\right)$$

Chapter 23

Solving Linear Equations

23.1 Introduction

Importance for the human sciences

Most problems in the human sciences are concerned with attempts to predict certain criterion variables from certain predictor variables. Typically we have measures on a number of predictor variables or attributes for a number of entities. We also have measures on one or more criterion variables for these same entities. We wish to determine the relationships between the predictor variables and the criterion variables so that for other entities for whom only predictor variables are available we can estimate criterion variables. In the human sciences we usually assume that the relationships between the predictor variables or certain known functions of them and the criterion variables are linear. Therefore, the problem of prediction is usually that of finding a vector of weights which, when used as a postmultiplier of the matrix of predictor variables, gives the best estimate of a criterion variable. If we have more than one criterion variable, we have a vector of prediction weights corresponding to each criterion, or a matrix of predictor vectors.

The fundamental matrix equation

We may let X_{Nn} be a matrix of predictor measures of n variables for each of N entities, Z_{Nm} be a matrix of criterion measures of m different criteria for each of the N entities, a_{nm} be a matrix of vectors of predictor weights for the m different criteria. Then the typical problem can be stated in the form

$$X_{Nn}a_{nm} - Z_{Nm} = \epsilon_{Nm} \qquad (23.1.1)$$

where a_{nm} is to be solved for, so that the elements of ϵ_{Nm} will be as close to zero as possible. In many problems having to do with the analysis of data in the human sciences, we also work with secondary or derived matrices in which we have equations of the type given in Eq. (23.1.1), where X and Z are known and a is to be solved for.

Therefore, the general problem with which we are concerned in this chapter is the solution of a set of linear equations in a specified number of unknowns. In matrix notation the general problem consists in solving for a in the equation

$$Xa - Z = \epsilon \qquad (23.1.2)$$

where X and Z are known and not null. Now according to our practice we let X be the vertical order, or in the limiting case, a square matrix. Equation (23.1.2) represents the case of the number of scalar equations equal to the number of unknowns and also the case of more equations than unknowns. If the number of unknowns is greater than the number of scalar equations, instead of Eq. (23.1.2) we may write

$$X'a - Z = \epsilon \qquad (23.1.3)$$

For both Eq. (23.1.2) and (23.1.3), X may be either basic or nonbasic.

The four possible conditions

We shall now consider solutions for a under each of the following four conditions of X: (1) It is square and basic; (2) It is vertical and basic; (3) It is horizontal and basic; (4) It is nonbasic. Traditionally the mathematicians have emphasized the fact that an exact solution for a such that ϵ is 0 is possible only when X is square and basic. The statisticians have developed the "method of least squares" for the case when X is vertical and basic. For the case of more unknowns than equations, the mathematicians have been largely content to emphasize that an infinite number of solutions exist, while for the case of X nonbasic they have been even less helpful.

The general inverse solution

We find that the concept of the general inverse developed in Chapter 19 can be applied to arrive at a solution for each of the four cases. We recall the basic structure of X,

$$X = P\Delta Q' \qquad (23.1.4)$$

and its general inverse

$$Y' = Q\Delta^{-1}P' \qquad (23.1.5)$$

We shall then define the optimal solution for a in the case of Eq. (23.1.2) as

$$a = Y'Z \qquad (23.1.6)$$

and in the case of Eq. (23.1.3) as

$$a = YZ \qquad (23.1.7)$$

Equations (23.1.6) and (23.1.7) apply when X is either basic or nonbasic. Substituting Eq. (23.1.5) in Eqs. (23.1.6) and (23.1.7) we have respectively

$$a = Q\Delta^{-1}P'Z \qquad (23.1.8)$$

$$a = P\Delta^{-1}Q'Z \qquad (23.1.9)$$

23.2   The Solution When X Is Square and Basic

The traditional case

When X is square and basic, and Z is a vector, we have the traditional case in which the number of scalar equations and unknowns is equal. The mathematicians call X here a nonsingular matrix. In no case, however, would we be justified in working with a data matrix that is square. As was pointed out in Chapter 1, a data matrix, to give scientifically useful results, must have many more entities than attributes. However, we have many examples of derived or secondary matrices in which X is square. Actually, as you will recall from Chapter 19 on the inverse of a matrix, we require the inverses of symmetric basic matrices in order to get the general inverses of matrices that are not square, not basic, or neither. Correlation or covariance matrices are examples of symmetric matrices that are usually also basic.

Properties of the solution

It is easy to show that if X is basic and square, the value of a given by Eq. (23.1.6) is an exact solution so that ϵ is 0. We substitute Eqs. (23.1.4) and (23.1.8) in Eq. (23.1.2) and get

$$P\Delta Q'Q\,\Delta^{-1}P'Z - Z = \epsilon \qquad (23.2.1)$$

or

$$PP'Z - Z = \epsilon \qquad (23.2.2)$$

But since X is square and basic, P is square, and from Eq. (23.2.2),

$$\epsilon = Z - Z = 0 \qquad (23.2.3)$$

The more conventional way to solve for a when X is square and basic is simply to write

$$Xa = Z \qquad (23.2.4)$$

then premultiply Eq. (23.2.4) by X^{-1} to get

$$X^{-1}Xa = X^{-1}Z \qquad (23.2.5)$$

or

$$a = X^{-1}Z \qquad (23.2.6)$$

But from (23.1.4) we have

$$X^{-1} = [P\Delta Q']^{-1} \qquad (23.2.7)$$

or

$$X^{-1} = Q\Delta^{-1}P' \qquad (23.2.8)$$

Substituting Eq. (23.2.8) in Eq. (23.2.6) gives Eq. (23.1.8).

We have chosen, however, to start with the defined solution of a given by Eq. (23.1.8) because of the interesting and useful generalizations we can extend to the other types of X matrices.

23.3   The Solution When X Is Vertical and Basic

The method of least squares

When X is vertical and basic, we have the typical case of a data matrix with more entities than unknowns. It is for this case that the method of least squares was developed to provide a solution for a. This solution guarantees that the sum of the squares of the deviations of the predicted Z's from the observed Z's is a minimum, or that the trace of $\epsilon'\epsilon$ in Eq. (23.1.2) is a minimum. The method of least squares derived by rather complicated methods of the calculus gives as the solution of a

$$a = (X'X)^{-1}X'Z \qquad (23.3.1)$$

You have seen, however, in Chapter 19, that the general inverse of X is given by

$$Y' = (X'X)^{-1}X' \qquad (23.3.2)$$

Properties of the solution

We now show that the solution for a given by Eq. (23.1.8) does (1) give a minimum for the trace of $\epsilon'\epsilon$, and also (2) gives an ϵ that is orthogonal to X, or

$$X'\epsilon = 0 \qquad (23.3.3)$$

First we prove Eq. (23.3.3). We substitute Eqs. (23.1.4) and (23.1.8) in Eq. (23.1.2) to get

$$P\Delta Q'Q\Delta^{-1}P'Z - Z = \epsilon \qquad (23.3.4)$$

From (23.3.4)

$$(PP' - I)Z = \epsilon \qquad (23.3.5)$$

Premultiplying both sides of Eq. (23.3.5) by X'

$$X'(PP' - I)Z = X'\epsilon \qquad (23.3.6)$$

Substituting Eq. (23.1.4) in the left side of Eq. (23.3.6)

$$Q\Delta P'(PP' - I)Z = X'\epsilon \qquad (23.3.7)$$

$$\text{or}\quad (Q\Delta P'PP' - Q\Delta P')Z = X'\epsilon$$

$$\text{or}\quad (Q\Delta P' - Q\Delta P')Z = X'\epsilon$$

or Eq. (23.3.3), $X'\epsilon = 0$, which is what we set out to prove.

Now to show that the trace of $\epsilon'\epsilon$ is a minimum we need only prove for any particular column vector $\epsilon_{.i}$ that $\epsilon'_{.i}\epsilon_{.i}$ is a minimum. Without loss of generality, therefore, we assume that Z and ϵ in Eq. (23.1.4) are vectors.

First assume that instead of a given by (23.1.8) we take some other vector b, where

$$b = a + E \qquad (23.3.8)$$

and write

$$Xb - Z = \epsilon_b \qquad (23.3.9)$$

We wish to prove now that

$$\epsilon'_b\epsilon_b > \epsilon'\epsilon \qquad (23.3.10)$$

for any nonvanishing E in Eq. (23.3.8). First we get from Eq. (23.3.5) for the case of ϵ a vector.

$$\epsilon'\epsilon = Z'(PP' - I)(PP' - I)Z \qquad (23.3.11)$$

or

$$\epsilon'\epsilon = Z'(I - PP')Z \qquad (23.3.12)$$

Next we substitute the right side of Eq. (23.1.4), for X in (23.3.9). We also substitute the right side of Eq. (23.1.8) for a in Eq. (23.3.8). Using this result for b in Eq. (23.3.9) gives

$$P\Delta Q'(Q\Delta^{-1}P'Z + E) - Z = \epsilon_b \qquad (23.3.13)$$

From (23.3.13)

$$PP'Z + P\Delta Q'E - Z = \epsilon_b$$

or

$$(PP' - I)Z + P\Delta Q'E = \epsilon_b \qquad (23.3.14)$$

From Eq. (23.3.14) we get

$$\epsilon'_b\epsilon_b = [Z'(PP' - I) + E'Q\Delta P'][(PP' - I)Z + P\Delta Q'E] \qquad (23.3.15)$$

or

$$\epsilon'_b\epsilon_b = Z'(I - PP')Z + 2Z'(PP' - I)P\Delta Q'E + (E'Q\Delta)(\Delta Q'E) \qquad (23.3.16)$$

But

$$Z'(PP' - I)P\Delta Q'E = Z'(P\Delta Q' - P\Delta Q')E = 0 \qquad (23.3.17)$$

Substituting Eq. (23.3.17) in Eq. (23.3.16)

$$\epsilon'_b\epsilon_b = Z'(I - PP')Z + (E'Q\Delta)(\Delta Q'E) \qquad (23.3.18)$$

Subtracting Eq. (23.3.12) from Eq. (23.3.18)

$$\epsilon'_b\epsilon_b - \epsilon'\epsilon = (E'Q\Delta)(\Delta Q'E) \qquad (23.3.19)$$

But the right of Eq. (23.3.19) is the minor product moment of a vector, therefore a sum of squares and positive, hence

$$\epsilon_b'\epsilon_b > \epsilon'\epsilon \tag{23.3.10}$$

which is the inequality we set out to prove.

23.4 The Case of the Horizontal Basic Matrix

An infinite number of solutions

In the case of Eq. (23.1.3) we have more unknowns than scalar equations and if X is basic, there is an infinite number of solutions for a that satisfy the equation so that ϵ is 0. This is very easy to prove if we substitute Eqs. (23.1.4) and (23.1.9) in Eq. (23.1.3) to get

$$Q\Delta P'P\Delta^{-1}Q'Z - Z = \epsilon \tag{23.4.1}$$

From Eq. (23.4.1)

$$QQ'Z - Z = \epsilon \tag{23.4.2}$$

But since X is assumed basic, Q is square and therefore from Eq. (23.4.2)

$$\epsilon = Z - Z = 0 \tag{23.4.3}$$

We could also have substituted Eq. (23.3.2) in Eq. (23.1.7) to get

$$a = X(X'X)^{-1}Z \tag{23.4.4}$$

and then substituted Eq. (23.4.4) in Eq. (23.1.3) to get at once

$$X'X(X'X)^{-1}Z - Z = \epsilon \tag{23.4.5}$$

$$\text{or} \quad Z - Z = \epsilon$$

An optimal solution

While it is true that an infinite number of a's exist to satisfy Eq. (23.1.3), other than Eq. (23.1.9), this solution does have a very interesting property of its own. We can show that in the solution given by Eq. (23.1.9) $a'a$ is a minimum. For any other solution, say

$$X'b = Z \tag{23.4.6}$$

we can prove that

$$b'b > a'a \tag{23.4.7}$$

We let

$$b = (a + E) \tag{23.4.8}$$

Substituting Eq. (23.4.8) in Eq. (23.4.6), we have

$$X'(a + E) = Z \tag{23.4.9}$$

or

$$X'a + X'E = Z \tag{23.4.10}$$

But from Eqs. (23.1.3), (23.4.3) and (23.4.10), we see that

$$X'E = 0 \qquad (23.4.11)$$

Now from Eq. (23.4.8)

$$b'b = a'a + 2a'E + E'E \qquad (23.4.12)$$

From Eq. (23.1.9)

$$a'E = Z'Q\Delta^{-1}P'E \qquad (23.4.13)$$

From Eqs. (23.1.4) and (23.4.11)

$$Q\Delta P'E = 0 \qquad (23.4.14)$$

Premultiplying Eq. (23.4.14) by $\Delta^{-1}Q'$ gives

$$P'E = 0 \qquad (23.4.15)$$

Substituting Eq. (23.4.15) in Eq. (23.4.13),

$$a'E = 0 \qquad (23.4.16)$$

Substituting Eq. (23.4.16) in Eq. (23.4.12),

$$b'b = a'a + E'E \qquad (23.4.17)$$

But since the second term on the right of Eq. (23.4.17) is a sum of squares, we have proved the inequality in Eq. (23.4.6), which we set out to prove.

23.5   The Solution When X Is Not Basic

Applications of the nonbasic case

It is unusual to find a data matrix that is not basic. Typically the derived matrices whose inverses we require are minor product moments of data matrices, or else derived from these. In such cases, the resulting symmetric matrices are basic. Usually, therefore, in the human sciences the matrices whose inverses we require are basic. Some of the more advanced developments of factor analysis, however, may involve solutions in which the X matrices are not basic. Also in certain extensions of the theory of least squares that assume errors of measurement in the predictor as well as the criterion variables, we may have a derived X matrix that is not basic.* Both because of the practical applications and to make the theory of solutions for linear equations more general, it is well to extend the solutions for a in Eqs. (23.1.8) and (23.1.9) to the case of X nonbasic. Our development will consider in detail only the case of X, a vertical matrix, but the same procedures apply to the case of X', a horizontal matrix.

---

*See Horst, Paul, *et al.*, "The Prediction of Personal Adjustment," *Social Science Research Council, Bull.* 48, 1941, pp. 436–444.

Properties of the solution

It is easy to show that the solution for a given by Eq. (23.1.8) gives a minimum for both $a'a$ and $\epsilon'\epsilon$, and that X is orthogonal to ϵ. We can use exactly the same proof to show that X is orthogonal to ϵ that we used in Eqs. (23.3.3) through (23.3.7), since none of these equations assume that Q is square. For the same reason, we can also use Eqs. (23.3.8) through (23.3.19) to prove that $\epsilon'\epsilon$ is a minimum.

Similarly we can use Eqs. (23.4.8) through (23.4.17) to prove that $a'a$ is a minimum, since none of these equations assume that Q is square.

It should be noted that none of these proofs depend on the width of a and Z. If a and Z are not vectors, then instead of referring to $a'a$ and $\epsilon'\epsilon$ we talk about their traces.

A method of solving for the general inverse of X, nonbasic, is given in Chapter 19. There you recall that we let

$$X = uv' \qquad (23.5.1)$$

where u and v' are basic matrices whose common order defines the rank of X. Then according to the solution given for the general inverse we have as the solution for a

$$a = V(v'v)^{-1}(u'u)^{-1}u'Z \qquad (23.5.2)$$

Solution if rank of X is known

One method for finding matrices u and v is given in Chapters 15 and 16, on the rank of a matrix. An approach, if the rank of X is known, is suggested in Chapter 22. Let us now indicate the left and right arbitrary multipliers of X as G_L and G_R respectively. We know that if the width of G_L and the height of G_R are equal to the rank of X, the Y of Chapter 22, Eq. (23.1.2) must be 0. Therefore, using this equation, we get

$$X = XG_R(G_LXG_R)^{-1}G_LX \qquad (23.5.3)$$

Suppose now we let

$$u = XG_R(G_LXG_R)^{-1} \qquad (23.5.4)$$

$$v' = G_LX \qquad (23.5.5)$$

Then substituting Eqs. (23.5.4) and (23.5.5) in Eq. (23.5.2),

$$Y = XG_R(G_LXG_R)^{-1}[(G_R'X'G_L')^{-1}G_R'X'XG_R(G_LXG_R)^{-1}]^{-1}(G_LXX'G_L')^{-1}G_LX \qquad (23.5.6)$$

which reduces to

$$Y' = X'G_L'(G_LXX'G_L')^{-1}(G_LXG_R)(G_R'X'XG_R)^{-1}G_R'X' \qquad (23.5.7)$$

We can show that Eq. (23.5.7) reduces to Eq. (23.1.5) by substituting in Eq. (23.5.7) the right side of Eq. (23.1.4).

$$Y' = Q\Delta P'G'_L(G_LP\Delta^2P'G'_L)^{-1}(G_LP\Delta Q'G_R)(G'_RQ\Delta^2Q'G_R)^{-1}G'_RQ\Delta P' \quad (23.5.8)$$

Now G_LP, $Q'G_R$ and their transposes are all square basic matrices. Suppose then we let

$$k = G_LP \quad (23.5.9)$$

$$h = Q'G_R \quad (23.5.10)$$

We substitute Eqs. (23.5.9) and (23.5.10) in (23.5.8), and remove parentheses to get

$$Y' = Q\Delta k'k'^{-1}\Delta^{-2}k^{-1}k\Delta hh^{-1}\Delta^{-2}h'^{-1}h'\Delta P'$$

or $\quad Y' = Q\Delta\Delta^{-2}\Delta\Delta^{-2}\Delta P'$

or $\quad Y' = Q\Delta^{-1}P'$

which is what we set out to prove.

Now G_R and G_L may be any matrices such that G_LXG_R has a regular inverse. In particular then suppose the rank of X is r and we partition X into

$$X = \begin{bmatrix} X_{rr} & X_{rt} \\ X_{sr} & X_{st} \end{bmatrix} \quad (23.5.11)$$

where $r + s$ is the height of X and $r + t$ its width. Let us now take

$$G_L = [I_{rr} \mid 0_{rs}] \quad (23.5.12)$$

and

$$G_R = \begin{bmatrix} I_{rr} \\ - \\ 0_{tr} \end{bmatrix} \quad (23.5.13)$$

From Eqs. (23.5.11) and (23.5.13)

$$XG_R = \begin{bmatrix} X_{rr} \\ X_{sr} \end{bmatrix} \quad (23.5.14)$$

From Eqs. (23.5.11) and (23.5.12)

$$G_LX = [X_{rr} \mid X_{rt}] \quad (23.5.15)$$

From Eqs. (23.5.12) and (23.5.14)

$$G_LXG_R = X_{rr} \quad (23.5.16)$$

From Eq. (23.5.16) we see that if X_{rr} is basic, Eqs. (23.5.12) and (23.5.13) may be used as the arbitrary left and right multipliers of X. In any case, if X_{rr} is not basic, rows and columns can always be permuted so that in

$$\pi_LX\pi_R = Z \quad (23.5.17)$$

we will have Z_{rr} basic. Then the following will be applicable to Z rather than X.

If now we let

$$X_{Nr} = \begin{bmatrix} X_{rr} \\ \hline X_{sr} \end{bmatrix} \qquad (23.5.18)$$

and

$$X_{rn} = [X_{rr} \mid X_{rt}] \qquad (23.5.19)$$

then using Eqs. (23.5.16), (23.5.18) and (23.5.19) in Eq. (23.5.7), we have, because of (23.5.14) and (23.5.15),

$$a = X'_{rn}(X_{rn}X'_{rn})^{-1}X_{rr}(X'_{Nr}X_{Nr})^{-1}X'_{Nr}Z \qquad (23.5.20)$$

Equation (23.5.20) shows how the general inverse of a matrix of rank r may be expressed as a function of the elements in its first r rows and columns provided X_{rr} is basic. If X_{rr} is not basic, it is necessary to choose those r rows and columns that do give a basic submatrix Z_{rr} in Eq. (23.5.17).

It should be noted that in some practical problems basic factors of X may already be known. In this case Eq. (23.5.2) may be used directly to get the general inverse. Sometimes, the basic structure itself is known, in which case the general inverse is computed directly from Eq. (23.1.5).

23.6 The Identity Matrix and the Product of a Matrix by its General Inverse

By definition, the product of a square basic matrix by its inverse is the identity matrix. But what relationship does the product of any matrix by its general inverse bear to the identity matrix? This problem might be considered a special case of the situation described in the foregoing sections, in which now Z is the identity matrix. Then let us consider the matrix

$$X = P\Delta Q' \qquad (23.6.1)$$

We indicate the general inverse of X by

$$Y' = Q\Delta^{-1}P' \qquad (23.6.2)$$

which if $Z = I$, is the solution for a in Eq. (23.1.8) in section 23.1.

If X is vertical and basic

Assume that X is vertical and basic. Because of Eqs. (23.6.1) and (23.6.2)

$$Y'X = I \qquad (23.6.3)$$

We shall now show that in

$$XY' - I = E \qquad (23.6.4)$$

the trace of $E'E$ is a minimum. This we do by showing for any given $E_{.i}$ that

$$E'_{.i}E_{.i} \text{ is a minimum} \tag{23.6.5}$$

From Eq. (23.6.4) we have

$$XY_{i.} - e_i = E_{.i} \tag{23.6.6}$$

But from Eq. (23.6.2)

$$Y_{i.} = Q\Delta^{-1}P_{i.} \tag{23.6.7}$$

Substituting Eqs. (23.6.1) and (23.6.7) in Eq. (23.6.6)

$$P\Delta Q'Q\Delta^{-1}P_{i.} - e_i = E_{.i} \tag{23.6.8}$$

or

$$PP_{i.} - e_i = E_{.i} \tag{23.6.9}$$

Now from (23.6.9)

$$E'_{.i}E_{.i} = (P'_{i.}P' - e'_i)(PP_{i.} - e_i) \tag{23.6.10}$$

or

$$E'_{.i}E_{.i} = P'_{i.}P_{i.} - 2P'_{i.}P_{i.} + 1 \tag{23.6.11}$$

or

$$E'_{.i}E_{.i} = 1 - P'_{i.}P_{i.} \tag{23.6.12}$$

Now suppose we had let $Y_{i.}$ be some other vector, V_i, say, where

$$V_i = Y_{i.} + \epsilon_i \tag{23.6.13}$$

Then substituting Eq. (23.6.7) in Eq. (23.6.13) gives

$$V_i = Q\Delta^{-1}P_{i.} + \epsilon_i \tag{23.6.14}$$

Substituting Eq. (23.6.1) in Eq. (23.6.6) and using Eq. (23.6.14) instead of Eq. (23.6.7), we have

$$P\Delta Q'(Q\Delta^{-1}P_{i.} + \epsilon_i) - e_i = {}_VE_{.i} \tag{23.6.15}$$

or

$$(PP_{i.} - e_i + P\Delta Q'\epsilon_i) = {}_VE_{.i} \tag{23.6.16}$$

where ${}_VE_{.i}$ is the residual vector obtained by using V_i instead of $Y_{i.}$. From Eq. (23.6.16)

$${}_VE'_{.i}\,{}_VE_{.i} = (P'_{i.}P' - e'_i + \epsilon'_iQ\Delta P')(PP_{i.} - e_i + P\Delta Q'\epsilon_i) \tag{23.6.17}$$

From Eq. (23.6.17)

$${}_VE'_{.i}\,{}_VE_{.i} = 1 - P'_{i.}P_{i.} + 2[(P'_{i.}P' - e'_i)P\Delta Q'\epsilon_i] + \epsilon_iQ\Delta^2Q'\epsilon_i \tag{23.6.18}$$

But

$$(P'_{i.}P' - e'_i)(P\Delta Q'\epsilon_i) = P'_{i.}\Delta Q'\epsilon_i - P'_{i.}\Delta Q'\epsilon_i = 0 \tag{23.6.19}$$

Therefore Eq. (23.6.18) becomes

$$_vE'_{.i}\,_vE_{.i} = 1 - P'_{i.}P_{i.} + (\epsilon_iQ\Delta)(\Delta Q'\epsilon_i) \qquad (23.6.20)$$

From Eqs. (23.6.12) and (23.6.20)

$$_vE'_{.i}\,_vE_{.i} - E'_{.i}E_{.i} = (\epsilon_iQ\Delta)(\Delta Q'\epsilon_i) \qquad (23.6.21)$$

But the expression on the right of Eq. (23.3.2) is a sum of squares, and therefore positive. Thus, any vector other than $Y_{i.}$ given by Eq. (23.6.7) would yield a larger minor product moment than Eq. (23.6.13). This is true for all vectors of Y'. Therefore, the trace of $E'E$ is a minimum when Y' is given by Eq. (23.6.2).

If X is nonbasic

Next consider the case of X in Eq. (23.6.1) not basic. The demonstration in Eq. (23.6.4) through Eq. (23.6.21) would still hold so that the right of Eq. (23.6.2) would yield the minimum trace for the minor product moment of E in Eq. (23.6.4). But now let us commute the factors in the first term on the left of Eq. (23.6.4) to get

$$Y'X - I = F' \qquad (23.6.22)$$

or

$$X'Y - I = F \qquad (23.6.23)$$

We can show for X not basic that the trace of the minor product moment of F in Eq. (23.6.23) is a minimum for Y' given by Eq. (23.6.2). This we do by showing for any particular $F_{.i}$ that its minor product moment is a minimum. From Eq. (23.6.23) we have

$$X'Y_{.i} - e_i = F_{.i} \qquad (23.6.24)$$

But from Eq. (23.6.2)

$$Y_{.i} = P\Delta^{-1}Q_{i.} \qquad (23.6.25)$$

Substituting Eqs. (23.6.1) and (23.6.25) in Eq. (23.6.24)

$$Q\Delta P'P\Delta^{-1}Q_{i.} - e_i = F_{.i} \qquad (23.6.26)$$

or

$$QQ_{i.} - e_i = F_{.i} \qquad (23.6.27)$$

From (23.6.27)

$$F'_{.i}F_{.i} = (Q'_{i.}Q' - e'_i)(QQ_{i.} - e_i) \qquad (23.6.28)$$

or

$$F'_{.i}F_{.i} = Q'_{i.}Q_{i.} - 2\,Q'_{i.}Q_{i.} + 1 \qquad (23.6.29)$$

or

$$F'_{.i}F_{.i} = 1 - Q'_{i.}Q_{i.} \qquad (23.6.30)$$

We now proceed exactly as in Eqs. (23.6.13) through (23.6.21), except that we substitute Q for P and P for Q, so that corresponding to Eq. (23.6.21), we get

$$v F'_{.i} \, v F_{.i} - F'_{.i} F_{.i} = (\epsilon'_i P \Delta)(\Delta P' \epsilon_i) \qquad (23.6.31)$$

Equation (22.6.31) shows that any vector other than $Y_{.i}$ given by Eq. (23.6.25) would yield a larger minor product moment than Eq. (23.6.29) since the right side of Eq. (23.6.31) is a sum of squares.

Therefore, no matrix exists which when pre- or postmultiplied by X as defined in Eq. (23.6.1) gives a better fit to the identity matrix in the least square sense than the matrix Y' defined by Eq. (23.6.2). In the case of a square basic matrix, of course, this fit is perfect.

SUMMARY

1. Introduction:
   a. Most scientific problems in the human sciences are concerned with predicting criterion variables from predictor variables.
   b. If we let

   X_{Nn} be a matrix of known predictor measures of n variables for N entities,

   Z_{Nm} be a matrix of known criterion measures of m different criteria for each of the n entities, and

   a_{nm} be a matrix of unknown vectors of predictor weights for the m different criteria, then the fundamental matrix equation is

   $$X_{Nn} a_{nm} - Z_{Nm} = \epsilon_{Nm}$$

   where a_{nm} is to be determined so that

   $$tr \; \epsilon' \epsilon \text{ is a minimum}$$

   c. The X matrix may be one of four different types: (1) square and basic, (2) vertical and basic, (3) horizontal and basic, (4) nonbasic.
   d. The general inverse solution: If we let $X = P\Delta Q'$, and $Y' = Q\Delta^{-1}P'$, the general solution for a is given by

   $$(1) \; a = Y'Z \text{ for } Xa - Z = \epsilon$$
   $$(2) \; a = YZ \text{ for } X'a - Z = 0$$

2. When X is square and basic:
   a. This is the traditional mathematical case.
   b. The solution for a is $a = X^{-1}Z$.
   c. The solution is exact so that $\epsilon = 0$.

3. When X is vertical and basic:
   a. This is the usual case for data matrices in the social sciences.
   b. The general inverse solution for a is

   $$a = (X'X)^{-1}X'Z$$

c. The solution satisfies the conditions

$$(1) \qquad X'\epsilon = 0$$

$$(2) \ \epsilon'\epsilon = \text{minimum}$$

4. The horizontal basic matrix X':
   a. This case has an infinite number of solutions.
   b. The general inverse solution is

$$a = X(X'X)^{-1}Z$$

   c. The solution satisfies the conditions

$$(1) \qquad \epsilon = 0$$

$$(2) \ tr \ a'a \ = \text{a minimum}$$

5. When X is nonbasic:
   a. This case is encountered with derived matrices and in certain advanced problems in multivariate analysis.
   b. If the rank of X is given by $X = uv'$ the solution is

$$a = v(v'v)^{-1}(u'u)^{-1}u'Z$$

   c. The solution satisfies the conditions:

$$(1) \qquad X'\epsilon = 0$$

$$(2) \ tr \ \epsilon'\epsilon = \text{a minimum}$$

$$(3) \ tr \ a'a = \text{a minimum}$$

   d. If the rank r of X is known and we let

$$X = \begin{bmatrix} X_{rr} & X_{rt} \\ X_{sr} & X_{st} \end{bmatrix} \qquad X_{Nr} = \begin{bmatrix} X_{rr} \\ X_{sr} \end{bmatrix} \qquad X_{rn} = [X_{rr} \quad X_{rt}]$$

   where $r + s$ and $r + t$ are the height and width respectively of X, then if X_{rr} is basic, the general inverse solution of a is

$$a = X'_{rn}(X_{rn}X'_{rn})^{-1}X_{rr}(X'_{Nr}X_{Nr})^{-1}X'_{Nr}Z$$

6. The identity matrix and the product of a matrix by its general inverse: Let the basic structure of X be given by $X = P\Delta Q'$ and its general inverse by $Y' = Q\Delta^{-1}P'$.
   a. Assume X vertical and basic:
   (1) If $Y'X - I = E$, then $E = 0$ if Y' is the general inverse of X.
   (2) If $XY' - I = E$ then $X'E = 0$.

   and $tr \ E'E$ is a minimum if Y' is the general inverse of X.
   b. Assume X nonbasic:
   (1) If $XY' - I = E$, then $X'E = 0$ and $tr \ E'E$ is a minimum for Y' the general inverse of X.
   (2) If $X'Y - I = F$, then $XF = 0$ and $tr \ F'F$ is a minimum for Y' the general inverse of X.

Index

Index

505